MINIMAX UNDER TRANSPORTATION CONSTRAINS

Applied Optimization

Volume 27

Minimax Under Transportation Constrains

by

Vladimir Tsurkov
Computing Center,
Russian Academy of Sciences,
Moscow, Russia

and

Anatoli Mironov
Computing Center,
Russian Academy of Sciences,
Moscow, Russia

SPRINGER-SCIENCE+BUSINESS MEDIA, B.V.

A C.I.P. Catalogue record for this book is available from the Library of Congress.

ISBN 978-1-4613-6818-2 ISBN 978-1-4615-4060-1 (eBook)
DOI 10.1007/978-1-4615-4060-1

Printed on acid-free paper

Contents

Preface

Transportation problems belong to the domains mathematical programming and operations research. Transportation models are widely applied in various fields. Numerous concrete problems (for example, assignment and distribution problems, maximum-flow problem, etc.) are formulated as transportation problems. Some efficient methods have been developed for solving transportation problems of various types.

This monograph is devoted to transportation problems with minimax criteria. The classical (linear) transportation problem was posed several decades ago. In this problem, supply and demand points are given, and it is required to minimize the transportation cost. This statement paved the way for numerous extensions and generalizations.

In contrast to the original statement of the problem, we consider a minimax rather than a minimum criterion. In particular, a matrix with the minimal largest element is sought in the class of nonnegative matrices with given sums of row and column elements. In this case, the idea behind the minimax criterion can be interpreted as follows. Suppose that the shipment time from a supply point to a demand point is proportional to the amount to be shipped. Then, the minimax is the minimal time required to transport the total amount. It is a common situation that the decision maker does not know the tariff coefficients. In other situations, they do not have any meaning at all, and neither do nonlinear tariff objective functions. In such cases, the minimax interpretation leads to an effective solution.

The monograph is focused on the development of an algorithm for finding the minimax value and the corresponding minimax matrix. It is based on the construction of characteristic functions and equations, which determine the general properties of all matrices with a common bound on their elements. The bound is then interpreted as a minimax. The method is extended to the case of integer variables. An efficient (reduction) algorithm is constructed for calculating the integer matrix whose largest element is minimal. Minimax matrices are used in the theory of transportation polyhedrons. We describe a broad class of transportation matrices (uniform matrices), which make up the convex hull of a certain finite set of minimax matrices.

The monograph consists of five chapters. The presentation of theory is accompanied by detailed analysis of examples. Here, we briefly outline the contents of the book.

Chapter 1 plays an introductory role. A number of concrete transportation models with minimax criteria are formulated, and the cost functions minimized in Chapter 2 are constructed. We introduce characteristic functions and equations with a view to analyzing the plan matrices of an arbitrary transportation problem, whose matrix elements are bounded from above. An algorithm for calculating the elements of these matrices is constructed.

In Chapter 2, we present algorithms for solving minimax transportation problems. In the final analysis, when the basic criterion is employed, the solution is reduced to simple computational procedures for verifying linear inequalities and choosing the largest element. This approach is extended to the case of integer matrices. The so-called hereditarily minimax matrix is constructed, which provides solutions to problems with other minimax criteria.

In Chapter 3, the technique based on characteristic functions and equations is used to find admissible solutions to integer and boolean transportation problems. Unlike conventional methods, the algorithm constructs all admissible solutions. These solutions provide a basis for determining the vertices of transportation polyhedrons.

In Chapter 4, we introduce the concept of an extremal vector pair. By definition, a transportation problem specified by such a vector pair has a unique solution consisting of zeros and ones. This solution is called the extremal matrix. Matrices of this type determine the vertices of the convex hulls of matrix sets. Moreover, extremal matrices are constructed as solutions to all minimax transportation problems analyzed in the monograph.

In Chapter 5, we consider the convex hull spanned by the extremal vector pairs. We show that every extremal pair is a vertex of the hull. By invoking the well-known theorem about extreme points from functional analysis, it is established that each vector pair (which defines the classical transportation polyhedron) belongs to the hull. In particular, this result facilitates solution of some optimization problems posed on the sets of solutions to certain minimax problems.

Chapter 1

Transportation Models with Minimax Criteria and Preliminary Constructions

Foundations of theory of transportation problems were laid by Hitchcock [43]. Note, however, that some particular problems were examined earlier (see [39]). The scope of transportation models is very wide. We can mention problems related, for example, to design of production and shipments, to creation of computation and information systems, to allocation of resources and stocks, etc. [14, 27, 28, 42].

Classical transportation problems are generally placed in the field of linear programming, and a body of mathematics for solving them is well developed and known [9, 27, 40, 45]. Currently, one often has to invoke nonlinear transportation models deal with practical tasks [14, 20, 42]; however, even in such cases, linear approximation of functionals is sometimes used [2, 27]. In the general case, an arbitrary objective function is given on the set of matrices of the transportation polyhedron. Therefore, mathematical programming can be applied to the transportation problems [1, 15, 16, 19, 29, 38]. Depending on the type of the objective function, extremum transportation problems can be subsumed, for example, under convex [32, 35], dynamic [3, 18, 45–48], or discrete [36, 44] programming.

In this chapter, we introduce transportation models with minimax criteria. A mixed criterion—minimax—is widely used in operations research [10, 26, 33]; it is also used in decision theory, in the theory of active and hierarchical systems, and in matrix games [4–6, 30, 31].

The preface gives only one example of possible minimax criteria defined on transportation polyhedrons. Here is another example. The sum of the largest elements of all submatrices of a transportation matrix is a functional on the matrices of an arbitrary transportation polyhedron. The minimization of this functional is examined in Chapter 2. Here, we only note that the solution of this optimization problem is unique, and the minimum is attained on a hereditarily minimax matrix.

Chapter 1 is introductory. In Section 1, we consider a number of transportation models and construct objective functions to be minimized in Chapter 2.

In Section 2, we obtain necessary and sufficient conditions in the form of an irreducible system of inequalities that determine plan matrices for transportation problems with upper bounds on the elements. The set of such matrices is a special case of truncated transportation polyhedrons [11].

An algorithm for calculating elements of these matrices is developed that constructs any transportation matrix of this truncated polyhedron.

In Section 3, we introduce characteristic equations determined by the sums of elements of principal submatrices. By means of these equations, we establish conditions that make it possible to decompose a truncated transportation polyhedron into two polyhedrons of lower dimensions. This, as well as the technique of characteristic equations, is widely used in Chapters 2 and 4 for finding minimax matrices. In this chapter, the technique of characteristic equations gives us estimates for the number of integer matrices of a transportation polyhedron.

§1. Transportation Models, Uniform Matrices, and Characteristic Functions

In this section, we introduce various minimax functionals, some of which are meaningful. We also consider a hereditary property of transportation matrices—their uniformity. Uniform matrices are solutions to certain minimax problems, and the uniformity is used to evaluate the corresponding functionals.

We construct here a characteristic function depending on the components of a pair of nonnegative vectors and on a nonnegative constant. This function will be often used in what follows. For a transportation polyhedron, the nonnegativity of this function guarantees that the set of matrices of this polyhedron with a given upper bound on all elements is not empty. Here, the components of the first vector are assumed to be arranged in a nonascending order. The function under consideration is the left-hand side of the characteristic equations.

1.1. Uniform Matrices and Matrix Sets

Consider constraints for a closed transportation problem [9, 27, 40]. Denote by R^n_+ the set of all n-dimensional vectors with nonnegative components:

$$R^n_+ = \left\{ \mathbf{A} = (a_1, \ldots, a_n) \in R^n : a_i \geq 0, \quad 1 \leq i \leq n \right\}.$$

Let vectors $\mathbf{A} = (a_1, \ldots, a_n) \in R^n_+$ and $\mathbf{B} = (b_1, \ldots, b_m) \in R^m_+$ have equal sums of the components:

$$\sum_{i=1}^{n} a_i = \sum_{j=1}^{m} b_j. \tag{1.1}$$

Condition (1.1) is called the balance or closure condition for a system consisting of the pair of vectors $\mathbf{A}$ and $\mathbf{B}$. For vectors $\mathbf{A}$ and $\mathbf{B}$ with equal sums of components, we denote by $X = X(\mathbf{A}, \mathbf{B}) = (x_{ij})$, $1 \leq i \leq n$,

$1 \le j \le m$ a matrix that satisfies the following conditions:

$$\begin{aligned} &x_{ij} \ge 0, \quad 1 \le i \le n, \quad 1 \le j \le m, \\ &\sum_{j=1}^{m} x_{ij} = a_i, \quad 1 \le i \le n, \\ &\sum_{i=1}^{n} x_{ij} = b_j, \quad 1 \le j \le m. \end{aligned} \tag{1.2}$$

The matrix $X(\mathbf{A}, \mathbf{B})$ satisfying conditions (1.2) is called a *plan matrix* of the transportation problem or a *transportation matrix* determined by the vectors $\mathbf{A}$ and $\mathbf{B}$. The set of all matrices $\{X(\mathbf{A}, \mathbf{B})\}$ is denoted by $M(\mathbf{A}, \mathbf{B})$:

$$M(\mathbf{A}, \mathbf{B}) = \Big\{X(\mathbf{A}, \mathbf{B}) = (x_{ij}) : x_{ij} \ge 0,\ 1 \le i \le n,\ 1 \le j \le m, \\ \sum_{j=1}^{m} x_{ij} = a_i,\ 1 \le i \le n,\ \sum_{i=1}^{n} x_{ij} = b_j,\ 1 \le j \le m\Big\}.$$

The set of matrices $M(\mathbf{A}, \mathbf{B})$ is called *classical transportation polyhedron* [11].

The following assertions about the matrix sets $M(\mathbf{A}, \mathbf{B})$ are evident:

(a) $M(\mathbf{A}, \mathbf{B}) \ne \varnothing$ for any admissible vectors $\mathbf{A}$ and $\mathbf{B}$;

(b) the matrix set $M(\mathbf{A}, \mathbf{B})$ is convex [21] and bounded;

(c) if n_1 and m_1 denote the numbers of positive components of the vectors $\mathbf{A}$ and $\mathbf{B}$, respectively, then the set $M(\mathbf{A}, \mathbf{B})$ consists of only one matrix if and only if $\min(n_1, m_1) \le 1$.

In the set $M(\mathbf{A}, \mathbf{B})$, we single out the subset of matrices that possess a specific property.

Definition 1.1. A Matrix $X(\mathbf{A}, \mathbf{B}) = (x_{ij})$ is called *uniform* if its elements and the components of the vectors $\mathbf{A} = (a_1, \ldots, a_n) \in R_+^n$ and $\mathbf{B} = (b_1, \ldots, b_m) \in R_+^m$ satisfy the following conditions:

(a) if $a_i = a_p$ and $b_j = b_q$, then $x_{ij} = x_{pq}$, $1 \le i, p \le n$, $1 \le j, q \le m$;

(b) if $a_i \ge a_p$ and $b_j \ge b_q$, then $x_{ij} \ge x_{pq}$, $1 \le i, p \le n$, $1 \le j, q \le m$.

For the set of uniform matrices determined by vectors $\mathbf{A}$ and $\mathbf{B}$, we define

$$\widetilde{M}(\mathbf{A}, \mathbf{B}) = \{X(\mathbf{A}, \mathbf{B}) \in M(\mathbf{A}, \mathbf{B}) : X(\mathbf{A}, \mathbf{B}) \text{ is a uniform matrix}\}.$$

Clearly, the set of uniform matrices $\widetilde{M}(\mathbf{A},\mathbf{B})$ is a convex subset of the set $M(\mathbf{A},\mathbf{B})$. Below, we construct transportation models for which optimal solutions are attained on uniform matrices.

We denote by $\widetilde{\mathbf{A}} = (\tilde{a}_1,\dots,\tilde{a}_n)$ the vector obtained from the vector $\mathbf{A} = (a_1,\dots,a_n)$ by arranging its components in nonascending order ($\tilde{a}_i \geq \tilde{a}_{i+1}$, $1 \leq i \leq n-1$, if $n \geq 2$). Let $\mathbf{A} = (a_1,\dots,a_n) \in R^n_+$ and $\mathbf{B} = (b_1,\dots,b_m) \in R^m_+$. By permutating rows and columns, every matrix $X(\mathbf{A},\mathbf{B}) = (x_{ij})$ can be transformed into the matrix $\widetilde{X}(\widetilde{\mathbf{A}},\widetilde{\mathbf{B}}) = (\tilde{x}_{ij})$ such that

$$\sum_{j=1}^{m} \tilde{x}_{ij} = \tilde{a}_i, \quad 1 \leq i \leq n, \quad \sum_{i=1}^{n} \tilde{x}_{ij} = \tilde{b}_j, \quad 1 \leq j \leq m,$$

where $\widetilde{\mathbf{A}} = (\tilde{a}_1,\dots,\tilde{a}_n)$ and $\widetilde{\mathbf{B}} = (\tilde{b}_1,\dots,\tilde{b}_m)$. Without loss of generality, in many transportation problems, the components of the vectors $\mathbf{A}$ and $\mathbf{B}$ may be assumed to be ordered in nonascending order. The set of vectors whose components are arranged in nonascending order is denoted by

$$\overline{R}^n_+ = \left\{\mathbf{A} = (a_1,\dots,a_n) \in R^n_+ : a_i \geq a_{i+1}, \quad 1 \leq i \leq n-1\right\},$$

where $n \geq 2$.

Clearly, if $\mathbf{A} \in \overline{R}^n_+$ and $\mathbf{B} \in \overline{R}^m_+$, then $\widetilde{\mathbf{A}} = \mathbf{A}$, $\widetilde{\mathbf{B}} = \mathbf{B}$, and $X(\mathbf{A},\mathbf{B}) = X(\widetilde{\mathbf{A}},\widetilde{\mathbf{B}}) = \widetilde{X}(\mathbf{A},\mathbf{B})$.

Let us introduce two operations on pairs of vectors. For $\alpha \in R$, $\mathbf{A}$, $\mathbf{C} \in R^n$, and $\mathbf{B}$, $\mathbf{D} \in R^m$, we define $\alpha(\mathbf{A},\mathbf{B}) = (\alpha\mathbf{A},\alpha\mathbf{B})$ and $(\mathbf{A},\mathbf{B}) + (\mathbf{C},\mathbf{D}) = (\mathbf{A}+\mathbf{C},\mathbf{B}+\mathbf{D})$. Clearly, these operations on pairs of vectors are consistent with those on plan matrices for transportation problems. If $\alpha \geq 0$ and matrices $X(\mathbf{A},\mathbf{B})$ and $X(\mathbf{C},\mathbf{D})$ are of the same dimension, then $\alpha X(\mathbf{A},\mathbf{B}) = X(\alpha\mathbf{A},\alpha\mathbf{B})$ and $X(\mathbf{A},\mathbf{B}) + X(\mathbf{C},\mathbf{D}) = X(\mathbf{A}+\mathbf{C},\mathbf{B}+\mathbf{D})$.

In the literature devoted to the sets of matrices for transportation problems, the set $M(\mathbf{A},\mathbf{B})$ is referred to as the classical transportation polyhedron (of dimension n-by-m if $\mathbf{A} \in R^n_+$ and $\mathbf{B} \in R^m_+$) [11, 12]. Among the polyhedrons arising in linear programming problems, the transportation polyhedrons are the best studied ones. In particular, various results concerning the number of their vertices and edges were obtained, and asymptotic behavior of certain classes of these polyhedrons was investigated [13, 17]. A detailed discussion of transportation polyhedrons can be found in [11].

Definition 1.2. The set of uniform matrices $\widetilde{M}(\mathbf{A},\mathbf{B})$ is called a *uniform matrix set* or a *uniform transportation polyhedron.*

1.2. Examples of Transportation Models with Minimax Criteria

A standard interpretation of the vectors $\mathbf{A} = (a_1,\dots,a_n) \in R^n_+$ and $\mathbf{B} = (b_1,\dots,b_m) \in R^m_+$ and of the matrix $X(\mathbf{A},\mathbf{B})$ in closed transportation models is as follows [9, 27, 40]. There are n supply points and m demand points for a certain homogeneous product. The component a_i, $1 \le i \le n$, of the vector $\mathbf{A}$ is the amount produced at the ith supply point, and the component b_j, $1 \le j \le m$, of the vector $\mathbf{B}$ is the amount consumed at the jth demand point. In the closed transportation models, the total amount produced is to be consumed. Therefore, the vectors $\mathbf{A}$ and $\mathbf{B}$ satisfy the balance condition (1.1). Every supply point is connected with every demand point by a route. Let x_{ij}, $1 \le i \le n$, $1 \le j \le m$, be the amount to be shipped from the ith supply point to the jth demand point. Since all of the product is to be consumed, the matrix $X = (x_{ij})$ satisfies relation (1.2). The plan matrix $X = (x_{ij}) = X(\mathbf{A}, \mathbf{B})$ is called the transportation matrix.

The transportation model is designed for choosing the most efficient transportation matrix $X(\mathbf{A}, \mathbf{B})$, and the performance criterion is the total cost of shipments. When $\varphi_{ij}(x)$, $1 \le i \le n$, $1 \le j \le m$, are given functions of a real variable, the transportation problem is possed as a minimization (sometimes, maximization) for the functional

$$\Phi\left(X(\mathbf{A}, \mathbf{B})\right) = \sum_{i=1}^{n} \sum_{j=1}^{m} \varphi_{ij}(x_{ij}), \quad X(\mathbf{A}, \mathbf{B}) = (x_{ij}) \in M(\mathbf{A}, \mathbf{B}). \tag{1.3}$$

In the classical (linear) model, the functional to be minimized is

$$\Phi\left(X(\mathbf{A}, \mathbf{B})\right) = \sum_{i=1}^{n} \sum_{j=1}^{m} c_{ij} x_{ij}, \quad X(\mathbf{A}, \mathbf{B}) = (x_{ij}) \in M(\mathbf{A}, \mathbf{B}), \tag{1.4}$$

where (c_{ij}), $1 \le i \le n$, $1 \le j \le m$, is a tariff matrix (c_{ij} is the cost of transportation from the ith supply point to the jth demand point per unit amount).

The general transportation model (1.3), (1.4) is designed for choosing the plan matrix with minimum expenses. The expenses are interpreted not only as transportation costs [9, 27, 40] but also as the total transportation time, the time required for execution of all works, the sum of taxes and fines, and so on.

In a well-known transportation problem, the performance criterion for a plan is the maximum time required for a shipment [9, 40]: given a matrix (t_{ij}), $1 \le i \le n$, $1 \le j \le m$, where t_{ij} is the time required to carry the product from the ith point to the jth point (irrespective of the amount), minimize the objective function

$$T\left(X(\mathbf{A}, \mathbf{B})\right) = \max_{x_{ij} > 0} t_{ij}, \quad X(\mathbf{A}, \mathbf{B}) = (x_{ij}) \in M(\mathbf{A}, \mathbf{B}). \tag{1.5}$$

Problem (1.5) arises, for example, when a perishable product is to be shipped.

The function $T\left(X(\mathbf{A},\mathbf{B})\right)$ is not linear with respect to variables x_{ij}; however, the problem of its minimization can be reduced to a sequence of ordinary (linear) transportation problems [9, 40]. An algorithm for minimization of this function based on successive solving maximum-flow problems has been constructed in [40].

In contrast to this model, we construct here an algorithm of minimization of the maximum time required for a shipment under the condition that t_{ij} is proportional to x_{ij}: $t_{ij} = tx_{ij}$, $1 \le i \le n$, $1 \le j \le m$, $t > 0$. Generally, the matrix that minimizes the function

$$\Phi\left(X(\mathbf{A},\mathbf{B}),t\right) = t\max_{i,j} x_{ij}, \quad X(\mathbf{A},\mathbf{B}) = (x_{ij}) \in M(\mathbf{A},\mathbf{B}), \tag{1.6}$$

is not unique, and there always exists a uniform matrix

$$X^*(\mathbf{A},\mathbf{B}) = (x^*_{ij}) \in \widetilde{M}(\mathbf{A},\mathbf{B})$$

such that

$$\min_{X(\mathbf{A},\mathbf{B})\in M(\mathbf{A},\mathbf{B})} \Phi\left(X(\mathbf{A},\mathbf{B}),t\right) = t \min_{X(\mathbf{A},\mathbf{B})\in M(\mathbf{A},\mathbf{B})} \max_{i,j} x_{ij} = t\max_{i,j} x^*_{ij}.$$

Let us introduce the following functionals:

$$\Phi_1\left(X(\mathbf{A},\mathbf{B})\right) = \max_{i,j} x_{ij}, \tag{1.7}$$

$$\Phi_2\left(X(\mathbf{A},\mathbf{B})\right) = \sum_{j=1}^{m} \max_{i} x_{ij}, \tag{1.8}$$

$$\Phi_3\left(X(\mathbf{A},\mathbf{B})\right) = \sum_{j=1}^{m} \max_{i} x_{ij} + \sum_{i=1}^{n} \max_{j} x_{ij}, \tag{1.9}$$

$$\Phi_4\left(X(\mathbf{A},\mathbf{B})\right) = \max_{i,j} x_{ij} + \sum_{j=1}^{m} \max_{i} x_{ij}, \tag{1.10}$$

$$\Phi_5\left(X(\mathbf{A},\mathbf{B})\right) = \max_{i,j} x_{ij} + \sum_{j=1}^{m} \max_{i} x_{ij} + \sum_{i=1}^{n} \max_{j} x_{ij}, \tag{1.11}$$

where $\mathbf{A} \in R^n_+$, $\mathbf{B} \in R^m_+$, and $X(\mathbf{A},\mathbf{B}) = (x_{ij}) \in M(\mathbf{A},\mathbf{B})$. It is evident that the objective functions (1.7)–(1.11) satisfy the relations

$$\Phi_3\left(X(\mathbf{A},\mathbf{B})\right) = \Phi_2\left(X(\mathbf{A},\mathbf{B})\right) + \Phi_2\left(X(\mathbf{B},\mathbf{A})\right),$$

$$\Phi_4\left(X(\mathbf{A},\mathbf{B})\right) = \Phi_1\left(X(\mathbf{A},\mathbf{B})\right) + \Phi_2\left(X(\mathbf{A},\mathbf{B})\right),$$

$$\Phi_5\left(X(\mathbf{A},\mathbf{B})\right) = \Phi_1\left(X(\mathbf{A},\mathbf{B})\right) + \Phi_3\left(X(\mathbf{A},\mathbf{B})\right).$$

For different functionals $\Phi_i\left(X(\mathbf{A},\mathbf{B})\right)$, the quantities to be minimized are as follows: the largest element of the matrix ($i = 1$), the sum of the largest elements of all columns of the matrix ($i = 2$), the sum of the largest elements of all columns and rows of the matrix ($i = 3$), the sum of the largest element of the matrix and the largest elements of all columns and rows ($i = 5$).

The matrices that minimize functionals (1.7)–(1.11) are not unique, and, as shown below, for any nonnegative vectors $\mathbf{A}$ and $\mathbf{B}$ with equal sums of components, there exists a uniform matrix $X^*(\mathbf{A},\mathbf{B}) \in \widetilde{M}(\mathbf{A},\mathbf{B})$ (common for all $\Phi_i\left(X(\mathbf{A},\mathbf{B})\right)$, $1 \le i \le 5$) that minimizes all these functionals:

$$\min_{X(\mathbf{A},\mathbf{B})\in M(\mathbf{A},\mathbf{B})} \Phi_i\left(X(\mathbf{A},\mathbf{B})\right) = \Phi_i\left(X^*(\mathbf{A},\mathbf{B})\right), \quad 1 \le i \le 5.$$

Sometimes, the matrix $X^*(\mathbf{A},\mathbf{B})$ is a unique solution to the problem

$$\Phi_i\left(X(\mathbf{A},\mathbf{B})\right) \to \min, \quad X(\mathbf{A},\mathbf{B}) \in M(\mathbf{A},\mathbf{B}),$$

if $i = 1, 4$, or 5. We formulate necessary and sufficient conditions for vectors $\mathbf{A}$ and $\mathbf{B}$ under which the maximum of functionals (1.7), (1.10), and (1.11) is attained on a unique matrix. We also present an algorithm of minimization of the functional $\Phi_1\left(X(\mathbf{A},\mathbf{B})\right)$ over the class of integer matrices for integer vectors $\mathbf{A}$ and $\mathbf{B}$.

Clearly, the problem of minimization of the time required for shipments (1.6) is equivalent to the minimization problem for functional (1.7):

$$\Phi\left(X(\mathbf{A},\mathbf{B}),t\right) = t\Phi_1\left(X(\mathbf{A},\mathbf{B})\right).$$

The minimization of the objective function $\Phi_5\left(X(\mathbf{A},\mathbf{B})\right)$, as well as any function (1.7)–(1.11), can be interpreted in terms of game theory [30, 31]. Consider the following model. Let $t > 0$ be a constant, and let $X(\mathbf{A},\mathbf{B}) = (x_{ij})$ be a plan of amounts to be shipped. Suppose that, for all admissible values of i and j, the sums of taxes paid by the producers and consumers to "local" budgets are

$$h'_i = t\max_j x_{ij} \quad \text{and} \quad h''_j = t\max_i x_{ij},$$

respectively, and that the total tax paid to the government is

$$h = t\max_{i,j} x_{ij}.$$

Then, the total of the taxes paid by the producers and consumers is given by

$$H\left(X(\mathbf{A},\mathbf{B})\right) = t\Phi_5\left(X(\mathbf{A},\mathbf{B})\right).$$

Since each producer and consumer wants to pay less tax, a conflict arises between the subjects of financial relations (producers, consumers, and tax inspectors). If the producers and consumers come to an agreement and form a coalition against tax inspectors, then the functional $H\left(X(\mathbf{A},\mathbf{B})\right) = t\Phi_5\left(X(\mathbf{A},\mathbf{B})\right)$ can be minimized.

Consider the following probabilistic problem statement in terms of model (1.6). Let $\mathbf{A} \in R_+^n$, $\mathbf{B} \in R_+^m$, $t > 0$, and let $X(\mathbf{A},\mathbf{B}) = (x_{ij})$ be a plan matrix for the amounts to be shipped. Then, $tX(\mathbf{A},\mathbf{B}) = (tx_{ij})$ is the matrix of the times required to carry the product from the ith producer to the jth consumer, $1 \le i \le n$, $1 \le j \le m$. Denote by $\mathfrak{M}(X)$ the set of all submatrices of the matrix $X = X(\mathbf{A},\mathbf{B})$. It is well known that the number of all subsets of any set consisting of n elements is 2^n (including the empty subset). Since every submatrix is determined by arbitrary subsets of rows and columns, the number of all submatrices in $\mathfrak{M}(X)$ is $|\mathfrak{M}(X)| = (2^n - 1)(2^m - 1)$. Here, $|Q|$ denotes the number of elements in a finite set Q.

Suppose submatrices Y are chosen from $\mathfrak{M}(X)$ with equal probabilities, and the criterion for every choice is the value $t\Phi_1(Y)$ of the largest element of tY. Since the probability distribution is uniform on the set of submatrices from $\mathfrak{M}(X)$, the probability of choosing an arbitrary submatrix Y from $\mathfrak{M}(X)$ is $p(Y) = 1/\left((2^n - 1)(2^m - 1)\right)$. Hence,

$$m\left(X(\mathbf{A},\mathbf{B}),t\right) = \frac{t}{(2^n - 1)(2^m - 1)} \sum_{Y \in \mathfrak{M}(X)} \Phi_1(Y)$$

is the expected value of the maximum time required for a shipment on the set of submatrices in $\mathfrak{M}(X)$, where $X = X(\mathbf{A},\mathbf{B})$ is some matrix in $M(\mathbf{A},\mathbf{B})$. Thus, we arrive at the problem

$$m\left(X(\mathbf{A},\mathbf{B}),t\right) \to \min, \quad X(\mathbf{A},\mathbf{B}) \in M(\mathbf{A},\mathbf{B}).$$

In Chapter 2, we construct a matrix $\overset{*}{X}(\mathbf{A},\mathbf{B}) = (\overset{*}{x}_{ij})$ that minimizes the expectation:

$$\min_{X(\mathbf{A},\mathbf{B}) \in M(\mathbf{A},\mathbf{B})} m\left(X(\mathbf{A},\mathbf{B}),t\right) = m(\overset{*}{X}(\mathbf{A},\mathbf{B}),t).$$

It is also shown that this problem has a unique solution.

Let us eliminate all trivial matrices from the set $\mathfrak{M}(X)$: let $\mathfrak{M}_0(X)$ be the set of submatrices of $X = X(\mathbf{A},\mathbf{B})$ of dimension greater than one. Clearly, the number of one-element submatrices is nm, and the sum of all elements of $X(\mathbf{A},\mathbf{B})$ is equal to the sum S of the components of $\mathbf{A}$ or $\mathbf{B}$. Then, the matrix $\overset{*}{X}(\mathbf{A},\mathbf{B})$ minimizes the expectation

$$m_0\left(X(\mathbf{A},\mathbf{B}),t\right) = \frac{t}{(2^n - 1)(2^m - 1) - nm} \sum_{Y \in \mathfrak{M}_0(X)} \Phi_1(Y):$$

$$\min_{X(\mathbf{A},\mathbf{B})\in M(\mathbf{A},\mathbf{B})} m_0(X(\mathbf{A},\mathbf{B}),t) = m_0(\overset{*}{X}(\mathbf{A},\mathbf{B}),t).$$

It is worth noting that

$$\sum_{Y\in\mathfrak{M}_0(X)} \Phi_1(Y) = \sum_{Y\in\mathfrak{M}(X)} \Phi_1(Y) - S.$$

In the next chapter, minimization algorithms are constructed for the following objective functions:

$$\Phi_6\left(X(\mathbf{A},\mathbf{B})\right) = \sum_{Y\in\mathfrak{M}(X(\mathbf{A},\mathbf{B}))} \Phi_2(Y), \tag{1.12}$$

$$\Phi_7\left(X(\mathbf{A},\mathbf{B})\right) = \sum_{Y\in\mathfrak{M}(X(\mathbf{A},\mathbf{B}))} \Phi_3(Y), \tag{1.13}$$

$$\Phi_8\left(X(\mathbf{A},\mathbf{B})\right) = \sum_{Y\in\mathfrak{M}(X(\mathbf{A},\mathbf{B}))} \Phi_1(Y), \tag{1.14}$$

$$\Phi_9\left(X(\mathbf{A},\mathbf{B})\right) = \sum_{Y\in\mathfrak{M}(X(\mathbf{A},\mathbf{B}))} \Phi_1(Y) + \sum_{Y\in\mathfrak{M}(X(\mathbf{A},\mathbf{B}))} \Phi_2(Y), \tag{1.15}$$

$$\Phi_{10}\left(X(\mathbf{A},\mathbf{B})\right) = \sum_{Y\in\mathfrak{M}(X(\mathbf{A},\mathbf{B}))} \Phi_1(Y) + \sum_{Y\in\mathfrak{M}(X(\mathbf{A},\mathbf{B}))} \Phi_3(Y), \tag{1.16}$$

where $\mathfrak{M}\left(X(\mathbf{A},\mathbf{B})\right)$ is the set of all submatrices of the matrix $X(\mathbf{A},\mathbf{B}) \in M(\mathbf{A},\mathbf{B})$. It is evident that

$$\begin{aligned} \Phi_9\left(X(\mathbf{A},\mathbf{B})\right) &= \Phi_8\left(X(\mathbf{A},\mathbf{B})\right) + \Phi_6\left(X(\mathbf{A},\mathbf{B})\right), \\ \Phi_{10}\left(X(\mathbf{A},\mathbf{B})\right) &= \Phi_8\left(X(\mathbf{A},\mathbf{B})\right) + \Phi_7\left(X(\mathbf{A},\mathbf{B})\right). \end{aligned} \tag{1.17}$$

Note that, in what follows, functional (1.7) is also minimized on the class of integer transportation matrices defined by a pair of integer vectors:

$$\Phi_Z\left(X(\mathbf{A},\mathbf{B})\right) = \max_{i,j} x_{ij}, \quad X(\mathbf{A},\mathbf{B}) = (x_{ij}) \in M(\mathbf{A},\mathbf{B}), \tag{1.18}$$

where $\mathbf{A} = (a_1,\ldots,a_n) \in R_+^n$, $a_i \in Z$, $1 \le i \le n$, $\mathbf{B} = (b_1,\ldots,b_m) \in R_+^m$, $b_j \in Z$, $1 \le j \le m$, $x_{ij} \in Z$, $1 \le i \le n$, $1 \le j \le m$.

It is easy to see that the minimization problem for the expected value can be reduced to minimization of functional (1.14): $m\left(X(\mathbf{A},\mathbf{B}),t\right) = t\Phi_8\left(X(\mathbf{A},\mathbf{B})\right)$.

Consider an arbitrary matrix set $M(\mathbf{A},\mathbf{B})$. For any matrices (x'_{ij}), $(x''_{ij}) \in M(\mathbf{A},\mathbf{B})$ and any α, $0 \le \alpha \le 1$, we have $(x_{ij}) = \alpha(x'_{ij}) + (1-\alpha)(x''_{ij}) \in M(\mathbf{A},\mathbf{B})$ (i.e., $M(\mathbf{A},\mathbf{B})$ is a convex set), and it is clear that

$$\begin{aligned} \max_{i,j} x_{ij} &\le \alpha \max_{i,j} x'_{ij} + (1-\alpha) \max_{i,j} x''_{ij}, \\ \max_{k} x_{ij} &\le \alpha \max_{k} x'_{ij} + (1-\alpha) \max_{k} x''_{ij}, \end{aligned} \tag{1.19}$$

where $k = 1$ or $k = j$. Inequalities (1.19) show that the functions

$$\max_{i,j} x_{ij}, \quad \max_{i} x_{ij}, \quad \max_{j} x_{ij}$$

are convex. Therefore, functionals (1.7)–(1.16) are convex as well.

1.3. Truncated Transportation Polyhedrons

Consider the subset of matrices from $M(\mathbf{A}, \mathbf{B})$ whose elements do not exceed $c \ge 0$:

$$M(\mathbf{A}, \mathbf{B}; c) = \left\{ X(\mathbf{A}, \mathbf{B}) = (x_{ij}) \in M(\mathbf{A}, \mathbf{B}) : x_{ij} \le c,\ 1 \le i \le n,\ 1 \le j \le m \right\},$$

where $\mathbf{A} \in R_+^n$ and $\mathbf{B} \in R_+^m$.

The following propositions about the matrix set $M(\mathbf{A}, \mathbf{B}; c)$ are obviously true:

(a) $M(\mathbf{A}, \mathbf{B}; 0) \ne \varnothing$ if and only if $\mathbf{A}$ and $\mathbf{B}$ are zero vectors;

(b) if $c < d$, then $M(\mathbf{A}, \mathbf{B}; c) \subseteq M(\mathbf{A}, \mathbf{B}; d)$;

(c) if $a = \max\{a_i : 1 \le i \le n\}$, $b = \max\{b_j : 1 \le j \le m\}$, and $d \ge \min(a, b)$, then $X(\mathbf{A}, \mathbf{B}) \in M(\mathbf{A}, \mathbf{B}; d)$ and $M(\mathbf{A}, \mathbf{B}) = M(\mathbf{A}, \mathbf{B}; d)$;

(d) the fulfilment of the inequalities $\max\{a_i : 1 \le i \le n\} \le cm$ and $\max\{b_j : 1 \le j \le m\} \le cn$ is necessary for the set $M(\mathbf{A}, \mathbf{B}; c)$ to be nonempty.

In the next section, we formulate necessary and sufficient conditions for $X(\mathbf{A}, \mathbf{B}) \in M(\mathbf{A}, \mathbf{B}; c)$.

In the literature devoted to transportation polyhedrons, the matrix set $M(\mathbf{A}, \mathbf{B}, c)$ is subsumed under the class of truncated transportation polyhedrons. Let vectors $\mathbf{A} \in R_+^n$ and $\mathbf{B} \in R_+^m$ and a nonnegative matrix (c_{ij}), $1 \le i \le n$, $1 \le j \le m$, be given. The subset of matrices $M\left(\mathbf{A}, \mathbf{B}; (c_{ij})\right)$ from $M(\mathbf{A}, \mathbf{B})$ defined by the condition

$$M\left(\mathbf{A}, \mathbf{B}; (c_{ij})\right) = \Big\{ X(\mathbf{A}, \mathbf{B}) = (x_{ij}) \in M(\mathbf{A}, \mathbf{B}) : x_{ij} \le c_{ij}, \\ 1 \le i \le n,\ 1 \le j \le m \Big\},$$

is referred to as the n-by-m *truncated transportation polyhedron* [11].

The following conditions quarantee that $M\left(\mathbf{A}, \mathbf{B}; (c_{ij})\right) \ne \varnothing$ as a corollary to Gale's theorem [7, 11, 41]. If vectors $\mathbf{A} = (a_1, \dots, a_n) \in R_+^n$ and $\mathbf{B} =$

$(b_1, \ldots, b_m) \in R_+^m$ have equal sums of the components and the matrix (c_{ij}), $1 \le i \le n$, $1 \le j \le m$, is nonnegative, then the truncated transportation polyhedron $M\left(\mathbf{A}, \mathbf{B}; (c_{ij})\right)$ is not empty if and only if

$$\sum_{j=1}^{m} \min\Big(b_j, \sum_{i \in I} c_{ij}\Big) \ge \sum_{i \in I} a_i \tag{1.20}$$

for any subset of indices $I \subseteq \{1, \ldots, n\}$.

In particular, when $c_{ij} = c$, $1 \le i \le n$, $1 \le j \le m$, in (1.20), we obtain necessary and sufficient conditions for the nonemptiness of $M(\mathbf{A}, \mathbf{B}; c)$. In this case, the number of inequalities to be checked is reduced. Here, it is convenient to assume that the components of the first vector $\mathbf{A}$ are arranged in nonascending order. Since

$$\sum_{i \in I} c_{ij} = c|I|$$

for $c_{ij} = c$, $1 \le i \le n$, $1 \le j \le m$, where $|I|$ is the number of indices in the set I, the following assertion is valid: if vectors $\mathbf{A} \in R_+^n$ and $\mathbf{B} \in R_+^n$ satisfy balance condition (1.1) and $c \ge 0$, then the transportation polyhedron $M(\mathbf{A}, \mathbf{B}; c)$ is not empty if and only if

$$\sum_{j=1}^{m} \min(b_j, ck) \ge \sum_{i=1}^{k} \tilde{a}_i \tag{1.21}$$

for any $k \in Z$, $1 \le k \le n$, where $\widetilde{\mathbf{A}} = (\tilde{a}_1, \ldots \tilde{a}_n)$ is the vector derived from the vector $\mathbf{A}$ by arranging its components in nonascending order.

For vectors $\mathbf{A} = (a_1, \ldots, a_n) \in R_+^n$ and $\mathbf{B} = (b_1, \ldots, b_m) \in R_+^m$ (generally, not satisfying the balance condition) and numbers c and k, where $c \ge 0$ and $k \in Z$, $1 \le k \le n$, consider the function

$$\delta_k(\mathbf{A}, \mathbf{B}; c) = \sum_{j=1}^{m} b_j - \sum_{b_j \ge ck} (b_j - ck) - \sum_{i=1}^{k} a_i, \tag{1.22}$$

called the *characteristic* function. In certain cases, $k = 0$ in (1.22); then, $\delta_0(\mathbf{A}, \mathbf{B}; c) = 0$. It is evident that

$$\sum_{j=1}^{m} b_j - \sum_{b_j \ge ck} (b_j - ck) = \sum_{j=1}^{m} \min(b_j, ck).$$

Therefore, if $\mathbf{A} \in \overline{R}_+^n$ and the vectors $\mathbf{A}$ and $\mathbf{B}$ satisfy condition (1.1), then the inequalities $\delta_k(\mathbf{A}, \mathbf{B}; c) \ge 0$ and (1.21) are equivalent. Hence, the following proposition is valid: when the vectors $\mathbf{A} \in R_+^n$ and $\mathbf{B} \in R_+^m$ have equal sums of the components and $c \geq 0$, then $M(\mathbf{A}, \mathbf{B}; c) \neq \varnothing$ if and only if

$$\delta_k(\widetilde{\mathbf{A}}, \mathbf{B}; c) \geq 0 \tag{1.23}$$

for all k, $1 \leq k \leq n$ [7, 23–25].

We prove this proposition will be proved in Section 2 when constructing an algorithm for calculation of a matrix from $M(\mathbf{A}, \mathbf{B}; c)$ (note that all dependent inequalities in (1.23) will be eliminated).

Conditions (1.23) constitute a system of n inequalities some of which can be dependent; hence, the number of inequalities in this system can be reduced. A consistent system of inequalities is referred to as irreducible [49] if it does not contain dependent inequalities. The properties of function (1.22) determine an irreducible system of inequalities and a condition that ensures that $M(\mathbf{A}, \mathbf{B}; c) \neq \varnothing$.

We will also use the following equivalent expressions for function (1.22):

$$\delta_k(\mathbf{A}, \mathbf{B}; c) = \sum_{j=1}^{m} \min(b_j, ck) - \sum_{i=1}^{k} a_i, \tag{1.24}$$

$$\delta_k(\mathbf{A}, \mathbf{B}; c) = \sum_{b_j < ck} b_j + ckl_k(\mathbf{B}; c) - \sum_{i=1}^{k} a_i, \tag{1.25}$$

where $l_k(\mathbf{B}; c)$ is the number of components of the vector $\mathbf{B} = (b_1, \ldots, b_m)$ that are not less than ck:

$$l_k(\mathbf{B}; c) = \left|\{j : b_j \geq ck,\ 1 \leq j \leq m\}\right| .$$

Note that the system of inequalities (1.23) also determines a criterion for the existence of an integer transportation matrix ($\mathbf{A}$ and $\mathbf{B}$ are integer vectors).

§2. Transportation Matrices with Bounded Elements

In this section, we present two algorithms for construction of transportation matrices with a common upper bound on all elements. One of them constructs a certain matrix (if the given truncated transportation polyhedron is not empty). By means of the other algorithm, an arbitrary matrix from an admissible set can be constructed.

For an arbitrary pair of vectors that determine a nonempty transportation polyhedron and for any positive bound, a system of inequalities is constructed that does not contain reduntant inequalities. A criterion for the existence of a transportation matrix with a given upper bound on all elements is formulated. These results are extended to the case of integer matrices.

2.1. Characteristic Functions for Partitioned Vector Pairs

The algorithm for calculation of the elements of a matrix $X(\mathbf{A},\mathbf{B})$ from $M(\mathbf{A},\mathbf{B};c) \neq \varnothing$, where $\mathbf{A} \in R_+^n$, $n \geq 2$, and $\mathbf{B} \in R_+^m$, is based on the partition of vectors $\mathbf{A}$ and $\mathbf{B}$ into vector pairs $\mathbf{A}' \in R_+^{n'}$, $\mathbf{B}^* \in R_+^m$ and $\mathbf{A}'' \in R_+^{n''}$, $\mathbf{B}^{**} \in R_+^m$, $n'+n''=n$ and on constructing matrices $X(\mathbf{A}',\mathbf{B}^*) \in M(\mathbf{A}',\mathbf{B}^*;c)$ and $X(\mathbf{A}'',\mathbf{B}^{**}) \in M(\mathbf{A}'',\mathbf{B}^{**};c)$ that make up the matrix $X(\mathbf{A},\mathbf{B})$.

Let $\mathbf{A} = (a_1,\ldots,a_n) \in R_+^n$, $n \geq 2$, $\mathbf{B} = (b_1,\ldots,b_m) \in R_+^m$, $c \geq 0$, $p \in Z$, $1 \leq p \leq n-1$. Here, the balance condition for the vectors $\mathbf{A}$ and $\mathbf{B}$ is not assumed to hold. Let us construct vector pairs $\mathbf{A}' = (a'_1,\ldots,a'_{n-p})$, $\mathbf{B}' = (b'_1,\ldots,b'_m)$ and $\mathbf{A}'' = (a''_1,\ldots,a''_p)$, $\mathbf{B}'' = (b''_1,\ldots,b''_m)$ with components defined by the following relations:

$$a'_i = a_{p+i}, \quad 1 \leq i \leq n-p,$$
$$b'_j = \begin{cases} b_j - cp, & 1 \leq j \leq m, \quad b_j \geq cp, \\ 0, & 1 \leq j \leq m, \quad b_j < cp; \end{cases} \tag{2.1}$$

$$a''_i = a_i, \quad 1 \leq i \leq p,$$
$$b''_j = \begin{cases} cp, & 1 \leq j \leq m, \quad b_j \geq cp, \\ b_j, & 1 \leq j \leq m, \quad b_j < cp. \end{cases} \tag{2.2}$$

It is easy to see that $\mathbf{A}' \in R_+^{n-p}$, $\mathbf{B}' \in R_+^m$, $\mathbf{A}'' \in R_+^p$, $\mathbf{B}'' \in R_+^m$, the vector $\mathbf{B}$ is the sum of $\mathbf{B}'$ and $\mathbf{B}''$ ($\mathbf{B} = \mathbf{B}' + \mathbf{B}''$), and the vector $\mathbf{A}$ is the union of the components of the vectors $\mathbf{A}'$ and $\mathbf{A}''$.

To construct an algorithm for calculating the elements of a matrix from $M(\mathbf{A},\mathbf{B};c)$, we need two lemmas relating values of the characteristic function (1.22) of two vectors $\mathbf{A}$ and $\mathbf{B}$ (the balance condition is not assumed) to the values of the same function considered as one of the vector pairs $\mathbf{A}'$, $\mathbf{B}'$ and $\mathbf{A}''$, $\mathbf{B}''$ with components defined in (2.1) and (2.2).

Lemma 2.1. *Let* $\mathbf{A} = (a_1,\ldots,a_n) \in R_+^n$, $\mathbf{B} = (b_1,\ldots,b_m) \in R_+^m$ *where* $n \geq 2$ *and* $c \geq 0$, $p \in Z$, $1 \leq p \leq n-1$. *Then, function* (1.22) *of vectors* $\mathbf{A}$, $\mathbf{B}$ *and* $\mathbf{A}'$, $\mathbf{B}'$ *with components defined in* (2.1) *satisfies the equation*

$$\delta_k(\mathbf{A}',\mathbf{B}';c) = \delta_{p+k}(\mathbf{A},\mathbf{B};c) - \delta_p(\mathbf{A},\mathbf{B};c) \tag{2.3}$$

for any k, $1 \leq k \leq n-p$.

Proof. Consider function (1.22) of vectors $\mathbf{A}'$ and $\mathbf{B}'$ for an arbitrary $k \in Z$, $1 \leq k \leq n-p$:

$$\delta_k(\mathbf{A}',\mathbf{B}';c) = \sum_{j=1}^{m} b'_j - \sum_{b'_j \geq ck} (b'_j - ck) - \sum_{i=1}^{k} a'_i.$$

It follows from (2.1) that

$$\sum_{i=1}^{k} a'_i = \sum_{i=p+1}^{p+k} a_i,$$

$$\sum_{b_j \geq ck} (b'_j - ck) = \sum_{b_j \geq c(p+k)} (b'_j - c(p+k)),$$

$$\sum_{j=1}^{m} b'_j = \sum_{b_j \geq cp} b'_j = \sum_{b_j \geq cp} (b_j - cp) = \sum_{b_j \geq cp} b_j - cpl_p(\mathbf{B};c),$$

where $l_p(\mathbf{B};c)$ is defined in (1.25). Therefore, function (1.22) of $\mathbf{A}'$ and $\mathbf{B}'$ written in terms of the components of $\mathbf{A}$ and $\mathbf{B}$ takes the form

$$\delta_k(\mathbf{A}',\mathbf{B}';c) = \sum_{b_j \geq cp} b_j - cpl_p(\mathbf{B};c) - \sum_{b_j \geq c(p+k)} (b_j - c(p+k)) - \sum_{i=p+1}^{p+k} a_i.$$

Applying the obvious identities

$$\sum_{b_j \geq cp} b_j = \sum_{j=1}^{m} b_j - \sum_{b_j < cp} b_j, \quad \sum_{i=p+1}^{p+k} a_i = \sum_{i=1}^{p+k} a_i - \sum_{i=1}^{p} a_i,$$

to the function $\delta_k(\mathbf{A}',\mathbf{B}';c)$, we find that

$$\delta_k(\mathbf{A}',\mathbf{B}';c) =$$

$$\sum_{j=1}^{m} b_j - \sum_{b_j < cp} b_j - cpl_p(\mathbf{B};c) - \sum_{b_j \geq c(p+k)} (b_j - c(p+k)) - \sum_{i=1}^{p+k} a_i + \sum_{i=1}^{p} a_i =$$

$$\sum_{j=1}^{m} b_j - \sum_{b_j \geq c(p+k)} (b_j - c(p+k)) - \sum_{i=1}^{p+k} a_i - \Big(\sum_{b_j < cp} b_j + cpl_p(\mathbf{B};c) - \sum_{i=1}^{p} a_i\Big) =$$

$$\delta_{p+k}(\mathbf{A},\mathbf{B};c) - \Big(\sum_{j=1}^{m} b_j - \sum_{b_j \geq cp} (b_j - cp) - \sum_{i=1}^{p} a_i\Big) =$$

$$\delta_{p+k}(\mathbf{A},\mathbf{B};c) - \delta_p(\mathbf{A},\mathbf{B};c).$$

Thus, $\delta_k(\mathbf{A}',\mathbf{B}';c) = \delta_{p+k}(\mathbf{A},\mathbf{B};c) - \delta_p(\mathbf{A},\mathbf{B};c)$, and the lemma is proved.

Lemma 2.2. *Let* **A**, **B**, c, *and* p *be the same as in Lemma 2.1. Then, function (1.22) of the vectors* **A**, **B** *and* **A**″, **B**″ *with components defined in (2.2) satisfies the equation*

$$\delta_k(\mathbf{A}'', \mathbf{B}''; c) = \delta_k(\mathbf{A}, \mathbf{B}; c) \tag{2.4}$$

for any k, $1 \le k \le p$.

Proof. For any $k \in Z$, $1 \le k \le p$, the components of the vectors **A**, **B** and **A**″, **B**″ satisfy the equations (see (2.2))

$$\sum_{i=1}^{k} a_i'' = \sum_{i=1}^{k} a_i,$$

$$\sum_{j=1}^{m} b_j'' = \sum_{j=1}^{m} b_j - \sum_{b_j \ge cp} (b_j - cp),$$

$$\sum_{b_j'' \ge ck} (b_j'' - ck) = \sum_{b_j \ge ck} (b_j - ck) - \sum_{b_j \ge cp} (b_j - cp).$$

Therefore,

$$\delta_k(\mathbf{A}'', \mathbf{B}''; c) = \sum_{j=1}^{m} b_j'' - \sum_{b_j'' \ge ck} (b_j'' - ck) - \sum_{i=1}^{k} a_i'' =$$

$$\sum_{j=1}^{m} b_j - \sum_{b_j \ge cp} (b_j - cp) - \Big(\sum_{b_j \ge ck} (b_j - ck) - \sum_{b_j \ge cp} (b_j - cp)\Big) - \sum_{i=1}^{k} a_i =$$

$$\sum_{j=1}^{m} b_j - \sum_{b_j \ge ck} (b_j - ck) - \sum_{i=1}^{k} a_i = \delta_k(\mathbf{A}, \mathbf{B}; c),$$

which proves the lemma.

Suppose that pair of vectors **A** and **B** satisfies the balance conditions (1.1), nonnegativity conditions for function (1.22), and the assumptions of Lemmas 2.1. and 2.2. Then the pair of vectors $\mathbf{A} \in R_+^n$ and $\mathbf{B} \in R_+^m$ can be split into vector pairs $\mathbf{A}' \in R_+^{n'}$, $\mathbf{B}^* \in R_+^m$ and $\mathbf{A}'' \in R_+^{n''}$, $\mathbf{B}^{**} \in R_+^m$, $n' + n'' = n$, for which the balance conditions and nonnegativity conditions for function (1.22) are also valid. The following plays a basic role in the construction of the algorithm for calculating the elements of a matrix from $M(\mathbf{A}, \mathbf{B}; c)$ and in the proof of sufficiency of conditions (1.23) for the nonemptiness of $M(\mathbf{A}, \mathbf{B}; c)$.

Lemma 2.3. *Suppose that for vectors* $\mathbf{A} = (a_1, \ldots, a_n) \in R_+^n$, $n \geq 2$, *and* $\mathbf{B} = (b_1, \ldots, b_m) \in R_+^m$ *satisfying the balance condition*

$$\sum_{i=1}^{n} a_i = \sum_{j=1}^{m} b_j,$$

and for $c \geq 0$ *and* $p \in Z$, $1 \leq p \leq n-1$, *such that* $\delta_k(\mathbf{A}, \mathbf{B}; c) \geq 0$ *for any* k, $1 \leq k \leq n$, *and*

$$\delta_p(\mathbf{A}, \mathbf{B}; c) = \min \left\{ \delta_k(\mathbf{A}, \mathbf{B}; c) : 1 \leq k \leq n-1 \right\},$$

vector pairs $\mathbf{A}' = (a'_1, \ldots, a'_{n-p}) \in R_+^{n-p}$, $\mathbf{B}' = (b'_1, \ldots, b'_m) \in R_+^m$ *and* $\mathbf{A}'' = (a''_1, \ldots, a''_p) \in R_+^p$, $\mathbf{B}'' = (b''_1, \ldots, b''_m) \in R_+^m$ *are constructed that satisfy conditions* (2.1) *and* (2.2), *respectively. Then, the following propositions are valid:*

(a) *If* $\delta_p(\mathbf{A}, \mathbf{B}; c) = 0$, *then the vector pairs* $\mathbf{A}'$, $\mathbf{B}^*$ *and* $\mathbf{A}''$, $\mathbf{B}^{**}$, *where* $\mathbf{B}^* = (b_1^*, \ldots, b_m^*) = \mathbf{B}'$ *and* $\mathbf{B}^{**} = (b_1^{**}, \ldots, b_m^{**}) = \mathbf{B}''$, *satisfy the balance conditions*

$$\sum_{i=1}^{n-p} a'_i = \sum_{j=1}^{m} b_j^*, \quad \sum_{i=1}^{p} a''_i = \sum_{j=1}^{m} b_j^{**}, \tag{2.5}$$

and the inequalities

$$\delta_k(\mathbf{A}', \mathbf{B}^*; c) \geq 0, \quad \delta_k(\mathbf{A}'', \mathbf{B}^{**}; c) \geq 0 \tag{2.6}$$

for $k \in Z$, *where, respectively,* $1 \leq k \leq n-p$ *and* $1 \leq k \leq p$.

(b) *Suppose that* $\delta_p(\mathbf{A}, \mathbf{B}; c) \geq 0$ *and a vector* $(\beta_1, \ldots, \beta_m)$ *from* R_+^m *satisfies the conditions*

$$\begin{aligned} &\sum_{j=1}^{m} \beta_j = \delta_p(\mathbf{A}, \mathbf{B}; c), \\ &\max\{b'_j + \beta_j : 1 \leq j \leq m\} \leq c(n-p), \\ &\min\{b''_j - \beta_j : 1 \leq j \leq m\} \geq 0. \end{aligned} \tag{2.7}$$

Then, the vector pairs $\mathbf{A}'$, $\mathbf{B}^* = (b_1^*, \ldots, b_m^*)$ *and* $\mathbf{A}''$, $\mathbf{B}^{**} = (b_1^{**}, \ldots, b_m^{**})$, *where*

$$\begin{aligned} b_j^* &= b'_j + \beta_j, \quad 1 \leq j \leq m, \\ b_j^{**} &= b''_j - \beta_j, \quad 1 \leq j \leq m, \end{aligned}$$

satisfy balance conditions (2.5) *and nonnegativity conditions* (2.6) *for function* (1.22).

(c) *If* $\mathbf{A} \in \overline{R}^n_+$, *i.e., the components of the vector* $\mathbf{A}$ *are arranged in nonascending order, then there exists a vector* $(\beta_1, \ldots, \beta_m)$ *that satisfies conditions* (2.7). *In particular, when* $\beta_{m+1} = 0$, *conditions* (2.7) *are satisfied by the vector* $(\beta_1, \ldots, \beta_m)$ *whose components are defined as*

$$\beta_{m-i} = \min \left\{ \begin{array}{l} cp, \\ cn - cp, \\ b_{m-i}, \\ cn - b_{m-i}, \\ \delta_p(\mathbf{A}, \mathbf{B}; c) - \sum\limits_{j=m-i+1}^{m+1} \beta_j \end{array} \right\}, \quad 0 \le i \le m-1. \tag{2.8}$$

Proof. We apply equations (1.24), (2.3), and (2.4) to the vectors and to the characteristic function (1.22).

(a) In view of (1.24), the condition $\delta_p(\mathbf{A}, \mathbf{B}; c) = 0$ implies that

$$\sum_{i=1}^{p} a_i = \sum_{j=1}^{m} \min(b_j, cp).$$

Since the components of $\mathbf{B}^{**}$ and $\mathbf{A}''$ are defined as

$$b_j^{**} = b_j'' = \min(b_j, cp) = \begin{cases} cp, & 1 \le j \le m, \quad b_j \ge cp, \\ b_j, & 1 \le j \le m, \quad b_j < cp, \end{cases}$$
$$a_i'' = a_i, \quad 1 \le i \le p,$$

the vectors $\mathbf{A}''$ and $\mathbf{B}^{**}$ make up a closed system:

$$\sum_{i=1}^{p} a_i'' = \sum_{j=1}^{m} b_j^{**}.$$

The vector pairs $\mathbf{A}$, $\mathbf{B}$; $\mathbf{A}'$, $\mathbf{B}^*$; and $\mathbf{A}''$, $\mathbf{B}^{**}$ satisfy the equations

$$\sum_{i=1}^{n-p} a_i' = \sum_{i=1}^{n} a_i - \sum_{i=1}^{p} a_i'', \quad \sum_{j=1}^{m} b_j^* = \sum_{j=1}^{m} b_j - \sum_{j=1}^{m} b_j^{**}.$$

Therefore, the balance condition for $\mathbf{A}$ and $\mathbf{B}$ implies that

$$\sum_{i=1}^{n-p} a_i' = \sum_{j=1}^{m} b_j^*,$$

and (2.5) is proved.

Applying relations (2.3) and (2.4) to the pairs $\mathbf{A}$, $\mathbf{B}$; $\mathbf{A}'$, $\mathbf{B}^* = \mathbf{B}'$; and $\mathbf{A}''$, $\mathbf{B}^{**} = \mathbf{B}''$ and taking into account the nonnegativity of the function $\delta_k(\mathbf{A},\mathbf{B};c)$, $1 \le k \le n$, and the equality $\delta_p(\mathbf{A},\mathbf{B};c) = 0$, we find that conditions (2.6) are valid:

$$\delta_k(\mathbf{A}',\mathbf{B}^*;c) = \delta_{p+k}(\mathbf{A},\mathbf{B};c) \ge 0, \quad 1 \le k \le n-p,$$

$$\delta_k(\mathbf{A}'',\mathbf{B}^{**};c) = \delta_k(\mathbf{A},\mathbf{B};c) \ge 0, \quad 1 \le k \le p.$$

(b) This proposition is a generalization of (a). If $\delta_p(\mathbf{A},\mathbf{B};c) = 0$, then the vector $(\beta_1,\ldots,\beta_m)$ is zero.

It is evident that $\mathbf{B}^*, \mathbf{B}^{**} \in R^m_+$ (see (2.7)). By definition (2.2) of $\mathbf{B}''$ and expression (1.24) for function (1.22) it holds that

$$\sum_{j=1}^{m} b''_j = \sum_{j=1}^{m} \min(b_j, cp) = \sum_{i=1}^{p} a_i + \delta_p(\mathbf{A},\mathbf{B};c).$$

Then, using the first condition in (2.7), we obtain the balance condition for the vectors $\mathbf{A}''$ and $\mathbf{B}^{**}$:

$$\sum_{j=1}^{m} b^{**}_j = \sum_{j=1}^{m} b''_j - \delta_p(\mathbf{A},\mathbf{B};c) = \sum_{i=1}^{p} a_i = \sum_{i=1}^{p} a''_i.$$

Therefore, the fact that vectors $\mathbf{A}$ and $\mathbf{B}$ make up a closed system implies that $\mathbf{A}'$ and $\mathbf{B}^*$ satisfy equation (2.5):

$$\sum_{j=1}^{m} b^*_j = \sum_{j=1}^{m} b'_j + \delta_p(\mathbf{A},\mathbf{B};c) = \sum_{j=1}^{m} b_j - \sum_{j=1}^{m} b''_j + \delta_p(\mathbf{A},\mathbf{B};c) =$$
$$\sum_{j=1}^{m} b_j - \sum_{i=1}^{p} a_i = \sum_{i=p+1}^{n} a_i = \sum_{i=1}^{n-p} a'_i.$$

To prove inequality (2.6), we must analyse the behavior of function (1.22) when the components of the second vector are changed.

Lemma 2.4. *Suppose that* $\mathbf{A} \in R^n_+$, $\mathbf{B} = (b_1,\ldots,b_m) \in R^m_+$, $q \in Z$, $1 \le q \le m$, $\alpha, \beta \in R$, $\alpha \ge 0$, $0 \le \beta \le b_q$, *and vectors* $\mathbf{C} = (c_1,\ldots,c_m)$ *and* $\mathbf{D} = (d_1,\ldots,d_m)$ *are derived from the vector* $\mathbf{B}$ *according to the following rules:*

$$c_j = \begin{cases} b_j, & 1 \le j \le m, \quad j \ne q, \\ b_j + \alpha, & j = q, \end{cases} \qquad d_j = \begin{cases} b_j, & 1 \le j \le m, \quad j \ne q, \\ b_q - \beta, & j = q. \end{cases}$$

Then, the values of function (1.22) *for the vector pairs* **A**, **B**; **A**, **C**; *and* **A**, **D** *are related to each other by the following equations*:

$$\delta_k(\mathbf{A},\mathbf{C};c) = \begin{cases} \delta_k(\mathbf{A},\mathbf{B};c), & b_q \geq ck, \\ \delta_k(\mathbf{A},\mathbf{B};c) + ck - b_q, & ck - \alpha \leq b_q < ck, \\ \delta_k(\mathbf{A},\mathbf{B};c) + \alpha, & b_q < ck - \alpha; \end{cases}$$
$$\delta_k(\mathbf{A},\mathbf{D};c) = \begin{cases} \delta_k(\mathbf{A},\mathbf{B};c) - \beta, & b_q < ck, \\ \delta_k(\mathbf{A},\mathbf{B};c) - ck - \beta + b_q, & ck \leq b_q < ck + \beta, \\ \delta_k(\mathbf{A},\mathbf{B};c), & b_q \geq ck + \beta. \end{cases} \tag{2.9}$$

The proof follows immediately from the definition of function (1.22).

Consider the vectors $\mathbf{A}'$, $\mathbf{B}'$, and $\mathbf{B}^*$. Since the components of $\mathbf{B}^*$ are not less than the corresponding components of $\mathbf{B}'$, it follows from (2.9) that $\delta_k(\mathbf{A}',\mathbf{B}^*;c) \geq \delta_k(\mathbf{A}',\mathbf{B}';c)$, $1 \leq k \leq n-p$. Formula (2.3) relates values of function (1.22) of vectors **A** and **B** to those of the same function of vectors $\mathbf{A}'$ and $\mathbf{B}'$: $\delta_k(\mathbf{A}',\mathbf{B}';c) = \delta_{p+k}(\mathbf{A},\mathbf{B};c) - \delta_p(\mathbf{A},\mathbf{B};c)$. Since $\delta_p(\mathbf{A},\mathbf{B};c)$ is the minimum value of the function $\delta_k(\mathbf{A},\mathbf{B};c)$ for $1 \leq k \leq n-1$, we have

$$\delta_k(\mathbf{A}',\mathbf{B}^*;c) \geq \delta_k(\mathbf{A}',\mathbf{B}';c) = \delta_{p+k}(\mathbf{A},\mathbf{B};c) - \delta_p(\mathbf{A},\mathbf{B};c) \geq 0$$

for any k, $1 \leq k \leq n-p$.

Consider the vectors $\mathbf{A}''$, $\mathbf{B}''$, and $\mathbf{B}^{**}$. It follows from (2.9) that the decrease of the jth component of the second vector by β_j results in a decrease of no more than β_j in function (1.22). The components of the vectors $\mathbf{B}''$ and $\mathbf{B}^{**}$ satisfy the equations $b_j^{**} = b_j'' - \beta_j$, $1 \leq j \leq m$. Therefore, it follows from the first equality in (2.7) that, when the vector $\mathbf{B}''$ is replaced by the vector $\mathbf{B}^{**}$, the decrease in function (1.22) is not greater than $\delta_p(\mathbf{A},\mathbf{B};c)$: $\delta_k(\mathbf{A}'',\mathbf{B}'';c) - \delta_k(\mathbf{A}'',\mathbf{B}^{**};c) \leq \delta_p(\mathbf{A},\mathbf{B};c)$. Taking into account (2.4) and the definition of $\delta_p(\mathbf{A},\mathbf{B};c)$, we find that

$$\delta_k(\mathbf{A}'',\mathbf{B}^{**};c) \geq \delta_k(\mathbf{A}'',\mathbf{B}'';c) - \delta_p(\mathbf{A},\mathbf{B};c) = \delta_k(\mathbf{A},\mathbf{B};c) - \delta_p(\mathbf{A},\mathbf{B};c) \geq 0$$

for any k, $1 \leq k \leq p$, which proves (2.6).

(c) Let $(\beta_1,\ldots,\beta_m)$ be defined by (2.8). Since **A** and **B** satisfy (1.1) and function (1.22) is nonnegative for any $k \in Z$, $1 \leq k \leq n$, we have

$$\delta_n(\mathbf{A},\mathbf{B};c) = \sum_{j=1}^{m} b_j - \sum_{b_j \geq cn} (b_j - cn) - \sum_{i=1}^{n} a_i = -\sum_{b_j \geq cn} (b_j - cn) \geq 0$$

and $b_j \leq cn$ for any j, $1 \leq j \leq m$. Hence,

$$\min\{cp, cn - cp, b_{m-i}, cn - b_{m-i}\} \geq 0, \quad 0 \leq i \leq m-1,$$

$$\min\left\{cp, cn - cp, b_m, cn - b_m, \delta_p(\mathbf{A}, \mathbf{B}; c)\right\} \geq 0,$$

and (2.8) implies that $\beta_m \geq 0$. Suppose that the vector $(\beta_1, \dots, \beta_m)$ has negative components, and let i $(1 \leq i \leq m-1)$ be the least index for which $\beta_{m-i} < 0$. Then, it follows from (2.8) that

$$\beta_{m-i} = \delta_p(\mathbf{A}, \mathbf{B}; c) - \sum_{j=m-i+1}^{m+1} \beta_j = \delta_p(\mathbf{A}, \mathbf{B}; c) - \sum_{j=m-i+2}^{m+1} \beta_j - \beta_{m-i+1} < 0. \quad (2.10)$$

On the other hand, (2.8) also implies that

$$\beta_{m-i+1} \leq \delta_p(\mathbf{A}, \mathbf{B}; c) - \sum_{j=m-i+2}^{m+1} \beta_j,$$

which contradicts (2.10). Hence, all components of $(\beta_1, \dots, \beta_m)$ are nonnegative; i.e., $(\beta_1, \dots, \beta_m) \in R_+^m$. Note here that if $\beta_k = 0$, then $\beta_j = 0$ for any j, $1 \leq j \leq k$.

Let us show that the vector $(\beta_1, \dots, \beta_m)$ constructed according to (2.8) satisfies conditions (2.7).

It follows from (2.8) that

$$\beta_j \leq \min\{cp, cn - cp, b_j, cn - b_j\}, \quad 1 \leq j \leq m.$$

Hence, it follows from definitions (2.1) and (2.2) of the vectors $\mathbf{B}'$ and $\mathbf{B}''$ that

$$b'_j + \beta_j = \begin{cases} b_j - cp + \beta_j \leq b_j - cp + cn - b_j = cn - cp, & b_j \geq cp, \\ \beta_j \leq cn - cp, \quad b_j < cp; \end{cases}$$

$$b''_j - \beta_j = \begin{cases} cp - \beta_j \geq cp - cp = 0, & b_j \geq cp, \\ b_j - \beta_j \geq b_j - b_j = 0, & b_j < cp. \end{cases}$$

for any j, $1 \leq j \leq m$. Thus, the two inequalities in (2.7) are proved.

Let us show that

$$\sum_{j=1}^{m} \beta_j = \delta_p(\mathbf{A}, \mathbf{B}; c).$$

It follows from (2.8) and from the nonnegativity of the components of $(\beta_1, \dots, \beta_m)$ that

$$\beta_j \leq \delta_p(\mathbf{A}, \mathbf{B}; c) - \sum_{j=2}^{m+1} \beta_j, \quad \delta_p(\mathbf{A}, \mathbf{B}; c) \geq \sum_{j=1}^{m} \beta_j,$$

where $\beta_{m+1} = 0$.

Suppose that

$$\delta_p(\mathbf{A}, \mathbf{B}; c) > \sum_{j=1}^{m} \beta_j. \tag{2.11}$$

In view of (2.8), the components of $(\beta_1, \ldots, \beta_m)$ satisfy the inequalities

$$\beta_{m-i} \leq \delta_p(\mathbf{A}, \mathbf{B}; c) - \sum_{j=m-i+1}^{m+1} \beta_j, \quad 0 \leq i \leq m-1.$$

Using assumption (2.11), we find that

$$\beta_{m-i} < \delta_p(\mathbf{A}, \mathbf{B}; c) - \sum_{j=m-i+1}^{m+1} \beta_j \tag{2.12}$$

for any i, $0 \leq i \leq m-1$; otherwise, the equality

$$\beta_{m-i_1} = \delta_p(\mathbf{A}, \mathbf{B}; c) - \sum_{j=m-i_1+1}^{m+1} \beta_j$$

for some i_1 would imply, in view of (2.8), that

$$\delta_p(\mathbf{A}, \mathbf{B}; c) = \sum_{j=m-i_1}^{m+1} \beta_j, \quad \beta_{m-i} = 0, \quad 0 \leq i \leq i_1 - 1$$

and $\delta_p(\mathbf{A}, \mathbf{B}; c) = \beta_1 + \cdots + \beta_m$, which contradicts (2.11).

Thus, assumption (2.11) leads to inequality (2.12) and the relation

$$\beta_{m-i} = \min\{cp, cn - cp, b_{m-i}, cn - b_{m-i}\}, \quad 0 \leq i \leq m-1$$

or, equivalently,

$$\beta_j = \min\{cp, cn - cp, b_j, cn - b_j\}, \quad 1 \leq j \leq m.$$

Consider the vectors $\mathbf{B}^*$, $\mathbf{B}'$, and $\mathbf{B}$. If $b_j \geq cp$, then $cn - cp \geq cn - b_j$ and $\beta_j = \min\{cp, cn - b_j\}$. In this case, it follows from (2.1) that

$$b_j^* = b_j' + \beta_j = b_j - cp + \min\{cp, cn - b_j\} = \min\left\{b_j, c(n-p)\right\}.$$

If $b_j < cp$, then $cn - cp < cn - b_j$ and $\beta_j = \min\{b_j, cn - cp\}$. Hence, in view of (2.1), we have

$$b_j^* = b_j' + \beta_j = 0 + \beta_j = \min\left\{b_j, c(n-p)\right\}.$$

Thus, $b_j^* = \min\{b_j, c(n-p)\}$ for any j, $1 \le j \le m$, and

$$\sum_{j=1}^{m} b_j^* = \sum_{j=1}^{m} \min\{b_j, c(n-p)\}. \qquad (2.13)$$

Applying (1.24) and (2.2) to the characteristic function $\delta_p(\mathbf{A}, \mathbf{B}; c)$, we obtain

$$\delta_p(\mathbf{A}, \mathbf{B}; c) = \sum_{j=1}^{m} \min(b_j, cp) - \sum_{i=1}^{p} a_i = \sum_{j=1}^{m} b_j'' - \sum_{i=1}^{p} a_i.$$

Then, assumption (2.11) results in the inequalities

$$\sum_{j=1}^{m} \beta_j < \sum_{j=1}^{m} b_j'' - \sum_{i=1}^{p} a_i, \quad \sum_{i=1}^{p} a_i < \sum_{j=1}^{m} b_j'' - \sum_{j=1}^{m} \beta_j,$$

and, in view of (2.2) and the definition of $\mathbf{B}^{**}$, the vectors $\mathbf{A}''$ and $\mathbf{B}^{**}$ do not satisfy the balance condition:

$$\sum_{i=1}^{p} a_i'' < \sum_{j=1}^{m} b_j^{**}.$$

Since the vectors $\mathbf{A}$ and $\mathbf{B}$ make up a closed system and (see (2.1) and the definition of $\mathbf{B}^*$)

$$\sum_{i=1}^{n-p} a_i' = \sum_{i=1}^{n} a_i - \sum_{i=1}^{p} a_i'', \quad \sum_{j=1}^{m} b_j^* = \sum_{j=1}^{m} b_j - \sum_{j=1}^{m} b_j^{**},$$

it holds that

$$\sum_{i=1}^{n-p} a_i' > \sum_{j=1}^{m} b_j^*;$$

i.e., the vectors $\mathbf{A}'$ and $\mathbf{B}^*$ do not satisfy (1.1).

Using (2.13), consider the value of function (1.22) (or (1.24)) for $k = n-p$:

$$\delta_{n-p}(\mathbf{A}, \mathbf{B}; c) = \sum_{j=1}^{m} \min\{b_j, c(n-p)\} - \sum_{i=1}^{n-p} a_i =$$

$$\sum_{j=1}^{m} b_j^* - \sum_{i=1}^{n-p} a_i' + \sum_{i=1}^{n-p} a_i' - \sum_{i=1}^{n-p} a_i < \sum_{i=1}^{n-p} a_i' - \sum_{i=1}^{n-p} a_i = \sum_{i=p+1}^{n} a_i - \sum_{i=1}^{n-p} a_i.$$

Since $\mathbf{A} \in \overline{R}^n_+$ ($a_i \geq a_{i+1}$, $1 \leq i \leq n-1$), the value $\delta_{n-p}(\mathbf{A}, \mathbf{B}; c)$ of the characteristic function satisfies the inequality

$$\delta_{n-p}(\mathbf{A}, \mathbf{B}; c) < \sum_{i=p+1}^{n} a_i - \sum_{i=1}^{n-p} a_i \leq 0.$$

Thus, assumption (2.11) implies that $\delta_{n-p}(\mathbf{A}, \mathbf{B}; c) < 0$, which contradicts the nonnegativity of the function $\delta_k(\mathbf{A}, \mathbf{B}; c)$ for any k, $1 \leq k \leq n$. Hence, the first condition in (2.7) is also satisfied. The lemma is proved.

2.2. Existence Criterion for Matrices with a Given Bound

We now formulate necessary and sufficient conditions for the existence of matrices of the truncated polyhedron $M(\mathbf{A}, \mathbf{B}; c)$. The proof of sufficiency is based on Lemma 2.3.

Suppose that vectors $\mathbf{A}$ and $\mathbf{B}$ satisfy conditions (1.1) and $c \leq 0$, where $\mathbf{A} = (a_1)$ is a 1-component vector and $\mathbf{B} = (b_1, \ldots, b_m) \in R^m_+$, and let $\delta_1(\mathbf{A}, \mathbf{B}; c) \geq 0$ (here, $n = 1$). Then, the set of matrices $M(\mathbf{A}, \mathbf{B})$ contains only one matrix $X(\mathbf{A}, \mathbf{B}) = (x_{ij})$, $i = 1$, $1 \leq j \leq m$, where $x_{1j} = b_j$, $1 \leq j \leq m$. It follows from (1.22) and (1.1) that

$$\begin{cases} \displaystyle\sum_{j=1}^{m} b_j - \sum_{b_j \geq c} (b_j - c) \geq a_1, \\ \displaystyle\sum_{j=1}^{m} b_j = a_1. \end{cases}$$

Hence,

$$-\sum_{b_j \geq c} (b_j - c) \geq 0,$$

which is possible only if $b_j \leq c$, $1 \leq j \leq m$. Since $x_{1j} = b_j$, $1 \leq j \leq m$, the matrix $(x_{1j}) \in M(\mathbf{A}, \mathbf{B}; c) = M(\mathbf{A}, \mathbf{B})$ and $M(\mathbf{A}, \mathbf{B}; c) \neq \varnothing$, which proves the sufficiency of condition (1.21) or (1.23) in the case of $n = 1$.

Let us now apply the method of mathematical induction.

Suppose that $M(\mathbf{A}, \mathbf{B}; c) \neq \varnothing$ if the number of components of $\mathbf{A}$ does not exceed $n-1$, $n \geq 2$, and the characteristic function (1.22) of $\widetilde{\mathbf{A}}$ and $\mathbf{B}$ is nonnegative for every admissible value of k, $k \in Z$.

Suppose that $\mathbf{A} \in R^n_+$ and $\mathbf{B} \in R^m_+$ satisfy the balance condition and $\delta_k(\widetilde{\mathbf{A}}, \mathbf{B}; c) \geq 0$ for any k, $1 \leq k \leq n$. Then we apply Lemma 2.3 to construct the pairs of vectors $\widetilde{\mathbf{A}}' = (\tilde{a}_{p+1}, \ldots, \tilde{a}_n) \in \overline{R}^{n-p}_+$, $\mathbf{B}^* \in R^m_+$ and $\widetilde{\mathbf{A}}'' = (\tilde{a}_1, \ldots, \tilde{a}_p) \in \overline{R}^p_+$, $\mathbf{B}^{**} \in R^m_+$ based on $\widetilde{\mathbf{A}}$ and $\mathbf{B}$ that satisfy the balance conditions and the inequalities $\delta_k(\widetilde{\mathbf{A}}', \mathbf{B}^*; c) \geq 0$ and $\delta_k(\widetilde{\mathbf{A}}'', \mathbf{B}^{**}; c) \geq 0$ for any k, where $1 \le k \le n-p$ and $1 \le k \le p$, respectively. By the induction hypothesis, $M(\widetilde{\mathbf{A}}', \mathbf{B}^*; c) \ne \varnothing$ and $M(\widetilde{\mathbf{A}}'', \mathbf{B}^{**}; c) \ne \varnothing$. Suppose that $X(\widetilde{\mathbf{A}}', \mathbf{B}^*) = (x'_{ij}) \in M(\widetilde{\mathbf{A}}', \mathbf{B}^*; c)$ and $X(\widetilde{\mathbf{A}}'', \mathbf{B}^{**}) = (x''_{ij}) \in M(\widetilde{\mathbf{A}}'', \mathbf{B}^{**}; c)$. Since $\mathbf{B} = \mathbf{B}^* + \mathbf{B}^{**}$ and $\widetilde{\mathbf{A}}$ is the union of the components of $\widetilde{\mathbf{A}}''$ and $\widetilde{\mathbf{A}}'$, the matrix (x_{ij}), where

$$x_{ij} = \begin{cases} x''_{ij}, & 1 \le i \le p, \quad 1 \le j \le m, \\ x'_{i-pj}, & p+1 \le i \le n, \quad 1 \le j \le m, \end{cases}$$

belongs to the matrix set $M(\widetilde{\mathbf{A}}, \mathbf{B}; c)$, and $M(\widetilde{\mathbf{A}}, \mathbf{B}; c) \ne \varnothing$. Since matrices from $M(\mathbf{A}, \mathbf{B}; c)$ and $M(\widetilde{\mathbf{A}}, \mathbf{B}; c)$ differ by the order of their rows, we have $M(\mathbf{A}, \mathbf{B}; c) \ne \varnothing$. Hence, balance condition (1.1) and the nonnegativity of function (1.22) are sufficient for $M(\mathbf{A}, \mathbf{B}; c)$ to be nonempty.

Suppose that $M(\mathbf{A}, \mathbf{B}; c) \ne \varnothing$ and $X(\mathbf{A}, \mathbf{B}) = (x_{ij}) \in M(\mathbf{A}, \mathbf{B}; c)$, where $\mathbf{A} \in R^n_+$ and $\mathbf{B} \in R^m_+$. Let $k \in Z$, $1 \le k \le n$. The balance condition for $\mathbf{A}$ and $\mathbf{B}$ is equivalent to the equation

$$\sum_{j=1}^{m} b_j - \sum_{i=1}^{k} a_i = \sum_{i>k} a_i.$$

Taking into account the transportation conditions (1.2) for the matrix $X(\mathbf{A}, \mathbf{B})$, we rewrite the right-hand side of this equation as

$$\sum_{i>k} a_i = \sum_{i>k} \sum_{j=1}^{m} x_{ij} = \sum_{j=1}^{m} \sum_{i>k} x_{ij} = \sum_{j=1}^{m} \left(b_j - \sum_{i=1}^{k} x_{ij} \right).$$

Then, the equation equivalent to the balance condition takes the form

$$\sum_{j=1}^{m} b_j - \sum_{i=1}^{k} a_i = \sum_{j=1}^{m} \left(b_j - \sum_{i=1}^{k} x_{ij} \right).$$

Since $x_{ij} \le c$ for any i, j, $1 \le i \le n$, $1 \le j \le m$, we have

$$\sum_{j=1}^{m} b_j - \sum_{i=1}^{k} a_i \ge \sum_{b_j \ge ck} \left(b_j - \sum_{i=1}^{k} x_{ij} \right) \ge \sum_{b_j \ge ck} (b_j - ck).$$

Hence,

$$\sum_{j=1}^{m} b_j - \sum_{i=1}^{k} a_i \ge \sum_{b_j \ge ck} (b_j - ck),$$

which proves that conditions (1.21) (or (1.23)) are necessary for the nonemptiness of $M(\mathbf{A}, \mathbf{B}; c)$.

Thus, we proved the sufficiency of conditions (1.21) (or (1.23)).

Lemma 2.5. *The set of matrices $M(\mathbf{A}, \mathbf{B}; c)$ is not empty for $\mathbf{A} \in R_+^n$ if and only if $\delta_k(\widetilde{\mathbf{A}}, \mathbf{B}; c) \geq 0$ for any k, $1 \leq k \leq n$.*

Note that, if $M(\mathbf{A}, \mathbf{B}; c) \neq \varnothing$, the characteristic function (1.22) is nonnegative for any arrangement of the components of the vector $\mathbf{A}$. It can be readily shown that the inequalities $\delta_k(\mathbf{A}, \mathbf{B}; c) \geq \delta_k(\widetilde{\mathbf{A}}, \mathbf{B}; c)$, $0 \leq k \leq n$ follow directly from formula (1.22). The requirement that the components of the first vector are arranged in nonascending order is important, which is illustrated by the following example.

Example 2.1. Let $\mathbf{A} = (1, 2, 3)$, $\mathbf{B} = (3, 3)$ and $c = 1$. Clearly, the sums of components of $\mathbf{A}$ and $\mathbf{B}$ are equal, and $\delta_1(\mathbf{A}, \mathbf{B}; 1) = \delta_2(\mathbf{A}, \mathbf{B}; 1) = 1$, $\delta_3(\mathbf{A}, \mathbf{B}; 1) = 0$. The assumption that $M(\mathbf{A}, \mathbf{B}; 1) \neq \varnothing$ implies that both columns consist of ones. Then, the sum of the elements of the first row is 2; however, the first component of $\mathbf{A}$ is 1. Hence, $M((1, 2, 3), (3, 3); 1) = \varnothing$.

2.3. Existence Criterion for a Transportation Matrix in the Form of an Irreducible Set of Inequalities

In what follows, we eliminate the dependent inequalities from those determined by Lemma 2.5. To do this, let us establish some properties of the characteristic function (1.22).

It can be readily shown that function (1.22) is certainly nonnegative for some $k \in Z$.

Lemma 2.6. *Let $\mathbf{A} = (a_1, \ldots, a_n) \in R_+^n$, $\mathbf{B} = (b_1, \ldots, b_m) \in R_+^m$, $c \geq 0$ and $k \in Z$, $1 \leq k \leq n$.*

(a) *If $\min\{b_j : 1 \leq j \leq m\} \geq ck$ and $a_i \leq cm$ for $i \leq k$, then $\delta_k(\mathbf{A}, \mathbf{B}; c) \geq 0$.*

(b) *If $\max\{b_j : 1 \leq j \leq m\} \leq ck$ and $\mathbf{A}$ and $\mathbf{B}$ satisfy the balance condition, then $\delta_k(\mathbf{A}, \mathbf{B}; c) \geq 0$.*

The proof is easy to derive from formula (1.24): if the assumptions in (a) hold, we have

$$\delta_k(\mathbf{A}, \mathbf{B}; c) = cmk - \sum_{i=1}^{k} a_i;$$

if the assumptions in (b) hold, we obtain

$$\delta_k(\mathbf{A},\mathbf{B};c) = \sum_{j=1}^{m} b_j - \sum_{i=1}^{k} a_i.$$

In the following lemma, bounds on the values of (1.22) as functions of consecutive values of $k \in Z$ are given.

Lemma 2.7. *Let* $\mathbf{A} = (a_1, \ldots, a_n) \in R^n_+$, $\mathbf{B} = (b_1, \ldots, b_m) \in R^m_+$, *and* $c \geq 0$, *where* $n \geq 2$. *Then, function* (1.22) *satisfies the inequalities*

$$\delta_k(\mathbf{A},\mathbf{B};c) \geq \delta_{k-1}(\mathbf{A},\mathbf{B};c) - a_k + cl_k(\mathbf{B};c), \quad k > 1, \tag{2.14}$$

$$\delta_k(\mathbf{A},\mathbf{B};c) \geq \delta_{k+1}(\mathbf{A},\mathbf{B};c) + a_{k+1} - cl_k(\mathbf{B};c), \quad k < n. \tag{2.15}$$

Proof. For $k > 1$, consider the difference $\delta_k(\mathbf{A},\mathbf{B};c) - \delta_{k-1}(\mathbf{A},\mathbf{B};c)$. Since

$$\sum_{b_j \geq c(k-1)} (b_j - c(k-1)) - \sum_{b_j \geq ck} (b_j - ck) =$$
$$\sum_{c(k-1) \leq b_j < ck} b_j + ckl_k(\mathbf{B};c) - c(k-1)l_{k-1}(\mathbf{B};c) =$$
$$\sum_{c(k-1) \leq b_j < ck} b_j + c(k-1)\left(l_k(\mathbf{B};c) - l_{k-1}(\mathbf{B};c)\right) + cl_k(\mathbf{B};c) \geq$$
$$c(k-1)\left(l_{k-1}(\mathbf{B};c) - l_k(\mathbf{B};c)\right) + c(k-1)\left(l_k(\mathbf{B};c) - l_{k-1}(\mathbf{B};c)\right) +$$
$$cl_k(\mathbf{B};c) = cl_k(\mathbf{B};c),$$

we have

$$\delta_k(\mathbf{A},\mathbf{B};c) - \delta_{k-1}(\mathbf{A},\mathbf{B};c) =$$
$$-a_k - \sum_{b_j \geq ck} (b_j - ck) + \sum_{b_j \geq c(k-1)} (b_j - c(k-1)) \geq -a_k + cl_k(\mathbf{B};c).$$

Hence, $\delta_k(\mathbf{A},\mathbf{B};c) \geq \delta_{k-1}(\mathbf{A},\mathbf{B};c) - a_k + cl_k(\mathbf{B};c)$.

For $k < n$,

$$\sum_{b_j \geq c(k+1)} (b_j - c(k+1)) - \sum_{b_j \geq ck} (b_j - ck) =$$
$$- \sum_{ck \leq b_j < c(k+1)} b_j - c(k+1)l_{k+1}(\mathbf{B};c) + ckl_k(\mathbf{B};c) =$$

$$- \sum_{ck \le b_j < c(k+1)} b_j - ck\left(l_{k+1}(\mathbf{B};c) - l_k(\mathbf{B};c)\right) - cl_{k+1}(\mathbf{B};c) \ge$$
$$-c(k+1)\left(l_k(\mathbf{B};c) - l_{k+1}(\mathbf{B};c)\right) - ck\left(l_{k+1}(\mathbf{B};c) - l_k(\mathbf{B};c)\right) - cl_{k+1}(\mathbf{B};c) = -cl_k(\mathbf{B};c).$$

Hence,

$$\delta_k(\mathbf{A},\mathbf{B};c) - \delta_{k+1}(\mathbf{A},\mathbf{B};c) = a_{k+1} - \sum_{b_j \ge ck}(b_j - ck) + \sum_{b_j \ge c(k+1)}(b_j - c(k+1)) \ge a_{k+1} - cl_k(\mathbf{B};c)$$

and $\delta_k(\mathbf{A},\mathbf{B};c) \ge \delta_{k+1}(\mathbf{A},\mathbf{B};c) + a_{k+1} - cl_k(\mathbf{B};c)$.

By Lemma 2.7 for certain values of $k \in Z$, the nonnegativity of function (1.22) can be derived from its nonnegativity for $k-1$ and $k+1$.

Lemma 2.8. *If* $\mathbf{A} = (a_1, \ldots, a_n) \in R_+^n$, $n \ge 2$, $\mathbf{B} \in R_+^m$, $c \ge 0$, *and* $a_k = a_{k+1}$, *then the inequalities* $\delta_{k-1}(\mathbf{A},\mathbf{B};c) \ge 0$ *and* $\delta_{k+1}(\mathbf{A},\mathbf{B};c) \ge 0$ *imply that* $\delta_k(\mathbf{A},\mathbf{B};c) \ge 0$.

Proof. Adding inequalities (2.14) and (2.15) together term by term and dividing the result by two, we obtain the required inequality.

The following result is also derived from inequalities (2.14) and (2.15).

Lemma 2.9. *If the assumptions of Lemma* 2.7 *are fulfilled and* $k \in Z$, $1 < k < n$, *then*

$$\delta_k(\mathbf{A},\mathbf{B};c) \ge \left(\delta_{k-1}(\mathbf{A},\mathbf{B};c) + \delta_{k+1}(\mathbf{A},\mathbf{B};c) + a_{k+1} - a_k\right)/2.$$

The following lemma is a generalization of Lemma 2.8.

Lemma 2.10. *Let* $\mathbf{A} = (a_1, \ldots, a_n) \in R_+^n$, $n \ge 2$, $\mathbf{B} = (b_1, \ldots, b_m) \in R_+^m$, $c \ge 0$, *and* $a_k = \cdots = a_{k+t}$, *where* $1 \le t \le n-k$. *Then the inequalities* $\delta_{k-1}(\mathbf{A},\mathbf{B};c) \ge 0$ *and* $\delta_{k+t}(\mathbf{A},\mathbf{B};c) \ge 0$ *imply that* $\delta_{k+q}(\mathbf{A},\mathbf{B};c) \ge 0$ *for any* q, $0 \le k \le t-1$.

Proof. It will suffice to prove that the assumptions of the lemma imply that either $\delta_k(\mathbf{A},\mathbf{B};c) \ge 0$ or $\delta_{k+t-1}(\mathbf{A},\mathbf{B};c) \ge 0$. This reduced the range of indices for which the nonnegativity of function (1.22) is to be established. Applying the reduced assertion t times, we will prove the lemma.

In view of the equalities $a_k = \cdots = a_{k+t}$, it follows from Lemma 2.9 that

$$2\delta_{k+i}(\mathbf{A},\mathbf{B};c) \ge \delta_{k+i-1}(\mathbf{A},\mathbf{B};c) + \delta_{k+i+1}(\mathbf{A},\mathbf{B};c), \quad 0 \le i \le t-1.$$

Adding these inequalities termwise together, we obtain

$$2\sum_{i=0}^{t-1}\delta_{k+i}(\mathbf{A},\mathbf{B};c) \geq \sum_{i=0}^{t-1}\delta_{k+i-1}(\mathbf{A},\mathbf{B};c) + \sum_{i=0}^{t-1}\delta_{k+i+1}(\mathbf{A},\mathbf{B};c),$$

which is equivalent to

$$\delta_k(\mathbf{A},\mathbf{B};c) + \delta_{k+t-1}(\mathbf{A},\mathbf{B};c) \geq \delta_{k-1}(\mathbf{A},\mathbf{B};c) + \delta_{k+t}(\mathbf{A},\mathbf{B};c).$$

Since the right-hand side of this inequality is nonnegative, at least one addend on the left-hand side is nonnegative. The lemma is proved.

For a vector $\mathbf{B} = (b_1, \ldots, b_m) \in R_+^m$ and numbers n and c, where $n \in Z$, $n \geq 1$, $c \geq 0$, and $\max\{b_j : 1 \leq j \leq m\} \leq cn$, we define:

$$k'(\mathbf{B}; n, c) = \max\left\{k \in Z : 0 \leq k \leq n,\ \min\{b_j : 1 \leq j \leq m\} \geq ck\right\},$$

$$k''(\mathbf{B}; n, c) = \min\left\{k \in Z : 0 \leq k \leq n,\ \max\{b_j : 1 \leq j \leq m\} \leq ck\right\},$$

$$k^*(\mathbf{B}; n, c) = k'(\mathbf{B}; n, c) + 1 \text{ and } k^{**}(\mathbf{B}; n, c) = k''(\mathbf{B}; n, c) - 1.$$

When vectors $\mathbf{A} = (a_1, \ldots, a_n) \in R_+^n$ and $\mathbf{B} \in R_+^m$ have equal sums of components and $\max\{a_i : 1 \leq i \leq n\} \leq cm$, Lemma 2.6 implies that the function $\delta_k(\mathbf{A},\mathbf{B};c)$ is nonnegative for $k \leq k'(\mathbf{B}; n, c)$ and $k \geq k''(\mathbf{B}; n, c)$. Clearly, $k'(\mathbf{B}; n, c) \leq k''(\mathbf{B}; n, c)$; moreover, $k'(\mathbf{B}; n, c) = k''(\mathbf{B}; n, c)$ if and only if there exists $k \in Z$ such that $b_j = ck$, $1 \leq j \leq m$. Note also that, if $b_j = b$, $1 \leq j \leq m$, then $k''(\mathbf{B}; n, c) - k'(\mathbf{B}; n, c) \leq 1$. If $k''(\mathbf{B}; n, c) - k'(\mathbf{B}; n, c) \geq 2$, we define

$$K(\mathbf{B},\mathbf{A}; n, c) = \left\{k^*(\mathbf{B}; n, c), k^{**}(\mathbf{B}; n, c)\right\} \cup \left\{k \in Z : k^*(\mathbf{B}; n, c) < k < k^{**}(\mathbf{B}; n, c),\ \tilde{a}_k > \tilde{a}_{k+1}\right\}.$$

The number of inequalities in Lemma 2.5 that quarantee the nonemptiness of $M(\mathbf{A},\mathbf{B};c)$ can be reduced in the general case. The inequality $\delta_k(\mathbf{A},\mathbf{B};c) \geq 0$ certainly holds for small and large $k \in Z$ by Lemma 2.6; it may not be verified for k such that $a_k = a_{k+1}$ (Lemma 2.10). Note that, if $\mathbf{A}$ and $\mathbf{B}$ have dimensions n and m, respectively, then the inequalities $a_i \leq cm$ and $b_j \leq cn$, where $1 \leq i \leq n$ and $1 \leq j \leq m$, must hold for $M(\mathbf{A},\mathbf{B};c)$ to be nonempty.

Finally, we arrive at the main result of this subsection.

Theorem 2.1. *Suppose that $c \geq 0$, the components of $\mathbf{A} = (a_1, \ldots, a_n) \in R_+^n$ and $\mathbf{B} = (b_1, \ldots, b_m) \in R_+^m$ are such that*

$$\sum_{i=1}^{n} a_i = \sum_{j=1}^{m} b_j$$

and $a_i \le cm$ *and* $b_j \le cn$, *where* $1 \le i \le n$, $1 \le j \le m$, *and the vector* $\widetilde{\mathbf{A}} = (\tilde{a}_1, \ldots, \tilde{a}_n)$ *belonging to* $\overline{R}^n_+$ *is derived from* $\mathbf{A}$ *by arranging its components in nonascending order.*

(a) *The matrix set* $M(\mathbf{A}, \mathbf{B}; c)$ *is nonempty if and only if either* $k''(\mathbf{B}; n, c) - k'(\mathbf{B}; n, c) \le 1$ *or* $k''(\mathbf{B}; n, c) - k'(\mathbf{B}; n, c) \ge 2$ *and*

$$\delta_k(\widetilde{\mathbf{A}}, \mathbf{B}; c) = \sum_{j=1}^{m} b_j - \sum_{b_j \ge ck} (b_j - ck) - \sum_{i=1}^{k} \tilde{a}_i \ge 0 \tag{2.16}$$

for any $k \in K(\mathbf{B}, \mathbf{A}; n, c)$.

(b) *Suppose that* $k''(\mathbf{B}; n, c) - k'(\mathbf{B}; n, c) \ge 2$, $M(\mathbf{A}, \mathbf{B}; c) \ne \varnothing$,*and* p *is the number of elements in the set* $K(\mathbf{B}, \mathbf{A}; n, c)$. *If* $m \ge p+2$, *then the system of inequalities* (2.16) *does not contain dependent inequalities.*

(c) *Suppose that an arbitrary real number* $c > 0$ *and arbitrary integers* $n \ge 2$, $p \ge 1$, $m \ge p + 2$, *and* k_q, $1 \le q \le p$, *are such that* $k_1 \ge 1$, $k_p < n$ *and, if* $p \ge 2$, $k_q < k_{q+1}$ *for* $1 \le q \le p - 1$. *Let* x_i *and* y_i, *where* $1 \le i \le n$, $1 \le j \le m$, *be nonnegative variables satisfying the inequalities*

$$x_1 \le cm, \quad x_i \ge x_{i+1}, \quad 1 \le i \le n - 1.$$

Let

$$x_{k_1+1} = x_{k_2},$$

for $p \ge 2$ *and*

$$\begin{aligned} &x_{k_q} > x_{k_q+1}, \quad x_{k_q+1} = x_{k_{q+1}}, \quad 2 \le q \le p - 1; \\ &c(k_1 - 1) \le \min\{y_j : 1 \le j \le m\} < ck_1; \\ &ck_p < \max\{y_j : 1 \le j \le m\} \le c(k_p + 1); \\ &\sum_{i=1}^{n} x_i = \sum_{j=1}^{m} y_j \end{aligned} \tag{2.17}$$

for $p \ge 3$.

Then, the set of inequalities

$$\begin{cases} \displaystyle\sum_{j=1}^{m} y_j - \sum_{y_j \ge ck_q} (y_j - ck_q) - \sum_{i=1}^{k_q} x_i \ge 0, \\ 1 \le q \le p, \end{cases} \tag{2.18}$$

is consistent and irreducible.

Prior to proving this theorem, we want to emphasize these facts:

(1) The system of inequalities (2.16) is consistent with respect to the variables $\tilde{a}_i$ and b_j since $M(\mathbf{A}, \mathbf{B}; c) \neq \varnothing$.

(2) For items (b) and (c) of the theorem, the two sets coincide:

$$\{k_q : 1 \le q \le p\} = \Big\{k \in Z : k^*(\mathbf{B}; n, c) \le k \le k^{**}(\mathbf{B}; n, c) \text{ and } \tilde{a}_k > \tilde{a}_{k+1}, \text{ if } k^*(\mathbf{B}; n, c) < k < k^{**}(\mathbf{B}; n, c)\Big\},$$

when $\tilde{a}_i = x_i$, $1 \le i \le n$ and $b_j = y_j$, $1 \le j \le m$.

(3) The inequality $m \ge p + 2$ plays a basic role in (b).

Let us consider the last fact in some detail. If $p = 1$, then system (2.16) does not contain inequalities when $m = 1$; when $m = 2$, it contains not more than one inequality. Then, the only inequality in (2.16) is independent since it does not result from the constraints. It is easy to see that the system under consideration contains no more than one inequality also in the case of $p \ge 2$ and $m \ge 2$.

Let $p \ge 2$ and $3 \le m \le p+1$. Define $k_1 = k^*(\mathbf{B}; n, c)$ and $k_p = k^{**}(\mathbf{B}; n, c)$, and suppose that $k_i \in Z$, $2 \le i \le p - 1$, satisfy the inequalities

$$k_1 < k_i < k_{i+1} < k_p, \quad 2 \le i \le p - 2,$$

$$\tilde{a}_{k_i} > \tilde{a}_{k_i+1}, \quad 2 \le i \le p - 1.$$

Then, (2.16) is equivalent to the system of inequalities

$$\begin{cases} \delta_k(\widetilde{\mathbf{A}}, \mathbf{B}; c) \ge 0, \\ k \in \{k_i : 1 \le i \le p\}. \end{cases}$$

When $m \le p$, the definitions of k_1 and k_p imply that there exists a number $q \in Z$, $1 \le q \le p-1$, such that $l_{k_q}(\mathbf{B}; c) = l_{k_{q+1}}(\mathbf{B}; c)$ for any vector $\mathbf{B} \in R_+^m$. Hence,

$$\delta_{k_q}(\widetilde{\mathbf{A}}, \mathbf{B}; c) - \delta_{k_{q+1}}(\widetilde{\mathbf{A}}, \mathbf{B}; c) = \big(a - cl_{k_q}(\mathbf{B}; c)\big)\left(k_{q+1} - k_q\right),$$

where $a = \tilde{a}_i$, $k_q + 1 \le i \le k_{q+1}$. This equation shows that for $a \ge cl_{k_q}(\mathbf{B}; c)$ the inequality $\delta_{k_{q+1}}(\widetilde{\mathbf{A}}, \mathbf{B}; c) \ge 0$ results in $\delta_{k_q}(\widetilde{\mathbf{A}}, \mathbf{B}; c) \ge 0$, and, for $a < cl_{k_q}(\mathbf{B}; c)$ the inequality $\delta_{k_{q+1}}(\widetilde{\mathbf{A}}, \mathbf{B}; c) \ge 0$ results from the inequality $\delta_{k_q}(\widetilde{\mathbf{A}}, \mathbf{B}; c) \ge 0$.

If $m = p + 1$, then inequalities (2.16) are independent only in the case when $\min\{b_j : 1 \le j \le m\} < c$, $\max\{b_j : 1 \le j \le m\} > ck_p$ and

$$l_{k_i}(\mathbf{B}; c) = m - i, \quad 1 \le i \le p = m - 1.$$

As stated in proposition (b) of Theorem 2.1, system (2.16) does not contain dependent inequalities since the condition $m \geq p+2$ implies that

$$\Big\{\mathbf{B} = (b_1, \ldots, b_m) \in R_+^m : \min\{b_j : 1 \leq j \leq m\} \geq 0,$$
$$\max\{b_j : 1 \leq j \leq m\} \leq cn,\ l_{k_q}(\mathbf{B}; c) > l_{k_{q+1}}(\mathbf{B}; c),\ 1 \leq q \leq p-1\Big\} \neq \varnothing.$$

Proof of Theorem 2.1.

(a) The validity of this proposition follows from Lemmas 2.5, 2.6, and 2.10.

(b) This proposition is proved by proving proposition (c) of the theorem.

(c) For nonnegative variables x_i and y_j, $1 \leq i \leq n$, $1 \leq j \leq m$, we set

$$x_i = \begin{cases} cm, & i \leq k_1 - 1, \\ c(m-1), & i = k_1, \\ c(m-q), & k_{q-1} < i \leq k_q, \text{ if } p \geq 2, \quad 2 \leq q \leq p, \\ c(m-p-1), & i = k_p + 1, \\ 0, & i > k_p + 1; \end{cases} \tag{2.19}$$

$$y_j = \begin{cases} c(k_1 - 1), & j = m, \\ ck_q, & j = m - q, \quad 1 \leq q \leq p, \\ c(k_p + 1), & j \leq m - p - 1. \end{cases} \tag{2.20}$$

It is obvious that the components of the vectors $(x_1, \ldots, x_n)$ and $(y_1, \ldots, y_m)$ satisfy the corresponding inequalities in (2.17). Note also that $(x_1, \ldots, x_n) \in \overline{R}_+^n$ and $(y_1, \ldots, y_m) \in \overline{R}_+^m$. Let us prove that the last inequality in (2.17) is true.

For the components of $(x_1, \ldots, x_n)$, we have

$$\sum_{j=1}^{n} x_i = cm(k_1 - 1) + c(m-1) + \sum_{q=2}^{p} c(m-q)(k_q - k_{q-1}) + c(m-p-1) =$$
$$c\Big(mk_1 + m - p - 2 + \sum_{q=2}^{p} (m-q)(k_q - k_{q-1})\Big).$$

Since $(y_1, \ldots, y_m) \in \overline{R}_+^m$, setting $y_{m+1} = 0$, we find that

$$\sum_{j=1}^{m} y_j = \sum_{q=0}^{m-1} (m-q)(y_{m-q} - y_{m-q+1}).$$

Since $y_j = c(k_p + 1)$ for $j \leq m - p - 1$, we have

$$\sum_{j=1}^{m} y_j = \sum_{q=0}^{p+1} (m-q)(y_{m-q} - y_{m-q+1}) =$$

$$m(y_m - y_{m+1}) + (m-1)(y_{m-1} - y_m) + \sum_{q=2}^{p} (m-q)(y_{m-q} - y_{m-q+1}) +$$

$$(m-p-1)(y_{m-p-1} - y_{m-p}) = mc(k_1 - 1) + (m-1)\left(ck_1 - c(k_1 - 1)\right) +$$

$$\sum_{q=2}^{p} (m-q)(ck_q - ck_{q-1}) + (m-p-1)\left(c(k_p + 1) - ck_p\right) =$$

$$c\Big(mk_1 + m - p - 2 + \sum_{q=2}^{p} (m-q)(k_q - k_{q-1})\Big) = \sum_{i=1}^{n} x_i.$$

Let us show that components of the vectors $(x_1, \ldots, x_n)$ and $(y_1, \ldots, y_m)$ satisfy the system of inequalities (2.18), which will imply that this system is consistent. We write the left-hand sides of inequalities (2.18), using (2.19) and (2.20):

$$\delta_{k_q}\left((x_1, \ldots, x_n), (y_1, \ldots, y_m); c\right) = \sum_{j=1}^{m} y_j - \sum_{y_j \geq ck_q} (y_j - ck_q) - \sum_{i=1}^{k_q} x_i =$$

$$\sum_{j=1}^{m} \min(y_j, ck_q) - \sum_{i=1}^{k_q} x_i = \delta_{k_q}.$$

When $q = 1$, we have

$$\delta_{k_1} = ck_1(m-1) + c(k_1 - 1) - cm(k_1 - 1) - c(m-1) = 0.$$

Assume that $2 \leq q \leq p$ when $p \geq 2$. Then,

$$\delta_{k_q} = ck_q(m-q) + \sum_{i=1}^{q-1} ck_i + c(k_1 - 1) - cm(k_1 - 1) -$$

$$c(m-1) - \sum_{i=2}^{q} c(m-i)(k_i - k_{i-1}) =$$

$$c\Big(k_q(m-q) + k_1 - mk_1 + \sum_{i=1}^{q-1} k_i - \sum_{i=2}^{q} (m-i)(k_i - k_{i-1})\Big).$$

We calculate the following sum:

$$\sum_{i=2}^{q}(m-i)(k_i-k_{i-1}) = m\sum_{i=2}^{q}(k_i-k_{i-1}) - \sum_{i=2}^{q} i(k_i-k_{i-1}) = m(k_q-k_1)-$$
$$\sum_{i=2}^{q} ik_i + \sum_{i=2}^{q} ik_{i-1} = m(k_q-k_1) - qk_q - \sum_{2\le i\le q-1} ik_i + \sum_{3\le i\le q} ik_{i-1} + 2k_1 =$$
$$m(k_q-k_1) - qk_q + 2k_1 - \sum_{2\le i\le q-1} ik_i + \sum_{2\le i\le q-1}(i+1)k_i =$$
$$m(k_q-k_1) - qk_q + 2k_1 + \sum_{2\le i\le q-1} k_i.$$

Hence,

$$\delta_{k_q} = c\Big(k_q(m-q)+k_1-mk_1+\sum_{i=1}^{q-1} k_i - m(k_q-k_1)+qk_q-2k_1-\sum_{2\le i\le q-1} k_i\Big) = 0.$$

Thus, the components of the vectors defined by (2.19) and (2.20) satisfy inequalities (2.18), which reduce to equalities in the case under consideration. Hence, system (2.18) is consistent.

Let us prove that system (2.18) does not contain dependent inequalities. To do this, we show that, for any $q' \in Z$, $1 \le q' \le p$, there exist vectors from $\overline{R}_+^n$ and $\overline{R}_+^m$ with components satisfying inequalities (2.17) and all inequalities in (2.18) for $q \ne q'$, whereas the corresponding inequality in (2.18) does not hold when $q = q'$.

Assume that $\alpha \in R$, $0 < \alpha < c$, and $q' \in Z$. Having the vectors constructed according to rules (2.19) and (2.20), consider vectors $(x'_1, \ldots, x'_n) \in \overline{R}_+^n$ and $(y'_1, \ldots, y'_m) \in \overline{R}_+^m$ with components defined as

$$x'_i = \begin{cases} x_i, & i \le k_{q'-1}, \\ x_i + \alpha/(k_{q'} - k_{q'-1}), & k_{q'-1} < i \le k_{q'}, \\ x_i, & i > k_{q'}; \end{cases}$$
$$y'_j = \begin{cases} y_{m-q'} + \alpha, & j = m - q', \\ y_j, \quad 1 \le j \le m, & j \ne m - q', \end{cases}$$

where $k_0 = k - 1$.

It is obvious that the components of these vectors satisfy all inequalities in (2.17).

Let us show that the number

$$\delta_{k_q}\left((x'_1, \ldots, x'_n), (y'_1, \ldots, y'_m); c\right)$$

is nonnegative (in fact, zero) for any q, $1 \leq q \leq p$, $q \neq q'$, if $p > 1$, and negative for $q = q'$. By the construction of the vectors $(x_1, \ldots, x_n)$, $(y_1, \ldots, y_m)$ and $(x'_1, \ldots, x'_n)$, $(y'_1, \ldots, y'_m)$, the following relations are valid:

$$\sum_{j=1}^{m} \min(y'_j, ck_q) = \begin{cases} \sum_{j=1}^{m} \min(y_j, ck_q), & q \leq q', \\ \sum_{j=1}^{m} \min(y_j, ck_q) + \alpha, & q > q'; \end{cases}$$

$$\sum_{i=1}^{k_q} x'_i = \begin{cases} \sum_{i=1}^{k_q} x_i, & q < q', \\ \sum_{i=1}^{k_q} x_i + \alpha, & q \geq q'. \end{cases}$$

Hence,

$$\delta_{k_q}\left((x'_1, \ldots, x'_n), (y'_1, \ldots, y'_m); c\right) = \\ \begin{cases} \delta_{k_q}\left((x_1, \ldots, x_n), (y_1, \ldots, y_m); c\right) - \alpha = -\alpha, & q = q', \\ \delta_{k_q}\left((x_1, \ldots, x_n), (y_1, \ldots, y_m); c\right) = 0, & q \neq q'. \end{cases}$$

Therefore, every inequality in (2.18) is independent (if $p = 1$, the only inequality in this system does not result from constraints (2.17)). This proves the irreducibility of system (2.18). The theorem is proved.

As an illustration to Theorem 2.1, consider the following example.

Example 2.2. Consider the vectors

$$\mathbf{A} = (36, 34, 33, 32, 28, 19, 8) \in \overline{R}^7_+$$

and

$$\mathbf{B} = (28, 21, 19, 18, 18, 18, 18, 18, 14, 10, 5, 3) \in \overline{R}^{12}_+,$$

with the sums of components equal to 190. Let us apply proposition (a) of Theorem 2.1 to find the minimal number of inequalities of the form $\delta_k(\mathbf{A}, \mathbf{B}; 4) \geq 0$ or $\delta_k(\mathbf{B}, \mathbf{A}; 4) \geq 0$ required to ensure that $M(\mathbf{A}, \mathbf{B}; 4)$ is not empty.

It can be readily shown that, for the vectors $\mathbf{A}$ and $\mathbf{B}$, $k'(\mathbf{B}; 7, 4) = 0$ and $k''(\mathbf{B}; 7, 4) = 7$. Since all components of $\mathbf{A}$ are different, the nonnegativity of function $\delta_k(\mathbf{A}, \mathbf{B}; 4)$ must be checked for every $k \in Z$, $1 \leq k \leq 6$, and the number of inequalities to be checked is six.

For the vectors $\mathbf{B}$ and $\mathbf{A}$, we have $k'(\mathbf{A}; 12, 4) = 2$ and $k''(\mathbf{A}; 12, 4) = 9$. In this case, $k^*(\mathbf{A}; 12, 4) = 3$ and $k^{**}(\mathbf{A}; 12, 4) = 8$; since $b_j = 18$, $4 \leq j \leq 8$, it is required to check only two inequalities: $\delta_3(\mathbf{B}, \mathbf{A}; 4) \geq 0$ and $\delta_8(\mathbf{B}, \mathbf{A}; 4) \geq 0$.

Applying (1.25), we obtain

$$\delta_3(\mathbf{B}, \mathbf{A}; 4) = 8 + 4 \cdot 3 \cdot 6 - 28 - 21 - 19 = 12 > 0,$$

$$\delta_8(\mathbf{B}, \mathbf{A}; 4) = 28 + 19 + 8 + 4 \cdot 8 \cdot 4 - 28 - 21 - 19 - 18 \cdot 4 = 7 > 0.$$

Hence, $M(\mathbf{A}, \mathbf{B}; 4) \neq \varnothing$ (or $M(\mathbf{B}, \mathbf{A}; 4) \neq \varnothing$). Note that direct calculation gives $\delta_8(\mathbf{B}, \mathbf{A}; 3) = -11 < 0$ and $M(\mathbf{A}, \mathbf{B}; 3) = \varnothing$.

The matrix set $M(\mathbf{A}, \mathbf{B}; c)$ in Example 2.2 contains integer matrices as well. In what follows, we will also consider integer vectors $\mathbf{A}$ and $\mathbf{B}$ and integer solutions. To do this, we define

$$Z_+^n = \left\{\mathbf{A} = (a_1, \ldots, a_n) \in R_+^n : a_i \in Z, \quad 1 \le i \le n\right\},$$

$$\overline{Z}_+^n = \left\{\mathbf{A} = (a_1, \ldots, a_n) \in \overline{R}_+^n : a_i \in Z, \quad 1 \le i \le n\right\},$$

and, for $\mathbf{A} \in Z_+^n$, $\mathbf{B} \in Z_+^m$, $c \ge 0$, $c \in Z$,

$$M_Z(\mathbf{A}, \mathbf{B}) = \Big\{X(\mathbf{A}, \mathbf{B}) = (x_{ij}) \in M(\mathbf{A}, \mathbf{B}) : x_{ij} \in Z,\ 1 \le i \le n, \\ 1 \le j \le m\Big\},$$

$$M_Z(\mathbf{A}, \mathbf{B}; c) = \Big\{X(\mathbf{A}, \mathbf{B}) = (x_{ij}) \in M_Z(\mathbf{A}, \mathbf{B}) : x_{ij} \le c,\ 1 \le i \le n, \\ 1 \le j \le m\Big\}.$$

Remark 2.1. When $\mathbf{A} \in Z_+^n$, $\mathbf{B} \in Z_+^m$, and $c \in Z$, $c \ge 0$, Lemmas 2.4 and 2.6–2.10 are obviously valid. All operations in Lemmas 2.1–2.3 result in integer vectors and integer values of function (1.22). In proposition (b) of Lemma 2.3, the vector $(\beta_1, \ldots, \beta_m)$ defined by (2.7) may be assumed integer, and Lemmas 2.1-2.3 are also valid. It can be readily shown that Lemma 2.5 is also valid for integer matrices.

By Remark 2.1 and Theorem 2.1, the following theorem on integer matrices is valid.

Theorem 2.2. *Suppose that $c \in Z$, $c \ge 0$, and vectors $\mathbf{A} = (a_1, \ldots, a_n) \in Z_+^n$ and $\mathbf{B} = (b_1, \ldots, b_m) \in Z_+^m$ satisfy the balance condition. The matrix set $M_Z(\mathbf{A}, \mathbf{B}; c)$ is nonempty if and only if $\delta_k(\widetilde{\mathbf{A}}, \mathbf{B}; c) \ge 0$ for any k, $1 \le k \le n$. Moreover, if $a_i \le cm$, $b_j \le cn$, $1 \le i \le n$, $1 \le j \le m$, it is sufficient condition for $M_Z(\mathbf{A}, \mathbf{B}; c)$ to be nonempty that either $k''(\mathbf{B}; n, c) - k'(\mathbf{B}; n, c) \le 1$ or $k''(\mathbf{B}; n, c) - k'(\mathbf{B}; n, c) \ge 2$ and $\delta_k(\widetilde{\mathbf{A}}, \mathbf{B}; c) \ge 0$ for every $k \in K(\mathbf{B}; n, c)$.*

The following assertion is important by itself.

Theorem 2.3. *Suppose that $c > 0$ is an arbitrary real number and $n \ge 2$, $p \ge 1$, $m \ge p + 2$, and k_q, $1 \le q \le p$, is arbitrary integers such that $k_1 \ge 1$,*

$k_p < n$. *Suppose also* $k_q < k_{q+1}$ *for* $1 \le q \le p-1$, *and* $p \ge 2$. *If nonnegative variables* x_i *and* y_j, $1 \le i \le n$, $1 \le j \le m$ *satisfy* (2.17), *then the system of equations*

$$\begin{cases} \sum_{j=1}^{m} y_j - \sum_{y_j \ge ck_q} (y_j - ck_q) - \sum_{i=1}^{k_q} x_i = 0, \\ 1 \le q \le p, \end{cases} \tag{2.21}$$

is consistent and irreducible.

This proposition was proved in proving proposition (b) of Theorem 2.1.

2.4. Algorithms for Constructing of Transportation Matrices

Given a vector $\mathbf{A} = (a_1, \dots, a_n)$ and numbers k_1, $k_2 \in Z$, $1 \le k_1 \le k_2 \le n$, denote by $\mathbf{A}_{[k_1,k_2]} = (a_{k_1}, \dots, a_{k_2})$ the subvector of $\mathbf{A}$ with $k_2 - k_1 + 1$ components. By definition, $\mathbf{A}_{[1,n]} = \mathbf{A}$ and $\mathbf{A}_{[k_1,k_1]} = (a_{k_1})$ is a 1-component vector. Denote by $\mathbf{B}^{[k_1,k_2]} = (b_1^{[k_1,k_2]}, \dots, b_m^{[k_1,k_2]})$ the vector that corresponds to the $\mathbf{A}_{[k_1,k_2]}$ and is obtained from $\mathbf{B} = (b_1, \dots, b_m)$ by the rules described in Algorithm 2.1. Note that, if $\mathbf{A} \in R_+^n$ and $\mathbf{B} \in R_+^m$ satisfy the balance condition and $\delta_k(\mathbf{A}, \mathbf{B}; c) \ge 0$, $1 \le k \le n$, then vectors $\mathbf{A}_{[k_1,k_2]}$ and $\mathbf{B}^{[k_1,k_2]}$ satisfy the following relations:

$$0 \le b_j^{[k_1,k_2]} \le b_j, \quad 1 \le j \le m,$$

$$\sum_{i=k_1}^{k_2} a_i = \sum_{j=1}^{m} b_j^{[k_1,k_2]},$$

$$\delta_k(\mathbf{A}_{[k_1,k_2]}, \mathbf{B}^{[k_1,k_2]}; c) \ge 0, \quad 1 \le k \le k_2 - k_1 + 1,$$

where

$$\delta_k(\mathbf{A}_{[p,q]}, \mathbf{B}; c) = \sum_{j=1}^{m} b_j - \sum_{b_j \ge ck} (b_j - ck) - \sum_{i=p}^{p+k-1} a_i.$$

Every matrix in $M(\mathbf{A}, \mathbf{B})$ can be obtained from the corresponding matrix of the set $M(\widetilde{\mathbf{A}}, \mathbf{B})$ by permuting of its rows. Therefore, without loss of generality, we can assume that the components of the vector $\mathbf{A}$ in the algorithm of construction of a matrix belonging to $M(\mathbf{A}, \mathbf{B}; c)$ are arranged in nonascending order.

Algorithm 2.1. Construction of a transportation matrix with a common bound on all elements.

Suppose vectors $\mathbf{A} = (a_1, \ldots, a_n) \in \overline{R}_+^n$ and $\mathbf{B} = (b_1, \ldots, b_m) \in R_+^m$ satisfy the balance condition, $c > 0$, and $\delta_k(\mathbf{A}, \mathbf{B}; c) \geq 0$ for any k, $1 \leq k \leq n$; i.e., $M(\mathbf{A}, \mathbf{B}; c) \neq \varnothing$.

Step 1. Set $k_1^{(1)} = 1$, $k_2^{(1)} = n$,

$$\mathbf{A}_{[k_1^{(1)}, k_2^{(1)}]} = \mathbf{A}_{[1,n]} = \mathbf{A} = (a_1, \ldots, a_n),$$

$$\mathbf{B}^{[k_1^{(1)}, k_2^{(1)}]} = (b_1^{[k_1^{(1)}, k_2^{(1)}]}, \ldots, b_m^{[k_1^{(1)}, k_2^{(1)}]}) = \mathbf{B} = (b_1, \ldots, b_m),$$

$t = 1$.

Step 2. Calculate

$$n^{(t)} = \max\left\{k_{i+1}^{(t)} - k_i^{(t)} : \mathbf{A}_{[k_i^{(t)}, k_{i+1}^{(t)}]},\ 1 \leq i \leq t\right\}.$$

Step 3. If $n^{(t)} = 0$, go to Step 10. If $n^{(t)} \geq 1$, go to the next step.

Step 4. Determine index q for which the vector $\mathbf{A}_{[k_q^{(t)}, k_{q+1}^{(t)}]}$, $k_{q+1}^{(t)} > k_q^{(t)}$ is constructed.

Step 5. For the vectors

$$\mathbf{A}_{[k_q^{(t)}, k_{q+1}^{(t)}]} = (a_{k_q^{(t)}}, \ldots, a_{k_{q+1}^{(t)}}), \quad \mathbf{B}^{[k_q^{(t)}, k_{q+1}^{(t)}]} = (b_1^{[k_q^{(t)}, k_{q+1}^{(t)}]}, \ldots, b_m^{[k_q^{(t)}, k_{q+1}^{(t)}]}),$$

choose a number $p^{(t)} \in Z$, $1 \leq p^{(t)} \leq k_{q+1}^{(t)} - k_q^{(t)}$, such that (see Lemma 2.3)

$$\begin{gathered}\delta_{p^{(t)}}\left(\mathbf{A}_{[k_q^{(t)}, k_{q+1}^{(t)}]}, \mathbf{B}^{[k_q^{(t)}, k_{q+1}^{(t)}]}; c\right) = \\ \min\left\{\delta_k\left(\mathbf{A}_{[k_q^{(t)}, k_{q+1}^{(t)}]}, \mathbf{B}^{[k_q^{(t)}, k_{q+1}^{(t)}]}; c\right) : 1 \leq k \leq k_{q+1}^{(t)} - k_q^{(t)}\right\}.\end{gathered} \tag{2.22}$$

Step 6. Set $\beta_{m+1}^{(t)} = 0$ and construct vector $(\beta_1^{(t)}, \ldots, \beta_m^{(t)})$ with components successively calculated (beginning with the last one) by rule (2.8) (see Lemma 2.3):

$$\beta_{m-i}^{(t)} = \min\left\{\begin{array}{l} cp^{(t)}, \\ c(k_{q+1}^{(t)} - k_q^{(t)} + 1) - cp^{(t)}, \\ b_{m-i}^{[k_q^{(t)}, k_{q+1}^{(t)}]}, \\ c(k_{q+1}^{(t)} - k_q^{(t)} + 1) - b_{m-i}^{[k_q^{(t)}, k_{q+1}^{(t)}]}, \\ \delta_{p^{(t)}}\left(\mathbf{A}_{[k_q^{(t)}, k_{q+1}^{(t)}]}, \mathbf{B}^{[k_q^{(t)}, k_{q+1}^{(t)}]}; c\right) - \displaystyle\sum_{j=m-i+1}^{m+1} \beta_j^{(t)} \end{array}\right\}, \tag{2.23}$$

$0 \leq i \leq m-1$.

Step 7. Using the number $p^{(t)}$, split the vector $\mathbf{A}_{[k_q^{(t)},k_{q+1}^{(t)}]}$ into two subvectors,

$$\mathbf{A}_{[k_q^{(t)},k_q^{(t)}+p^{(t)}-1]}, \quad \mathbf{A}_{[k_q^{(t)}+p^{(t)},k_{q+1}^{(t)}]}.$$

Step 8. Represent the vector $\mathbf{B}^{[k_q^{(t)},k_{q+1}^{(t)}]}$ as the sum of two vectors corresponding to the vectors $\mathbf{B}^*$ and $\mathbf{B}^{**}$ (Lemma 2.3),

$$\mathbf{B}^{[k_q^{(t)},k_q^{(t)}+p^{(t)}-1]}, \quad \mathbf{B}^{[k_q^{(t)}+p^{(t)},k_{q+1}^{(t)}]},$$

with components defined as

$$b_j^{[k_q^{(t)},k_q^{(t)}+p^{(t)}-1]} = \begin{cases} cp^{(t)} - \beta_j^{(t)}, & b_j^{[k_q^{(t)},k_{q+1}^{(t)}]} \geq cp^{(t)}, \\ b_j^{[k_q^{(t)},k_{q+1}^{(t)}]} - \beta_j^{(t)}, & b_j^{[k_q^{(t)},k_{q+1}^{(t)}]} < cp^{(t)}, \end{cases}$$

$$b_j^{[k_q^{(t)}+p^{(t)},k_{q+1}^{(t)}]} = \begin{cases} b_j^{[k_q^{(t)},k_{q+1}^{(t)}]} - cp^{(t)} + \beta_j^{(t)}, & b_j^{[k_q^{(t)},k_{q+1}^{(t)}]} \geq cp^{(t)}, \\ \beta_j^{(t)}, & b_j^{[k_q^{(t)},k_{q+1}^{(t)}]} < cp^{(t)}, \end{cases} \tag{2.24}$$

$$1 \leq j \leq m.$$

Step 9. Increase t by one, setting

$$k_i^{(t+1)} = \begin{cases} k_i^{(t)}, & 1 \leq i \leq q, \\ k_q^{(t)} + p^{(t)} - 1, & i = q+1, \\ k_q^{(t)} + p^{(t)}, & i = q+2, \\ k_{i-2}^{(t)}, & q+3 \leq i \leq 2(t+1), \end{cases}$$

and go to Step 2.

Step 10. Here, $t = n$, and the indices $k_q^{(n)}$ obtained, $1 \leq q \leq 2n$, satisfy the conditions $k_{q+1}^{(n)} = k_q^{(n)}$ when q is odd or $k_{q+1}^{(n)} = k_q^{(n)} + 1$ when q is even. Therefore, each 1-component vector $\mathbf{A}_{[k_q^{(n)},k_{q+1}^{(n)}]}$ constructed (q is odd) can be represented in the form

$$\mathbf{A}_{[k_q^{(n)},k_{q+1}^{(n)}]} = \mathbf{A}_{[i,i]} = (a_i), \quad q = 2i-1, \quad 1 \leq i \leq n.$$

The required matrix $X(\mathbf{A},\mathbf{B}) = (x_{ij})$ from $M(\mathbf{A},\mathbf{B};c)$ is

$$x_{ij} = b_j^{[i,i]}, \quad 1 \leq i \leq n, \quad 1 \leq j \leq m.$$

The algorithm is substantiated by Lemmas 2.3 and 2.5.

Remark 2.2.

(a) The basic operation in Algorithm 2.1 is the evaluation of function (1.22). For any subvector $\mathbf{A}_{[i_1,i_2]}$ of the vector $\mathbf{A} = \mathbf{A}_{[1,n]}$, the number of evaluations of function (1.22) is not greater than $i_2 - i_1$; for the subvectors $\mathbf{A}_{[i_1,i_3]}$ and $\mathbf{A}_{[i_3+1,i_2]}$, where $i_1 \le i_3 \le i_2$, the number of evaluations of this function does not exceed the sum $(i_3 - i_1) + (i_2 - i_3 - 1) = i_2 - i_1 - 1$. Therefore, the total number of evaluations of function (1.22) in this algorithm is obviously not greater than $n(n-1)/2$.

(b) In certain cases, a matrix from $M(\mathbf{A}, \mathbf{B}; c)$, $\mathbf{A} \in R_+^n$, $\mathbf{B} \in R_+^m$, can be constructed in a simpler way. For example, if

$$c \ge \min\Big\{\max\{a_i : 1 \le i \le n\}, \max\{b_j : 1 \le j \le m\}\Big\},$$

then $M(\mathbf{A}, \mathbf{B}; c) = M(\mathbf{A}, \mathbf{B})$. In this case, any matrix $X(\mathbf{A}, \mathbf{B}) = (x_{ij})$ in $M(\mathbf{A}, \mathbf{B}; c)$ can be constructed by successively inserting arbitrary nonnegative numbers in the rows $(x_{i1}, \ldots, x_{im})$, $1 \le i \le n$, and applying the following conditions:

$$\sum_{j=1}^{m} x_{ij} = a_i, \quad 1 \le i \le n, \quad \sum_{i=1}^{k} x_{ij} \le b_j, \quad 1 \le j \le m.$$

(c) If there exist indices i and j for vectors $\mathbf{A}$ and $\mathbf{B}$ such that $a_i = 0$ or $a_i = cm$ and $b_j = 0$ or $b_j = cn$, then the corresponding rows and columns of an arbitrary matrix in $M(\mathbf{A}, \mathbf{B}; c) \ne \varnothing$ consist of elements equal to 0 or c. In this case, prior to using Algorithm 2.1 to construct a matrix from $M(\mathbf{A}, \mathbf{B}; c)$, the number of components of the vectors can be reduced.

Remark 2.3. If $c \in Z$ and all components of $\mathbf{A}$ and $\mathbf{B}$ are integer, then, according to Remark 2.1, all operations in Algorithm 2.1 result in integer values and the matrix constructed is integer.

The following example illustrates the application of Algorithm 2.1.

Example 2.3. The vectors

$$\mathbf{A} = (18, 15, 14, 12, 9) \text{ and } \mathbf{B} = (14, 13, 12, 10, 9, 6, 4)$$

satisfy the balance condition:

$$18 + 15 + 14 + 12 + 9 = 14 + 13 + 12 + 10 + 9 + 6 + 4 = 68.$$

The values of the characteristic function (1.22) for $c = 3$ are obtained by using (1.25):

$$\delta_1(\mathbf{A},\mathbf{B};3) = 3\cdot 7 - 18 = 3,$$
$$\delta_2(\mathbf{A},\mathbf{B};3) = 4 + 3\cdot 2\cdot 6 - 33 = 7,$$
$$\delta_3(\mathbf{A},\mathbf{B};3) = 10 + 3\cdot 3\cdot 5 - 47 = 8,$$
$$\delta_4(\mathbf{A},\mathbf{B};3) = 29 + 3\cdot 4\cdot 3 - 59 = 6,$$
$$\delta_5(\mathbf{A},\mathbf{B};3) = 68 - 68 = 0.$$

Hence, $M_Z(\mathbf{A},\mathbf{B};3) \neq \varnothing$ by Theorem 2.2.

Let us construct an integer matrix $X(\mathbf{A},\mathbf{B}) = (x_{ij})$ belonging to $M_Z(\mathbf{A},\mathbf{B};3)$.

Set $k_1^{(1)} = 1$, $k_2^{(1)} = 5$, $\mathbf{A}_{[1,5]} = \mathbf{A}$, $\mathbf{B}^{[1,5]} = \mathbf{B}$.

For $t = 1$, we have $n^{(1)} = k_2^{(1)} - k_1^{(1)} = 4 > 0$.

Since only the vector $\mathbf{A}_{[1,5]}$ is constructed, $q = 1$ (Step 4).

The values of the function $\delta_k(\mathbf{A},\mathbf{B};3)$ were calculated earlier. Here, $p^{(1)}=1$ (see (2.22)) and

$$\delta_1(\mathbf{A}_{[1,5]},\mathbf{B}^{[1,5]};3) = 3.$$

Calculate $\beta_j^{(1)}$, $1 \le j \le 7$, using (2.23). Here,

$$\beta_7^{(1)} = \min\left\{\begin{array}{l} cp^{(1)} = 3, \\ c(k_2^{(1)} - k_1^{(1)} + 1) - cp^{(1)} = 12, \\ b_7^{[1,5]} = 4, \\ c(k_2^{(1)} - k_1^{(1)} + 1) - b_7^{[1,5]} = 11, \\ \delta_1(\mathbf{A}_{[1,5]},\mathbf{B}^{[1,5]};3) = 3 \end{array}\right\} = 3.$$

Since (by Lemma 2.3)

$$\sum_{j=1}^{7} \beta_j^{(1)} = \delta_1(\mathbf{A}_{[1,5]},\mathbf{B}^{[1,5]};3),$$

we have $\beta_j^{(1)} = 0$, $1 \le j \le 6$, $\beta_7^{(1)} = 3$.

The number $p^{(1)} = 1$ splits the vector $\mathbf{A}_{[1,5]}$ into two subvectors, $\mathbf{A}_{[1,1]} = (18)$ and $\mathbf{A}_{[2,5]} = (15,14,12,9)$.

Write $\mathbf{B}^{[1,5]}$ as the sum of two vectors, $\mathbf{B}^{[1,1]}$ and $\mathbf{B}^{[2,5]}$, with components defined by (2.24); since $b_j^{[1,5]} > cp^{(1)} = 3$, $1 \le j \le m$, we have

$$b_j^{[1,1]} = cp^{(1)} - \beta_j^{(1)}, \quad 1 \le j \le 7,$$

$$b_j^{[2,5]} = b_j^{[1,5]} - cp^{(1)} + \beta_j^{(1)}, \quad 1 \le j \le 7.$$

It is now clear that

$$\mathbf{B}^{[1,1]} = (3,3,3,3,3,3,0), \quad \mathbf{B}^{[2,5]} = (11,10,9,7,6,3,4).$$

Thus, one iteration of Algorithm 2.1 is implemented, and the vectors $\mathbf{A}_{[1,1]}$ and $\mathbf{B}^{[1,1]}$ will determine the first row of the matrix.

Increase t by one and set $k_1^{(2)} = k_1^{(1)} = 1$, $k_2^{(2)} = k_1^{(1)} + p^{(1)} - 1 = 1$, $k_3^{(2)} = k_1^{(1)} + p^{(1)} = 2$, $k_4^{(2)} = k_2^{(1)} = 5$. Then, $n^{(2)} = \max\{k_2^{(2)} - k_1^{(2)}, k_4^{(2)} - k_3^{(2)}\} = 3 > 0$. Here, for the vector $\mathbf{A}_{[2,5]}$, we have $k_4^{(2)} - k_3^{(2)} > 0$ and $q = 3$. Therefore, the vector pair $\mathbf{A}_{[2,5]} = (15,14,12,9)$, $\mathbf{B}^{[2,5]} = (11,10,9,7,6,3,4)$ is to be considered. Evaluate function (1.22) for these vectors and $c = 3$, where $k \in Z$, $1 \le k \le 3$:

$$\delta_1(\mathbf{A}_{[2,5]}, \mathbf{B}^{[2,5]}; 3) = 3 \cdot 7 - 15 = 6,$$

$$\delta_2(\mathbf{A}_{[2,5]}, \mathbf{B}^{[2,5]}; 3) = 7 + 3 \cdot 2 \cdot 5 - 29 = 6,$$

$$\delta_3(\mathbf{A}_{[2,5]}, \mathbf{B}^{[2,5]}; 3) = 20 + 3 \cdot 3 \cdot 3 - 41 = 6.$$

In the case under consideration, $p^{(2)}$ is any number satisfying the inequality $1 \le p^{(2)} \le 3$. Set $p^{(2)} = 1$.

Calculate $\beta_j^{(2)}$, $1 \le j \le 7$, using the fact that

$$\delta_1(\mathbf{A}_{[2,5]}, \mathbf{B}^{[2,5]}; 3) = \min\Big\{\delta_k(\mathbf{A}_{[2,5]}, \mathbf{B}^{[2,5]}; 3) : 1 \le k \le 3\Big\} = 6,$$

$$cp^{(2)} = 3, \quad k_4^{(2)} - k_3^{(2)} + 1 = 4, \quad c(k_4^{(2)} - k_3^{(2)} + 1) - cp^{(2)} = 9.$$

$$\beta_7^{(2)} = \min \left\{ \begin{array}{l} cp^{(2)} = 3, \\ c(k_4^{(2)} - k_3^{(2)} + 1) - cp^{(2)} = 9, \\ b_7^{[2,5]} = 4, \\ c(k_4^{(2)} - k_3^{(2)} + 1) - b_7^{[2,5]} = 8, \\ \delta_1(\mathbf{A}_{[2,5]}, \mathbf{B}^{[2,5]}; 3) = 6 \end{array} \right\} = 3,$$

$$\beta_6^{(2)} = \min \left\{ \begin{array}{l} 3, \\ 9, \\ b_6^{[2,5]} = 3, \\ 9 - b_6^{[2,5]} = 6, \\ \delta_1(\mathbf{A}_{[2,5]}, \mathbf{B}^{[2,5]}; 3) - \beta_7^{(2)} = 3 \end{array} \right\} = 3.$$

Since $\beta_6^{(2)} + \beta_7^{(2)} = \delta_1(\mathbf{A}_{[2,5]}, \mathbf{B}^{[2,5]}; 3)$, we have $\beta_j^{(2)} = 0$, $1 \le j \le 5$.

Let us split the vector $\mathbf{A}_{[2,5]}$ into two subvectors using the fact that $p^{(2)} = 1$:

$$\mathbf{A}_{[2,2]} = (15), \quad \mathbf{A}_{[3,5]} = (14, 12, 9).$$

The components of $\mathbf{B}^{[2,5]} = \mathbf{B}^{[2,2]} + \mathbf{B}^{[3,5]}$ satisfy the inequalities $b_j^{[2,5]} \geq cp^{(2)} = 3$, Therefore,

$$b_j^{[2,2]} = cp^{(2)} - \beta_j^{(2)}, \quad 1 \leq j \leq 7,$$

$$b_j^{[3,5]} = b_j^{[2,5]} - cp^{(2)} + \beta_j^{(2)}, \quad 1 \leq j \leq 7.$$

Hence,

$$\mathbf{B}^{[2,2]} = (3,3,3,3,3,0,0), \quad \mathbf{B}^{[3,5]} = (8,7,6,4,3,3,4).$$

The vectors $\mathbf{A}_{[2,2]}$ and $\mathbf{B}^{[2,2]}$ constructed by the second iteration determine the second row of the matrix.

Increasing t by one, we obtain $k_1^{(3)} = k_1^{(2)} = 1$, $k_2^{(3)} = k_2^{(2)} = 1$, $k_3^{(3)} = k_3^{(2)} = 2$, $k_4^{(3)} = k_3^{(2)} + p^{(2)} - 1 = 2$, $k_5^{(3)} = k_3^{(2)} + p^{(2)} = 3$, $k_6^{(3)} = k_4^{(2)} = 5$. Only the vector $\mathbf{A}_{[3,5]}$ is such that the inequality $k_6^{(3)} - k_5^{(3)} = 2 > 0$ holds. Hence, $q = 5$, and we now consider the pair of vectors $\mathbf{A}_{[3,5]} = (14, 12, 9)$ and $\mathbf{B}^{[3,5]} = (8,7,6,4,3,3,4)$. For these vectors, evaluate function (1.22) (where $c = 3$ and $1 \leq k \leq 2$):

$$\delta_1(\mathbf{A}_{[3,5]}, \mathbf{B}^{[3,5]}; 3) = 3 \cdot 7 - 14 = 7,$$
$$\delta_2(\mathbf{A}_{[3,5]}, \mathbf{B}^{[3,5]}; 3) = 14 + 3 \cdot 2 \cdot 3 - 26 = 6.$$

Thus, $p^{(3)} = 2$, and

$$\delta_2(\mathbf{A}_{[3,5]}, \mathbf{B}^{[3,5]}; 3) = \min\left\{\delta_k(\mathbf{A}_{[3,5]}, \mathbf{B}^{[3,5]}; 3) : 1 \leq k \leq 2\right\} = 6.$$

Calculate the components $\beta_j^{(3)}$, $1 \leq j \leq 7$. Here, $cp^{(3)} = 6$, $k_6^{(3)} - k_5^{(3)} + 1 = 3$, $c(k_6^{(3)} - k_5^{(3)} + 1) - cp^{(3)} = 3$. Therefore,

$$\beta_7^{(3)} = \min \left\{ \begin{array}{l} cp^{(3)} = 6, \\ c(k_6^{(3)} - k_5^{(3)} + 1) - cp^{(3)} = 3, \\ b_7^{[3,5]} = 4, \\ c(k_6^{(3)} - k_5^{(3)} + 1) - b_7^{[3,5]} = 5, \\ \delta_2(\mathbf{A}_{[3,5]}, \mathbf{B}^{[3,5]}; 3) = 6 \end{array} \right\} = 3.$$

Similarly, we obtain $\beta_6^{(3)} = 3$ and $\beta_j^{(3)} = 0$, $1 \leq j \leq 5$.

Let us split the vector $\mathbf{A}_{[3,5]}$ for $p^{(3)} = 2$:

$$\mathbf{A}_{[3,4]} = (14, 12), \quad \mathbf{A}_{[5,5]} = (9).$$

Calculate the components of vectors $\mathbf{B}^{[3,4]}$ and $\mathbf{B}^{[5,5]}$ ($\mathbf{B}^{[3,5]} = \mathbf{B}^{[3,4]} + \mathbf{B}^{[5,5]}$). If $b_j^{[3,5]} \geq cp^{(3)} = 6$, i.e., $1 \leq j \leq 3$, then

$$b_1^{[3,4]} = cp^{(3)} - \beta_1^{(3)} = 6, \quad b_1^{[5,5]} = b_1^{[3,5]} - cp^{(3)} + \beta_1^{(3)} = 2,$$

$$b_2^{[3,4]} = cp^{(3)} - \beta_2^{(3)} = 6, \quad b_2^{[5,5]} = b_2^{[3,5]} - cp^{(3)} + \beta_2^{(3)} = 1,$$

$$b_3^{[3,4]} = cp^{(3)} - \beta_3^{(3)} = 6, \quad b_3^{[5,5]} = b_3^{[3,5]} - cp^{(3)} + \beta_3^{(3)} = 0.$$

If $b_j^{[3,4]} < cp^{(3)} = 6$, then $4 \leq j \leq 7$ and

$$b_4^{[3,4]} = b_4^{[3,5]} - \beta_4^{(3)} = 4, \quad b_4^{[5,5]} = \beta_4^{(3)} = 0,$$

$$b_5^{[3,4]} = b_5^{[3,5]} - \beta_5^{(3)} = 3, \quad b_5^{[5,5]} = \beta_5^{(3)} = 0,$$

$$b_6^{[3,4]} = b_6^{[3,5]} - \beta_6^{(3)} = 0, \quad b_6^{[5,5]} = \beta_6^{(3)} = 3,$$

$$b_7^{[3,4]} = b_7^{[3,5]} - \beta_7^{(3)} = 1, \quad b_7^{[5,5]} = \beta_7^{(3)} = 3.$$

Hence,

$$\mathbf{B}^{[3,4]} = (6, 6, 6, 4, 3, 0, 1), \quad \mathbf{B}^{[5,5]} = (2, 1, 0, 0, 0, 3, 3).$$

Thus, the third iteration of the algorithm is implemented. The vectors $\mathbf{A}_{[5,5]}$ and $\mathbf{B}^{[5,5]}$ determine the fifth row of the matrix.

In the next iteration cycle, the pair of vectors $\mathbf{A}_{[3,4]} = (14, 12)$ and $\mathbf{B}^{[3,4]} = (6, 6, 6, 4, 3, 0, 1)$ is to be considered. They determine the two remaining rows of the matrix. There is only one possibility for function (1.22): $p^{(4)} = 1$ and

$$\delta_1(\mathbf{A}_{[3,4]}, \mathbf{B}^{[3,4]}; 3) = 1 + 3 \cdot 5 - 14 = 2.$$

Calculate the components of the vector $(\beta_1^{(4)}, \ldots, \beta_7^{(4)})$:

$$\beta_7^{(4)} = \min \left\{ \begin{array}{l} cp^{(4)} = 3, \\ c(4 - 3 + 1) - cp^{(4)} = 3, \\ b_7^{[3,4]} = 1, \\ c(4 - 3 + 1) - b_7^{[3,4]} = 5, \\ \delta_1(\mathbf{A}_{[3,4]}, \mathbf{B}^{[3,4]}; 3) = 2 \end{array} \right\} = 1,$$

$$\beta_6^{(4)} = \min \left\{ \begin{array}{l} 3, \\ 3, \\ b_6^{[3,4]} = 0, \\ 6 - b_6^{[3,4]} = 6, \\ \delta_1(\mathbf{A}_{[3,4]}, \mathbf{B}^{[3,4]}; 3) - \beta_7^{(4)} = 1 \end{array} \right\} = 0,$$

$$\beta_5^{(4)} = \min \left\{ \begin{array}{l} 3, \\ 3, \\ b_5^{[3,4]} = 3, \\ 6 - b_5^{[3,4]} = 3, \\ \delta_1(\mathbf{A}_{[3,4]}, \mathbf{B}^{[3,4]}; 3) - (\beta_6^{(4)} + \beta_7^{(4)}) = 1 \end{array} \right\} = 1,$$

and $\beta_j^{(4)} = 0$, $1 \le j \le 4$.

The vector $\mathbf{A}_{[3,4]}$ is split into two 1-component subvectors $\mathbf{A}_{[3,3]} = (14)$, $\mathbf{A}_{[4,4]} = (12)$.

The components of $\mathbf{B}^{[3,3]}$ and $\mathbf{B}^{[4,4]}$ determine the elements of the two remaining rows of the matrix. It follows from (2.24) that

$$b_1^{[3,3]} = 3,\ b_2^{[3,3]} = 3,\ b_3^{[3,3]} = 3,\ b_4^{[3,3]} = 3,\ b_5^{[3,3]} = 2,\ b_6^{[3,3]} = 0,\ b_7^{[3,3]} = 0;$$

$$b_1^{[4,4]} = 3,\ b_2^{[4,4]} = 3,\ b_3^{[4,4]} = 3,\ b_4^{[4,4]} = 1, b_5^{[4,4]} = 1,\ b_6^{[4,4]} = 0,\ b_7^{[4,4]} = 1.$$

Hence,

$$\mathbf{B}^{[3,3]} = (3,3,3,3,2,0,0), \quad \mathbf{B}^{[4,4]} = (3,3,3,1,1,0,1).$$

Each pair of vectors $\mathbf{A}_{[i,i]}$, $\mathbf{B}^{[i,i]}$, $1 \le i \le 5$, thus constructed determines the ith row of the matrix $X(\mathbf{A},\mathbf{B})$ from $M_Z(\mathbf{A},\mathbf{B};3)$:

$$X(\mathbf{A},\mathbf{B}) = \begin{pmatrix} 3 & 3 & 3 & 3 & 3 & 3 & 0 \\ 3 & 3 & 3 & 3 & 3 & 0 & 0 \\ 3 & 3 & 3 & 3 & 2 & 0 & 0 \\ 3 & 3 & 3 & 1 & 1 & 0 & 1 \\ 2 & 1 & 0 & 0 & 0 & 3 & 3 \end{pmatrix} \in M_Z(\mathbf{A},\mathbf{B};3).$$

The number of matrices that can be constructed by Algorithm 2.1 depends only on the values of q (Step 4) and $p^{(t)}$ (Step 5). In constructing the vector pair

$$\mathbf{A}_{[k_q^{(t)}, k_{q+1}^{(t)}]}, \quad \mathbf{B}^{[k_q^{(t)}, k_{q+1}^{(t)}]},$$

the vector $(\beta_1^{(t)}, \ldots, \beta_m^{(t)})$ is of basic importance; it corrects the balance condition and preserves the nonnegativity of function (1.22) for new pairs of vectors. The components of this vector are determined by the strict rules in (2.23), which are justified by formula (2.8) (proposition (c) of Lemma 2.3). The strict rules in (2.23) make it possible to construct only certain matrices by Algorithm 2.1. In order to construct an arbitrary matrix from $M(\mathbf{A},\mathbf{B};c)$ or $M_Z(\mathbf{A},\mathbf{B};c)$, the strict rules in (2.23), which are used in Algorithm 2.1 to calculate the components of the vector $(\beta_1^{(t)},\dots,\beta_m^{(t)}) = (\beta_1,\dots,\beta_m)$, should be changed. In proposition (b) of Lemma 2.3, relations (2.7) are given. When splitting the vector pair $\mathbf{A}$, $\mathbf{B}$ into $\mathbf{A}'$, $\mathbf{B}^*$ and $\mathbf{A}''$, $\mathbf{B}^{**}$, the vector $(\beta_1,\dots,\beta_m)$ must satisfy these relations while the balance condition is retained and the characteristic function (1.22) remains nonnegative.

Algorithm 2.2. Construction of arbitrary matrices from the sets $M(\mathbf{A},\mathbf{B};c)$ and $M_Z(\mathbf{A},\mathbf{B};c)$.

Suppose that $\mathbf{A} = (a_1,\dots,a_n) \in \overline{R}_+^n$, $\mathbf{B} = (b_1,\dots,b_m) \in R_+^m$, and either $M(\mathbf{A},\mathbf{B};c)$ or $M_Z(\mathbf{A},\mathbf{B};c)$ is nonempty. Let us replace Step 6 of Algorithm 2.1 by the following Step 6′, keeping all other steps unchanged.

Step 6′. Construct a vector $(\beta_1^{(t)},\dots,\beta_m^{(t)})$ that satisfies the relations

$$\begin{cases} \beta_j^{(t)} \geq 0, \\ \sum_{j=1}^{m} \beta_j^{(t)} = \delta_{p^{(t)}}(\mathbf{A}_{[k_q^{(t)},k_{q+1}^{(t)}]}, \mathbf{B}^{[k_q^{(t)},k_{q+1}^{(t)}]}; c), \\ b_j^{[k_q^{(t)},k_{q+1}^{(t)}]} - cp^{(t)} + \beta_j^{(t)} \leq c(k_{q+1}^{(t)} - k_q^{(t)} + 1), \quad b_j^{[k_q^{(t)},k_{q+1}^{(t)}]} \geq cp^{(t)}, \\ \beta_j^{(t)} \leq c(k_{q+1}^{(t)} - k_q^{(t)} + 1), \quad b_j^{[k_q^{(t)},k_{q+1}^{(t)}]} < cp^{(t)}, \\ cp^{(t)} - \beta_j^{(t)} \geq 0, \quad b_j^{[k_q^{(t)},k_{q+1}^{(t)}]} \geq cp^{(t)}, \\ b_j^{[k_q^{(t)},k_{q+1}^{(t)}]} - \beta_j^{(t)} \geq 0, \quad b_j^{[k_q^{(t)},k_{q+1}^{(t)}]} < cp^{(t)}, \\ 1 \leq j \leq m. \end{cases} \tag{2.25}$$

The replacement of Step 6 by Step 6′ is substantiated in proposition (b) of Lemma 2.3. Note that, if $\mathbf{A} \in \overline{Z}_+^n$, $\mathbf{B} \in Z_+^m$, $c \in Z$, and $\beta_j^{(t)} \in Z$, $1 \leq j \leq m$, Algorithm 2.2 constructs a matrix belonging to $M_Z(\mathbf{A},\mathbf{B},c)$. It is obvious that different solutions to system (2.25) correspond to different representations of $\mathbf{B}^{[k_q^{(t)},k_{q+1}^{(t)}]}$ as a sum of two vectors:

$$\mathbf{B}^{[k_q^{(t)},k_{q+1}^{(t)}]} = \mathbf{B}^{[k_q^{(t)},k_{q+1}^{(t)}+p^{(t)}-1]} + \mathbf{B}^{[k_q^{(t)}+p^{(t)},k_{q+1}^{(t)}]}.$$

Hence, different vectors $(\beta_1^{(t)},\dots,\beta_m^{(t)})$ satisfying (2.25) lead to different matrices from $M(\mathbf{A},\mathbf{B};c)$ (or $M_Z(\mathbf{A},\mathbf{B};c)$). It may happen that system (2.25) has a unique solution. Then, the solution is determined by the strict condition in (2.23). Recall that the foregoing analysis applies to $\mathbf{A} = (a_1, \ldots, a_n) \in \overline{R}^n_+$, i.e., $a_i \geq a_{i+1}$ for $1 \leq i \leq n-1$ if $n \geq 2$. Under the assumptions of Algorithms 2.1 and 2.2, system (2.25) has a solution, and the vector $(\beta_1^{(t)}, \ldots, \beta_m^{(t)})$ determined by (2.23) is a particular solution to (2.25).

Algorithm 2.2 constructs any matrix from the nonempty set $M(\mathbf{A}, \mathbf{B}; c)$ (or $M_Z(\mathbf{A}, \mathbf{B}; c)$), where $\mathbf{A} = (a_1, \ldots, a_n) \in \overline{R}^n_+$ and $\mathbf{B} = (b_1, \ldots, b_m) \in R^n_+$. Let $X(\mathbf{A}, \mathbf{B}) = (x_{ij})$ be an arbitrary matrix from $M(\mathbf{A}, \mathbf{B}; c)$ or $M_Z(\mathbf{A}, \mathbf{B}; c)$, and let integer p, $1 \leq p \leq n-1$, satisfy the equation

$$\delta_p(\mathbf{A}, \mathbf{B}; c) = \min \left\{ \delta_k(\mathbf{A}, \mathbf{B}; c) : 1 \leq k \leq n-1 \right\}.$$

The number p splits the matrix $X(\mathbf{A}, \mathbf{B})$ into two submatrices $X(\mathbf{A}', \mathbf{B}^*) = (x_{ij})$, $p+1 \leq i \leq n$, $1 \leq j \leq m$, and $X(\mathbf{A}'', \mathbf{B}^{**}) = (x_{ij})$, $1 \leq i \leq p$, $1 \leq j \leq m$, where

$$\mathbf{A}' = \mathbf{A}_{[p+1,n]} = (a_{p+1}, \ldots, a_n), \quad \mathbf{A}'' = \mathbf{A}_{[1,p]} = (a_1, \ldots, a_p),$$

$$\mathbf{B}^* = (b_1^*, \ldots, b_m^*) = \mathbf{B}^{[p+1,n]} = (b_1^{[p+1,n]}, \ldots, b_m^{[p+1,n]}),$$

$$\mathbf{B}^{**} = (b_1^{**}, \ldots, b_m^{**}) = \mathbf{B}^{[1,p]} = (b_1^{[1,n]}, \ldots, b_m^{[1,n]}),$$

$$b_j^* = \sum_{i=p+1}^{n} x_{ij}, \quad b_j^{**} = \sum_{i=1}^{p} x_{ij}, \quad 1 \leq j \leq m.$$

Given the vector $\mathbf{B}$, we construct two vectors $\mathbf{B}' = (b_1', \ldots, b_m')$ and $\mathbf{B}'' = (b_1'', \ldots, b_m'')$, with components defined in terms of p and c (see Lemmas 2.1-2.3):

$$b_j' = \begin{cases} b_j - cp, & b_j \geq cp, \\ 0, & b_j < cp, \end{cases}$$

$$b_j'' = \begin{cases} cp, & b_j \geq cp, \\ b_j, & b_j < cp, \end{cases}$$

$$1 \leq j \leq m.$$

It follows from the trivial equalities $b_j' + b_j'' = b_j$, $b_j^* + b_j^{**} = b_j$, $1 \leq j \leq m$, that

$$b_j^* - b_j' = b_j - b_j^{**} - b_j' = b_j - b_j' - b_j^{**} = b_j'' - b_j^{**}.$$

Define $\beta_j = b_j^* - b_j' = b_j'' - b_j^{**}$, $1 \leq j \leq m$. If $b_j < cp$, then $\beta_j = b_j^* - b_j' = b_j^* \geq 0$. Since $x_{ij} \leq c$ for all admissible values of i and j, we have

$$\beta_j = b_j^* - b_j' = b_j^* - b_j + cp = cp - b_j^{**} = cp - \sum_{i=1}^{p} x_{ij} \geq 0$$

for $b_j \geq cp$. Hence, $(\beta_1, \ldots, \beta_m) \in R_+^m$, and the last relations in (2.7) follow from the definitions of the vectors $\mathbf{B}^*$ and $\mathbf{B}^{**}$: $b_j^* = b_j' + \beta_j \leq c(n-p)$, $b_j^{**} = b_j'' - \beta_j \geq 0$, $1 \leq j \leq m$. The first relation in (2.7) follows from the definitions of $\mathbf{B}''$, $(\beta_1, \ldots, \beta_m)$, and the matrix $X(\mathbf{A}'', \mathbf{B}^{**})$:

$$\delta_p(\mathbf{A}, \mathbf{B}; c) = \sum_{i=1}^{m} \min(b_j, cp) - \sum_{i=1}^{p} a_i = \sum_{j=1}^{m} b_j'' - \sum_{j=1}^{m} b_j^{**} = \sum_{j=1}^{m} (b_j'' - b_j^{**}) = \sum_{j=1}^{m} \beta_j.$$

Thus, for an arbitrary matrix $X(\mathbf{A}, \mathbf{B}) \in M(\mathbf{A}, \mathbf{B}; c)$ and a number $p \in Z$, where $1 \leq p \leq n-1$ and

$$\delta_p(\mathbf{A}, \mathbf{B}; c) = \min\Big\{\delta_k(\mathbf{A}, \mathbf{B}; c) : 1 \leq k \leq n-1\Big\},$$

a vector $(\beta_1, \ldots, \beta_m) \in R_+^m$ is constructed that satisfies relations (2.7) and uniquely determines vectors $\mathbf{B}^*$ and $\mathbf{B}^{**}$ such that $M(\mathbf{A}', \mathbf{B}^*; c)$, $M(\mathbf{A}'', \mathbf{B}^{**}; c) \neq \varnothing$. The fact that system (2.25) is equivalent to (2.7) for any $t \in Z$, $1 \leq t \leq n-1$, implies that the following proposition is proved.

Theorem 2.4. *Let $M(\mathbf{A}, \mathbf{B}; c) \neq \varnothing$ or $M_Z(\mathbf{A}, \mathbf{B}; c) \neq \varnothing$, where $\mathbf{A} \in \overline{R}_+^n$. Then, any matrix $X(\mathbf{A}, \mathbf{B})$ in $M(\mathbf{A}, \mathbf{B}; c)$ or $M_Z(\mathbf{A}, \mathbf{B}; c)$ can be constructed by Algorithm* 2.2.

§3. Characteristic Equations for Truncated Transportation Polyhedrons

We consider arbitrary vector pairs that determine nonempty transportation polyhedrons. For transportation matrices with a common upper bound on their elements, identities are introduced that relate the elements of principal submatrices to the components of source vectors determining a transportation polyhedron. These identities are treated as equations in the unknown elements of the matrices and are called characteristic equations. On the basis of these equations, we develop a technique that is widely used in what follows for finding optimal solutions to problems with minimax functionals.

3.1. Construction of Characteristic Equations

Suppose that vectors $\mathbf{A} = (a_1, \ldots, a_n) \in R_+^n$ and $\mathbf{B} = (b_1, \ldots, b_m) \in R_+^m$ have equal sums of the components. Let $c \geq 0$, $k \in Z$, $1 \leq k \leq n$. Suppose also that $M(\mathbf{A}, \mathbf{B}; c) \neq \varnothing$ (or $M_Z(\mathbf{A}, \mathbf{B}; c) \neq \varnothing$ if $\mathbf{A} \in Z_+^n$, $\mathbf{B} \in Z_+^m$, $c \in Z$). Then, for an arbitrary matrix $X(\mathbf{A}, \mathbf{B}) = (x_{ij}) \in M(\mathbf{A}, \mathbf{B}; c)$, the numbers c and k determine one or two submatrices, $\overline{X} = \overline{X}(\mathbf{A}, \mathbf{B}; c, k)$ and $\overline{\overline{X}} = \overline{\overline{X}}(\mathbf{A}, \mathbf{B}; c, k)$, with elements defined as follows:

$$\overline{X} = (x_{ij}), \quad 1 \le i \le k, \quad b_j \ge ck;$$
$$\overline{\overline{X}} = (x_{ij}), \quad k < i \le n, \quad b_j < ck.$$

Denote by $s(X)$ the sum of elements of an arbitrary matrix X. Recall that $l_k(\mathbf{B}; c)$ is the number of components of $\mathbf{B}$ that are not less than ck (see (1.25)). If either $\overline{X}$ or $\overline{\overline{X}}$ does not exist, set $s(\overline{X}) = 0$ or $s(\overline{\overline{X}}) = 0$. It can be readily shown that the matrix $\overline{X}$ does not exist if $l_k(\mathbf{B}; c) = 0$; if $l_k(\mathbf{B}; c) = m$ or $k = n$, then the matrix $\overline{\overline{X}}$ does not exist. The matrix $\overline{X}$ exists for $k = n$ if and only if $\max\{b_j : 1 \le j \le m\} = cn$.

Suppose that $M(\mathbf{A}, \mathbf{B}; c) \ne \varnothing$ (or $M_Z(\mathbf{A}, \mathbf{B}; c) \ne \varnothing$). Consider the characteristic function (1.22) written in the equivalent form (1.25):

$$\delta_k(\mathbf{A}, \mathbf{B}; c) = \sum_{b_j < ck} b_j + ckl_k(\mathbf{B}; c) - \sum_{i=1}^{k} a_i.$$

For any matrix $X(\mathbf{A}, \mathbf{B}) = (x_{ij}) \in M(\mathbf{A}, \mathbf{B}; c)$, relations (1.2) entail the equations

$$\sum_{b_j < ck} b_j = \sum_{b_j < ck} \sum_{i=1}^{n} x_{ij},$$
$$\sum_{i=1}^{k} a_i = \sum_{i=1}^{k} \sum_{j=1}^{m} x_{ij}.$$

Hence,

$$\sum_{b_j < ck} b_j - \sum_{i=1}^{k} a_i = \sum_{b_j < ck} \sum_{i=1}^{k} x_{ij} + \sum_{b_j < ck} \sum_{i>k} x_{ij} - \sum_{i=1}^{k} \sum_{b_j \ge ck} x_{ij} - \sum_{i=1}^{k} \sum_{b_j < ck} x_{ij} =$$
$$\sum_{b_j < ck} \sum_{i>k} x_{ij} - \sum_{i=1}^{k} \sum_{b_j \ge ck} x_{ij}.$$

It follows from the definitions of matrices $\overline{X}(\mathbf{A}, \mathbf{B}; c, k)$ and $\overline{\overline{X}}(\mathbf{A}, \mathbf{B}; c, k)$ that

$$\sum_{b_j < ck} b_j - \sum_{i=1}^{k} a_i = s(\overline{\overline{X}}(\mathbf{A}, \mathbf{B}; c, k)) - s(\overline{X}(\mathbf{A}, \mathbf{B}; c, k)),$$

and we have the following assertion.

Theorem 3.1. *If $M(\mathbf{A},\mathbf{B};c) \neq \varnothing$ (or $M_Z(\mathbf{A},\mathbf{B};c) \neq \varnothing$), then, for an arbitrary $k \in Z$, $1 \leq k \leq n$, any matrix $X(\mathbf{A},\mathbf{B}) = (x_{ij})$ from $M(\mathbf{A},\mathbf{B};c)$ (or $M_Z(\mathbf{A},\mathbf{B};c)$) satisfies the equation*

$$\delta_k(\mathbf{A},\mathbf{B};c) = ckl_k(\mathbf{B};c) - s(\overline{X}(\mathbf{A},\mathbf{B};c,k)) + s(\overline{\overline{X}}(\mathbf{A},\mathbf{B};c,k)), \tag{3.1}$$

where $\overline{X}(\mathbf{A},\mathbf{B};c,k) = (x_{ij})$, $1 \leq i \leq k$, $b_j \geq ck$, $\overline{\overline{X}}(\mathbf{A},\mathbf{B};c,k) = (x_{ij})$, $k < i \leq n$, $b_j < ck$.

Relations (3.1) for $1 \leq k \leq n$ are identities for the set of matrices from $M(\mathbf{A},\mathbf{B};c)$, which will be treated as equations in the unknown elements of a matrix.

Definition 3.1. Equation (3.1) is called the *characteristic of a matrix sets* $M(\mathbf{A},\mathbf{B};c)$ or $M(\mathbf{A},\mathbf{B};c) \neq \varnothing$.

The variables $s(\overline{X}(\mathbf{A},\mathbf{B};c,k))$ and $s(\overline{\overline{X}}(\mathbf{A},\mathbf{B};c,k))$ in (3.1) are related by the equations

$$s(\overline{X}(\mathbf{A},\mathbf{B};c,k)) = \sum_{i=1}^{k} a_i - \sum_{b_j<ck} b_j + s(\overline{\overline{X}}(\mathbf{A},\mathbf{B};c,k)),$$

$$s(\overline{\overline{X}}(\mathbf{A},\mathbf{B};c,k)) = \sum_{i>k} a_i - \sum_{b_j\geq ck} b_j + s(\overline{\overline{X}}(\mathbf{A},\mathbf{B};c,k)).$$

Therefore, if $M(\mathbf{A},\mathbf{B};c) \neq \varnothing$, then the variables $s(\overline{X}(\mathbf{A},\mathbf{B};c,k))$ and $s(\overline{\overline{X}}(\mathbf{A},\mathbf{B};c,k))$ satisfy the following natural inequalities:

$$\begin{aligned} &\sum_{i=1}^{k} a_i - \sum_{b_j<ck} b_j \leq s(\overline{X}(\mathbf{A},\mathbf{B};c,k)) \leq \min\Big\{ckl_k(\mathbf{B};c), \sum_{i=1}^{k} a_i\Big\}, \\ &\sum_{i>k} a_i - \sum_{b_j\geq ck} b_j \leq s(\overline{\overline{X}}(\mathbf{A},\mathbf{B};c,k)) \leq \\ &\qquad\qquad \min\Big\{c(n-k)\left(m - l_k(\mathbf{B};c)\right), \sum_{b_j<ck} b_j, \sum_{i>k} a_i\Big\}. \end{aligned} \tag{3.2}$$

If $M(\mathbf{A},\mathbf{B};c) \neq \varnothing$ and the characteristic function (1.22) of A, B, and c is zero for a certain $k \in Z$, then the analysis of the matrix set is simplified. In this case, the elements of the principal submatrices are completely determined, and the truncated transportation polyhedron is decomposed into two subpolyhedrons of lesser dimensions. This is stated in the following proposition.

Theorem 3.2. *Suppose that* $\mathbf{A} \in R_+^n$ *and* $\mathbf{B} \in R_+^m$. *If* $M(\mathbf{A},\mathbf{B};c) \neq \varnothing$ *and* $\delta_p(\mathbf{A},\mathbf{B};c) = 0$ *for a certain* $p \in Z$, *where* $1 \leq p \leq n$ *and* $cp \leq \max\{b_j : 1 \leq j \leq m\}$, *then the elements of any matrix* $(x_{ij}) = X(\mathbf{A},\mathbf{B}) \in M(\mathbf{A},\mathbf{B};c)$ *are*

$$x_{ij} = \begin{cases} c, & 1 \leq i \leq p, \quad b_j \geq cp, \\ 0, & p < i \leq n, \quad b_j < cp. \end{cases} \tag{3.3}$$

Proof. Applying Theorem 3.1, we find that, for $k = p$, the characteristic equation (3.1) takes the form

$$cpl_p(\mathbf{B};c) - s(\overline{X}(\mathbf{A},\mathbf{B};c,p)) + s(\overline{\overline{X}}(\mathbf{A},\mathbf{B};c,p)) = 0.$$

Therefore, the assertion of the theorem follows from the equations $s(\overline{X}(\mathbf{A},\mathbf{B};c,p)) = cpl_p(\mathbf{B};c)$ and $s(\overline{\overline{X}}(\mathbf{A},\mathbf{B};c,p)) = 0$.

3.2. Examples of Application of Characteristic Equations

Suppose that $X(\mathbf{A},\mathbf{B}) = (x_{ij})$, $1 \leq i \leq n$, $1 \leq j \leq m$, is an arbitrary matrix from the set $M(\mathbf{A},\mathbf{B};c)$ and $\delta_p(\mathbf{A},\mathbf{B};c) = 0$, where $cp \leq \max\{b_j : 1 \leq j \leq m\}$. Denote by $X_1 = X_1(\mathbf{A},\mathbf{B};c,p)$ and $X_2 = X_2(\mathbf{A},\mathbf{B};c,p)$ the submatrices of the matrix $X(\mathbf{A},\mathbf{B})$ (if they exist) with elements defined by the relations

$$X_1 = (x_{ij}), \quad 1 \leq i \leq p, \quad b_j < cp,$$
$$X_2 = (x_{ij}), \quad p < i \leq n, \quad b_j \geq cp.$$

Let us construct new vectors based on $\mathbf{A} = (a_1, \dots, a_n) \in R_+^n$ and $\mathbf{B} = (b_1, \dots, b_m) \in R_+^m$, with components denoted by the same indices:

$$\begin{aligned} &\mathbf{A}_1 = \left(a_i - cl_p(\mathbf{B};c)\right), \quad 1 \leq i \leq p, \\ &\mathbf{B}_1 = (b_j), \quad b_j < cp \quad (\text{if } l_p(\mathbf{B};c) < m), \\ &\mathbf{A}_2 = (a_i), \quad p < i \leq n \quad (\text{if } p < n), \\ &\mathbf{B}_2 = (b_j - cp), \quad b_j \geq cp. \end{aligned} \tag{3.4}$$

It is obvious that $\mathbf{B}_1 \in R_+^{m-l_p(\mathbf{B};c)}$, $\mathbf{A}_2 \in R_+^{n-p}$, and $\mathbf{A}_1 \in R_+^p$, $\mathbf{B}_1 \in R_+^{l_p(\mathbf{B};c)}$ by Theorem 3.2. Since

$$s(\overline{X}) + s(X_1) = \sum_{i=1}^{p} a_i, \quad s(X_2) + s(\overline{\overline{X}}) = \sum_{i>p} a_i,$$

$$s(\overline{X}) + s(X_2) = \sum_{b_j \geq cp} b_j, \quad s(X_1) + s(\overline{\overline{X}}) = \sum_{b_j < cp} b_j,$$

it follows from (3.4) that

$$s(X_1) = \sum_{i=1}^{p} \left(a_i - cl_p(\mathbf{B};c)\right) = \sum_{b_j<cp} b_j,$$

$$s(X_2) = \sum_{b_j \geq cp} (b_j - cp) = \sum_{i>p} a_i;$$

i.e., the vector pairs $\mathbf{A}_1$, $\mathbf{B}_1$ and $\mathbf{A}_2$, $\mathbf{B}_2$ satisfy balance condition (1.1). It is also obvious that the elements of the submatrices X_1 and X_2 of the matrix $X(\mathbf{A},\mathbf{B})$ satisfy conditions (1.2) for the pairs of vectors $\mathbf{A}_1$, $\mathbf{B}_1$ and $\mathbf{A}_2$, $\mathbf{B}_2$, respectively; i.e., X_1 and X_2 are plan matrices for the vector pairs $\mathbf{A}_1$, $\mathbf{B}_1$ and $\mathbf{A}_2$, $\mathbf{B}_2$: $X_1 = X_1(\mathbf{A}_1,\mathbf{B}_1)$, $X_2 = X_2(\mathbf{A}_2,\mathbf{B}_2)$. Since $X(\mathbf{A},\mathbf{B}) \in M(\mathbf{A},\mathbf{B};c)$, then $X_1 \in M(\mathbf{A}_1,\mathbf{B}_1;c)$ and $X_2 \in M(\mathbf{A}_2,\mathbf{B}_2;c)$ (if matrices X_1 and X_2 exist).

We define the set of matrix pairs $(X_1(\mathbf{A},\mathbf{B};c,p), X_2(\mathbf{A},\mathbf{B};c,p))$ as follows:

$$M(\mathbf{A}_1,\mathbf{B}_1;c) \times M(\mathbf{A}_2,\mathbf{B}_2;c) = \begin{cases} \{(X_1,X_2) : X_1 \in M(\mathbf{A}_1,\mathbf{B}_1;c),\ X_2 \in M(\mathbf{A}_2,\mathbf{B}_2;c)\}, \\ \qquad\qquad l_p(\mathbf{B};c) < m, \quad p < n, \\ \{X_1 : X_1 \in M(\mathbf{A}_1,\mathbf{B}_1;c)\}, \quad p = n, \\ \{X_2 : X_2 \in M(\mathbf{A}_2,\mathbf{B}_2;c)\}, \quad l_p(\mathbf{B};c) = m. \end{cases}$$

If $M(\mathbf{A}_1,\mathbf{B}_1;c) \neq \varnothing$ and $M(\mathbf{A}_2,\mathbf{B}_2;c) \neq \varnothing$, the set $M(\mathbf{A}_1,\mathbf{B}_1;c) \times M(\mathbf{A}_2,\mathbf{B}_2;c)$ is the product of $M(\mathbf{A}_1,\mathbf{B}_1;c)$ and $M(\mathbf{A}_2,\mathbf{B}_2;c)$.

Thus, we have proved the following theorem.

Theorem 3.3. *Suppose that the assumptions of Theorem* 3.2 *hold. Then,*

(a) *there exists a one-to-one correspondence*

$$f : M(\mathbf{A},\mathbf{B};c) \to M(\mathbf{A}_1,\mathbf{B}_1;c) \times M(\mathbf{A}_2,\mathbf{B}_2;c)$$

between the sets $M(A,B;c)$ *and* $M(\mathbf{A}_1,\mathbf{B}_1;c) \times M(\mathbf{A}_2,\mathbf{B}_2;c)$; *moreover, if*

$$f\left(X(\mathbf{A},\mathbf{B})\right) = \begin{cases} (X_1(\mathbf{A},\mathbf{B};c,p), X_2(\mathbf{A},\mathbf{B};c,p)), \quad l_p(\mathbf{B};c) < m, \\ \qquad\qquad p < n, \\ X_1 = X_1(\mathbf{A},\mathbf{B};c,p), \quad p = n, \\ X_2 = X_2(\mathbf{A},\mathbf{B};c,p), \quad l_p(\mathbf{B};c) = m, \end{cases}$$

then every element of X_1 *and* X_2 *is equal to the corresponding element of the matrix* $X(\mathbf{A},\mathbf{B})$ (*the remaining elements of* $X(\mathbf{A},\mathbf{B})$ *are determined by* (3.3));

(b) *If* $\mathbf{A} \in Z_+^n$, $\mathbf{B} \in Z_+^m$, *and* $c \in Z$, *then the number of matrices in the set* $M_Z(\mathbf{A}, \mathbf{B}; c)$ *is given by*

$$\left|M_Z(\mathbf{A}, \mathbf{B}; c)\right| = \begin{cases} \left|M_Z(\mathbf{A}_1, \mathbf{B}_1; c)\right| \times \left|M_Z(\mathbf{A}_2, \mathbf{B}_2; c)\right|, & l_p(\mathbf{B}; c) < m, \\ & p < n, \\ \left|M_Z(\mathbf{A}_1, \mathbf{B}_1; c)\right|, & p = n, \\ \left|M_Z(\mathbf{A}_2, \mathbf{B}_2; c)\right|, & l_p(\mathbf{B}; c) = m. \end{cases}$$

Note that, if $p = n$ and $l_p(\mathbf{B}; c) = m$, matrices X_1 and X_2 do not exist. In this case, the set $M(\mathbf{A}, \mathbf{B}; c)$ consists of only one matrix, with elements equal to c.

In certain cases, assertion (a) of Theorem 3.3 makes it easy to construct the set of matrices from $M_Z(\mathbf{A}, \mathbf{B}; c)$ as this problem reduces to the construction of sets of matrices of lesser dimensions.

Example 3.1. Let us construct the matrix set $M_Z(\mathbf{A}, \mathbf{B}; c)$ for $\mathbf{A} = (11, 11, 11, 4, 3)$, $\mathbf{B} = (12, 12, 10, 3, 3)$, and $c = 3$.

Here, $\delta_3(\mathbf{A}, \mathbf{B}; 3) = 0$ and $l_3(\mathbf{B}; 3) = 3$. Applying (3.4), we find that $\mathbf{A}_1 = (2, 2, 2)$, $\mathbf{B}_1 = (3, 3)$ and $\mathbf{A}_2 = (4, 3)$, $\mathbf{B}_2 = (3, 3, 1)$. Both $M_Z(\mathbf{A}_1, \mathbf{B}_1; 3)$ and $M_Z(\mathbf{A}_2, \mathbf{B}_2; 3)$ can be easily constructed since any column of the matrix $X_1 = X_1(\mathbf{A}_1, \mathbf{B}_1)$ and any row of $X_2 = X_2(\mathbf{A}_2, \mathbf{B}_2)$ uniquely define these matrices:

$$M_Z(\mathbf{A}_1, \mathbf{B}_1; 3) = \left\{ \begin{pmatrix} 2 & 0 \\ 1 & 1 \\ 0 & 2 \end{pmatrix}, \begin{pmatrix} 2 & 0 \\ 0 & 2 \\ 1 & 1 \end{pmatrix}, \begin{pmatrix} 1 & 1 \\ 2 & 0 \\ 0 & 2 \end{pmatrix}, \begin{pmatrix} 0 & 2 \\ 2 & 0 \\ 1 & 1 \end{pmatrix}, \begin{pmatrix} 1 & 1 \\ 0 & 2 \\ 2 & 0 \end{pmatrix}, \begin{pmatrix} 0 & 2 \\ 1 & 1 \\ 2 & 0 \end{pmatrix}, \begin{pmatrix} 1 & 1 \\ 1 & 1 \\ 1 & 1 \end{pmatrix} \right\},$$

$$M_Z(\mathbf{A}_2, \mathbf{B}_2; 3) = \left\{ \begin{pmatrix} 0 & 3 & 1 \\ 3 & 0 & 0 \end{pmatrix}, \begin{pmatrix} 3 & 0 & 1 \\ 0 & 3 & 0 \end{pmatrix}, \begin{pmatrix} 1 & 2 & 1 \\ 2 & 1 & 0 \end{pmatrix}, \begin{pmatrix} 1 & 3 & 0 \\ 2 & 0 & 1 \end{pmatrix}, \begin{pmatrix} 2 & 1 & 1 \\ 1 & 2 & 0 \end{pmatrix}, \begin{pmatrix} 3 & 1 & 0 \\ 0 & 2 & 1 \end{pmatrix}, \begin{pmatrix} 2 & 2 & 0 \\ 1 & 1 & 1 \end{pmatrix} \right\}.$$

It follows from Theorems 3.2 and 3.3 that each matrix $X(\mathbf{A}, \mathbf{B})$ in $M_Z(\mathbf{A}, \mathbf{B}; c)$ has the form

$$X(\mathbf{A}, \mathbf{B}) = \begin{pmatrix} 3 & 3 & 3 & x_{14} & x_{15} \\ 3 & 3 & 3 & x_{24} & x_{25} \\ 3 & 3 & 3 & x_{34} & x_{35} \\ x_{41} & x_{42} & x_{43} & 0 & 0 \\ x_{51} & x_{52} & x_{53} & 0 & 0 \end{pmatrix},$$

where $\begin{pmatrix} x_{14} & x_{15} \\ x_{24} & x_{25} \\ x_{34} & x_{35} \end{pmatrix}$ and $\begin{pmatrix} x_{41} & x_{42} & x_{43} \\ x_{51} & x_{52} & x_{53} \end{pmatrix}$ are arbitrary matrices from the sets $M_Z(\mathbf{A}_1, \mathbf{B}_1; 3)$ and $M_Z(\mathbf{A}_2, \mathbf{B}_2; 3)$, respectively.

Applying assertion (b) of Theorem 3.3, we find that

$$|M_Z(\mathbf{A}, \mathbf{B}; 3)| = |M_Z(\mathbf{A}_1, \mathbf{B}_1; 3)| \times |M_Z(\mathbf{A}_2, \mathbf{B}_2; 3)| = 49.$$

In view of the obvious inequality $\delta_k(\widetilde{\mathbf{A}}, \mathbf{B}; c) \geq \delta_k(\mathbf{A}, \mathbf{B}; c)$ for the characteristic function, which holds for all admissible values of k, the analysis of truncated transportation polyhedrons is simplified when the components of the first vector are arranged in nonascending order. This is further illustrated by the constructions presented below.

We see that the unknowns in characteristic equations are the sums of elements of principal submatrices. The following problem can be posed: under what conditions do the solutions found allow us to recover the whole matrix?

Example 3.2. It is easy to check that $M_Z(\mathbf{A}, \mathbf{B}; 2) \neq \varnothing$ for $\mathbf{A} = \mathbf{B} = (8, 8, 7, 3, 2)$. Let $k = 2$. Then, $\delta_2(\mathbf{A}, \mathbf{B}; 2) = 1$, $l_2(\mathbf{B}; 2) = 3$, and any matrix $X(\mathbf{A}, \mathbf{B}) \in M_Z(\mathbf{A}, \mathbf{B}; 2)$ satisfies the equation

$$12 - s(\overline{X}(\mathbf{A}, \mathbf{B}; 2, 2)) + s(\overline{\overline{X}}(\mathbf{A}, \mathbf{B}; 2, 2)) = 1.$$

This characteristic equation has two integer solutions:

$$s(\overline{X}(\mathbf{A}, \mathbf{B}; 2, 2)) = 12, \quad s(\overline{\overline{X}}(\mathbf{A}, \mathbf{B}; 2, 2)) = 1$$

and

$$s(\overline{X}(\mathbf{A}, \mathbf{B}; 2, 2)) = 11, \quad s(\overline{\overline{X}}(\mathbf{A}, \mathbf{B}; 2, 2)) = 0.$$

If we set $k = 3$, then $\delta_3(\mathbf{A}, \mathbf{B}; 2) = 0$, $l_3(\mathbf{B}; 2) = l_2(\mathbf{B}; 2) = 3$, and any matrix $X(\mathbf{A}, \mathbf{B})$ from $M_Z(\mathbf{A}, \mathbf{B}; 2)$ satisfies the equation

$$18 - s\left(\overline{X}(\mathbf{A}, \mathbf{B}; 2, 3)\right) + s(\overline{\overline{X}}(\mathbf{A}, \mathbf{B}; 2, 3)) = 0.$$

This equation has the unique solution

$$s(\overline{X}(\mathbf{A}, \mathbf{B}; 2, 3)) = 18, \quad s(\overline{\overline{X}}(\mathbf{A}, \mathbf{B}; 2, 3)) = 0.$$

Hence, each element of $\overline{X}(\mathbf{A}, \mathbf{B}; 2, 3)$ is two. Since $\overline{X}(\mathbf{A}, \mathbf{B}; 2, 2)$ is the matrix block of $\overline{X}(\mathbf{A}, \mathbf{B}; 2, 3)$, each element of $\overline{X}(\mathbf{A}, \mathbf{B}; 2, 2)$ is also equal to two. Therefore, only one of the two solutions to equation (3.1) corresponds to any matrix $X(\mathbf{A}, \mathbf{B}) \in M_Z(\mathbf{A}, \mathbf{B}; 2)$ for $k = 2$: $s(\overline{X}(\mathbf{A}, \mathbf{B}; 2, 2)) = 12$, $s(\overline{\overline{X}}(\mathbf{A}, \mathbf{B}; 2, 2)) = 1$.

The following theorem states a sufficient condition for recovering the original matrix given a solution to the characteristic equation.

Theorem 3.4. *Suppose that* $\mathbf{A} = (a_1, \ldots, a_n) \in \overline{R}^n_+$, $\mathbf{B} = (b_1, \ldots, b_m) \in \overline{R}^m_+$, $c > 0$, $M(\mathbf{A}, \mathbf{B}; c) \neq \varnothing$, *and* $p \in Z$, *where* $1 \leq p \leq n-1$ *and* $\delta_p(\mathbf{A}, \mathbf{B}; c) = \min\left\{\delta_k(\mathbf{A}, \mathbf{B}; c) : 1 \leq k \leq n-1,\ ck \leq \max\{b_j : 1 \leq j \leq m\}\right\}$. *Then, for any solution* $\overline{s}_p$, $\overline{\overline{s}}_p$ *of the characteristic equation*

$$cpl_p(\mathbf{B}; c) - \overline{s}_p + \overline{\overline{s}}_p = \delta_p(\mathbf{A}, \mathbf{B}; c),$$

where $\overline{s}_p$ *and* $\overline{\overline{s}}_p$ *satisfy* (3.2),

$$\max\left\{0, cpl_p(\mathbf{B}; c) - \delta_p(\mathbf{A}, \mathbf{B}; c)\right\} \leq \overline{s}_p \leq \min\left\{cpl_p(\mathbf{B}; c), \sum_{i=1}^{p} a_i\right\},$$

$$\max\left\{0, \delta_p(\mathbf{A}, \mathbf{B}; c) - cpl_p(\mathbf{B}; c)\right\} \leq \overline{\overline{s}}_p \leq \min\left\{c(n-p)\left\{m - l_p(\mathbf{B}; c)\right\}, \sum_{b_j < cp} b_j, \sum_{i=p+1}^{n} a_i\right\},$$

the following assertions are valid.

(a) *There exists a matrix* $X(\mathbf{A}, \mathbf{B})$ *in* $M(\mathbf{A}, \mathbf{B}; c)$ *such that* $s(\overline{X}(\mathbf{A}, \mathbf{B}; c, p)) = \overline{s}_p$ *and* $s(\overline{\overline{X}}(\mathbf{A}, \mathbf{B}; c, p)) = \overline{\overline{s}}_p$ (*if* X *does not exist,* then $s(X) = 0$).

(b) *If arbitrary numbers* β_{ij}, $0 \leq \beta_{ij} \leq c$, *where* $1 \leq i \leq p$, $b_j \geq cp$ *or* $p+1 \leq i \leq n$, $b_j < cp$, *satisfy the relations*

$$\begin{aligned} &\sum_{i=1}^{p} \sum_{b_j \geq cp} \beta_{ij} = cpl_p(\mathbf{B}; c) - \overline{s}_p, \\ &\sum_{i=p+1}^{n} \sum_{b_j < cp} \beta_{ij} = \overline{\overline{s}}_p, \\ &\sum_{b_j < cp} \beta_{ij} \leq a_i, \quad i \geq p+1, \\ &\sum_{i=p+1}^{n} \beta_{ij} \leq b_j, \quad b_j < cp \end{aligned} \tag{3.5}$$

(*if* $\{b_j : 1 \leq j \leq m,\ b_j < cp\} = \varnothing$, *the last three sums, for which* $i \geq p+1$ *or* $b_j < cp$, *are assumed to be zero*), *then there exists the matrix* $X(\mathbf{A}, \mathbf{B}) = (x_{ij}) \in M(\mathbf{A}, \mathbf{B}; c)$ *with elements defined as*

$$x_{ij} = \begin{cases} c - \beta_{ij}, & 1 \leq i \leq p, \quad b_j \geq cp, \\ \beta_{ij}, & p+1 \leq i \leq n, \quad b_j < cp. \end{cases}$$

Proof. Since assertion (a) follows from assertion (b), it will suffice to prove (b).

Given vectors $\mathbf{A} = (a_1, \ldots, a_n)$ and $\mathbf{B} = (b_1, \ldots, b_m)$, let us consider the vectors $\mathbf{A}' = (a'_1, \ldots, a'_n)$ and $\mathbf{B}' = (b'_1, \ldots, b'_m)$ with components defined as

$$a'_i = \begin{cases} a_i + \sum\limits_{b_j \geq cp} \beta_{ij}, & 1 \leq i \leq p, \\ a_i - \sum\limits_{b_j < cp} \beta_{ij}, & p+1 \leq i \leq n, \end{cases}$$

$$b'_j = \begin{cases} b_j + \sum\limits_{i=1}^{p} \beta_{ij}, & b_j \geq cp, \\ b_j - \sum\limits_{i=p+1}^{n} \beta_{ij}, & b_j < cp. \end{cases} \tag{3.6}$$

It is clear that $\mathbf{A}' \in R^n_+$ and $\mathbf{B}' \in R^m_+$.

Let us show that $\delta_k(\widetilde{\mathbf{A}}', \mathbf{B}'; c) \geq 0$ and $\delta_p(\widetilde{\mathbf{A}}', \mathbf{B}'; c) = 0$, where the vector $\widetilde{\mathbf{A}}' = (\tilde{a}'_1, \ldots, \tilde{a}'_n)$ is derived from the vector $\mathbf{A}'$ by arranging its components in nonascending order.

It is obvious from (3.6) that

$$\sum_{i=1}^{p} \tilde{a}'_i = \sum_{i=1}^{p} a'_i, \quad l_p(\mathbf{B}'; c) = l_p(\mathbf{B}; c).$$

Hence,

$$\delta_p(\widetilde{\mathbf{A}}', \mathbf{B}'; c) = \delta_p(\mathbf{A}', \mathbf{B}'; c) = \sum_{b'_j < cp} b'_j + cpl_p(\mathbf{B}'; c) - \sum_{i=1}^{p} a'_i =$$
$$\sum_{b_j < cp} b_j - \sum_{i=1}^{p} \sum_{b_j \geq cp} \beta_{ij} + cpl_p(\mathbf{B}; c) - \sum_{i=1}^{p} a_i - \sum_{i=p+1}^{n} \sum_{b_j < cp} \beta_{ij} =$$
$$\delta_p(\mathbf{A}, \mathbf{B}; c) - \sum_{i=1}^{p} \sum_{b_j \geq cp} \beta_{ij} - \sum_{i=p+1}^{n} \sum_{b_j < cp} \beta_{ij}.$$

Applying (3.5), we obtain

$$\delta_p(\widetilde{\mathbf{A}}', \mathbf{B}'; c) = \delta_p(\mathbf{A}, \mathbf{B}; c) - cpl_p(\mathbf{B}; c) + \overline{s}_p - \overline{\overline{s}}_p = \delta_p(\mathbf{A}, \mathbf{B}; c) - \delta_p(\mathbf{A}, \mathbf{B}; c) = 0.$$

Let $k \in Z$, $1 \leq k \leq p-1$. Then, it follows from (3.6) that $l_p(\mathbf{B}; c) \leq l_k(\mathbf{B}'; c) \leq l_k(\mathbf{B}; c)$. The addends on the right-hand side of the expression for the characteristic function

$$\delta_k(\widetilde{\mathbf{A}}', \mathbf{B}'; c) = \sum_{b'_j < ck} b'_j + ckl_k(\mathbf{B}'; c) - \sum_{i=1}^{k} \tilde{a}'_i$$

satisfy the following relations

$$\sum_{b'_j<ck} b'_j = \sum_{b_j<ck} b_j + \sum_{\substack{b'_j < ck \\ b_j \geq ck}} b'_j - \sum_{i=p+1}^{n} \sum_{b_j<ck} \beta_{ij} =$$
$$\sum_{b_j<ck} b_j + \sum_{\substack{b'_j < ck \\ b_j \geq ck}} b_j - \sum_{i=p+1}^{n} \sum_{\substack{b'_j < ck \\ b_j \geq ck}} \beta_{ij} - \sum_{i=p+1}^{n} \sum_{b_j<ck} \beta_{ij} \geq$$
$$\sum_{b_j<ck} b_j + \sum_{\substack{b'_j < ck \\ b_j \geq ck}} b_j - \sum_{i=p+1}^{n} \sum_{b_j<cp} \beta_{ij},$$

$$ckl_k(\mathbf{B}'; c) = ckl_k(\mathbf{B}; c) - ck\left(l_k(\mathbf{B}; c) - l_k(\mathbf{B}'; c)\right),$$

$$\sum_{i=1}^{k} \tilde{a}'_i = \sum_{i=1}^{k} a'_i + \sum_{i=k+1}^{p} a'_i - \sum_{i=k+1}^{p} \tilde{a}'_i =$$
$$\sum_{i=1}^{k} a_i + \sum_{i=1}^{k} \sum_{b_j \geq cp} \beta_{ij} + \sum_{i=k+1}^{p} a_i + \sum_{i=k+1}^{p} \sum_{b_j \geq cp} \beta_{ij} - \sum_{i=k+1}^{p} \tilde{a}'_i =$$
$$\sum_{i=1}^{k} a_i + \sum_{i=1}^{p} \sum_{b_j \geq cp} \beta_{ij} + \sum_{i=k+1}^{p} a_i - \sum_{i=k+1}^{p} \tilde{a}'_i.$$

The obvious inequalities

$$\sum_{\substack{b'_j < ck \\ b_j \geq ck}} b_j \geq ck\left(l_k(\mathbf{B}; c) - l_k(\mathbf{B}'; c)\right), \quad \sum_{i=k+1}^{p} a_i \leq \sum_{i=k+1}^{p} \tilde{a}_i$$

entail

$$\sum_{b'_j<ck} b'_j + ckl_k(\mathbf{B}'; c) \geq \sum_{b_j<ck} b_j + ckl_k(\mathbf{B}; c) - \sum_{i=p+1}^{n} \sum_{b_j<cp} \beta_{ij},$$

$$\sum_{i=1}^{k} \tilde{a}'_i \leq \sum_{i=1}^{k} a_i + \sum_{i=1}^{p} \sum_{b_j \geq cp} \beta_{ij}.$$

Hence, the characteristic function (1.22) satisfies the inequality

$$\delta_k(\widetilde{\mathbf{A}}', \mathbf{B}'; c) \geq$$

$$\sum_{b_j<ck} b_j + ckl_k(\mathbf{B}; c) - \sum_{i=1}^{k} a_i - \sum_{i=p+1}^{n} \sum_{b_j<cp} \beta_{ij} - \sum_{i=1}^{p} \sum_{b_j\geq cp} \beta_{ij} =$$

$$\delta_k(\mathbf{A}, \mathbf{B}; c) - \sum_{i=p+1}^{n} \sum_{b_j<cp} \beta_{ij} - \sum_{i=1}^{p} \sum_{b_j\geq cp} \beta_{ij}.$$

Since

$$\delta_p(\mathbf{A}, \mathbf{B}; c) = \min\Big\{\delta_k(\mathbf{A}, \mathbf{B}; c) : 1 \leq k \leq n-1,\ ck \leq \max\{b_j : 1 \leq j \leq m\}\Big\},$$

it follows from (3.5) that

$$\delta_k(\widetilde{\mathbf{A}}', \mathbf{B}'; c) \geq \delta_k(\mathbf{A}, \mathbf{B}; c) - cpl_p(\mathbf{B}; c) + \overline{s}_p - \overline{\overline{s}}_p = \delta_k(\mathbf{A}, \mathbf{B}; c) - \delta_p(\mathbf{A}, \mathbf{B}; c) \geq 0.$$

Let $k \in Z$, $p+1 \leq k \leq n$. Then $l_k(\mathbf{B}; c) \leq l_k(\mathbf{B}'; c) \leq l_p(\mathbf{B}; c)$. Let us consider separately each term on the right-hand side of the function

$$\delta_k(\widetilde{\mathbf{A}}', \mathbf{B}'; c) = \sum_{b'_j<ck} b'_j + ckl_k(\mathbf{B}'; c) - \sum_{i=1}^{k} \tilde{a}'_1 :$$

$$\sum_{b'_j<ck} b'_j = \sum_{b_j<ck} b_j - \sum_{\substack{b_j < ck \\ b'_j \geq ck}} b_j + \sum_{i=1}^{p} \sum_{\substack{b_j \geq cp \\ b'_j < ck}} \beta_{ij} - \sum_{i=p+1}^{n} \sum_{b_j<cp} \beta_{ij},$$

$$ckl_k(\mathbf{B}'; c) = ckl_k(\mathbf{B}; c) + ck\left(l_k(\mathbf{B}'; c) - l_k(\mathbf{B}; c)\right),$$

$$\sum_{i=1}^{k} \tilde{a}'_i = \sum_{i=1}^{p} \tilde{a}'_i + \sum_{i=p+1}^{k} \tilde{a}'_i = \sum_{i=1}^{p} a'_i + \sum_{i=p+1}^{k} \tilde{a}'_i =$$

$$\sum_{i=1}^{p} a_i + \sum_{i=1}^{p} \sum_{b_j\geq cp} \beta_{ij} + \sum_{i=p+1}^{k} a'_i + \sum_{i=p+1}^{k} \tilde{a}'_i - \sum_{i=p+1}^{k} a'_i =$$

$$\sum_{i=1}^{p} a_i + \sum_{i=1}^{p} \sum_{b_j\geq cp} \beta_{ij} + \sum_{i=p+1}^{k} a_i - \sum_{i=p+1}^{k} \sum_{b_j<cp} \beta_{ij} + \sum_{i=p+1}^{k} \tilde{a}'_i - \sum_{i=p+1}^{k} a'_i =$$

$$\sum_{i=1}^{k} a_i + \sum_{i=1}^{p} \sum_{b_j\geq cp} \beta_{ij} + \sum_{i=p+1}^{k} \tilde{a}'_i - \sum_{i=p+1}^{k} a_i.$$

It is easy to show that the following inequalities hold:

$$\sum_{\substack{b_j < ck \\ b'_j \geq ck}} b_j \leq ck\left(l_k(\mathbf{B}'; c) - l_k(\mathbf{B}; c)\right), \quad \sum_{i=p+1}^{k} \tilde{a}'_i \leq \sum_{i=p+1}^{k} a_i.$$

Hence,

$$\delta_k(\widetilde{\mathbf{A}}', \mathbf{B}; c) \geq$$

$$\sum_{b_j < ck} b_j + ckl_k(\mathbf{B}; c) - \sum_{i=p+1}^{n} \sum_{b_j < cp} \beta_{ij} - \sum_{i=1}^{k} a_i - \sum_{i=1}^{p} \sum_{b_j \geq cp} \beta_{ij} =$$

$$\delta_k(\mathbf{A}, \mathbf{B}; c) - \sum_{i=p+1}^{n} \sum_{b_j < cp} \beta_{ij} - \sum_{i=1}^{p} \sum_{b_j \geq cp} \beta_{ij}.$$

Therefore, taking into account (3.5) and the definition of $\delta_p(\mathbf{A}, \mathbf{B}; c)$, we find that

$$\delta_k(\widetilde{\mathbf{A}}', \mathbf{B}'; c) \geq \delta_k(\mathbf{A}, \mathbf{B}; c) - \delta_p(\mathbf{A}, \mathbf{B}; c) \geq 0.$$

Thus, $\delta_k(\widetilde{\mathbf{A}}', \mathbf{B}'; c) \geq 0$ for any k, $1 \leq k \leq n$. Therefore, $M(\mathbf{A}', \mathbf{B}'; c) \neq \varnothing$ by Lemma 2.5. Since $\delta_p(\mathbf{A}', \mathbf{B}'; c) = 0$, the assumptions of Theorem 3.2 are valid. Hence, there exists a matrix $X(\mathbf{A}', \mathbf{B}') = (x'_{ij}) \in M(\mathbf{A}', \mathbf{B}'; c)$ with elements satisfying (3.3):

$$x'_{ij} = \begin{cases} c, & 1 \leq i \leq p, \quad b_j \geq cp, \\ 0, & p+1 \leq i \leq n, \quad b_j < cp. \end{cases}$$

Let us construct $X = (x_{ij})$, $1 \leq i \leq n$, $1 \leq j \leq m$, with the elements

$$x_{ij} = \begin{cases} c - \beta_{ij}, & 1 \leq i \leq p, \quad b_j \geq cp, \\ x'_{ij}, & 1 \leq i \leq p, \quad b_j < cp, \\ \beta_{ij}, & p+1 \leq i \leq n, \quad b_j < cp, \\ x'_{ij}, & p+1 \leq i \leq n, \quad b_j \geq cp. \end{cases}$$

Applying (3.6) to the rows and columns of X, we obtain

$$\sum_{j=1}^{m} x_{ij} = \sum_{j=1}^{m} x'_{ij} - \sum_{b_j \geq cp} \beta_{ij} = a'_i - \sum_{b_j \geq cp} \beta_{ij} = a_i, \quad 1 \leq i \leq p,$$

$$\sum_{j=1}^{m} x_{ij} = \sum_{j=1}^{m} x'_{ij} + \sum_{b_j < cp} \beta_{ij} = a'_i + \sum_{b_j < cp} \beta_{ij} = a_i, \quad p+1 \leq i \leq n,$$

$$\sum_{i=1}^{n} x_{ij} = \sum_{i=1}^{n} x'_{ij} - \sum_{i=1}^{p} \beta_{ij} = b'_j - \sum_{i=1}^{p} \beta_{ij} = b_j, \quad b_j \geq cp,$$

$$\sum_{i=1}^{n} x_{ij} = \sum_{i=1}^{n} x'_{ij} + \sum_{i=p+1}^{n} \beta_{ij} = b'_j + \sum_{i=p+1}^{n} \beta_{ij} = b_j, \quad b_j < cp.$$

Thus, the matrix $X \in M(\mathbf{A}, \mathbf{B}; c)$ satisfies all assumptions of proposition (b) of the theorem.

Theorem 3.4 can be used to estimate the number of matrices in the set $M_Z(\mathbf{A}, \mathbf{B}; c)$.

Example 3.3. Consider a matrix set $M_Z(\mathbf{A}, \mathbf{B}; 2)$, where

$$\mathbf{A} = (14, 14, 13, 10, 8, 4, 3) \in \overline{R}^7_+ \quad \text{and} \quad \mathbf{B} = (12, 12, 10, 10, 8, 6, 4, 2, 2) \in \overline{R}^9_+.$$

It is easy to show that function (1.22) of the vectors $\mathbf{A}$ and $\mathbf{B}$ and the number $c = 2$ is nonnegative for all admissible values of $k \in Z$, and $\delta_4(\mathbf{A}, \mathbf{B}; 2) = \min\left\{\delta_k(\mathbf{A}, \mathbf{B}; 2) : 1 \le k \le 6,\ 2k \le \max\{b_j : 1 \le j \le 9\}\right\} = 3$. Let us apply Theorem 3.4 to estimate $\left|M_Z(\mathbf{A}, \mathbf{B}; 2)\right|$.

Here, $l_4(\mathbf{B}; 2) = 5$ and, since $\mathbf{B} \in \overline{R}^9_+$, $\{j : b_j \ge 8\} = \{j : 1 \le j \le 5\}$ and $\{j : b_j < 8\} = \{j : 6 \le j \le 9\}$. Therefore, the characteristic equation (3.1) for $k = 4$ has the form

$$40 - s(\overline{X}(\mathbf{A}, \mathbf{B}; 2, 4)) + s(\overline{\overline{X}}(\mathbf{A}, \mathbf{B}; 2, 4)) = 3,$$

where $\overline{X} = \overline{X}(\mathbf{A}, \mathbf{B}; 2, 4) = (x_{ij})$, $1 \le i \le 4$, $1 \le j \le 5$, $\overline{\overline{X}} = \overline{\overline{X}}(\mathbf{A}, \mathbf{B}; 2, 4) = (x_{ij})$, $5 \le i \le 7$, $6 \le j \le 9$. Define $\overline{s} = s(\overline{X})$ and $\overline{\overline{s}} = s(\overline{\overline{X}})$. It follows from proposition (a) of Theorem 3.4 that, for any solution to the equation $40 - \overline{s} + \overline{\overline{s}} = 3$ that satisfies (3.2), there exists a matrix $X(\mathbf{A}, \mathbf{B}) \in M(\mathbf{A}, \mathbf{B}; 2)$ such that $s(\overline{X}) = \overline{s}$ and $s(\overline{\overline{X}}) = \overline{\overline{s}}$. Since we consider integer matrices, the characteristic equation has four integer solutions $\overline{s}$, $\overline{\overline{s}}$:

$$40,\ 3; \quad 39,\ 2; \quad 38,\ 1; \quad 37,\ 0.$$

By proposition (b) of Theorem 3.4, the lower estimate for the number of matrices in $M_Z(\mathbf{A}, \mathbf{B}; 2)$ is the number r of solutions to four systems of equations:

$$\begin{cases} \displaystyle\sum_{i=1}^{4} \sum_{j=1}^{5} \beta_{ij} = 37 + q, \\ \displaystyle\sum_{i=5}^{7} \sum_{j=6}^{9} \beta_{ij} = q, \end{cases} \qquad q \in Z, \quad 0 \le q \le 3,$$

with $\beta_{ij} \in Z$, $0 \le \beta_{ij} \le 2$, where $1 \le i \le 4$, $1 \le j \le 5$ or $5 \le i \le 7$, $6 \le j \le 9$. Denote by $r(q)$ the number of solutions to this system for $q \in Z$, $0 \le q \le 3$, and by $r_1(q)$ and $r_2(q)$ the numbers of solutions to the first and second equations of the system, respectively. Since the equations in each system are independent, we have $r(q) = r_1(q)r_2(q)$, and

$$r = \sum_{q=0}^{3} r_1(q)r_2(q).$$

The numbers $r_k(q)$, $1 \le k \le 2$, $0 \le q \le 3$, can be easily calculated. The numbers of variables in the first and second equations are 20 and 12, respectively. Therefore,

$$\begin{aligned} r_1(0) &= C_{20}^3 + 2C_{20}^2 = 1520, & r_2(0) &= 1, \\ r_1(1) &= C_{20}^2 + C_{20}^1 = 210, & r_2(1) &= C_{12}^1 = 12, \\ r_1(2) &= C_{20}^1 = 20, & r_2(2) &= C_{12}^2 + C_{12}^1 = 78, \\ r_1(3) &= 1, & r_2(3) &= C_{12}^3 + 2C_{12}^2 = 352. \end{aligned}$$

Performing simple calculations, we find that r=5952 and $|M_Z(\mathbf{A}, \mathbf{B}; 2)| \geq$ 5952.

The results of this chapter are based on [22–25].

References to Chapter 1

[1] Alekseev V. M., Tikhomirov V. M., and Fomin S. V., *Optimal'noe upravlenie* (Optimal Control), Moscow: Nauka, 1979.

[2] Astakhov N. D., Fedoseev A. V., and Chan Van Khung, *O primenenii uskorennykh algoritmov resheniya zadach transportnogo tipa* (On the Use of Accelerated Algorithms for Solving Transportation Problems), Moscow: Computing Center of Russ. Acad. Sci., 1994.

[3] Bellman R. *Dynamic Programming*, Princeton: Princeton Univ. Press, 1957.

[4] Burkov V. N, *Matematicheskie osnovy teorii aktivnykh sistem* (Mathematical Foundations of the Theory of Active Systems), Moscow: Nauka, 1977.

[5] Burkov V. N. and Opoitsev V. N., Metagame Approach to Hierarchical Systems Control, *Avtom. Telemekh.*, 1974, no. 1, pp. 103–114.

[6] *Teoretiko-igrovye voprosy prinyatiya reshenii* (Game-Theoretic Problems in Decision Making), Vorob'eva, N.N., Ed., Leningrad: Nauka,1978.

[7] Gale D, *A Theorem on Flow in Networks*, *Pacific J. Math.*, 1957, vol. 7, pp. 1073–1082.

[8] Gnedenko B. V., *Kurs teorii veroyatnostei* (A Course in Probability Theory), Moscow: Fizmatgiz, 1969.

[9] Gol'shtein E. G. and Yudin D. B, *Zadachi lineinogo programmirovania transportnogo tipa* (Transportation Problems in Linear Programming), Moscow: Nauka, 1969.

[10] Dem'yaninov V. F. and Malozemov V. N., *Vvedenie v minimaks* (Introduction to Minimax), Moscow: Nauka, 1972.

[11] Emelichev V. A. and Kovalev M. M., *Mnogogranniki, grafy, optimizatsiya* (Polyhedrons, Graphs, and Optimization), Moscow: Nauka, 1981.

[12] Emelichev V. A. and Kononenko A. M., On a Class of Transportation Polyhedrons, *Izv. Belarus. Akad. Nauk.*, Ser. Fiz. Mat., 1971, no. 3, pp. 21–23.

[13] Emelichev V. A., Kravtsov M. K., and Krachkovski A. P., On Some Classes of Transportation Polyhedrons, *Dokl. Akad. Nauk SSSR*, 1978, vol. 241, no. 3, pp. 34–37.

[14] Zolotov A. V., Kim K. V., Mgeryan G. M., and Khachaturov V. R., *Algoritmicheskoe i programmnoe obespechenie resheniya zadach operativnogo planirovaniya transportnykh perevozok* (Algorithms and Software for Short-Term Planning of Shipments), Moscow: Computing Center of Russ. Acad. Sci., 1989.

[15] Ioffe A. D. and Tikhomirov V. M., *Teoriya ekstremal'nykh zadach* (Theory of Extremum Value Problems), Moscow: Nauka, 1974.

[16] Karmanov V. G., *Matematicheskoe programmirovanie* (Mathematical programming), Moscow: Nauka, 1986.

[17] Kononenko A. M. and Trukhanovskii N. N., On Transport Polyhedrons with Maximal Number of Vertices, *Izv. Belarus. Akad. Nauk, Ser. Fiz. Mat.*, 1978, no. 5, pp. 23–25.

[18] Krasnoshchekov P. S. and Petrov A. A., *Printsipy postroeniya modelei* (Principles of Model Construction), Moscow: Mosk. Gos. Univ., 1983.

[19] Kuznetsov A. V., Sakovich V. A., and Kholod N. N., *Vysshaya matematika. Matematicheskoe programmirovanie* (Higher Mathematics: Mathematical programming), Minsk: Vysshaya Shkola, 1994.

[20] Kuhn H. V. and Tucker A. W., *Nonlinear Programming, Proc. of the Second Berkeley Symp. on Math. Statistics and Probability*, Berkeley: Univ. of California Press, 1951, pp. 481–492.

[21] Marcus M. and Mink H., *Matrix Theory and Matrix Inequalities*, Boston: Allyn and Bacon, 1964. Translated under the title *Obzor po teorii matrits i matrichnykh neravenstv,* Moscow: Nauka, 1972.

[22] Mironov A. A. and Tsurkov V. I., A Class of Distributed Problems with Minimax Criteria, *Dokl. Akad. Nauk*, 1992, vol. 336, no. 1, pp. 35–38.

[23] Mironov A. A. and Tsurkov V. I., Transportation and Network Problems with Minimax Criteria, *Zh. Vychisl. Mat. Mat. Fiz.*, 1995, vol. 35, no. 1, pp. 24–45.

[24] Mironov A. A. and Tsurkov V. I., Transportation Problems with Minimax Criteria, *Avtom. Telemekh.*, 1995, no. 12, pp. 109–118.

[25] Mironov A. A. and Tsurkov V. I., Transportation Problems with Minimax Criteria, *Dokl. Akad. Nauk*, 1996, vol. 346, no. 2, pp. 1–4.

[26] *Sovremennoe sostoyanie teorii issledovaniya operatsii* (Modern Theory of Operations Research), Moiseev, N.N., Ed., Moscow: Nauka, 1979.

[27] Moder J. and Elmaghraby S. S., *Handbook of Operations Research*, New York: Van Nostrand, 1978. Translated under the title *Issledovanie operatsii. Metodologicheskie osnovy i matematicheskie metody*, vol. 1, Moscow: Mir, 1981.

[28] Moder J. and Elmaghraby S. S., *Handbook of Operations Research*, New York: Van Nostrand, 1978. Translated under the title *Issledovanie operatsii. Modeli i primeneniya*, vol. 2, Moscow: Mir, 1981.

[29] Mukhacheva E. A. and Rubinshtein G. Sh., *Matematicheskoe programmirovanie* (Mathematical Programming), Novosibirsk: Nauka, 1977.

[30] Neyman J. and Morgenshtein O., *Theory of Games and Economic Behavior*, Princeton University Press, Princeton, New Jersey, 1953. Translated under the title *Teoriya igr i economicheskoe povedenie*, Moscow: Nauka, 1970.

[31] Owen G., *Game Theory*, Philadelphia: Saunders, 1968. Translated under the title *Teoriya igr*, Moscow: Mir, 1971.

[32] Panos M. Pardalos, *Minimax and Applications*, University of Minnesota and Institute of Applied Mathematics, Beijing: Academia Sinica, 1995.

[33] Pshenichnyi B. N., *Vypuklyi analiz i ekstremal'nye zadachi* (Convex Analysis and Extremum Value Problems), Moscow: Nauka, 1980.

[34] Riordan J., *An Introduction to Combinatorial Analysis*, New York: Wiley, 1958. Translated under the title *Vvedenie v kombinatornyi analiz*, Moscow: Inostr. Literatura, 1963.

[35] Rockafellar R. T., *Convex Analysis*, Princeton: Princeton Univ. Press, 1970. Translated under the title *Vypuklyi analiz*, Moscow: Mir, 1973.

[36] Saaty T., *Optimization in Integers and Related Extremal Problems*, New York: McGraw-Hill, 1970. Translated under the title *Tselochislennye metody optimizatsii i svyazannye s nimi ekstremal'nye problemy*, Moscow: Mir, 1973.

[37] Sachkov V. N., *Vvedenie v kombinatornye metody diskretnoi matematiki* (Introduction to Combinatorial Methods of Discrete Mathematics), Moscow: Nauka, 1982.

[38] Sukharev A. G., Timokhov A. V., and Fedorov V. V., *Kurs metodov optimizatsii* (Lectures on Optimization Methods), Moscow: Nauka, 1986.

[39] Tolstoi A. N., Elimination of Inefficient Shipments in Planning, *Sotsialisticheskii Transport*, 1939, no. 9, pp. 28–51.

[40] Trius E. B., *Zadachi matematicheskogo programmirovaniya transportnogo tipa* (Transportation Problems in Mathematical Programming), Moscow: Sovetskoe Radio, 1967.

[41] Ford L. R. and Fulkerson D., *Flows in Networks*, Princeton: Princeton Univ. Press, 1962. Translated under the title *Potoki v setyakh*, Moscow: Mir, 1966.

[42] Khachaturov V.R. and Montlevich V.M., *Reshenie nelineinykh proizvodstvenno-transportnykh zadach s nedelimymi potrebitelyami* (Solution of Nonlinear Production Transportation Problems with Indivisible Consumers), Moscow: Computing Center of Russ. Acad. Sci., 1987.

[43] Hitchcock F. L., Distribution of a Product from Several Sources to Numerous Localities, *J. Math. Phys.*, 1941, vol. 20, pp. 224–230.

[44] Hu T., *Integer Programming and Network Flows*, Reading (USA): Addison-Wesley, 1969. Translated under the title *Tselochislennoe programmirovanie i potoki v setyakh,* Moscow: Mir, 1974.

[45] Tsurkov V. I., *Dekompozitsiya v zadachakh bol'shoi razmernosti* (Decomposition in Large-Scale Problems), Moscow: Nauka, 1981.

[46] Tsurkov V. I., *Dinamicheskie zadachi bol'shoi razmernosti* (Dynamic Large-Scale Problems), Moscow: Nauka, 1988.

[47] Tsurkov V. I. and Litvinichev I. S., *Optimizatsiya i issledovanie operatsii. Dekompozitsiya v dinamicheskikh zadachakh s perekrestnymi svyazyami* (Optimization and Operations Research: Decomposition in Dynamic Problems with Crossed Links), vol. 1, Moscow: Nauka, 1994.

[48] Tsurkov V. I. and Litvinichev I. S., *Optimizatsiya i issledovanie operatsii. Dekompozitsiya v dinamicheskikh zadachakh s perekrestnymi svyazyami* (Optimization and Operations Research: Decomposition in Dynamic Problems with Crossed Links), vol. 2, Moscow: Nauka, 1994.

[49] Chernikov S. N., *Lineinye neravenstva* (Linear Inequalities), Moscow: Nauka, 1968.

Chapter 2

Minimax Criteria and Solution Methods for Transportation Problems

Transportation problems have been the subject of extensive literature [4, 6–8, 10, 21, 26]. Whereas effective accurate methods have been developed for linear transportation problems, no general accurate optimization methods for nonlinear problems are available to this day. Optimization of (convex or concave) objective functions by means of convex programming is the most advanced field of nonlinear studies [1, 23–25]. Dynamic programming methods are also applicable under certain conditions [3, 9, 27]. Optimization may lead to good results when functionals are approximated by piecewise linear functions [2, 20]. Minimax problems commonly arise in operations research [9, 19, 22].

In this chapter, we develop a method for solving transportation problems with the minimax criteria formulated in Chapter 1. The method, based on the characteristic equations and functions introduced above, is finally reduced to simple computational procedures.

This chapter consists of three sections. In Section 1, we construct an algorithm for solving the basic problem of minimization of the largest element for the transportation polyhedron matrices. We give a uniqueness criterion for the solution. The algorithm is reduced to solving linear equations for single unknowns (their number is not greater than the number of matrix elements) and to checking some simple inequalities.

In solving transportation problems with other minimax criteria, a fundamental role is played by the concept of a hereditarily minimax matrix considered in Section 2, i.e., the matrix whose every submatrix solves the problem with the basic minimax criterion. In that section, we also establish the uniqueness of the hereditarily minimax matrix in an arbitrary transportation polyhedron and show that the matrix is uniform.

In Section 3, we present algorithms for solving problems with the remaining minimax criteria. We also prove that the optimal solutions are hereditarily minimax matrices. The method is reduced to repeated application of the basic algorithm for minimax calculation.

§1. The Basic Algorithm and Uniqueness Conditions for the Solution of the Minimization Problem for the Largest Element of a Transportation Matrix

In this section, we present the details of construction of an algorithm for calculating the minimax value in a transportation problem. The performance of the method is illustrated by examples. As a rule, the solution matrix is not unique. We formulate and prove necessary and sufficient conditions for the uniqueness of the solution. We give formulas for calculating the elements of the minimax matrix. We also discuss the case of an integer transportation matrix.

1.1. Minimax Values and Minimax Matrices

Suppose the vectors $\mathbf{A} = (a_1, \ldots, a_n) \in R^n_+$ and $\mathbf{B} = (b_1, \ldots, b_m) \in R^m_+$ are such that the sums of their components are equal. Consider functional (1.7) from Chapter 1, defined on a set of matrices from $M(\mathbf{A}, \mathbf{B})$:

$$\Phi_1\left(X(\mathbf{A}, \mathbf{B})\right) = \max_{i,j} x_{ij}, \quad (x_{ij}) = X(\mathbf{A}, \mathbf{B}) \in M(\mathbf{A}, \mathbf{B}).$$

Let us calculate the minimal value of the functional and determine the corresponding matrix.

Definition 1.1. The quantity

$$c(\mathbf{A}, \mathbf{B}) = \min_{(x_{ij}) \in M(\mathbf{A},\mathbf{B})} \max_{i,j} x_{ij} = \min_{X(\mathbf{A},\mathbf{B}) \in M(\mathbf{A},\mathbf{B})} \Phi_1\left(X(\mathbf{A}, \mathbf{B})\right)$$

is the *minimax* (*minimax value*) *of the set of matrices* $M(\mathbf{A}, \mathbf{B})$.

Since $M\left(\mathbf{A}, \mathbf{B}; c(\mathbf{A}, \mathbf{B})\right) \neq \varnothing$ and $M(\mathbf{A}, \mathbf{B}; c) = \varnothing$ if $c < c(\mathbf{A}, \mathbf{B})$, it is obvious that the following assertion is true for the minimax value.

Theorem 1.1. $c(\mathbf{A}, \mathbf{B}) = \min\left\{c : M(\mathbf{A}, \mathbf{B}; c) \neq \varnothing\right\}$.

The minimax can be expressed in terms of the characteristic function given by (1.22) in Chapter 1. By invoking equation (1.24) from Chapter 1, it can be readily shown that $\delta_k(\mathbf{A}, \mathbf{B}; c)$ is a continuous function when $c \geq 0$ for every $k \in Z$ where it is defined, and a nondecreasing function; moreover, if $\mathbf{A} \in R^n_+$ and $\mathbf{B} \in R^m_+$, then, for any $k \in Z$ $(1 \leq k \leq n)$, $\delta_k(\mathbf{A}, \mathbf{B}; c)$ is a strictly increasing function when $0 \leq c \leq \max\{b_j/k : 1 \leq j \leq m\}$ and a constant,

$$\delta_k(\mathbf{A}, \mathbf{B}; c) = \sum_{j=1}^{m} b_j - \sum_{i=1}^{k} a_i,$$

when $c \geq \max\{b_j/k : 1 \leq j \leq m\}$. Since the conditions $M(\mathbf{A}, \mathbf{B}; c) \neq \varnothing$ and $\delta_k(\widetilde{\mathbf{A}}, \mathbf{B}; c) \geq 0$ are equivalent for any $k \in Z$ $(1 \leq k \leq n)$ by Lemma 2.5 from Chapter 1, we have the following equivalent lemmas.

Lemma 1.1. *Suppose vectors $\mathbf{A} \in R^n_+$ and $\mathbf{B} \in R^m_+$ satisfy the balance condition, and $c \geq 0$. Then, $c = c(\mathbf{A}, \mathbf{B})$ if and only if the function given by* (1.22) *in Chapter 1 satisfies the following conditions:*

(a) $\delta_k(\widetilde{\mathbf{A}}, \mathbf{B}; c) \geq 0$ *for any* $k \in Z$ $(1 \leq k \leq n)$;

(b) *there exists* $p \in Z$ $(1 \leq p \leq n,\ cp \leq \max\{b_j : 1 \leq j \leq m\})$ *such that* $\delta_p(\widetilde{\mathbf{A}}, \mathbf{B}; c) = 0$.

Lemma 1.1′. *If $M(\mathbf{A}, \mathbf{B}; c) \neq \varnothing$, where $\mathbf{A} \in R^n_+$ and $\mathbf{B} \in R^m_+$, then $c = c(\mathbf{A}, \mathbf{B})$ if and only if there exists $p \in Z$ $(1 \leq p \leq n,\ cp \leq \max\{b_j : 1 \leq j \leq m\})$ such that $\delta_p(\widetilde{\mathbf{A}}, \mathbf{B}; c) = 0$.*

Definition 1.2. Every matrix in the set $M\left(\mathbf{A}, \mathbf{B}; c(\mathbf{A}, \mathbf{B})\right)$ is called the *minimax matrix.*

It is clear that $M(\mathbf{A}, \mathbf{B}; c) = \varnothing$ when $c < c(\mathbf{A}, \mathbf{B})$ and $M(\mathbf{A}, \mathbf{B}; c) = M(\mathbf{A}, \mathbf{B})$ when $c \geq \min\left\{\max\{a_i : 1 \leq i \leq n\}, \max\{b_j : 1 \leq j \leq m\}\right\}$, where $\mathbf{A} = (a_1, \ldots, a_n) \in R^n_+$ and $\mathbf{B} = (b_1, \ldots, b_m) \in R^m_+$. Note that if the number of positive components of either vector is not greater than one, then the set $M(\mathbf{A}, \mathbf{B})$ contains only one matrix and $M\left(\mathbf{A}, \mathbf{B}; c(\mathbf{A}, \mathbf{B})\right) = M(\mathbf{A}, \mathbf{B})$.

Theorem 3.2 in Chapter 1 provides formulas for calculating the elements of admissible solutions to the truncated transportation problem. Using these constructions, we now formulate rules for calculating the elements of the minimax matrix.

Theorem 1.2. *Let $X(\mathbf{A}, \mathbf{B}) = (x_{ij})$ be the minimax matrix, where $\mathbf{A} \in R^n_+$ and $\mathbf{B} \in R^m_+$.*

(a) *If p and q are such indices that $a_p = \max\{a_i : 1 \leq i \leq n\}$ and $b_q = \max\{b_j : 1 \leq j \leq m\}$, then $x_{pq} = c(\mathbf{A}, \mathbf{B})$.*

(b) *The elements of the matrix $\widetilde{X}(\widetilde{\mathbf{A}}, \mathbf{B}) = (\tilde{x}_{ij})$ obtained from $X(\mathbf{A}, \mathbf{B})$ by permuting its rows ($\widetilde{\mathbf{A}} \in \overline{R}^n_+$) are*

$$\tilde{x}_{ij} = \begin{cases} c(\mathbf{A}, \mathbf{B}), & \textit{if there exists } k \in Z, \\ & \textit{such that } k \geq i,\ ck \leq b_j \textit{ and } \delta_k(\widetilde{\mathbf{A}}, \mathbf{B}; c) = 0, \\ 0, & \textit{if there exists } k \in Z, \\ & \textit{such that } k < i,\ ck > b_j \textit{ and } \delta_k(\widetilde{\mathbf{A}}, \mathbf{B}; c) = 0. \end{cases}$$

Proof. Recall Theorem 3.2 from Chapter 1.

(a) If $X(\mathbf{A},\mathbf{B}) \in M\left(\mathbf{A},\mathbf{B};c(\mathbf{A},\mathbf{B})\right)$, then Lemmas 1.1 and 1.1′ imply that there exists $p \in Z$ $(1 \le p \le n,\ c(\mathbf{A},\mathbf{B})p \le \max\{b_j : 1 \le j \le m\} = b_q)$ such that $\delta_p(\widetilde{\mathbf{A}},\mathbf{B};c) = 0$, where the vector $\widetilde{\mathbf{A}} = (\tilde{a}_1,\dots,\tilde{a}_n)$ is constructed by arranging the components of $\mathbf{A} = (a_1,\dots,a_n)$ in nonascending order. Therefore, $a_p = \tilde{a}_1$, and condition (3.3) from Chapter 1 guarantees that $\tilde{x}_{1q} = c(\mathbf{A},\mathbf{B})$ for any matrix $\widetilde{X}(\widetilde{\mathbf{A}},\mathbf{B}) = (\tilde{x}_{ij}) \in M(\widetilde{\mathbf{A}},\mathbf{B};c(\mathbf{A},\mathbf{B}))$. Consequently, $x_{pq} = \tilde{x}_{1q} = c(\mathbf{A},\mathbf{B})$.

(b) This assertion is proved in a similar manner.

Example 1.1. In Example 3.1 from Chapter 1, we have the minimax value $c(\mathbf{A},\mathbf{B}) = 3$, and each $X(\mathbf{A},\mathbf{B})$ thus constructed is an integer minimax matrix. The set of all minimax matrix defined by the vectors $\mathbf{A} = (11,11,11,4,3)$ and $\mathbf{B} = (12,12,10,3,3)$ can be expressed as

$$M(\mathbf{A},\mathbf{B};3) = \left\{ \begin{pmatrix} 3 & 3 & 3 & x_{14} & x_{15} \\ 3 & 3 & 3 & x_{24} & x_{25} \\ 3 & 3 & 3 & x_{34} & x_{35} \\ x_{41} & x_{42} & x_{43} & 0 & 0 \\ x_{51} & x_{52} & x_{53} & 0 & 0 \end{pmatrix} : \right.$$

$$\left. \begin{pmatrix} x_{14} & x_{15} \\ x_{24} & x_{25} \\ x_{34} & x_{35} \end{pmatrix} \in M(\mathbf{A}_1,\mathbf{B}_1;3), \quad \begin{pmatrix} x_{41} & x_{42} & x_{43} \\ x_{51} & x_{52} & x_{53} \end{pmatrix} \in M(\mathbf{A}_2,\mathbf{B}_2;3) \right\},$$

where $\mathbf{A}_1 = (2,2,2)$, $\mathbf{B}_1 = (3,3)$, $\mathbf{A}_2 = (4,3)$ and $\mathbf{B}_2 = (3,3,1)$. Here, we have a nonunique solution.

1.2. Uniqueness Conditions for the Minimax Matrix of a Transportation Polyhedron

The restrictiveness of the constraints imposed on the elements of matrices from $M(\mathbf{A},\mathbf{B};c)$ depends on the components of $\mathbf{A}$ and $\mathbf{B}$ and the value of c. For prescribed vectors $\mathbf{A}$ and $\mathbf{B}$ (each having at least two positive components), the constraints become less restrictive as the value of c is decreased, and the matrix set $M(\mathbf{A},\mathbf{B};c)$ may contain only one matrix.

We have the following simple propositions.

Lemma 1.2. *Suppose the set $M(\mathbf{A},\mathbf{B};c)$, where $\mathbf{A} \in R^n_+$ and $\mathbf{B} \in R^m_+$, contains only the matrix $X(\mathbf{A},\mathbf{B}) = (x_{ij})$, and $c' = \min\{\max\{a_i : 1 \le i \le n\}, \max\{b_j : 1 \le j \le m\}\}$, $c'' = \max\{x_{ij} : 1 \le i \le n,\ 1 \le j \le m\}$. Then, $X(\mathbf{A},\mathbf{B})$ is the minimax matrix. Moreover,*

(a) *if the either $\mathbf{A}$ or $\mathbf{B}$ has not more than one positive component, then $c \ge c' = c'' = c(\mathbf{A},\mathbf{B})$ and $M(\mathbf{A},\mathbf{B};c) = M(\mathbf{A},\mathbf{B})$;*

(b) *if both* $\mathbf{A}$ *and* $\mathbf{B}$ *have at least two positive components, then* $c = c(\mathbf{A}, \mathbf{B}) = c'' < c'$ *and* $M(\mathbf{A}, \mathbf{B}; c) \subset M(\mathbf{A}, \mathbf{B}) = M(\mathbf{A}, \mathbf{B}; c')$.

Consider a vector $\mathbf{B} = (b_1, \ldots, b_m) \in R_+^m$ ($c \geq 0$, $k \in Z$, $k \geq 0$) and denote by $l(\mathbf{B}; c, k)$ the number of its components that are greater than ck: $l(\mathbf{B}; c, k) = \left|\{j : 1 \leq j \leq m,\ b_j > ck\}\right|$. If $\mathbf{B} \in \overline{R}_+^m$, then $l(\mathbf{B}; c, k) = \max\{j : 1 \leq j \leq m,\ b_j > ck\}$ when $b_1 > ck$ and $l(\mathbf{B}; c, k) = 0$ when $b_1 \leq ck$. Assuming that $\mathbf{A} \in \overline{R}_+^n$ and $\mathbf{B} \in \overline{R}_+^m$ (which does not restrict the generality of the analysis), use the function $\delta_k(\mathbf{A}, \mathbf{B}; c)$ and values of $l_k(\mathbf{B}; c)$ and $l(\mathbf{B}; c, k)$ that are necessary and sufficient for $\left|M(\mathbf{A}, \mathbf{B}; c)\right| = 1$. The basic theorem is stated as follows.

Theorem 1.3. *Suppose* $\mathbf{A} = (a_1, \ldots, a_n) \in \overline{R}_+^n$ *and* $\mathbf{B} = (b_1, \ldots, b_m) \in \overline{R}_+^m$ *satisfy the balance condition,* $c \geq 0$ *and* $\delta_k(\mathbf{A}, \mathbf{B}; c) \geq 0$ *for any* $k \in Z$ $(1 \leq k \leq n)$. *Set* $p_0 = 0$ *and* $\delta_{p_0}(\mathbf{A}, \mathbf{B}; c) = 0$ *and denote by* p_i $(p_i \in Z,\ 0 \leq i \leq t)$ *the values for which* $cp_i \leq b_1$ *and* $\delta_{p_i}(\mathbf{A}, \mathbf{B}; c) = 0$ *so that* $p_{i-1} < p_i$ *for* $1 \leq i \leq t$ if $t \geq 1$. *The matrix set* $M(\mathbf{A}, \mathbf{B}; c)$ *contains only one matrix if and only if*

$$l(\mathbf{B}; c, p_t) \leq 1$$

or

$$l(\mathbf{B}; c, p_t) > 1, \quad p_t + 1 = \max\{i : 1 \leq i \leq n,\ a_i > 0\},$$

and, if $t \geq 1$, *then, for every pair of* p_{i-1} *and* p_i $(1 \leq i \leq t)$, *either*

(a) $p_i - p_{i-1} = 1$,

or

(b) $p_i - p_{i-1} > 1$, $l(\mathbf{B}; c, p_{i-1}) - l_{p_i}(\mathbf{B}; c) \leq 1$.

The proof of the theorem should be preceded by the following remarks.

(A) The relations $\delta_{p_i}(\mathbf{A}, \mathbf{B}; c) = 0$ and $cp_i \leq b_1$ determine the matrix elements, which are either c or zero:

$$x_{pq} = \begin{cases} c, & 1 \leq p \leq p_i, \quad 1 \leq q \leq l_{p_i}(\mathbf{B}; c), \\ 0, & p > p_i, \quad q > l_{p_i}(\mathbf{B}; c). \end{cases}$$

As shown below,

(B) $p_i - p_{i-1} = 1$ is equivalent to an unambiguous definition of the elements of the p_ith row of the matrix;

(C) when $p_i - p_{i-1} > 1$, $l(\mathbf{B}; c, p_{i-1}) - l_{p_i}(\mathbf{B}; c) = 1$ is equivalent to an unambiguous definition of the elements of column $l(\mathbf{B}; c, p_{i-1})$ of the matrix.

Proof of Theorem 1.3.

Necessity.

Suppose $|M(\mathbf{A},\mathbf{B};c)| = 1$. Then, by Lemma 1.2, the only matrix in $M(\mathbf{A},\mathbf{B};c)$ is a minimax matrix.

Consider the trivial case $|M(\mathbf{A},\mathbf{B})| = 1$.

Let us show that the conditions of the theorem are necessary when $m = 1$; $m > 1$, $b_2 = 0$; $n = 1$; and $n > 1$, $a_2 = 0$. Then, the minimax value is given by $c(\mathbf{A},\mathbf{B}) = \min(a_1, b_1)$.

When $m = 1$, or $m > 1$ and $b_2 = 0$, we have $c(\mathbf{A},\mathbf{B}) = a_1$. If $c > c(\mathbf{A},\mathbf{B})$, then, for every $k \in Z$ $(1 \le k \le n)$, it holds that

$$\delta_k(\mathbf{A},\mathbf{B};c) = \begin{cases} ck - \sum_{i=1}^{k} a_i, & b_1 \ge ck, \\ b_i - \sum_{i=1}^{k} a_i, & b_1 < ck. \end{cases}$$

Therefore, $\delta_k(\mathbf{A},\mathbf{B};c) > 0$ if $ck \le b_1$. Consequently, $t = 0$, $p_0 = 0$, and $l(\mathbf{B};c,p_0) = l(\mathbf{B};c,0) \le 1$ (note that $l(\mathbf{B};c,0) = 0$ when $a_1 = b_1 = 0$). If $c = c(\mathbf{A},\mathbf{B})$, then $\delta_1(\mathbf{A},\mathbf{B};c) = 0$, $t \ge 1$, and $t = \max\{i : a_i = a_1,\ 1 \le i \le n\}$. Therefore, $\delta_i(\mathbf{A},\mathbf{B};c) = 0$, $ci \le b_1$ for any i $(1 \le i \le t)$. It is obvious that $l(\mathbf{B};c,i) \le 1$ for any i $(1 \le i \le t)$. Hence, setting $p_i = i$ $(1 \le i \le t)$, we obtain $p_i - p_{i-1} = 1$ for any i $(1 \le i \le t)$ and $l(\mathbf{B};c,p_t) \le 1$.

When $n = 1$, or $n > 1$ and $a_2 = 0$, we have $c(\mathbf{A},\mathbf{B}) = b_1$ and, since $c \ge c(\mathbf{A},\mathbf{B})$, $l(\mathbf{B};c,1) = 0$. Suppose $c > c(\mathbf{A},\mathbf{B})$. Then, $t = 0$ and $p_t = p_0 = 0$. Therefore, $l(\mathbf{B};c,p_t) = 0$ when $b_1 = 0$, and $p_t + 1 = \max\{i : a_i > 0,\ 1 \le i \le n\}$ when $b_1 > 0$. For $c = c(\mathbf{A},\mathbf{B})$, we have $\delta_k(\mathbf{A},\mathbf{B};c) = 0$ for any k $(1 \le k \le n)$, but the inequality $ck \le b_1$ is valid only for $k = 1$. Therefore, $t = 1$, $p_1 - p_0 = 1$, and $l(\mathbf{B};c,p_t) = l(\mathbf{B};c,1) = 0$.

Consider the case when $M(\mathbf{A},\mathbf{B})$ contains an infinite number of matrices.

Let us show that the conditions of the theorem are necessary when $n, m > 1$, and a_2, $b_2 > 0$. In this case, $c = c(\mathbf{A},\mathbf{B}) < \min(a_1, b_1)$ by Lemmas 1.1 and 1.2, and there exists $p \in Z$ $(1 \le p \le n)$ such that $cp \le b_1$ and $\delta_p(\mathbf{A},\mathbf{B};c) = 0$. Consequently, $t \ge 1$. Let us prove that $l(\mathbf{B};c,p_{i-1}) - l_{p_i}(\mathbf{B};c) = 1$ if $p_i - p_{i-1} > 1$ $(1 \le i \le t)$.

Consider the only matrix $X(\mathbf{A},\mathbf{B}) = (x_{ij})$ $(1 \le i \le n,\ 1 \le j \le m)$ in the set $M(\mathbf{A},\mathbf{B};c)$, and let p_{i-1} and p_i be the consecutive values for which $\delta_{p_{i-1}}(\mathbf{A},\mathbf{B};c) = \delta_{p_i}(\mathbf{A},\mathbf{B};c) = 0$ and $p_i - p_{i-1} > 1$. Recall that, if $\mathbf{B} \in \overline{R}^m_+$, $l_k(\mathbf{B};c) = \max\{j : 1 \le j \le m,\ b_j \ge ck\}$ when $b_1 \ge ck$, and $l_k(\mathbf{B};c) = 0$ when $b_1 < ck$. Theorem 1.2 determines the values of the elements of $X(\mathbf{A},\mathbf{B})$:

$$x_{pq} = \begin{cases} c, & 1 \le p \le p_{i-1}, \quad 1 \le q \le l_{p_{i-1}}(\mathbf{B}; c), \\ c, & 1 \le p \le p_i, \quad 1 \le q \le l_{p_i}(\mathbf{B}; c), \\ 0, & p_{i-1} + 1 \le p \le n, \quad q > l_{p_{i-1}}(\mathbf{B}; c), \\ 0, & p > p_i, \quad q > l_{p_i}(\mathbf{B}; c). \end{cases} \tag{1.1}$$

It is obvious that $l(\mathbf{B}; c, p_{i-1}) \ge l_{p_i}(\mathbf{B}; c)$. To show that the strict inequality is valid here, assume that $l(\mathbf{B}; c, p_{i-1}) = l_{p_i}(\mathbf{B}; c)$ and $p \in Z$, where $p_{i-1} + 1 \le p \le p_i - 1$. Since $l_{p_i}(\mathbf{B}; c) \le l_p(\mathbf{B}; c) \le l(\mathbf{B}; c, p_{i-1})$, we have $l_p(\mathbf{B}; c) = l_{p_i}(\mathbf{B}; c)$, and (1.1) implies that

$$\sum_{b_j < cp} b_j = \sum_{b_j < cp_i} b_j, \quad \sum_{k=p+1}^{p_i} a_k = c(p_i - p) l_p(\mathbf{B}; c).$$

Therefore,

$$\delta_p(\mathbf{A}, \mathbf{B}; c) = \sum_{b_j < cp} b_j + cpl_p(\mathbf{B}; c) - \sum_{k=1}^{p} a_k =$$

$$\sum_{b_j < cp} b_j + cp_i l_{p_i}(\mathbf{B}; c) - c(p_i - p) l_{p_i}(\mathbf{B}; c) - \sum_{k=1}^{p_i} a_k + \sum_{k=p+1}^{p_i} a_k =$$

$$\delta_{p_i}(\mathbf{A}, \mathbf{B}; c) = 0.$$

However, this contradicts the assumption that p_{i-1} and p_i are the adjacent values for which the function defined by (1.22) in Chapter 1 vanishes. Consequently, $l(\mathbf{B}; c, p_{i-1}) > l_{p_i}(\mathbf{B}; c)$. We now proceed with proving Theorem 1.3.

Next, consider a submatrix $X(\mathbf{C}, \mathbf{D}) = (x_{pq})$, $p_{i-1}+1 \le p \le p_i$, $l_{p_i}(\mathbf{B}; c) + 1 \le q \le l(\mathbf{B}; c, p_{i-1})$ of the matrix $X(\mathbf{A}, \mathbf{B}) = (x_{pq})$ $(1 \le p \le n, 1 \le q \le m)$, where

$$\mathbf{C} = (c_{p_{i-1}+1}, \ldots, c_{p_i}), \quad c_p = \sum_{q=l_{p_i}(\mathbf{B};c)+1}^{l(\mathbf{B};c,p_{i-1})} x_{pq}, \quad p_{i-1} + 1 \le p \le p_i,$$

$$\mathbf{D} = (d_{l_{p_i}(\mathbf{B};c)+1}, \ldots, d_{l(\mathbf{B};c,p_{i-1})}), \quad d_q = \sum_{p=p_{i-1}+1}^{p_i} x_{pq},$$

$$l_{p_i}(\mathbf{B}; c) + 1 \le q \le l(\mathbf{B}; c, p_{i-1}).$$

It is clear that $X(\mathbf{C}, \mathbf{D}) \in M(\mathbf{C}, \mathbf{D}; c)$. By applying (1.1), it can be readily shown that

$$c_p = a_p - cl_{p_i}(\mathbf{B}; c), \quad p_{i-1} + 1 \le p \le p_i,$$

$$d_p = b_q - cp_{i-1}, \quad l_{p_i}(\mathbf{B};c) + 1 \le q \le l(\mathbf{B};c,p_{i-1}),$$

and, since $a_i \ge a_{i+1}$ $(1 \le i \le n-1)$ and $b_j \ge b_{j+1}$ $(1 \le j \le m-1)$, it holds that

$$\mathbf{C} \in \overline{R}_+^{p_i - p_{i-1}}, \quad \mathbf{D} \in \overline{R}_+^{l(\mathbf{B};c,p_{i-1}) - l_{p_i}(\mathbf{B};c)}.$$

To show that all components of the vector $\mathbf{C}$ are positive, suppose that $c_{p_i} = 0$. Then, (1.1) implies that

$$a_{p_i} = cl_{p_i}(\mathbf{B};c), \quad l_{p_i}(\mathbf{B};c) = l_{p_i-1}(\mathbf{B};c), \quad \sum_{b_j < cp_i} b_j = \sum_{b_j < c(p_i-1)} b_j.$$

Using these equalities, calculate $\delta_{p_i-1}(\mathbf{A},\mathbf{B};c)$:

$$\delta_{p_i-1}(\mathbf{A},\mathbf{B};c) = \sum_{b_j < c(p_i-1)} b_j + c(p_i-1)l_{p_i-1}(\mathbf{B};c) - \sum_{i=1}^{p_i-1} a_i =$$

$$\sum_{b_j < cp_i} b_j + c(p_i-1)l_{p_i}(\mathbf{B};c) - \sum_{i=1}^{p_i} a_i + a_{p_i} =$$

$$\sum_{b_j < cp_i} b_j + cp_i l_{p_i}(\mathbf{B};c) - \sum_{i=1}^{p_i} a_i - cl_{p_i}(\mathbf{B};c) + a_{p_i} = \delta_{p_i}(\mathbf{A},\mathbf{B};c) = 0.$$

This contradicts the inequality $\delta_{p_i-1}(\mathbf{A},\mathbf{B};c) > 0$. Thus, $c_{p_i} > 0$ and the components of $\mathbf{C}$ are positive. Moreover, the components of $\mathbf{D}$ are positive by (1.1) and definition of $l(\mathbf{B};c,p_{i-1})$.

The dimensions of $X(\mathbf{C},\mathbf{D})$ are $(p_i - p_{i-1})$-by-$\big(l(\mathbf{B};c,p_{i-1}) - l_{p_i}(\mathbf{B};c)\big)$, where $p_i - p_{i-1} \ge 2$ and $l(\mathbf{B};c,p_{i-1}) - l_{p_i}(\mathbf{B};c) \ge 1$. We now prove that $l(\mathbf{B};c,p_{i-1}) - l_{p_i}(\mathbf{B};c) = 1$. Suppose $l(\mathbf{B};c,p_{i-1}) - l_{p_i}(\mathbf{B};c) > 1$. Since $X(\mathbf{A},\mathbf{B})$ is the only matrix in $M(\mathbf{A},\mathbf{B};c)$, the matrix set $M(\mathbf{C},\mathbf{D};c)$ also contains only one matrix, $X(\mathbf{C},\mathbf{D})$. Then, by Lemmas 1.1 and 1.2, $X(\mathbf{C},\mathbf{D})$ is a minimax matrix, and there exists $k \in Z$ $(1 \le k \le p_i - p_{i-1})$ such that

$$ck \le d_{l_{p_i}+1}(\mathbf{B};c), \quad \delta_k(\mathbf{C},\mathbf{D};c) = 0, \quad l_k(\mathbf{D};c) \ge 1.$$

Note that $k < p_i - p_{i-1}$ in the case considered here, because the equality $k = p_i - p_{i-1}$ is inconsistent with the definition of $l_{p_i}(\mathbf{B};c)$. Theorem 1.2 and Theorem 3.2 from Chapter 1 determine the values of certain elements of $X(\mathbf{C},\mathbf{D})$ (and $X(\mathbf{A},\mathbf{B})$):

$$x_{pq} = \begin{cases} c, & p_{i-1} + 1 \le p \le p_{i-1} + k, \\ & l_{p_i}(\mathbf{B};c) + 1 \le q \le l_{p_i}(\mathbf{B};c) + l_k(\mathbf{D};c), \\ 0, & p_{i-1} + k + 1 \le p \le p_i, \\ & l_{p_i}(\mathbf{B};c) + l_k(\mathbf{D};c) + 1 \le q \le l(\mathbf{B};c,p_{i-1}). \end{cases} \tag{1.2}$$

Combining (1.1) and (1.2), we readily obtain

$$x_{pq} = \begin{cases} c, & 1 \leq p \leq p_i + k, \quad 1 \leq q \leq l_{p_i}(\mathbf{B}; c) + l_k(\mathbf{D}; c), \\ 0, & p_i + k + 1 \leq p \leq n, \quad q > l_{p_i}(\mathbf{B}; c) + l_k(\mathbf{D}; c). \end{cases} \tag{1.3}$$

Let us use (1.3) to express the sums

$$\sum_{b_j < c(p_i+k)} b_j, \quad \sum_{p=1}^{p_i+k} a_p$$

in terms of the elements of $X(\mathbf{A}, \mathbf{B}) = (x_{pq})$ $(1 \leq p \leq n,\ 1 \leq q \leq m)$.

$$\sum_{b_j < c(p_i+k)} b_j = \sum_{p=1}^{n} \sum_{q > l_{p_i}(\mathbf{B};c) + l_k(\mathbf{D};c)} x_{pq} = \sum_{p=1}^{p_i+k} \sum_{q > l_{p_i}(\mathbf{B};c) + l_k(\mathbf{D};c)} x_{pq},$$

$$\sum_{p=1}^{p_i+k} a_p = \sum_{p=1}^{p_i+k} \sum_{q=1}^{m} x_{pq} = \sum_{p=1}^{p_i+k} \sum_{q=1}^{l_{p_i}(\mathbf{B};c)+l_k(\mathbf{D};c)} x_{pq} + \sum_{p=1}^{p_i+k} \sum_{q > l_{p_i}(\mathbf{B};c) + l_k(\mathbf{D};c)} x_{pq} =$$

$$c(p_i + k)\left(l_{p_i}(\mathbf{B}; c) + l_k(\mathbf{D}; c)\right) + \sum_{p=1}^{p_i+k} \sum_{q > l_{p_i}(\mathbf{B};c) + l_k(\mathbf{D};c)} x_{pq}.$$

Since $l_{p_i+k}(\mathbf{B}; c) = l_{p_i}(\mathbf{B}; c) + l_k(\mathbf{D}; c)$,

$$\delta_{p_i+k}(\mathbf{A}, \mathbf{B}; c) = \sum_{b_j < c(p_i+k)} b_j + c(p_i + k) l_{p_i+k}(\mathbf{B}; c) - \sum_{p=1}^{p_i+k} a_p =$$

$$\sum_{p=1}^{p_i+k} \sum_{q > l_{p_i}(\mathbf{B};c) + l_k(\mathbf{D};c)} x_{pq} + c(p_i + k) l_{p_i+k}(\mathbf{B}; c) - c(p_i + k) l_{p_i+k}(\mathbf{B}; c) -$$

$$\sum_{p=1}^{p_i+k} \sum_{q > l_{p_i}(\mathbf{B};c) + l_k(\mathbf{D};c)} x_{pq} = 0.$$

Since the equality $\delta_{p_i+k}(\mathbf{A}, \mathbf{B}; c) = 0$ contradicts the assumption that p_{i-1} and p_i are consecutive values for which the function defined by (1.22) in Chapter 1 vanishes, the assumption that $l(\mathbf{B}; c, p_{i-1}) - l_{p_i}(\mathbf{B}; c) > 1$ is not valid. Thus, $l(\mathbf{B}; c, p_{i-1}) - l_{p_i}(\mathbf{B}; c) = 1$.

It now remains to show that either $l(\mathbf{B}; c, p_t) \leq 1$ or $p_t + 1 = \max\{a_i : 1 \leq i \leq n,\ a_i > 0\}$, where $p_t = \max\{k : 1 \leq k \leq n,\ ck \leq b_1,\ \delta_k(\mathbf{A}, \mathbf{B}; c) = 0\}$ and $t = \left|\{k : 1 \leq k \leq n,\ ck \leq b_1,\ \delta_k(\mathbf{A}, \mathbf{B}; c) = 0\}\right| \geq 1$. Suppose $l(\mathbf{B}; c, p_t) > 1$. Then, it is obvious that $n > p_t$ and $a_{p_t+1} > 0$. Consider a submatrix

$X(\mathbf{C},\mathbf{D}) = (x_{pq})$, $p_t+1 \le p \le n$, $1 \le q \le l(\mathbf{B};c,p_t)$ of the matrix $X(\mathbf{A},\mathbf{B}) = (x_{pq})$ $(1 \le p \le n,\ 1 \le q \le m)$, where

$$\mathbf{C} = (c_{p_i},\dots,c_n), \quad c_p = \sum_{q=1}^{l(\mathbf{B};c,p_t)} x_{pq}, \quad p_t+1 \le p \le n,$$

$$\mathbf{D} = (d_1,\dots,d_{l(\mathbf{B};c,p_t)}), \quad d_p = \sum_{p=p_t+1}^{n} x_{pq}, \quad 1 \le q \le l(\mathbf{B};c,p_t).$$

Note that $X(\mathbf{C},\mathbf{D}) \in M(\mathbf{C},\mathbf{D};c)$ and, by Theorem 1.2 (or Theorem 3.2 from Chapter 1),

$$x_{pq} = \begin{cases} c, & 1 \le p \le p_t, \quad 1 \le q \le l_{p_t}(\mathbf{B};c), \\ 0, & p_t+1 \le p \le n, \quad q > l_{p_t}(\mathbf{B};c). \end{cases} \tag{1.4}$$

Therefore, the components of the vectors $\mathbf{C}$ and $\mathbf{D}$ are expressed as

$$c_p = a_p, \quad p_t+1 \le p \le n,$$

$$d_q = b_q - cp_t, \quad 1 \le q \le l(\mathbf{B};c,p_t) \le l_{p_t}(\mathbf{B};c).$$

It is obvious that the components of $\mathbf{C}$ and $\mathbf{D}$ are nonnegative and are arranged in nondescending order; moreover, the definition of $l(\mathbf{B};c,p_t)$ implies that the components of a vector $\mathbf{D}$ of dimension $l(\mathbf{B};c,p_t) > 1$ are positive. Let us now show that the number of positive components of $\mathbf{C}$ is not greater than one.

Suppose $n > p_t+1$ and $a_{p_t+2} > 0$. Since the set $M(\mathbf{A},\mathbf{B};c)$ contains only the matrix $X(\mathbf{A},\mathbf{B})$, the set $M(\mathbf{C},\mathbf{D};c)$ contains only the matrix $X(\mathbf{C},\mathbf{D})$. By Lemmas 1.1 and 1.2, $X(\mathbf{C},\mathbf{D})$ is a minimax matrix, and there exists $k \in Z$ $(1 \le k \le n-p_t)$ such that $ck \le d_1$ and

$$\delta_k(\mathbf{C},\mathbf{D};c) = \sum_{d_q<ck} d_q + ckl_k(\mathbf{D};c) - \sum_{p=p_t+1}^{k} c_p = 0,$$

where $1 \le l_k(\mathbf{D};c) \le l(\mathbf{B};c,p_t)$. Theorem 1.2 and Theorem 3.2 from Chapter 1 determine the values of the elements of $X(\mathbf{C},\mathbf{D}) = (x_{pq})$, $p_t+1 \le p \le n$, $1 \le q \le l(\mathbf{B};c,p_t)$:

$$x_{pq} = \begin{cases} c, & p_t+1 \le p \le p_t+k, \quad 1 \le q \le l_k(\mathbf{D};c), \\ 0, & p > p_t+k, \quad l_k(\mathbf{D};c)+1 \le q \le l(\mathbf{B};c,p_t). \end{cases} \tag{1.5}$$

Combining (1.4) and (1.5), we obtain

$$x_{pq} = \begin{cases} c, & 1 \le p \le p_t+k, \quad 1 \le q \le l_k(\mathbf{D};c), \\ 0, & p > p_t+k, \quad q > l_k(\mathbf{D};c). \end{cases} \tag{1.6}$$

Since $l_{p_t+k}(\mathbf{B};c) = l_k(\mathbf{D};c)$, (1.6) determines the value of $\delta_{p_t+k}(\mathbf{A},\mathbf{B};c)$:

$$\delta_{p_t+k}(\mathbf{A},\mathbf{B};c) = \sum_{b_j<c(p_t+k)} b_j + c(p_t+k)l_{p_t+k}(\mathbf{B};c) - \sum_{p=1}^{p_t+k} a_p =$$

$$\sum_{p=1}^{p_t+k} \sum_{q>l_{p_t+k}(\mathbf{B};c)} x_{pq} + c(p_t+k)l_{p_t+k}(\mathbf{B};c) - c(p_t+k)l_{p_t+k}(\mathbf{B};c) -$$

$$\sum_{p=1}^{p_t+k} \sum_{q>l_{p_t+k}(\mathbf{B};c)} x_{pq} = 0.$$

The equality $\delta_{p_t+k}(\mathbf{A},\mathbf{B};c) = 0$ is inconsistent with the definitions of p_t and t. Therefore, the assumption that $a_{p_t+2} > 0$ is not valid, and $p_t + 1 = \max\{i : 1 \le i \le n,\ a_i > 0\}$ if $l(\mathbf{B};c,p_t) > 1$. The necessity is thus proved.

Sufficiency.

Consider a matrix set $M(\mathbf{A},\mathbf{B};c) \neq \varnothing$, where $\mathbf{A} \in \overline{R}^n_+$ and $\mathbf{B} \in \overline{R}^m_+$. Suppose that

$$\delta_0(\mathbf{A},\mathbf{B};c) = 0, \quad t = \left|\{k : 1 \le k \le n,\ ck \le b_1,\ \delta_k(\mathbf{A},\mathbf{B};c) = 0\}\right|$$

and $p_i \in Z$ $(0 \le i \le t)$ are the values for which $\delta_{p_i}(\mathbf{A},\mathbf{B};c) = 0$; suppose also that $p_{i-1} < p_i$ for $1 \le i \le t$. Let us show that $M(\mathbf{A},\mathbf{B};c)$ contains only one matrix if either condition (1) or condition (2) of Theorem 1.3 is satisfied. To do this, we prove that these conditions uniquely determine all elements of the matrix $X(\mathbf{A},\mathbf{B}) = (x_{ij})$ from $M(\mathbf{A},\mathbf{B};c)$.

If $t = 0$, we have either $l(\mathbf{A},\mathbf{B};0) \le 1$ or $1 = \max\{i : 1 \le i \le n,\ a_i > 0\}$ when $l(\mathbf{A},\mathbf{B};c) > 1$. Suppose $l(\mathbf{A},\mathbf{B};0) \le 1$. Then, the vector $\mathbf{B}$ is such that either $m = 1$ or $m > 1$ and $b_2 = 0$. Therefore, $X(\mathbf{A},\mathbf{B})$ is defined by its first column, which is identical with the vector $\mathbf{A}$:

$$x_{ij} = \begin{cases} a_i, & 1 \le i \le n,\ j = 1, \\ 0, & 1 \le i \le n,\ j > 1. \end{cases}$$

If $l(\mathbf{A},\mathbf{B};c) > 1$, the vector $\mathbf{A}$ is such that either $n = 1$ or $n > 1$ and $a_2 = 0$. Accordingly, $X(\mathbf{A},\mathbf{B})$ is defined by its first row, which is identical with the vector $\mathbf{B}$:

$$x_{ij} = \begin{cases} b_j, & i = 1,\ 1 \le j \le m, \\ 0, & i > 1,\ 1 \le j \le m. \end{cases}$$

Suppose that $t \ge 1$. Then, there exists $p \in Z$ such that $cp \le b_1$ and $\delta_p(\mathbf{A},\mathbf{B};c) = 0$. Therefore, every $X(\mathbf{A},\mathbf{B}) = (x_{pq})$ $(1 \le p \le n,\ 1 \le q \le m)$ from $M(\mathbf{A},\mathbf{B},c)$ is a minimax matrix by Lemma 1.1 or 1.1′, $c = c(\mathbf{A},\mathbf{B})$, and some elements of the matrix are uniquely determined by Theorem 3.2

from Chapter 1. Applying that theorem, we now show that the conditions given above uniquely determine all elements of $X(\mathbf{A},\mathbf{B}) = (x_{pq})$ and, therefore, $X(\mathbf{A},\mathbf{B})$ is the only matrix in the set $M(\mathbf{A},\mathbf{B},c)$. Let us evaluate an arbitrary element x_{pq} of $X(\mathbf{A},\mathbf{B})$ as depending on p and q. For p, we have two possibilities: either $p_{i-1} < p < p_i$ for a certain i $(1 \le i \le t)$, or $p > p_t$ (provided that $n > p_t$).

Consider the case $p_{i-1} < p < p_i$.

If $p_i - p_{i-1} = 1$, then $p = p_i$ and $p - 1 = p_{i-1}$. Since $\delta_p(\mathbf{A},\mathbf{B};c) = \delta_{p-1}(\mathbf{A},\mathbf{B};c) = 0$, we apply Theorem 3.2 from Chapter 1 to the elements of the pth row of $X(\mathbf{A},\mathbf{B})$ to obtain

$$x_{pq} = \begin{cases} c, & 1 \le q \le l_p(\mathbf{B};c), \\ b_q - c(p-1), & l_p(\mathbf{B};c) < q \le l_{p-1}(\mathbf{B};c), \\ 0, & q > l_{p-1}(\mathbf{B};c), \end{cases}$$

(note that $x_{pq} = 0$ for $l(\mathbf{B};c,p-1) < q \le l_{p-1}(\mathbf{B};c)$).

Suppose now that $p_i - p_{i-1} > 1$. Then, $l(\mathbf{B};c,p_{i-1}) - l_{p_i}(\mathbf{B};c) = 1$, and we apply Theorem 3.2 from Chapter 1 to show that

$$x_{pq} = \begin{cases} c, & 1 \le q \le l_{p_i}(\mathbf{B};c), \\ a_p - c l_{p_i}(\mathbf{B};c), & q = l_{p_i}(\mathbf{B};c) + 1 = l(\mathbf{B};c,p_{i-1}), \\ 0, & q > l(\mathbf{B};c,p_{i-1}). \end{cases}$$

Consider the case $p > p_t$, $n > p_t$. We have $l(\mathbf{B};c,p_t) \le 1$ or $p_t + 1 = \max\{i : 1 \le i \le n,\ a_i > 0\}$ if $l(\mathbf{B};c,p_t) > 1$. Theorem 3.2 from Chapter 1 (or Theorem 1.2) determines the values of the elements of the lowest rows of $X(\mathbf{A},\mathbf{B})$:

$$x_{pq} = \begin{cases} a_p, & p_t + 1 \le p \le n,\ q = 1, \\ 0, & p_t + 1 \le p \le n,\ q > 1, \end{cases}$$

when $l(\mathbf{B};c,p_t) \le 1$ ($x_{pq} = 0$ for $p_t + 1 \le p \le n$ and $q \ge 1$ if $l(\mathbf{B};c,p_t) = 0$), and

$$x_{pq} = \begin{cases} b_q - cp_t, & p = p_t + 1, \quad 1 \le q \le l(\mathbf{B};c,p_t), \\ 0, & p = p_t + 1, \quad q > l(\mathbf{B};c,p_t), \\ 0, & p > p_t + 1, \quad 1 \le q \le m, \end{cases}$$

when $l(\mathbf{B};c,p_t) > 1$ and $p_t+1 = \max\{a_i : 1 \le i \le n,\ a_i > 0\}$. These relations determine all elements of the matrix $X(\mathbf{A},\mathbf{B})$ $(1 \le i \le n,\ 1 \le j \le m)$. Consequently, $|M(\mathbf{A},\mathbf{B};c)| = 1$, which proves the sufficiency and completes the proof of Theorem 1.3.

As a corollary to Theorem 1.3, we have an analogous proposition concerning the case when the transportation polyhedron contains an infinite number of matrices.

Theorem 1.4. *Suppose* $M(\mathbf{A},\mathbf{B};c) \neq \varnothing$ *and* $X(\mathbf{A},\mathbf{B}) \in M(\mathbf{A},\mathbf{B};c)$, *where* $\mathbf{A} = (a_1,\dots,a_n) \in \overline{R}^n_+$ $(n \geq 2)$, $\mathbf{B} = (b_1,\dots,b_m) \in \overline{R}^m_+$ $(m \geq 2)$, $a_2 > 0$, *and* $b_2 > 0$. *Set* $p_0 = 0$, $\delta_{p_0}(\mathbf{A},\mathbf{B};c) = 0$, *and* $t = |\{k \in Z : 1 \leq k \leq n,\ \delta_k(\mathbf{A},\mathbf{B};c) = 0,\ ck \leq b_1\}|$. *A matrix* $X(\mathbf{A},\mathbf{B}) = (x_{pq})$ $(1 \leq p \leq n,\ 1 \leq q \leq m)$ *is the only one contained in the set* $M(\mathbf{A},\mathbf{B};c)$ *if and only if* $t \geq 1$. *Moreover, if* p_i $(p_i \in Z$ *and* $0 \leq i \leq t)$ *are the only consecutive values such that* $cp_i \leq b_1$ *and* $\delta_{p_i}(\mathbf{A},\mathbf{B};c) = 0$, *then* $l(\mathbf{B};c,p_{i-1}) - l_{p_i}(\mathbf{B};c) \leq 1$ *corresponds to* $p_i - p_{i-1} > 1$ $(1 \leq i \leq t)$ *and* $p_t + 1 = \max\{i : 1 \leq i \leq n,\ a_i > 0\}$ *corresponds to* $l(\mathbf{B};c,p_t) > 1$. *When these relations hold, the elements of the only matrix* $X(\mathbf{A},\mathbf{B}) = (x_{pq})$ *in* $M(\mathbf{A},\mathbf{B};c)$ *are expressed as*

$$x_{pq} = \begin{cases} c, & p_{i-1} < p \leq p_i, \quad 1 \leq q \leq l_{p_i}(\mathbf{B};c), \quad 1 \leq i \leq t; \\ b_q - c(p-1), & p_{i-1} < p \leq p_i, \quad p_i - p_{i-1} = 1, \\ & l_p(\mathbf{B};c) < q \leq l_{p-1}(\mathbf{B};c), \quad 1 \leq i \leq t; \\ a_p - cl_{p_i}(\mathbf{B};c), & p_{i-1} < p \leq p_i, \quad p_i - p_{i-1} > 1, \\ & q = l_{p_i}(\mathbf{B};c) + 1, \quad 1 \leq i \leq t; \\ a_p, & p_t < p \leq n, \quad l(\mathbf{B};c,p_t) \leq 1, \quad q = 1; \\ b_q - cp_t, & p = p_t + 1, \quad l(\mathbf{B};c,p_t) > 1, \\ & 1 \leq q \leq l(\mathbf{B};c,p_t); \\ 0, & p = p_i, \quad p_i - p_{i-1} = 1, \quad q > l_{p-1}(\mathbf{B};c), \\ & 1 \leq i \leq t; \\ 0, & p_{i-1} < p \leq p_i, \quad p_i - p_{i-1} > 1, \\ & q > l(\mathbf{B};c,p_{i-1}), \quad 1 \leq i \leq t; \\ 0, & p_t < p \leq n, \quad l(\mathbf{B};c,p_t) \leq 1, \quad 2 \leq q \leq m; \\ 0, & p = p_t + 1, \quad l(\mathbf{B};c,p_t) > 1, \quad l(\mathbf{B};c,p_t) < q \leq m; \\ 0, & p_t + 1 < p \leq n, \quad l(\mathbf{B};c,p_t) > 1, \quad 1 \leq q \leq m; \end{cases} \tag{1.7}$$

where $c = c(\mathbf{A},\mathbf{B})$.

Note that, under the assumptions of Theorem 1.4, the only (minimax) matrix in the set $M\left(\mathbf{A},\mathbf{B};c(\mathbf{A},\mathbf{B})\right)$ can be constructed by applying Theorem 3.2 from Chapter 1 for every $p \in \{p_i : 1 \leq i \leq t\}$. The remaining elements of the matrix are evaluated automatically (see Example 1.2).

Example 1.2. Consider the vectors

$$\mathbf{A} = (24, 17, 17, 16, 12, 8, 8, 2, 2) \in \overline{R}_+^9$$

and

$$\mathbf{B} = (25, 21, 19, 14, 13, 8, 2, 2, 1, 1) \in \overline{R}_+^{10}.$$

It is easy to check the balance condition for these vectors, and the set $M(\mathbf{A}, \mathbf{B})$ obviously contains an infinite number of matrices. Let us evaluate the function defined by (1.22) in Chapter 1 for $c = 3$: $\delta_1(\mathbf{A}, \mathbf{B}; 3) = 0$, $\delta_2(\mathbf{A}, \mathbf{B}; 3) = 1$, $\delta_3(\mathbf{A}, \mathbf{B}; 3) = 1$, $\delta_4(\mathbf{A}, \mathbf{B}; 3) = 0$, $\delta_5(\mathbf{A}, \mathbf{B}; 3) = 0$, $\delta_6(\mathbf{A}, \mathbf{B}; 3) = 1$, $\delta_7(\mathbf{A}, \mathbf{B}; 3) = 0$, $\delta_8(\mathbf{A}, \mathbf{B}; 3) = 1$, $\delta_9(\mathbf{A}, \mathbf{B}; 3) = 0$. Thus, $M(\mathbf{A}, \mathbf{B}; 3) \neq \varnothing$. In this example, $\mathbf{A} \in \overline{R}_+^9$ and $\mathbf{B} \in \overline{R}_+^{10}$; therefore, $M_Z(\mathbf{A}, \mathbf{B}; 3) \neq \varnothing$.

Let us now apply Theorem 1.4 to show that the matrix set $M(\mathbf{A}, \mathbf{B}; 3)$ contains only one matrix (which belongs to $M_Z(\mathbf{A}, \mathbf{B}; 3)$) and construct the matrix by calculating all parameters required. Here, $t = 4$, $p_0 = 0$, $p_1 = 1$, $p_2 = 4$, $p_3 = 5$, and $p_4 = 7$. Recall that, when $\mathbf{B} = (b_1, \ldots, b_m) \in \overline{R}_+^m$, we have ($k \in Z$, $k \geq 0$)

$$l(\mathbf{B}; c, k) = \begin{cases} \max\{j : 1 \leq j \leq m,\ b_j > ck\}, & b_1 > ck, \\ 0, \quad b_1 \leq ck, \end{cases}$$

$$l_k(\mathbf{B}; c) = \begin{cases} \max\{j : 1 \leq j \leq m,\ b_j \geq ck\}, & b_1 \geq ck, \\ 0, \quad b_1 < ck. \end{cases}$$

Calculate the values of $p_i - p_{i-1}$ and $l(\mathbf{B}; 3, p_{i-1}) - l_{p_i}(\mathbf{B}; 3)$, where $1 \leq i \leq t = 4$:

$$p_i - p_{i-1} = \begin{cases} 1 - 0 = 1, & i = 1, \\ 4 - 1 = 3, & i = 2, \\ 5 - 4 = 1, & i = 3, \\ 7 - 5 = 2, & i = 4, \end{cases}$$

$$l(\mathbf{B}; 3, p_{i-1}) - l_{p_i}(\mathbf{B}; 3) = \begin{cases} 10 - 6 = 4, & i = 1, \\ 6 - 5 = 1, & i = 2, \\ 5 - 3 = 2, & i = 3, \\ 3 - 2 = 1, & i = 4. \end{cases}$$

It is obvious that $p_i - p_{i-1} > 1$ only for $i = 2$ and $i = 4$. For these values of i, $l(\mathbf{B}; 3, p_{i-1}) - l_{p_i}(\mathbf{B}; 3) = 1$. Since we have $l(\mathbf{B}; 3, p_t) = l(\mathbf{B}; 3, p_4) = l(\mathbf{B}; 3, 7) = 1$, the matrix set $M(\mathbf{A}, \mathbf{B}; 3)$ contains only one (minimax) matrix by Theorem 1.4.

Because the expressions for matrix elements given by (1.7) are cumbersome, we apply Theorem 3.2 from Chapter 1 to construct the matrix $X(\mathbf{A}, \mathbf{B}) = (x_{ij}) \in M(\mathbf{A}, \mathbf{B}; 3)$ for all $p \in \{1, 4, 5, 7\} = \{p_i : 1 \leq i \leq 4\}$:

$$x_{ij} = \begin{cases} 3, & i=1, \quad 1 \le j \le 6, \\ 0, & 2 \le i \le 9, \quad 7 \le j \le 10, \end{cases} \qquad x_{ij} = \begin{cases} 3, & 1 \le i \le 4, \quad 1 \le j \le 5, \\ 0, & 5 \le i \le 9, \quad 6 \le j \le 10, \end{cases}$$

$$x_{ij} = \begin{cases} 3, & 1 \le i \le 5, \quad 1 \le j \le 3, \\ 0, & 6 \le i \le 9, \quad 4 \le j \le 10, \end{cases} \qquad x_{ij} = \begin{cases} 3, & 1 \le i \le 7, \quad 1 \le j \le 2, \\ 0, & 8 \le i \le 9, \quad 3 \le j \le 10. \end{cases}$$

The required matrix is written as

$$\begin{array}{c} \\ 24 \\ 17 \\ 17 \\ 16 \\ 12 \\ 8 \\ 8 \\ 2 \\ 2 \end{array} \begin{array}{c} \begin{array}{cccccccccc} 25 & 21 & 19 & 14 & 13 & 8 & 2 & 2 & 1 & 1 \end{array} \\ \begin{pmatrix} 3 & 3 & 3 & 3 & 3 & 3 & x_{17} & x_{18} & x_{19} & x_{1\,10} \\ 3 & 3 & 3 & 3 & 3 & x_{26} & 0 & 0 & 0 & 0 \\ 3 & 3 & 3 & 3 & 3 & x_{36} & 0 & 0 & 0 & 0 \\ 3 & 3 & 3 & 3 & 3 & x_{46} & 0 & 0 & 0 & 0 \\ 3 & 3 & 3 & x_{54} & x_{55} & 0 & 0 & 0 & 0 & 0 \\ 3 & 3 & x_{63} & 0 & 0 & 0 & 0 & 0 & 0 & 0 \\ 3 & 3 & x_{73} & 0 & 0 & 0 & 0 & 0 & 0 & 0 \\ x_{81} & x_{82} & 0 & 0 & 0 & 0 & 0 & 0 & 0 & 0 \\ x_{91} & x_{92} & 0 & 0 & 0 & 0 & 0 & 0 & 0 & 0 \end{pmatrix} \end{array} = X(\mathbf{A}, \mathbf{B}) \in M(\mathbf{A}, \mathbf{B}; 3).$$

On the left of each row and on top of each column, we write the respective sums of their elements, which are equal to the corresponding components of the vectors $\mathbf{A}$ and $\mathbf{B}$. Then, it is clear that $x_{17} = x_{18} = 2$, $x_{19} = x_{1\,10} = 1$, $x_{26} = x_{36} = 2$, $x_{46} = 1$, $x_{54} = 2$, $x_{55} = 1$, $x_{63} = x_{73} = 2$, $x_{81} = x_{91} = 2$, $x_{82} = x_{92} = 0$. Therefore,

$$X(\mathbf{A}, \mathbf{B}) = \begin{pmatrix} 3 & 3 & 3 & 3 & 3 & 3 & 2 & 2 & 1 & 1 \\ 3 & 3 & 3 & 3 & 3 & 2 & 0 & 0 & 0 & 0 \\ 3 & 3 & 3 & 3 & 3 & 2 & 0 & 0 & 0 & 0 \\ 3 & 3 & 3 & 3 & 3 & 1 & 0 & 0 & 0 & 0 \\ 3 & 3 & 3 & 2 & 1 & 0 & 0 & 0 & 0 & 0 \\ 3 & 3 & 2 & 0 & 0 & 0 & 0 & 0 & 0 & 0 \\ 3 & 3 & 2 & 0 & 0 & 0 & 0 & 0 & 0 & 0 \\ 2 & 0 & 0 & 0 & 0 & 0 & 0 & 0 & 0 & 0 \\ 2 & 0 & 0 & 0 & 0 & 0 & 0 & 0 & 0 & 0 \end{pmatrix}.$$

Now, we formulate simpler sufficient conditions for the uniqueness of the minimax matrix.

Theorem 1.5. *Suppose vectors* $\mathbf{A} = (a_1, \ldots, a_n) \in \overline{R}^n_+$ *and* $\mathbf{B} = (b_1, \ldots, b_m) \in \overline{R}^n_+$, *where* n, $m \ge 0$ *and* a_2, $b_2 > 0$, *satisfy the balance condition, and* $\delta_k\left(\mathbf{A}, \mathbf{B}; c(\mathbf{A}, \mathbf{B})\right) = 0$ *for any* k $(1 \le k \le n)$. *Then, the matrix set* $M(\mathbf{A}, \mathbf{B})$ *contains only one minimax matrix* $X(\mathbf{A}, \mathbf{B})$ $(X(\mathbf{A}, \mathbf{B}) \in M\left(\mathbf{A}, \mathbf{B}; c(\mathbf{A}, \mathbf{B})\right)$, $\left|M\left(\mathbf{A}, \mathbf{B}; c(\mathbf{A}, \mathbf{B})\right)\right| = 1)$, *whose elements can be evaluated by applying Theorem* 3.2 *from Chapter* 1 *for all* $p \in Z$ $(1 \le p \le n-1)$.

This proposition is a corollary to Theorem 1.4.

Example 1.3. The vectors

$$\mathbf{A} = (37, 22, 12) \in \overline{R}^3_+ \text{ and } \mathbf{B} = (12, 12, 11, 9, 7, 6, 5, 3, 3, 2, 1) \in \overline{R}^{11}_+$$

satisfy the balance condition. It is easy to check that $\delta_k(\mathbf{A}, \mathbf{B}; 4) = 0$ for every k $(1 \le k \le 3)$. By Theorem 1.4 (or 1.5), $M(\mathbf{A}, \mathbf{B})$ contains only one minimax matrix (with the minimax value $c(\mathbf{A}, \mathbf{B}) = 4$). Applying Theorem 3.2 from Chapter 1, construct the matrix $X(\mathbf{A}, \mathbf{B}) = (x_{ij})$ with $1 \le i \le 3$ and $1 \le j \le 11$. Calculate two auxiliary parameters, $l_1(\mathbf{B}; 4) = 7$ and $l_2(\mathbf{B}; 4) = 4$.

For $p = 1$,

$$x_{ij} = \begin{cases} 4, & i = 1, \quad 1 \le j \le 7, \\ 0, & 2 \le i \le 3, \quad 8 \le j \le 11. \end{cases}$$

It is readily checked that the calculated elements determine the remaining entries in the first row: $x_{18} = x_{19} = 3$, $x_{1\,10} = 2$, $x_{1\,11} = 1$.

For $p = 2$

$$x_{ij} = \begin{cases} 4, & 1 \le i \le 2, \quad 1 \le j \le 4, \\ 0, & i = 3, \quad 5 \le j \le 11. \end{cases}$$

These elements determine the remaining entries in the second row: $x_{25} = 3$, $x_{26} = 2$, $x_{27} = 1$. The last row of $X(\mathbf{A}, \mathbf{B})$ is determined automatically. We obtain

$$X(\mathbf{A}, \mathbf{B}) = \begin{pmatrix} 4 & 4 & 4 & 4 & 4 & 4 & 4 & 3 & 3 & 2 & 1 \\ 4 & 4 & 4 & 4 & 3 & 2 & 1 & 0 & 0 & 0 & 0 \\ 4 & 4 & 3 & 1 & 0 & 0 & 0 & 0 & 0 & 0 & 0 \end{pmatrix}.$$

Theorems 1.3 and 1.4 give necessary and sufficient conditions — and Theorem 1.5 gives only sufficient conditions — for the uniqueness of the solution of a minimization problem with the functional defined by (1.7) in Chapter 1.

Theorem 1.6. *Suppose* $\mathbf{A} = (a_1, \ldots, a_n) \in R^n_+$ *and* $\mathbf{B} = (b_1, \ldots, b_m) \in R^m_+$ *satisfy the balance condition, and* $\widetilde{\mathbf{A}} = (\tilde{a}_1, \ldots, \tilde{a}_n) \in \overline{R}^n_+$ *and* $\widetilde{\mathbf{B}} = (\tilde{b}_1, \ldots, \tilde{b}_m) \in \overline{R}^m_+$ *are constructed by arranging the components of* $\mathbf{A}$ *and* $\mathbf{B}$, *respectively, in nonascending order. Then,*

(a) *equation*

$$\Phi_1\left(X(\mathbf{A}, \mathbf{B})\right) = c(\mathbf{A}, \mathbf{B}), \quad X(\mathbf{A}, \mathbf{B}) \in M(\mathbf{A}, \mathbf{B}) \tag{1.8}$$

has a unique solution if and only if the vectors $\widetilde{\mathbf{A}}$ *and* $\widetilde{\mathbf{B}}$ *satisfy the conditions of Theorem* 1.3;

(b) *if* $n, m \ge 2$ *and* $\tilde{a}_2, \tilde{b}_2 > 0$, *then equation* (1.8) *has a unique solution if and only if the vectors* $\widetilde{\mathbf{A}}$ *and* $\widetilde{\mathbf{B}}$ *satisfy the assumptions of Theorem* 1.4;

(c) *if the vectors* $\widetilde{\mathbf{A}}$ *and* $\widetilde{\mathbf{B}}$ *satisfy the assumptions of Theorem* 1.5, *then equation* (1.8) *has a unique solution.*

1.3. An Algorithm for Calculating a Minimax Value

Let us calculate the minimax value $c(\mathbf{A}, \mathbf{B})$ that minimizes the functional defined by (1.7) in Chapter 1 for vectors $\mathbf{A} \in R_+^n$ and $\mathbf{B} \in R_+^m$ characterized by equal sums of their components:

$$\min \Phi_1 \left(X(\mathbf{A}, \mathbf{B})\right) = c(\mathbf{A}, \mathbf{B}), \quad X(\mathbf{A}, \mathbf{B}) \in M(\mathbf{A}, \mathbf{B}).$$

It is clear that $c(\mathbf{A}, \mathbf{B}) = 0$ if and only if $\mathbf{A}$ and $\mathbf{B}$ are zero vectors, and $c(\mathbf{A}, \mathbf{B}) = c(\widetilde{\mathbf{A}}, \widetilde{\mathbf{B}})$. Consider $\mathbf{B} = (b_1, \ldots, b_m) \in \overline{R}_+^m$, where $b_1 > 0$. Single out the indices of those components of $\mathbf{B}$ whose values are greater than those of the components next in order (if $b_1 > b_m$). Denote by m' the index of the last positive component of $\mathbf{B}$ ($b_{m'} > 0$ and $b_{m'+1} = 0$ if $m > m'$): $T(\mathbf{B}) = \{k \in Z : 1 \le k \le m - 1,\ b_k > b_{k+1}\} \cup \{m'\}$. Define $t = |T(\mathbf{B})|$ and $k_i \in T(\mathbf{B})$ $(1 \le i \le t)$ so that $k_i < k_{i+1}$ for $1 \le i \le t-1$ (if $t > 1$). Thus, the elements of the set $T(\mathbf{B})$ are numbered in ascending order.

For nonzero vectors $\mathbf{A} = (a_1, \ldots, a_n) \in \overline{R}_+^n$ and $\mathbf{B} = (b_1, \ldots, b_m) \in \overline{R}_+^m$ and a pair of numbers, $k \in Z$ and $k_i \in T(\mathbf{B})$, where $1 \le k \le n$ and $1 \le i \le t$, consider the system of relations

$$\begin{cases} \sum_{j=1}^{m} b_j - \sum_{j=1}^{k_i} \left(b_j - c(k, k_i)k\right) - \sum_{i=1}^{k} a_i = 0, \\ c(k, k_i)k \le b_{k_i}, \\ c(k, k_i)k > b_{k_i+1}, \quad i \ne t, \end{cases} \qquad (1.9)$$
$$1 \le k \le n, \quad 1 \le i \le t,$$

with the only unknown quantity $c(k, k_i)$. If there exists a solution $c(k, k_i)$ to this system, then it is unique, because the first equation of the system is linear in $c(k, k_i)$. If system (1.9) has no solutions, then set $c(k, k_i) = 0$.

Let us prove two auxiliary propositions.

Lemma 1.3. *Suppose* $\mathbf{A} = (a_1, \ldots, a_n) \in R_+^n$, $\mathbf{B} = (b_1, \ldots, b_m) \in \overline{R}_+^m$, $b_1 > 0$, *and* $\delta_k(\mathbf{A}, \mathbf{B}; c) \ge 0$ *for some* $k \in Z$ $(1 \le k \le n)$. *Then,* $c \ge c(k, k_i)$ *for any* $k_i \in T(\mathbf{B})$, *where* $c(k, k_i)$ *is the solution of* (1.9).

Proof. If $c(k, k_i)$ is the unique solution of (1.9), then

$$\sum_{j=1}^{k_i} \left(b_j - c(k, k_i)k\right) = \sum_{b_j \ge c(k, k_i)k} \left(b_j - c(k, k_i)k\right).$$

Therefore,

$$\sum_{j=1}^{m} b_j - \sum_{b_j \geq c(k,k_i)k} (b_j - c(k,k_i)k) - \sum_{i=1}^{k} a_i = 0.$$

Since the function defined by (1.22) in Chapter 1 is such that

$$\sum_{j=1}^{m} b_j - \sum_{b_j \geq ck} (b_j - ck) - \sum_{i=1}^{k} a_i \geq 0,$$

the two relations above can be combined to obtain

$$\sum_{b_j \geq ck} (b_j - ck) \leq \sum_{b_j \geq c(k,k_i)k} (b_j - c(k,k_i)k),$$

which is possible only if $c \geq c(k, k_i)$.

Lemma 1.4. *For nonzero vectors* $\mathbf{A} \in \overline{R}_+^n$ *and* $\mathbf{B} \in \overline{R}_+^m$, *where* $M(\mathbf{A},\mathbf{B}) \neq \varnothing$, $c(\mathbf{A},\mathbf{B}) \geq \max\{c(k,k_i) : 1 \leq k \leq n,\ k_i \in T(\mathbf{B})\}$.

The proof follows from Lemma 1.3.

If vectors **A** and **B** satisfy the assumptions of Lemma 1.4, then the minimax $c(\mathbf{A},\mathbf{B})$ and the characteristic function defined by (1.22) in Chapter 1 are such that $\delta_k(\mathbf{A},\mathbf{B};c(\mathbf{A},\mathbf{B})) \geq 0$ for any k $(1 \leq k \leq n)$ and $\delta_p(\mathbf{A},\mathbf{B};c(\mathbf{A},\mathbf{B})) = 0$ for a certain p $(1 \leq p \leq n,\ cp \leq b_1)$ (see Lemma 1.1). Since

$$\delta_p(\mathbf{A},\mathbf{B};c(\mathbf{A},\mathbf{B})) = \sum_{j=1}^{m} b_j - \sum_{j=1}^{l_p(\mathbf{B};c(\mathbf{A},\mathbf{B}))} (b_j - c(\mathbf{A},\mathbf{B})p) - \sum_{i=1}^{p} a_i,$$

there exists an index i such that $l_p(\mathbf{B};c(\mathbf{A},\mathbf{B})) = k_i \in T(\mathbf{B})$, $c(\mathbf{A},\mathbf{B})p \leq b_{k_i}$, and $c(\mathbf{A},\mathbf{B})p > b_{k_i+1}$ if $i \neq t$. Accordingly, $c(\mathbf{A},\mathbf{B}) = c(p,k_i)$. Combining this inequality with Lemma 1.4, we obtain the theorem about the minimax value.

The following proposition follows from Lemmas 1.3 and 1.4 and provides a clue to constructing the minimax value.

Theorem 1.7. *For nonzero vectors* $\mathbf{A} \in \overline{R}_+^n$ *and* $\mathbf{B} \in \overline{R}_+^m$ *with equal sums of components*, $c(\mathbf{A},\mathbf{B}) = \max\{c(k,k_i) : 1 \leq k \leq n,\ k_i \in T(\mathbf{B})\}$.

By Theorem 1.7 and the system of relations (1.9), the method works as follows.

Algorithm 1.1. Calculation of the minimax value $c(\mathbf{A},\mathbf{B})$.

Suppose vectors $\mathbf{A} = (a_1, \ldots, a_n) \in R_+^n$ and $\mathbf{B} = (b_1, \ldots, b_m) \in R_+^m$ satisfy conditions (1.1) from Chapter 1.

Step 1. Arranging the components of $\mathbf{A}$ and $\mathbf{B}$ in nonascending order, construct vectors $\widetilde{\mathbf{A}} = (\tilde{a}_1, \ldots, \tilde{a}_n)$ and $\widetilde{\mathbf{B}} = (\tilde{b}_1, \ldots, \tilde{b}_m)$.

Step 2. For the vector $\widetilde{\mathbf{B}} = (\tilde{b}_1, \ldots, \tilde{b}_m)$, construct the ordered set of indices $T(\widetilde{\mathbf{B}}) = \{k_i : 1 \leq i \leq t\}$. Then, $\tilde{b}_{k_t+1} = 0$ for $k_t < m$, and if $t > 1$, then $\tilde{b}_{k_i} > \tilde{b}_{k_i+1}$, where $1 \leq i \leq t-1$.

Step 3. For every pair of integers k and k_i, where $1 \leq k \leq n$ and $k_i \in T(\widetilde{\mathbf{B}})$, find the solution $c(k, k_i)$ to system (1.9); if the system does not have a solution, set $c(k, k_i) = 0$.

Step 4. Calculate $\max\left\{c(k, k_i) : 1 \leq k \leq n,\ k_i \in T(\widetilde{\mathbf{B}})\right\}$.

Step 5. Set $c(\mathbf{A},\mathbf{B}) = \max\left\{c(k, k_i) : 1 \leq k \leq n,\ k_i \in T(\widetilde{\mathbf{B}})\right\}$.

For integer vectors $\mathbf{A}$ and $\mathbf{B}$, the minimax value can be calculated in the class of integer plan matrices of the transportation problem by applying Theorem 1.7 and Algorithm 1.1.

Definition 1.3. Suppose that vectors $\mathbf{A} \in Z_+^n$ and $\mathbf{B} \in Z_+^m$ are characterized by equal sums of components. The quantity

$$c_Z(\mathbf{A},\mathbf{B}) = \min_{(x_{ij}) \in M_Z(\mathbf{A},\mathbf{B})} \left\{\max\left\{x_{ij} : 1 \leq i \leq n,\ 1 \leq j \leq m,\ (x_{ij}) \in M_Z(\mathbf{A},\mathbf{B})\right\}\right\}$$

is said to be the *integer minimax of the matrix set* $M_Z(\mathbf{A},\mathbf{B})$.

The minimax $c_Z(\mathbf{A},\mathbf{B})$ is the minimal value of the objective function defined by (1.18) in Chapter 1:

$$c_Z(\mathbf{A},\mathbf{B}) = \min\left\{\Phi_Z\left(X(\mathbf{A},\mathbf{B})\right) : X(\mathbf{A},\mathbf{B}) \in M_Z(\mathbf{A},\mathbf{B})\right\}.$$

Definition 1.4. The matrices from the set

$$M_Z\left(\mathbf{A},\mathbf{B}; c_Z(\mathbf{A},\mathbf{B})\right)$$

are called *minimax matrices in the class of integer matrices from* $M_Z(\mathbf{A},\mathbf{B})$.

As indicated above, the characteristic function $\delta_k(\mathbf{A},\mathbf{B}; c)$ is a strictly increasing function of the argument c ($0 \leq c \leq \max\{b_j/k : 1 \leq j \leq m\}$) when $\mathbf{A}$, $\mathbf{B}$, and k are held constant. Therefore, Theorem 1.2 from Chapter 1 entails the following theorem and algorithm for integer matrices.

Theorem 1.8. *For nonzero vectors* $\mathbf{A} \in \overline{Z}_+^n$ *and* $\mathbf{B} \in \overline{Z}_+^m$ *with equal sums of components,* $c_Z(\mathbf{A},\mathbf{B}) = \max\left\{]c(k, k_i)[: 1 \leq k \leq n,\ k_i \in T(\mathbf{B})\right\}$, *where* $]a[$ *is the least integer that is not less than* a.

Algorithm 1.2. Calculation of the minimax value in the class of integer matrices.

Suppose vectors $\mathbf{A} \in Z_+^n$ and $\mathbf{B} \in Z_+^m$ satisfy the balance condition.

Step 1. Apply Algorithm 1.1 to calculate the minimax $c(\mathbf{A}, \mathbf{B})$.

Step 2. Calculate the integer minimax as $c_Z(\mathbf{A}, \mathbf{B}) =]c(\mathbf{A}, \mathbf{B})[$.

Example 1.4. Let us apply Algorithms 1.1 and 1.2 to calculate the minimax values $c(\mathbf{A}, \mathbf{B})$ and $c_Z(\mathbf{A}, \mathbf{B})$, where $\mathbf{A} = (10, 18, 18, 6) \in Z_+^4$ and $\mathbf{B} = (12, 4, 8, 12, 4, 12) \in Z_+^6$. The sum of components of either vector is 52, and the balance condition is satisfied.

Arranging the components of $\mathbf{A}$ and $\mathbf{B}$ in nonascending order, construct the vectors $\widetilde{\mathbf{A}} = (18, 18, 10, 6) \in \overline{Z}_+^4$ and $\widetilde{\mathbf{B}} = (12, 12, 12, 8, 4, 4) \in \overline{Z}_+^6$ (Step 1 of Algorithm 1.1).

For the vector $\widetilde{\mathbf{B}}$, single out the indices of those components whose values are greater than those of the components that are next in order and also single out the last index. Here, $T(\widetilde{\mathbf{B}}) = \{3, 4, 6\}$ and $t = 3$ (Step 2 of Algorithm 1.1).

At Step 3 of Algorithm 1.1, construct a system of relations (1.9) for each pair of integers k and k_i ($1 \le k \le 4$, $k_i \in T(\widetilde{\mathbf{B}})$) and determine its solution, or set $c(k, k_i) = 0$ if the system has no solutions. Write out and solve the resulting systems, setting $c(k, k_i) = c_{kk_i}$.

$$\begin{cases} 52 - 3(12 - c_{13}) - 18 = 0, \\ 8 < c_{13} \le 12, \end{cases} \qquad \begin{cases} 52 - 3(12 - c_{14}) - (8 - c_{14}) - 18 = 0, \\ 4 < c_{14} \le 8, \end{cases}$$
$$c_{13} = 0; \qquad c_{14} = 0;$$

$$\begin{cases} 52 - 3(12 - c_{16}) - (8 - c_{16}) - 2(4 - c_{16}) - 18 = 0, \\ c_{16} \le 4, \end{cases}$$
$$c_{16} = 3;$$

$$\begin{cases} 52 - 3(12 - c_{23}) - 36 = 0, \\ 8 < 2c_{23} \le 12, \end{cases} \qquad \begin{cases} 52 - 3(12 - 2c_{24}) - (8 - c_{24}) - 36 = 0, \\ 4 < 2c_{24} \le 8, \end{cases}$$
$$c_{23} = 0; \qquad c_{24} = 7/2;$$

$$\begin{cases} 52 - 3(12 - 2c_{26}) - (8 - 2c_{26}) - 2(4 - 2c_{26}) - 36 = 0, \\ 2c_{26} \le 4, \end{cases}$$
$$c_{26} = 0;$$

$$\begin{cases} 52 - 3(12 - 3c_{33}) - 46 = 0, \\ 8 < 3c_{33} \le 12, \end{cases} \qquad \begin{cases} 52 - 3(12 - 3c_{34}) - (8 - 3c_{34}) - 46 = 0, \\ 4 < 3c_{34} \le 8, \end{cases}$$
$$c_{33} = 10/3; \qquad c_{34} = 0;$$

$$\begin{cases} 52 - 3(12 - 3c_{36}) - (8 - 3c_{36}) - 2(4 - 3c_{36}) - 46 = 0, \\ 3c_{36} \le 4, \end{cases}$$
$$c_{36} = 0$$

$$\begin{cases} 52 - 3(12 - 4c_{43}) - 52 = 0, \\ 8 < 4c_{43} \le 12, \end{cases} \quad \begin{cases} 52 - 3(12 - 4c_{44}) - (8 - 4c_{44}) - 52 = 0, \\ 4 < 4c_{44} \le 8, \end{cases}$$
$$c_{43} = 3; \qquad c_{44} = 0$$

$$\begin{cases} 52 - 3(12 - 4c_{46}) - (8 - 4c_{46}) - 2(4 - 4c_{46}) - 52 = 0, \\ 4c_{46} \le 4, \end{cases}$$
$$c_{46} = 0;$$

It is obvious that $\max\{c(k, k_i) : 1 \le k \le 4, k_i \in T(\widetilde{\mathbf{B}})\} = 7/2$; therefore, the minimax values are $c(\mathbf{A}, \mathbf{B}) = 7/2$ and $c_Z(\mathbf{A}, \mathbf{B}) = 4$. These are examples of minimax matrices from $M(\mathbf{A}, \mathbf{B}; 7/2)$ and $M_Z(\mathbf{A}, \mathbf{B}; 4)$:

$$\begin{array}{c} \\ 10 \\ 18 \\ 18 \\ 6 \end{array} \begin{array}{c} \begin{array}{cccccc} 12 & 4 & 8 & 12 & 4 & 12 \end{array} \\ \begin{pmatrix} 3 & 0 & 1 & 3 & 0 & 3 \\ 3,5 & 2 & 3,5 & 3,5 & 2 & 3,5 \\ 3,5 & 2 & 3,5 & 3,5 & 2 & 3,5 \\ 2 & 0 & 0 & 2 & 0 & 2 \end{pmatrix} \end{array} \qquad \begin{array}{c} \\ 10 \\ 18 \\ 18 \\ 6 \end{array} \begin{array}{c} \begin{array}{cccccc} 12 & 4 & 8 & 12 & 4 & 12 \end{array} \\ \begin{pmatrix} 2 & 2 & 0 & 2 & 2 & 2 \\ 4 & 1 & 4 & 4 & 1 & 4 \\ 4 & 1 & 4 & 4 & 1 & 4 \\ 2 & 0 & 0 & 2 & 0 & 2 \end{pmatrix} \end{array}$$

$$X(\mathbf{A}, \mathbf{B}) \in M(\mathbf{A}, \mathbf{B}; 7/2) \qquad X(\mathbf{A}, \mathbf{B}) \in M_Z(\mathbf{A}, \mathbf{B}; 4)$$

Note that in this example the components of **A** and **B** are such that one can construct a matrix from $M_Z(\mathbf{A}, \mathbf{B})$ that has a row and a column in which every element equals 3. However, another row, characterized by the sum of elements equal to 18, must contain an element that is not less than 5:

$$\begin{array}{c} \\ 10 \\ 18 \\ 18 \\ 6 \end{array} \begin{array}{c} \begin{array}{cccccc} 12 & 4 & 8 & 12 & 4 & 12 \end{array} \\ \begin{pmatrix} 3 & 0 & 1 & 3 & 0 & 3 \\ 3 & 3 & 3 & 3 & 3 & 3 \\ 3 & 1 & 4 & 5 & 1 & 4 \\ 3 & 0 & 0 & 1 & 0 & 2 \end{pmatrix} \end{array}$$

$$X(\mathbf{A}, \mathbf{B}) \in M(\mathbf{A}, \mathbf{B}; 7/2)$$

Let us now summarize the results obtained. For vectors $\mathbf{A} \in R_+^n$ and $\mathbf{B} \in R_+^m$, denote any matrix from the matrix sets $M(\mathbf{A}, \mathbf{B}; c(\mathbf{A}, \mathbf{B}))$ or $M_Z(\mathbf{A}, \mathbf{B}; c_Z(\mathbf{A}, \mathbf{B}))$ by $X^*(\mathbf{A}, \mathbf{B}) = (x_{ij}^*)$ or $X_Z^*(\mathbf{A}, \mathbf{B}) = (x_{ij}^*)$, respectively $(1 \le i \le n,\ 1 \le j \le m)$. Then,

$$c(\mathbf{A}, \mathbf{B}) = \min_{X(\mathbf{A},\mathbf{B}) \in M(\mathbf{A},\mathbf{B})} \Phi_1(X(\mathbf{A}, \mathbf{B})) = \min_{X(\mathbf{A},\mathbf{B}) \in M(\mathbf{A},\mathbf{B})} \max\{x_{ij} : (x_{ij}) = X(\mathbf{A}, \mathbf{B})\} = \max\{x_{ij}^* : (x_{ij}^*) = X^*(\mathbf{A}, \mathbf{B})\} = \Phi_1(X^*(\mathbf{A}, \mathbf{B})),$$

$$c_Z(\mathbf{A}, \mathbf{B}) = \min_{X(\mathbf{A},\mathbf{B}) \in M_Z(\mathbf{A},\mathbf{B})} \Phi_Z(X(\mathbf{A}, \mathbf{B})) = \min_{X(\mathbf{A},\mathbf{B}) \in M_Z(\mathbf{A},\mathbf{B})} \max\{x_{ij} : (x_{ij}) = X(\mathbf{A}, \mathbf{B})\} = \max\{x_{ij}^* : (x_{ij}^*) = X_Z^*(\mathbf{A}, \mathbf{B})\} = \Phi_Z(X^*(\mathbf{A}, \mathbf{B})),$$

where

$$X^*(\mathbf{A}, \mathbf{B}) \in M(\mathbf{A}, \mathbf{B}; c(\mathbf{A}, \mathbf{B})) \quad \text{and} \quad X_Z^*(\mathbf{A}, \mathbf{B}) \in M(\mathbf{A}, \mathbf{B}; c(\mathbf{A}, \mathbf{B})).$$

Theorems 1.3 and 1.4 give necessary and sufficient conditions — and Theorem 1.5 contains sufficient conditions — for the matrix set $M\left(\mathbf{A}, \mathbf{B}; c(\mathbf{A}, \mathbf{B})\right)$ to contain only one matrix. In other words, these theorems state the uniqueness conditions for the solution to the problem of minimization of an objective function $\Phi_1\left(X(\mathbf{A}, \mathbf{B})\right)$, $X(\mathbf{A}, \mathbf{B}) \in M(\mathbf{A}, \mathbf{B})$. If the minimax value $c(\mathbf{A}, \mathbf{B})$ is calculated for a transportation polyhedron, and $\left|M\left(\mathbf{A}, \mathbf{B}; c(\mathbf{A}, \mathbf{B})\right)\right| = 1$, then a unique minimax matrix $X^*(\mathbf{A}, \mathbf{B})$ can be constructed by applying Theorem 3.2 from Chapter 1 or formula (1.7) (by arranging the components of $\mathbf{A}$ and $\mathbf{B}$ in nonascending order).

In the general case, minimax matrices from $M(\mathbf{A}, \mathbf{B})$ and $M_Z(\mathbf{A}, \mathbf{B})$ should be constructed by applying Algorithm 1.1 from this chapter and Algorithm 2.1 from Chapter 1 and, respectively, Algorithm 1.2 from this chapter and Algorithm 2.1 from Chapter 1. If Algorithm 2.2 from Chapter 1 is applied instead of Algorithm 2.1 from Chapter 1, any minimax matrix from the sets $M\left(\mathbf{A}, \mathbf{B}; c(\mathbf{A}, \mathbf{B})\right)$ and $M_Z\left(\mathbf{A}, \mathbf{B}; c_Z(\mathbf{A}, \mathbf{B})\right)$ can be constructed.

In the next section, we apply Algorithm 1.1 and Theorem 3.2 from Chapter 1 to construct a minimax matrix having important special properties.

§2. Hereditary Properties of Transportation Matrices

Every submatrix of a transportation matrix is a transportation matrix. Thus, the property of being a transportation matrix is the simplest hereditary property of transportation matrices. Uniformity is another hereditary property of transportation matrices.

In this section, we establish yet another hereditary property of matrices. In a transportation polyhedron, there exists a matrix whose submatrices are minimax matrices. This matrix is found to be unique and uniform and to minimize the functionals defined by formulas (1.7)–(1.16) in Chapter 1.

2.1. Hereditarily Minimax Matrices

Let $X = (x_{ij})$ $(1 \le i \le n,\ 1 \le j \le m)$ be an arbitrary matrix with nonnegative entries. The matrix determines vectors $\mathbf{A}(X) = (a_1, \ldots, a_n) \in R^n_+$ and $\mathbf{B}(X) = (b_1, \ldots, b_m) \in R^m_+$, where

$$a_i = \sum_{j=1}^{m} x_{ij}, \quad 1 \le i \le n, \quad b_j = \sum_{i=1}^{n} x_{ij}, \quad 1 \le j \le m;$$

equivalently, $X \in M\left(\mathbf{A}(X), \mathbf{B}(X)\right)$ and $X = X\left(\mathbf{A}(X), \mathbf{B}(X)\right)$. Define $c(X) = \max\{x_{ij} : 1 \le i \le n,\ 1 \le j \le m\}$. Then, $X \in M\left(\mathbf{A}(X), \mathbf{B}(X); c(X)\right)$. It is clear that the minimax property of $X\left(\mathbf{A}(X), \mathbf{B}(X)\right)$ in the matrix set $M\left(\mathbf{A}(X), \mathbf{B}(X)\right)$ is equivalent to the condition $c(X) = c\left(\mathbf{A}(X), \mathbf{B}(X)\right)$. Let

p and q be positive numbers such that $1 \le p \le n$ and $1 \le q \le m$. For the matrix X and numbers p and q, we denote by $i_1, \ldots, i_p$ and $j_1, \ldots, j_q$ the strictly increasing sequences of integer numbers satisfying the conditions $1 \le i_r \le n$, $1 \le j_t \le m$, $1 \le r \le p$, and $1 \le t \le q$. The matrix $Y = (x_{i_r j_t})$, where $1 \le r \le p$ and $1 \le t \le q$, is called a submatrix of $X = (x_{ij})$ $(1 \le i \le n,$ $1 \le j \le m)$.

Definition 2.1. A matrix $X = X\left(\mathbf{A}(X), \mathbf{B}(X)\right) = (x_{ij})$ $(1 \le i \le n,$ $1 \le j \le m)$ is said to be a *hereditarily minimax matrix* if any submatrix $Y = Y\left(\mathbf{C}(Y), \mathbf{D}(Y)\right) = (x_{i_r j_t})$ $(1 \le r \le p,$ $1 \le t \le q)$ of X is a minimax matrix of the set $M\left(\mathbf{C}(Y), \mathbf{D}(Y)\right)$.

Note that the minimax property of the submatrix Y can be expressed as follows: $Y \in M\left(\mathbf{C}(Y), \mathbf{D}(Y); c\left(\mathbf{C}(Y), \mathbf{D}(Y)\right)\right)$ or $c(Y) = c\left(\mathbf{C}(Y), \mathbf{D}(Y)\right)$.

The matrices that have the form of column or row vectors with nonnegative components are the simplest hereditarily minimax matrices. The matrices whose elements are nonnegative and equal to each other are also hereditarily minimax matrices. One can easily verify that these are minimax matrices:

$$\begin{pmatrix} 2 & 2 & 1 & 1 & 1 \\ 2 & 2 & 1 & 0 & 0 \end{pmatrix}, \quad \begin{pmatrix} 3 & 3 & 3 & 2 & 2 & 1 & 1 \\ 3 & 3 & 3 & 2 & 2 & 1 & 1 \end{pmatrix}.$$

If a matrix set $M\left(\mathbf{A}, \mathbf{B}; c(\mathbf{A}, \mathbf{B})\right)$ contains only one matrix (Theorems 1.3, 1.4, 1.5), then the matrix (which can be constructed by applying Theorem 3.2 from Chapter 1 or using formula (1.7) if $M(\mathbf{A}, \mathbf{B})$ is an infinite matrix set) is also a hereditarily minimax one. Note that Definition 2.1 implies that if $Y = Y\left(\mathbf{C}(Y), \mathbf{D}(Y)\right)$ is a submatrix of a hereditarily minimax matrix $X = X\left(\mathbf{A}(X), \mathbf{B}(X)\right)$, then $Y = Y\left(\mathbf{C}(Y), \mathbf{D}(Y)\right)$ is a hereditarily minimax matrix.

Lemma 2.1. *If vectors* $\mathbf{A}$ *and* $\mathbf{B}$ *have nonnegative components whose sums are equal, there exists a hereditarily minimax matrix* $X(\mathbf{A}, \mathbf{B})$.

A constructive proof of this lemma is contained in Algorithm 2.1 and Lemma 2.2.

The construction of a hereditarily minimax matrix is based on repeated application of Algorithm 1.1 and Theorem 3.2 from Chapter 1 (or Theorem 1.2). Without loss of generality, one may assume that the components of the vectors considered are arranged in nonascending order.

Algorithm 2.1. Construction of a hereditarily minimax matrix.

Suppose $\mathbf{A} = (a_1, \ldots, a_n) \in \overline{R}^n_+$ and $\mathbf{B} = (b_1, \ldots, b_m) \in \overline{R}^m_+$ satisfy the balance condition.

Step 1. Apply Algorithm 1.1 to calculate the minimax value $c(\mathbf{A}, \mathbf{B})$.

Step 2. Choose an arbitrary (say, the least) value of $k \in Z$ $(1 \le k \le n)$ such that $c(\mathbf{A},\mathbf{B})k \le b_1$ and $\delta_k\left(\mathbf{A},\mathbf{B};c(\mathbf{A},\mathbf{B})\right) = 0$ (Lemma 1.1).

Step 3. Applying Theorem 3.2 from Chapter 1, calculate certain elements of the desired matrix $X(\mathbf{A},\mathbf{B}) = (x_{ij})$ from $M\left(\mathbf{A},\mathbf{B};c(\mathbf{A},\mathbf{B})\right)$ $(1 \le i \le n,$ $1 \le j \le m)$ as follows:

$$x_{ij} = \begin{cases} c(\mathbf{A},\mathbf{B}), & 1 \le i \le k, \quad 1 \le j \le l_k\left(\mathbf{B};c(\mathbf{A},\mathbf{B})\right), \\ 0, & k < i \le n, \quad l_k\left(\mathbf{B};c(\mathbf{A},\mathbf{B})\right) < j \le m. \end{cases}$$

The elements x_{ij} of the matrix $X(\mathbf{A},\mathbf{B})$ and number k thus calculated define a submatrix $\overline{X} = \overline{X}\left(\mathbf{A},\mathbf{B};c(\mathbf{A},\mathbf{B}),k\right) = (x_{ij})$ $(1 \le i \le k,\ 1 \le j \le l_k\left(\mathbf{B};c(\mathbf{A},\mathbf{B})\right))$ whose elements are equal to $c(\mathbf{A},\mathbf{B})$ and (if $k < n$ and $l_k\left(\mathbf{B};c(\mathbf{A},\mathbf{B})\right) < m$) a zero submatrix $\overline{\overline{X}} = \overline{\overline{X}}\left(\mathbf{A},\mathbf{B};c(\mathbf{A},\mathbf{B}),k\right) = (x_{ij})$, where $k < i \le n$ and $l_k\left(\mathbf{B};c(\mathbf{A},\mathbf{B})\right) < j \le m$.

Step 4. Using the matrix $\overline{X}$ (if $l_k\left(\mathbf{B};c(\mathbf{A},\mathbf{B})\right) < m$), construct the pair of vectors $\mathbf{A}' = (a'_1,\dots,a'_k)$ and $\mathbf{B}' = (b'_{l_k(\mathbf{B};c(\mathbf{A},\mathbf{B}))+1},\dots,b'_m)$, where

$$\begin{aligned} a'_i &= a_i - c(\mathbf{A},\mathbf{B})l_k\left(\mathbf{B};c(\mathbf{A},\mathbf{B})\right), \quad 1 \le i \le k, \\ b'_j &= b_j, \quad l_k\left(\mathbf{B};c(\mathbf{A},\mathbf{B})\right) < j \le m. \end{aligned} \tag{2.1}$$

Using the submatrix $\overline{\overline{X}}$ (if $k < n$), construct the pair of vectors $\mathbf{A}'' = (a''_{k+1},\dots,a''_n)$ and $\mathbf{B}'' = (b''_1,\dots,b''_{l_k(\mathbf{B};c(\mathbf{A},\mathbf{B}))})$, where

$$\begin{aligned} a''_i &= a_i, \quad k < i \le n, \\ b''_j &= b_j - c(\mathbf{A},\mathbf{B})k, \quad 1 \le j \le l_k\left(\mathbf{B};c(\mathbf{A},\mathbf{B})\right). \end{aligned} \tag{2.2}$$

By setting $c = c(\mathbf{A}, \mathbf{B})$ and $l = l_k(\mathbf{B}; c)$, the result can be represented as the following matrix:

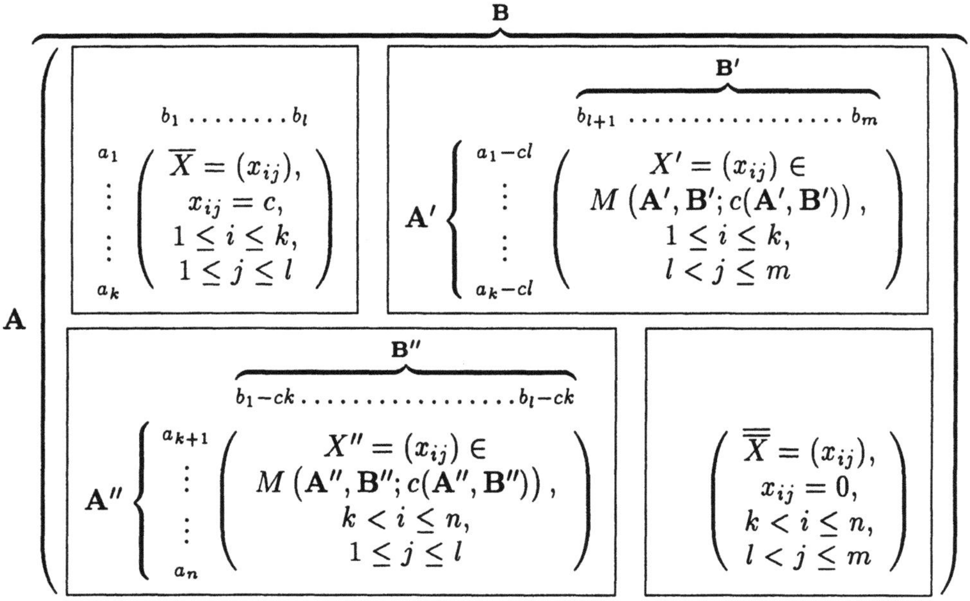

In view of the condition imposed on the components of $\mathbf{A}$ and $\mathbf{B}$, it is clear that the newly constructed (as well as existing) vectors with components defined by rules (2.1) and (2.2) satisfy the following conditions:

$$\mathbf{A}' \in \overline{R}^k_+, \quad \mathbf{B}' \in \overline{R}^{m-l_k(\mathbf{B};c(\mathbf{A},\mathbf{B}))}_+ \text{ (if } l_k\left(\mathbf{B}; c(\mathbf{A},\mathbf{B})\right) < m),$$

$$\mathbf{A}'' \in \overline{R}^{n-k}_+ \text{ (if } k < n), \quad \mathbf{B}'' \in \overline{R}^{l_k(\mathbf{B};c(\mathbf{A},\mathbf{B}))}_+,$$

$$\sum_{i=1}^{k} a'_i = \sum_{j>l_k(\mathbf{B};c(\mathbf{A},\mathbf{B}))} b'_j, \quad \sum_{i>k} a''_i = \sum_{j=1}^{l_k(\mathbf{B};c(\mathbf{A},\mathbf{B}))} b''_j.$$

Denote the submatrices of $X(\mathbf{A}, \mathbf{B})$ whose elements have not been calculated by $X' = (x_{ij})$ $(1 \le i \le k,\ l_k\left(\mathbf{B}; c(\mathbf{A},\mathbf{B})\right) < j \le m)$ and $X'' = (x_{ij})$ $(k < i \le n,\ 1 \le j \le l_k\left(\mathbf{B}; c(\mathbf{A},\mathbf{B})\right))$; note that X' and X'' do not exist when $l_k\left(\mathbf{B}; c(\mathbf{A},\mathbf{B})\right) = m$ and $k = n$, respectively. It is obvious that the pairs $\mathbf{A}'$, $\mathbf{B}'$ and $\mathbf{A}''$, $\mathbf{B}''$ define X' and X'', respectively: $X' \in M(\mathbf{A}', \mathbf{B}')$, $X'' \in M(\mathbf{A}'', \mathbf{B}'')$.

Some elements of X' and X'' may remain undefined. Apply Steps 1–4 of the procedure just described consecutively to the pairs of vectors $\mathbf{A}'$, $\mathbf{B}'$ and $\mathbf{A}''$, $\mathbf{B}''$; etc.

By executing Steps 1–4 for the second time, the minimax values $c(\mathbf{A}', \mathbf{B}')$ and $c(\mathbf{A}'', \mathbf{B}'')$ are calculated (obviously, $c(\mathbf{A}', \mathbf{B}')$, $c(\mathbf{A}'', \mathbf{B}'') \le c(\mathbf{A}, \mathbf{B})$), and some (if not all) elements x_{ij} of the minimax matrix $X' \in M(\mathbf{A}', \mathbf{B}'; c(\mathbf{A}', \mathbf{B}'))$ are evaluated as $c(\mathbf{A}', \mathbf{B}')$ or 0, and those of the minimax matrix $X'' \in M(\mathbf{A}'', \mathbf{B}''; c(\mathbf{A}'', \mathbf{B}''))$ are evaluated as $c(\mathbf{A}'', \mathbf{B}'')$ or 0. Every time Steps 1–4 are further executed, the number of pairs of vectors considered is increased by a factor of not more than two, and some elements of the minimax matrices corresponding to these vectors are evaluated. Algorithm 2.1 is completed by evaluating the last element of the matrix $X(\mathbf{A}, \mathbf{B}) \in M(\mathbf{A}, \mathbf{B}; c(\mathbf{A}, \mathbf{B}))$.

Lemma 2.2. *If vectors* $\mathbf{A} = (a_1, \ldots, a_n) \in \overline{R}_+^n$ *and* $\mathbf{B} = (b_1, \ldots, b_m) \in \overline{R}_+^m$ *are characterized by equal sums of their components, the matrix* $X(\mathbf{A}, \mathbf{B})$ *constructed by applying Algorithm* 2.1 *is a hereditarily minimax one.*

Proof. By construction, $X(\mathbf{A}, \mathbf{B}) \in M(\mathbf{A}, \mathbf{B}; c(\mathbf{A}, \mathbf{B}))$; consequently, $X(\mathbf{A}, \mathbf{B})$ is a minimax matrix of the set $M(\mathbf{A}, \mathbf{B})$. We now show that any submatrix Y of $X(\mathbf{A}, \mathbf{B})$ is a minimax matrix.

Algorithm 2.1 is designed to execute Steps 1–4 in cycles until new pairs of vectors are obtained for further analysis. Let us rename the vectors $\mathbf{A}$ and $\mathbf{B}$: $\mathbf{A}_1^1 = \mathbf{A}$ and $\mathbf{B}_1^1 = \mathbf{B}$. By executing Steps 1–4 for the first time, not more than two pairs of vectors are formed: $\mathbf{A}_2^1$, $\mathbf{B}_2^1$ and $\mathbf{A}_2^2$, $\mathbf{B}_2^2$ (if such pairs of vectors do not exist, then all elements of $X(\mathbf{A}, \mathbf{B})$ are evaluated by executing Steps 1–4 for the first time). These pairs of vectors determine the minimax values $c(\mathbf{A}_2^1, \mathbf{B}_2^1)$ and $c(\mathbf{A}_2^2, \mathbf{B}_2^2)$ for the submatrices $X_2^1 = X_2^1(\mathbf{A}_2^1, \mathbf{B}_2^1)$ and $X_2^2 = X_2^2(\mathbf{A}_2^2, \mathbf{B}_2^2)$ of the matrix $X(\mathbf{A}, \mathbf{B})$, such that $X_2^1 \in M(\mathbf{A}_2^1, \mathbf{B}_2^1; c(\mathbf{A}_2^1, \mathbf{B}_2^1))$ and $X_2^2 \in M(\mathbf{A}_2^2, \mathbf{B}_2^2; c(\mathbf{A}_2^2, \mathbf{B}_2^2))$ (note that the notation used here corresponds to that in Algorithm 2.1: $\mathbf{A}' = \mathbf{A}_2^1$, $\mathbf{B}' = \mathbf{B}_2^1$, $\mathbf{A}'' = \mathbf{A}_2^2$, $\mathbf{B}'' = \mathbf{B}_2^2$, $X' = X_2^1$, $X'' = X_2^2$). After Algorithm 2.1 has been applied for the kth time, not more than 2^{k-1} pairs of vectors $\mathbf{A}_k^i$, $\mathbf{B}_k^i$ are considered, where $1 \leq i \leq s$ and $s \leq 2^{k-1}$. Each vector pair $\mathbf{A}_k^i$, $\mathbf{B}_k^i$ determines a minimax value $c(\mathbf{A}_k^i, \mathbf{B}_k^i)$ and, therefore, the elements of the submatrix $X_k^i = X_k^i(\mathbf{A}_k^i, \mathbf{B}_k^i)$ of $X(\mathbf{A}, \mathbf{B})$, because X_k^i is a minimax matrix from the set $M(\mathbf{A}_k^i, \mathbf{B}_k^i; c(\mathbf{A}_k^i, \mathbf{B}_k^i))$.

Suppose $Y = (x_{i_r j_t})$ $(1 \leq i_r \leq n,\ 1 \leq j_t \leq m,\ 1 \leq r \leq p,\ 1 \leq t \leq q)$ is an arbitrary submatrix formed by p rows and q columns of the matrix $X(\mathbf{A}, \mathbf{B})$ constructed by means of Algorithm 2.1. Let us now prove that Y is a minimax matrix. Denote the largest element of c by Y. If $c = 0$, then Y is a minimax matrix. Suppose $c > 0$. Then, $c = c(\mathbf{A}_k^i, \mathbf{B}_k^i)$ for certain k and i. Among the elements of Y equal to $c = c(\mathbf{A}_k^i, \mathbf{B}_k^i)$, choose the one that is calculated by the algorithm with the lowest number k. This element, $x_{i^* j^*}$, is then common to both $X_k^i = X_k^i(\mathbf{A}_k^i, \mathbf{B}_k^i)$ and Y. Since the element $x_{i^* j^*}$ equals $c(\mathbf{A}_k^i, \mathbf{B}_k^i)$ for any matrix from $M(\mathbf{A}_k^i, \mathbf{B}_k^i; c(\mathbf{A}_k^i, \mathbf{B}_k^i))$ (Theorem 3.2 from Chapter 1), Y is a minimax matrix. This completes the proof of the lemma and proves Lemma 2.1.

2.2. Uniqueness of the Hereditarily Minimax Matrix of a Transportation Polyhedron

Step 2 of Algorithm 2.1 involves an uncertainty in choosing the number k. Despite the possible nonuniqueness of the choice of k, the algorithm leads to a unique matrix. Moreover, the following proposition is true.

Lemma 2.3. *For every pair of vectors with equal sums of components, the matrix set $M(\mathbf{A},\mathbf{B})$ contains not more than one hereditarily minimax matrix.*

Proof. Without loss of generality, consider $\mathbf{A} \in \overline{R}^n_+$ and $\mathbf{B} \in \overline{R}^m_+$. Apply mathematical induction on the number of components of $\mathbf{A}$. When $n = 1$, the assertion of the lemma is obvious. Suppose the lemma is true for $n \leq t$ and set $n = t+1$. Let $X_1 = X_1(\mathbf{A},\mathbf{B}) = (x_{ij})$ and $X_2 = X_2(\mathbf{A},\mathbf{B}) = (x'_{ij})$ $(1 \leq i \leq t+1$ and $1 \leq j \leq m)$ are hereditarily minimax matrices, where $\mathbf{A} = (a_1, \ldots, a_{t+1}) \in \overline{R}^{t+1}_+$ and $\mathbf{B} = (b_1, \ldots, b_m) \in \overline{R}^m_+$. Since X_1 and X_2 are minimax matrices, it holds that $x_{11} = x'_{11} = c(\mathbf{A},\mathbf{B})$ (by Theorem 3.2 from Chapter 1 or Theorem 1.2), and $X_1, X_2 \in M\left(\mathbf{A},\mathbf{B}; c(\mathbf{A},\mathbf{B})\right)$. By Lemma 1.1, there exists $p \in Z$ $(1 \leq p \leq t+1)$ such that $c(\mathbf{A},\mathbf{B})p \leq b_1$ and $\delta_p\left(\mathbf{A},\mathbf{B}; c(\mathbf{A},\mathbf{B})\right) = 0$. Applying Theorem 3.2 from Chapter 1 to X_1 and X_2, we obtain $x_{ij} = x'_{ij} = c(\mathbf{A},\mathbf{B})$ for $1 \leq i \leq p$, $1 \leq j \leq l_p\left(\mathbf{B}; c(\mathbf{A},\mathbf{B})\right)$ and $x_{ij} = x'_{ij} = 0$ for $p < i \leq t+1$, $l_p\left(\mathbf{B}; c(\mathbf{A},\mathbf{B})\right) < j \leq m$.

Suppose $p < t+1$. Consider the following submatrices of the matrices X_1 and X_2: $X'_1 = (x_{ij})$ and $X'_2 = (x'_{ij})$, where $1 \leq i \leq p$ and $l_p\left(\mathbf{B}; c(\mathbf{A},\mathbf{B})\right) < j \leq m$ (if $l_p\left(\mathbf{B}; c(\mathbf{A},\mathbf{B})\right) = m$, then X'_1 and X'_2 do not exist), and $X''_1 = (x_{ij})$ and $X''_2 = (x'_{ij})$, where $p+1 \leq i \leq t+1$ and $1 \leq j \leq l_p\left(\mathbf{B}; c(\mathbf{A},\mathbf{B})\right)$. It is obvious that the former pair of submatrices, X'_1 and X'_2 (if they exist), is determined by the vectors

$$\mathbf{A}' = (a'_1, \ldots, a'_p) \in \overline{R}^p_+ \text{ and } \mathbf{B}' = (b'_{l_p(\mathbf{B};c(\mathbf{A},\mathbf{B}))+1}, \ldots, b'_m) \in \overline{R}^{m-l_p(\mathbf{B};c(\mathbf{A},\mathbf{B}))}_+,$$

where

$$a'_i = a_i - c(\mathbf{A},\mathbf{B})l_p\left(\mathbf{B}; c(\mathbf{A},\mathbf{B})\right), \quad 1 \leq i \leq p,$$

$$b'_j = b_j, \quad l_p\left(\mathbf{B}; c(\mathbf{A},\mathbf{B})\right) + 1 \leq j \leq m;$$

and the latter pair, X''_1 and X''_2, is determined by $\mathbf{A}'' = (a''_{p+1}, \ldots, a''_{t+1}) \in \overline{R}^{t+1-p}_+$ and $\mathbf{B}'' = (b''_1, \ldots, b''_{l_p(\mathbf{B};c(\mathbf{A},\mathbf{B}))}) \in \overline{R}^{l_p(\mathbf{B};c(\mathbf{A},\mathbf{B}))}_+$, where

$$a''_i = a_i, \quad p+1 \leq i \leq t+1,$$

$$b''_j = b_j - c(\mathbf{A},\mathbf{B})p, \quad 1 \leq j \leq l_p\left(\mathbf{B}; c(\mathbf{A},\mathbf{B})\right):$$

$$X_1' = X_1'(\mathbf{A}', \mathbf{B}'), \quad X_2' = X_2'(\mathbf{A}', \mathbf{B}'), \quad X_1'' = X_1''(\mathbf{A}'', \mathbf{B}''),$$
$$X_2'' = X_2''(\mathbf{A}'', \mathbf{B}''), \quad X_1', X_2' \in M(\mathbf{A}', \mathbf{B}'), \quad X_1'', X_2'' \in M(\mathbf{A}'', \mathbf{B}'').$$

Since every submatrix of a hereditarily minimax matrix is a minimax and hereditarily minimax matrix, X_1' and X_2' are hereditarily minimax matrices from $M(\mathbf{A}', \mathbf{B}')$, and X_1'' and X_2'' are hereditarily minimax matrices from $M(\mathbf{A}'', \mathbf{B}'')$. By the induction hypothesis, $X_1' = X_2'$ and $X_1'' = X_2''$. Thus, $X_1(\mathbf{A}, \mathbf{B}) = X_2(\mathbf{A}, \mathbf{B})$.

Suppose $p = t + 1$. Consider the case $l_p(\mathbf{B}; c(\mathbf{A}, \mathbf{B})) < m$ (otherwise, $X_1(\mathbf{A}, \mathbf{B}) = X_2(\mathbf{A}, \mathbf{B})$). The value of $l_p(\mathbf{B}; c(\mathbf{A}, \mathbf{B}))$ determines the submatrices of X_1 and X_2: $X_1' = (x_{ij})$, $X_2' = (x'_{ij})$ $(1 \le i \le t+1,\ l_p(\mathbf{B}; c(\mathbf{A}, \mathbf{B})) + 1 \le j \le m)$. It is clear that $X_1' = X_1'(\mathbf{A}', \mathbf{B}')$ and $X_2' = X_2'(\mathbf{A}', \mathbf{B}')$, where $\mathbf{A}' = (a_1', \ldots, a_{t+1}') \in \overline{R}_+^{t+1}$, $\mathbf{B}' = (b'_{l_p(\mathbf{B};c(\mathbf{A},\mathbf{B}))} + 1, \ldots, b_m') \in \overline{R}_+^{m - l_p(\mathbf{B};c(\mathbf{A},\mathbf{B}))}$, and

$$a_i' = a_i - c(\mathbf{A}, \mathbf{B}) l_p(\mathbf{B}; c(\mathbf{A}, \mathbf{B})), \quad 1 \le i \le t+1,$$
$$b_j' = b_j, \quad l_p(\mathbf{B}; c(\mathbf{A}, \mathbf{B})) + 1 \le j \le m.$$

Since X_1 and X_2 are hereditarily minimax matrices, the submatrices X_1' and X_2' are also hereditarily minimax matrices, and $X_1', X_2' \in M(\mathbf{A}', \mathbf{B}'; c(\mathbf{A}', \mathbf{B}'))$. Applying Theorem 3.2 from Chapter 1 to the vectors $\mathbf{A}'$ and $\mathbf{B}'$ and the minimax value $c(\mathbf{A}', \mathbf{B}')$, we obtain $X_1' = X_2'$ (by the induction hypothesis in the case $p < t + 1$); otherwise, the first $l_p(\mathbf{B}'; c(\mathbf{A}', \mathbf{B}'))$ rows of X_1' and X_2' are identical (when $p = t+1$). However, it is now obvious that the hereditarily minimax matrices $X_1'(\mathbf{A}', \mathbf{B}')$ and $X_2'(\mathbf{A}', \mathbf{B}')$ are identical in the latter case as well. Consequently, $X_1(\mathbf{A}, \mathbf{B}) = X_2(\mathbf{A}, \mathbf{B})$, and the matrix set $M(\mathbf{A}, \mathbf{B})$ contains only one minimax matrix. This completes the proof of the lemma.

Hereditarily minimax matrices have another property (see Definition 1.1 in Chapter 1).

Lemma 2.4. *A hereditarily minimax matrix (a matrix constructed by means of Algorithm* 2.1) *is a uniform matrix.*

The proof is readily derived from the construction of Algorithm 2.1 and the proof of Lemma 2.2.

Starting from the following theorem, which follows from Lemmas 2.1–2.4, Algorithm 2.1, and Theorems 1.3 and 1.4, the asterisk denotes a hereditarily minimax matrix.

Theorem 2.1. *Suppose that vectors* $\mathbf{A} \in R_+^n$ *and* $\mathbf{B} \in R_+^m$ *satisfy the balance condition. The matrix set* $M(\mathbf{A}, \mathbf{B})$ *contains strictly one hereditarily* *minimax matrix* $\overset{*}{X}(\mathbf{A},\mathbf{B}) = (\overset{*}{x}_{ij})$ $(1 \le i \le n,\ 1 \le j \le m)$, *which is uniform. When a matrix* $\overset{*}{X}(\widetilde{\mathbf{A}},\widetilde{\mathbf{B}})$ *is constructed by permuting rows and columns in* $\overset{*}{X}(\mathbf{A},\mathbf{B})$ ($\widetilde{\mathbf{A}}$ *and* $\widetilde{\mathbf{B}}$ *are obtained by arranging the components of* $\mathbf{A}$ *and* $\mathbf{B}$, *respectively, in nonascending order*), *it is defined by Algorithm* 2.1.

Example 2.1. Let us apply Algorithm 2.1 to construct the hereditarily minimax matrix $\overset{*}{X}(\mathbf{A},\mathbf{B}) = (\overset{*}{x}_{ij})$ $(1 \le i \le 10,\ 1 \le j \le 13)$, where

$$\mathbf{A} = (79, 74, 61, 56, 19, 19, 16, 3, 3, 2) \in \overline{R}^{10}_{+},$$

$$\mathbf{B} = (52, 44, 44, 40, 40, 34, 21, 21, 15, 8, 8, 3, 2) \in \overline{R}^{13}_{+}.$$

The notation employed here is the same as in Algorithm 2.1 and proof of Lemma 2.2.

Consider Algorithm 2.1 as applied to the vectors $\mathbf{A}^1_1 = \mathbf{A}$ and $\mathbf{B}^1_1 = \mathbf{B}$. By means of Algorithm 1.1, determine the minimax value $c(\mathbf{A}^1_1,\mathbf{B}^1_1) = c(\mathbf{A},\mathbf{B}) = 8$, $c(\mathbf{A}^1_1,\mathbf{B}^1_1)\cdot 4 = 32 \le b_1 = 52$, $\delta_4(\mathbf{A}^1_1,\mathbf{B}^1_1;8) = 0$, and $l_4(\mathbf{B};8) = 6$. Then, Theorem 3.2 from Chapter 1 (or Theorem 1.2) gives the following elements of the minimax matrix $\overset{*}{X}(\mathbf{A},\mathbf{B})$:

$$\overset{*}{x}_{ij} = \begin{cases} 8, & 1 \le i \le 4, \quad 1 \le j \le 6, \\ 0, & 5 \le i \le 10, \ 7 \le j \le 13. \end{cases}$$

Thus, we have determined 66 of the 130 elements of $\overset{*}{X}(\mathbf{A},\mathbf{B})$. The unknown elements make up two submatrices of $\overset{*}{X}(\mathbf{A},\mathbf{B})$: $X^1_2 = (\overset{*}{x}_{ij})$ $(1 \le i \le 4,\ 7 \le j \le 13)$ and $X^2_2 = (\overset{*}{x}_{ij})$ $(5 \le i \le 10,\ 1 \le j \le 6)$. These submatrices are defined by the pairs of vectors $\mathbf{A}^1_2$, $\mathbf{B}^1_2$ and $\mathbf{A}^2_2$, $\mathbf{B}^2_2$, whose components are calculated by (2.1) and (2.2).

$$X^1_2 = X^1_2(\mathbf{A}^1_2,\mathbf{B}^1_2), \quad \mathbf{A}^1_2 = (31, 26, 13, 8), \quad \mathbf{B}^1_2 = (21, 21, 15, 8, 8, 3, 2),$$

$$X^2_2 = X^1_2(\mathbf{A}^2_2,\mathbf{B}^2_2), \quad \mathbf{A}^2_2 = (19, 19, 16, 3, 3, 2), \quad \mathbf{B}^2_2 = (20, 12, 12, 8, 8, 2).$$

We now apply Algorithm 2.1 to the two pairs of vectors $\mathbf{A}^1_2$, $\mathbf{B}^1_2$ and $\mathbf{A}^2_2$, $\mathbf{B}^2_2$. By means of Algorithm 1.1, we calculate the minimax values and the values for which the function defined by (1.22) in Chapter 1 vanishes: $c(\mathbf{A}^1_2,\mathbf{B}^1_2) = 6$, $c(\mathbf{A}^1_2,\mathbf{B}^1_2)\cdot 2 = 12 < 21$, $\delta_2(\mathbf{A}^1_2,\mathbf{B}^1_2;6) = 0$, $l_2(\mathbf{B}^1_2;6) = 3$, $c(\mathbf{A}^2_2,\mathbf{B}^2_2) = 4$, $c(\mathbf{A}^2_2,\mathbf{B}^2_2)\cdot 3 = 12 < 20$, $\delta_3(\mathbf{A}^2_2,\mathbf{B}^2_2;4) = 0$, $l_3(\mathbf{B}^2_2;4) = 3$. Then, Theorem 3.2 from Chapter 1 determines the following elements of X^1_2 and X^2_2 (which are also elements of $\overset{*}{X}(\mathbf{A},\mathbf{B})$):

$$\overset{*}{x}_{ij} = \begin{cases} 6, & 1 \le i \le 2, \ 7 \le j \le 9, \\ 0, & 3 \le i \le 4, \ 10 \le j \le 13, \end{cases}$$

$$\overset{*}{x}_{ij} = \begin{cases} 4, & 5 \le i \le 7, \quad 1 \le j \le 3, \\ 0, & 8 \le i \le 10, \quad 4 \le j \le 6. \end{cases}$$

Now, we have determined $66 + 14 + 18 = 98$ elements of the matrix $\overset{*}{X}(\mathbf{A}, \mathbf{B})$.

The elements of the desired matrix $\overset{*}{X}(\mathbf{A}, \mathbf{B})$ that remain unknown make up four submatrices:

$$X_3^1 = (\overset{*}{x}_{ij}), \quad 1 \le i \le 2, \quad 10 \le j \le 13,$$

$$X_3^2 = (\overset{*}{x}_{ij}), \quad 3 \le i \le 4, \quad 7 \le j \le 9,$$

$$X_3^3 = (\overset{*}{x}_{ij}), \quad 5 \le i \le 7, \quad 4 \le j \le 6,$$

$$X_3^4 = (\overset{*}{x}_{ij}), \quad 8 \le i \le 10, \quad 1 \le j \le 3.$$

These submatrices are determined by four pairs of vectors, whose components are calculated by (2.1) and (2.2):

$$X_3^1 = X_3^1(\mathbf{A}_3^1, \mathbf{B}_3^1), \quad \mathbf{A}_3^1 = (13, 8), \quad \mathbf{B}_3^1 = (8, 8, 3, 2),$$
$$X_3^2 = X_3^2(\mathbf{A}_3^2, \mathbf{B}_3^2), \quad \mathbf{A}_3^2 = (13, 8), \quad \mathbf{B}_3^2 = (9, 9, 3),$$
$$X_3^3 = X_3^3(\mathbf{A}_3^3, \mathbf{B}_3^3), \quad \mathbf{A}_3^3 = (7, 7, 4), \quad \mathbf{B}_3^3 = (8, 8, 2),$$
$$X_3^4 = X_3^4(\mathbf{A}_3^4, \mathbf{B}_3^4), \quad \mathbf{A}_3^4 = (3, 3, 2), \quad \mathbf{B}_3^4 = (8, 0, 0).$$

Next, we apply Algorithm 2.1 (for the third time) to the four pairs of vectors constructed. Note that the vector $\mathbf{B}_3^4$ in the last pair has only one positive component; therefore, the matrix X_3^4 is determined solely by the vector $\mathbf{A}_3^4$, which constitutes its first column: $\overset{*}{x}_{81} = \overset{*}{x}_{91} = 3$, $\overset{*}{x}_{101} = 2$, and $\overset{*}{x}_{ij} = 0$ for $8 \le i \le 10$ and $2 \le j \le 3$. Thus, we have determined all elements of $\overset{*}{X}(\mathbf{A}, \mathbf{B})$ for $8 \le i \le 10$. Applying Algorithm 1.1 to the remaining three pairs of vectors, we obtain

$$c(\mathbf{A}_3^1, \mathbf{B}_3^1) = 4, \;\; c(\mathbf{A}_3^1, \mathbf{B}_3^1) \cdot 2 = 8 \le 8, \;\; \delta_2(\mathbf{A}_3^1, \mathbf{B}_3^1; 4) = 0, \;\; l_2(\mathbf{B}_3^1; 4) = 2,$$

$$c(\mathbf{A}_3^2, \mathbf{B}_3^2) = 5, \;\; c(\mathbf{A}_3^2, \mathbf{B}_3^2) \cdot 1 = 5 < 9, \;\; \delta_1(\mathbf{A}_3^2, \mathbf{B}_3^2; 5) = 0, \;\; l_1(\mathbf{B}_3^2; 5) = 2,$$

$$c(\mathbf{A}_3^3, \mathbf{B}_3^3) = 3, \;\; c(\mathbf{A}_3^3, \mathbf{B}_3^3) \cdot 2 = 6 < 8, \;\; \delta_2(\mathbf{A}_3^3, \mathbf{B}_3^3; 3) = 0, \;\; l_1(\mathbf{B}_3^3; 3) = 2.$$

Theorem 3.2 from Chapter 1 defines the following elements of the matrices X_3^1, X_3^2, and X_3^3 and, therefore, the elements of $\overset{*}{X}(\mathbf{A}, \mathbf{B})$:

$$\overset{*}{x}_{ij} = 4, \quad 1 \le i \le 2, \quad 10 \le j \le 11,$$

$$\overset{*}{x}_{ij} = \begin{cases} 5, & i = 3, \quad 7 \le j \le 8, \\ 0, & i = 4, \quad j = 9, \end{cases}$$

$$\overset{*}{x}_{ij} = \begin{cases} 3, & 5 \le i \le 6, \quad 4 \le j \le 5, \\ 0, & i = 7, \quad j = 6. \end{cases}$$

Now, we have calculated $98+4+3+5+9 = 119$ elements of $\overset{*}{X}(\mathbf{A},\mathbf{B})$. The elements of $\overset{*}{X}(\mathbf{A},\mathbf{B})$ that remain unknown make up only five submatrices $(5 < 2^3)$, because the submatrix X_3^4 is completely defined, and the unknown elements of X_3^1 make up only one submatrix:

$$X_4^1 = X_4^1(\mathbf{A}_4^1, \mathbf{B}_4^1),\ \mathbf{A}_4^1 = (5,0),\ \mathbf{B}_4^1 = (3,2),\ X_4^1 = (\overset{*}{x}_{ij}),\ 1 \le i \le 2,\ 12 \le j \le 13,$$

$$X_4^2 = X_4^2(\mathbf{A}_4^2, \mathbf{B}_4^2),\ \mathbf{A}_4^2 = (3),\ \mathbf{B}_4^2 = (3),\ X_4^2 = (\overset{*}{x}_{ij}),\ i = 3,\ j = 9,$$

$$X_4^3 = X_4^3(\mathbf{A}_4^3, \mathbf{B}_4^3),\ \mathbf{A}_4^3 = (8),\ \mathbf{B}_4^3 = (4,4),\ X_4^3 = (\overset{*}{x}_{ij}),\ i = 4,\ 7 \le j \le 8,$$

$$X_4^4 = X_4^4(\mathbf{A}_4^4, \mathbf{B}_4^4),\ \mathbf{A}_4^4 = (1,1),\ \mathbf{B}_4^4 = (2),\ X_4^4 = (\overset{*}{x}_{ij}),\ 5 \le i \le 6,\ j = 6,$$

$$X_4^5 = X_4^5(\mathbf{A}_4^5, \mathbf{B}_4^5),\ \mathbf{A}_4^5 = (4),\ \mathbf{B}_4^5 = (2,2),\ X_4^5 = (\overset{*}{x}_{ij}),\ i = 7,\ 4 \le j \le 5,$$

where the components of vectors are calculated by (2.1) and (2.2).

When the algorithm is applied for the fourth time, the remaining elements of $\overset{*}{X}(\mathbf{A},\mathbf{B})$ are calculated directly, because each new pair of vectors contains a single-component one:

$$\overset{*}{x}_{1\,12} = 3, \quad \overset{*}{x}_{1\,13} = 2, \quad \overset{*}{x}_{2\,12} = \overset{*}{x}_{2\,13} = 0, \quad \overset{*}{x}_{39} = 3,$$

$$\overset{*}{x}_{47} = \overset{*}{x}_{48} = 4, \quad \overset{*}{x}_{56} = \overset{*}{x}_{66} = 1, \quad \overset{*}{x}_{74} = \overset{*}{x}_{75} = 2.$$

Thus, we have calculated all elements of $\overset{*}{X}(\mathbf{A},\mathbf{B})$:

	52	44	44	40	40	34	21	21	15	8	8	3	2
79	8	8	8	8	8	8	6	6	6	4	4	3	2
74	8	8	8	8	8	8	6	6	6	4	4	0	0
61	8	8	8	8	8	8	5	5	3	0	0	0	0
56	8	8	8	8	8	8	4	4	0	0	0	0	0
19	4	4	4	3	3	1	0	0	0	0	0	0	0
19	4	4	4	3	3	1	0	0	0	0	0	0	0
16	4	4	4	2	2	0	0	0	0	0	0	0	0
3	3	0	0	0	0	0	0	0	0	0	0	0	0
3	3	0	0	0	0	0	0	0	0	0	0	0	0
2	2	0	0	0	0	0	0	0	0	0	0	0	0

$$\overset{*}{X}(\mathbf{A},\mathbf{B}) \in M(\mathbf{A},\mathbf{B};8)$$

2.3. Minimization by Means of Uniform Minimax Matrices

In what follows, we show that optimal solutions to transportation problems with some minimax functionals are given by uniform minimax matrices.

Let $\mathbf{A} = (a_1, \ldots, a_n) \in R_+^n$ and $\mathbf{B} = (b_1, \ldots, b_m) \in R_+^m$ be arbitrary vectors with equal sums of components, and p, q $(1 \le p \le n,\ 1 \le q \le m)$ be such integers that $a_p = \max\{a_i : 1 \le i \le n\}$ and $b_q = \max\{b_j : 1 \le j \le m\}$. For an arbitrary matrix $X(\mathbf{A},\mathbf{B}) = (x_{ij})$ from $M(\mathbf{A},\mathbf{B})$ $(1 \le i \le n,\ 1 \le j \le m)$, consider objective functions (1.8) and (1.9) defined in Chapter 1. It is obvious that

$$\Phi_2\left(X(\mathbf{A},\mathbf{B})\right) = \sum_{j=1}^{m} \max_i x_{ij} \ge \sum_{j=1}^{m} x_{pj} = a_p,$$

$$\Phi_2\left(X(\mathbf{B},\mathbf{A})\right) = \sum_{i=1}^{n} \max_j x_{ij} \ge \sum_{i=1}^{n} x_{qi} = b_q,$$

$$\Phi_3\left(X(\mathbf{A},\mathbf{B})\right) = \sum_{j=1}^{m} \max_i x_{ij} + \sum_{i=1}^{n} \max_j x_{ij} \ge \sum_{j=1}^{m} x_{pj} + \sum_{i=1}^{n} x_{iq} = a_p + b_q.$$

Therefore,

$$\begin{aligned}
&\min_{X(\mathbf{A},\mathbf{B}) \in M(\mathbf{A},\mathbf{B})} \Phi_2\left(X(\mathbf{A},\mathbf{B})\right) \ge \max\{a_i : 1 \le i \le n\},\\
&\min_{X(\mathbf{B},\mathbf{A}) \in M(\mathbf{B},\mathbf{A})} \Phi_2\left(X(\mathbf{B},\mathbf{A})\right) \ge \max\{b_j : 1 \le j \le m\},\\
&\min_{X(\mathbf{A},\mathbf{B}) \in M(\mathbf{A},\mathbf{B})} \Phi_3\left(X(\mathbf{A},\mathbf{B})\right) \ge \max\{a_i : 1 \le i \le n\} + \max\{b_j : 1 \le j \le m\}.
\end{aligned} \tag{2.3}$$

Let $X(\mathbf{A},\mathbf{B})$ be a uniform matrix (the set of uniform matrices, $\widetilde{M}(\mathbf{A},\mathbf{B})$, is nonempty, because the hereditarily minimax matrix $\overset{*}{X}(\mathbf{A},\mathbf{B})$ is uniform). Then, by Definition 1.1 from Chapter 1,

$$\max_i x_{ij} = x_{pj}, \quad \max_j x_{ij} = x_{iq}, \quad 1 \le i \le n, \quad 1 \le j \le m.$$

Therefore,

$$\begin{aligned}
&\Phi_2\left(X(\mathbf{A},\mathbf{B})\right) = \sum_{j=1}^{m} x_{pj} = a_p, \quad X(\mathbf{A},\mathbf{B}) = (x_{ij}) \in \widetilde{M}(\mathbf{A},\mathbf{B}),\\
&\Phi_2\left(X(\mathbf{B},\mathbf{A})\right) = \sum_{i=1}^{n} x_{qi} = b_q, \quad X(\mathbf{B},\mathbf{A}) = (x_{ij}) \in \widetilde{M}(\mathbf{B},\mathbf{A}),\\
&\Phi_3\left(X(\mathbf{A},\mathbf{B})\right) = \sum_{j=1}^{m} x_{pj} + \sum_{i=1}^{n} x_{iq} = a_p + b_q, \quad X(\mathbf{A},\mathbf{B}) = (x_{ij}) \in \widetilde{M}(\mathbf{A},\mathbf{B}).
\end{aligned} \tag{2.4}$$

Remark 2.1. Suppose vectors $\mathbf{A} = (a_1, \ldots, a_n) \in R_+^n$ and $\mathbf{B} = (b_1, \ldots, b_m) \in R_+^m$ are such that $a_1 + \cdots + a_n = b_1 + \cdots + b_m > 0$. The matrix $X(\mathbf{A},\mathbf{B}) = (x_{ij})$ whose elements are defined by the condition

$$x_{ij} = \frac{a_i b_j}{\sum_{j=1}^{m} b_j} = \frac{a_i b_j}{\sum_{i=1}^{n} a_i}, \quad 1 \le i \le n, \quad 1 \le j \le m,$$

demonstrates that $M(\mathbf{A},\mathbf{B}) \neq \varnothing$. This matrix is uniform, and $\widetilde{M}(\mathbf{A},\mathbf{B}) \neq \varnothing$.

If the formula for the elements of matrices $X'(A',B')$ and $X''(A'',B'')$ is applied after Steps 1–4 of Algorithm 2.1, then a minimax uniform matrix will be constructed. This matrix minimizes functionals (1.10) and (1.11) from Chapter 1 (Theorem 2.3).

It follows from (2.3) and (2.4) that any uniform matrix minimizes functionals (1.8) and (1.9) defined in Chapter 1. Thus we have the following proposition.

Theorem 2.2. *Suppose vectors* $\mathbf{A} = (a_1,\dots,a_n) \in R_+^n$ *and* $\mathbf{B} = (b_1,\dots,b_m) \in R_+^m$ *satisfy the balance condition. Then any uniform matrix* $\overline{X}(\mathbf{A},\mathbf{B})$ *minimizes the functionals* $\Phi_2\left(X(\mathbf{A},\mathbf{B})\right)$ *and* $\Phi_3\left(X(\mathbf{A},\mathbf{B})\right)$:

$$\begin{aligned} \min_{X(\mathbf{A},\mathbf{B})\in M(\mathbf{A},\mathbf{B})} \Phi_2\left(X(\mathbf{A},\mathbf{B})\right) &= \Phi_3\left(\overline{X}(\mathbf{A},\mathbf{B})\right) = \\ &\max\{a_i : 1 \le i \le n\}, \\ \min_{X(\mathbf{A},\mathbf{B})\in M(\mathbf{A},\mathbf{B})} \Phi_3\left(X(\mathbf{A},\mathbf{B})\right) &= \Phi_3\left(\overline{X}(\mathbf{A},\mathbf{B})\right) = \\ &\max\{a_i : 1 \le i \le n\} + \max\{b_j : 1 \le j \le m\}, \end{aligned} \tag{2.5}$$

where $\overline{X}(\mathbf{A},\mathbf{B}) \in \widetilde{M}(\mathbf{A},\mathbf{B})$.

Since $\overset{*}{X}(\mathbf{A},\mathbf{B}) \in \widetilde{M}(\mathbf{A},\mathbf{B})$, the objective functions defined by (1.8) and (1.9) in Chapter 1 are minimized, in particular, by the hereditarily minimax matrix $\overset{*}{X}(\mathbf{A},\mathbf{B})$.

Any minimax matrix $X^*(\mathbf{A},\mathbf{B}) = (x_{ij}^*)$ from $M\left(\mathbf{A},\mathbf{B};c(\mathbf{A},\mathbf{B})\right)$ minimizes objective function (1.7) from Chapter 1:

$$\min_{X(\mathbf{A},\mathbf{B})\in M(\mathbf{A},\mathbf{B})} \Phi_1\left(X(\mathbf{A},\mathbf{B})\right) = \Phi_1\left(X^*(\mathbf{A},\mathbf{B})\right) = c(\mathbf{A},\mathbf{B}), \tag{2.6}$$

where $X^*(\mathbf{A},\mathbf{B}) \in M\left(\mathbf{A},\mathbf{B};c(\mathbf{A},\mathbf{B})\right)$ and, in particular, $\Phi_1(\overset{*}{X}(\mathbf{A},\mathbf{B})) = c(\mathbf{A},\mathbf{B})$.

As a corollary to (2.5) and (2.6), we have a proposition concerning the minimization of functionals (1.10) and (1.11) defined in Chapter 1,

$$\Phi_4\left(X(\mathbf{A},\mathbf{B})\right) = \max_{i,j} x_{ij} + \sum_{j=1}^{m} \max_{i} x_{ij} = \Phi_1\left(X(\mathbf{A},\mathbf{B})\right) + \Phi_2\left(X(\mathbf{A},\mathbf{B})\right),$$

$$\Phi_5\left(X(\mathbf{A},\mathbf{B})\right) = \max_{i,j} x_{ij} + \sum_{j=1}^{m} \max_{i} x_{ij} + \sum_{i=1}^{n} \max_{j} x_{ij} = \Phi_1\left(X(\mathbf{A},\mathbf{B})\right) + \Phi_3\left(X(\mathbf{A},\mathbf{B})\right),$$

where $X(\mathbf{A},\mathbf{B}) = (x_{ij}) \in M(\mathbf{A},\mathbf{B})$.

Theorem 2.3. *Suppose vectors* $\mathbf{A} = (a_1,\ldots,a_n) \in R_+^n$ *and* $\mathbf{B} = (b_1,\ldots,b_m) \in R_+^m$ *are characterized by equal sums of components. Any minimax, uniform matrix* $\overline{X}^*(\mathbf{A},\mathbf{B})$ *from the set* $M\left(\mathbf{A},\mathbf{B};c(\mathbf{A},\mathbf{B})\right) \cap \widetilde{M}(\mathbf{A},\mathbf{B})$ *minimizes the functionals* $\Phi_4\left(X(\mathbf{A},\mathbf{B})\right)$ *and* $\Phi_5\left(X(\mathbf{A},\mathbf{B})\right)$; *and*

$$\begin{aligned}\min_{X(\mathbf{A},\mathbf{B})\in M(\mathbf{A},\mathbf{B})} \Phi_4\left(X(\mathbf{A},\mathbf{B})\right) &= \Phi_4\left(\overline{X}^*(\mathbf{A},\mathbf{B})\right) = \\ &c(\mathbf{A},\mathbf{B}) + \max\{a_i : 1 \le i \le n\}, \\ \min_{X(\mathbf{A},\mathbf{B})\in M(\mathbf{A},\mathbf{B})} \Phi_5\left(X(\mathbf{A},\mathbf{B})\right) &= \Phi_5\left(\overline{X}^*(\mathbf{A},\mathbf{B})\right) = \\ c(\mathbf{A},\mathbf{B}) + \max\{a_i &: 1 \le i \le n\} + \max\{b_j : 1 \le j \le m\},\end{aligned} \tag{2.7}$$

where $\overline{X}^*(\mathbf{A},\mathbf{B}) \in M\left(\mathbf{A},\mathbf{B};c(\mathbf{A},\mathbf{B})\right) \cap \widetilde{M}(\mathbf{A},\mathbf{B})$; *in particular, the hereditarily minimax matrix* $\overset{*}{X}(\mathbf{A},\mathbf{B})$ *also minimizes these functionals*:

$$\Phi_4(\overset{*}{X}(\mathbf{A},\mathbf{B})) = c(\mathbf{A},\mathbf{B}) + \max\{a_i : 1 \le i \le n\},$$

$$\Phi_5(\overset{*}{X}(\mathbf{A},\mathbf{B})) = c(\mathbf{A},\mathbf{B}) + \max\{a_i : 1 \le i \le n\} + \max\{b_j : 1 \le j \le m\}.$$

2.4. Properties of Uniform Matrices

Now, we introduce the concept of uniformity of a transportation matrix with respect to its rows and columns. We establish a relation between this concept and the concept of uniformity previously employed. We formulate a functional criterion for uniformity of a matrix, which is used to minimize functionals (1.12) and (1.13) from Chapter 1.

Let $X = (x_{ij})$ ($1 \le i \le n$, $1 \le j \le m$) be an arbitrary matrix with nonnegative elements that defines vectors $\mathbf{A} = \mathbf{A}(X) \in R_+^n$ and $\mathbf{B} = \mathbf{B}(X) \in R_+^m$ (the components of $\mathbf{A}$ and $\mathbf{B}$ are the sums of elements in the rows and columns of X, respectively). Thus, $X = X(\mathbf{A},\mathbf{B})$. Suppose that the matrix is uniform; that is, $X = X(\mathbf{A},\mathbf{B}) \in \widetilde{M}(\mathbf{A},\mathbf{B})$. The submatrix formed by removing any row or column from X is also uniform. This can be readily demonstrated by rearranging rows and columns in $X = X(\mathbf{A}, \mathbf{B})$ so as to form a matrix $\widetilde{X} = \widetilde{X}(\widetilde{\mathbf{A}}, \widetilde{\mathbf{B}})$, where the vectors $\widetilde{\mathbf{A}}$ and $\widetilde{\mathbf{B}}$ are constructed by arranging the components of $\mathbf{A}$ and $\mathbf{B}$ in nonascending order (it is clear that uniformity is invariant with respect to permutation of rows or columns of a matrix). Thus, uniformity is a hereditary property of matrices.

Lemma 2.5. *If a transportation matrix X is uniform, then its every submatrix Y is also uniform.*

Let us analyze the behavior of functionals (1.8) and (1.9) from Chapter 1 as functions of a submatrix of an arbitrary matrix $X = (x_{ij})$ with nonnegative elements. Let $X = X(\mathbf{A}, \mathbf{B})$, where $\mathbf{A} = \mathbf{A}(X) = (a_1, \dots, a_n) \in R^n_+$, $\mathbf{B} = \mathbf{B}(X) = (b_1, \dots, b_m) \in R^m_+$, and $Y = (x_{i_r j_t})$ $(1 \le r \le p,\ 1 \le t \le q)$, be an arbitrary submatrix of X, with p rows and q columns. If $Y = Y(\mathbf{C}, \mathbf{D})$, where $\mathbf{C} = \mathbf{C}(Y) = (c_{i_1}, \dots, c_{i_p})$ and $\mathbf{D} = \mathbf{D}(Y) = (d_{j_1}, \dots, d_{j_q})$, then the definitions of functionals (1.8) and (1.9) from Chapter 1 imply that

$$\Phi_2\left(Y(\mathbf{C}, \mathbf{D})\right) = \sum_{t=1}^{q} \max_r x_{i_r j_t} \ge \max\{c_{i_r} : 1 \le r \le p\},$$

$$\Phi_3\left(Y(\mathbf{C}, \mathbf{D})\right) = \sum_{t=1}^{q} \max_r x_{i_r j_t} + \sum_{r=1}^{p} \max_t x_{i_r j_t} \ge \max\{c_{i_r} : 1 \le r \le p\} + \max\{d_{j_t} : 1 \le t \le q\};$$

moreover, if $Y(\mathbf{C}, \mathbf{D})$ is uniform, then (see (2.4))

$$\begin{aligned} \Phi_2\left(Y(\mathbf{C}, \mathbf{D})\right) &= \max\{c_{i_r} : 1 \le r \le p\}, \\ \Phi_3\left(Y(\mathbf{C}, \mathbf{D})\right) &= \max\{c_{i_r} : 1 \le r \le p\} + \max\{d_{j_t} : 1 \le t \le q\}. \end{aligned} \tag{2.8}$$

Thus, we have the following lemma.

Lemma 2.6. *If $X(\mathbf{A}, \mathbf{B})$ is a uniform matrix, then relations* (2.8) *are valid for its every submatrix $Y(\mathbf{C}, \mathbf{D})$.*

The proof follows from relations (2.3) and (2.4) combined with Lemma 2.5.

In what follows, we show that the validity of (2.8) for functional (1.9) from Chapter 1 as a function of any submatrix $Y(\mathbf{C}, \mathbf{D})$ of $X(\mathbf{A}, \mathbf{B})$ is sufficient for the matrix $X(\mathbf{A}, \mathbf{B})$ to be uniform. However, the validity of (2.8) for functional (1.8) from Chapter 1 as a function of any submatrix of a given matrix does not entail its uniformity.

Example 2.2. Consider matrices $X(\mathbf{A},\mathbf{B})$ and $X(\mathbf{B},\mathbf{A})$, where $\mathbf{A} = (5,5,2)$ and $\mathbf{B} = (6,6)$:

$$\begin{array}{c} \begin{matrix} & 6 & 6 \\ 5 & 2 & 3 \\ 5 & 2 & 3 \\ 2 & 2 & 0 \end{matrix} \\ X(\mathbf{A},\mathbf{B}) \end{array} \qquad \begin{array}{c} \begin{matrix} & 5 & 5 & 2 \\ 6 & 2 & 2 & 2 \\ 6 & 3 & 3 & 0 \end{matrix} \\ X(\mathbf{B},\mathbf{A}) \end{array}$$

It is obvious that these matrices are not uniform. In $X(\mathbf{A},\mathbf{B})$, the uniformity with respect to rows is violated, and the first relation in (2.8) can be readily checked for the matrix and its every submatrix. In $X(\mathbf{B},\mathbf{A})$, the uniformity with respect to columns is violated, and the first relation in (2.8) is not valid for the matrix itself: $\Phi_2\left(X(\mathbf{B},\mathbf{A})\right) = 8 > \max\{b_j : 1 \le j \le 2\} = 6$.

Note that any matrix that has the form of a row vector or column vector is a uniform matrix.

To examine the uniformity of a matrix, it is sufficient to consider each row and column of the matrix separately. Let $X = (x_{ij})$ be a matrix with nonnegative elements, such that $X = X(\mathbf{A},\mathbf{B})$, where $\mathbf{A} = (a_1,\ldots,a_n) \in R_+^n$ and $\mathbf{B} = (b_1,\ldots,b_m) \in R_+^m$. A row of X with a number p $(1 \le p \le n)$ is said to be *uniform with respect to the vector* $\mathbf{B}$ if the relations $b_k \ge b_j$ or $b_k = b_j$ entail $x_{pk} \ge x_{pj}$ or $x_{pk} = x_{pj}$, respectively. A column of X with a number q $(1 \le q \le m)$ is said to be *uniform with respect to the vector* $\mathbf{A}$ if the relations $a_k \ge a_i$ or $a_k = a_i$ entail $x_{kq} \ge x_{iq}$ or $x_{kq} = x_{iq}$, respectively.

Lemma 2.7. *A matrix $X(\mathbf{A},\mathbf{B})$ is uniform if and only if its every row is uniform with respect to $\mathbf{B}$ and its every column is uniform with respect to $\mathbf{A}$.*

Proof.

Necessity follows from the definition of uniformity of a matrix.

Sufficiency. Suppose that the rows of $X(\mathbf{A},\mathbf{B}) = (x_{ij})$ are uniform with respect to $\mathbf{B} = (b_1,\ldots,b_m) \in R_+^m$ and its columns are uniform with respect to $\mathbf{A} = (a_1,\ldots,a_n) \in R_+^n$. If the components of $\mathbf{A}$ and $\mathbf{B}$ are such that $a_p \ge a_q$, $b_r \ge b_t$ or $a_p = a_q$, $b_r = b_t$, then the uniformity of the rows or columns implies, respectively, that $x_{pr} \ge x_{pt}$, $x_{pt} \ge x_{qt}$, and $x_{pr} \ge x_{qt}$ or $x_{pr} = x_{pt}$, $x_{pt} = x_{qt}$, and $x_{pr} = x_{qt}$. This is equivalent to the definition of the uniformity of $X(\mathbf{A},\mathbf{B})$.

If a matrix X is not uniform, then (2.4) may be valid for functional (1.9) defined in Chapter 1. However, there exists a submatrix of this matrix for which the value of functional (1.9) from Chapter 1 is inconsistent with (2.4).

Example 2.3. For the matrix $X(\mathbf{A},\mathbf{B})$, where $\mathbf{A} = (10,5,5) \in \overline{R}_+^3$ and $\mathbf{B} = (6,4,4,4,2) \in \overline{R}_+^5$,

$$\begin{array}{c} \\ 10 \\ 5 \\ 5 \end{array} \begin{array}{c} \begin{array}{ccccc} 6 & 4 & 4 & 4 & 2 \end{array} \\ \begin{pmatrix} 2 & 2 & 2 & 2 & 2 \\ 2 & 2 & 0 & 1 & 0 \\ 2 & 0 & 2 & 1 & 0 \end{pmatrix} \end{array}$$

$$X(\mathbf{A}, \mathbf{B})$$

it holds that

$$\Phi_3\left(X(\mathbf{A}, \mathbf{B})\right) = \max\{a_i : 1 \le i \le 3\} + \max\{b_j : 1 \le j \le 5\} = 16.$$

It is obvious that $X(\mathbf{A}, \mathbf{B})$ is not uniform. Removing the first row and column from $X(\mathbf{A}, \mathbf{B})$, we obtain a submatrix $Y(\mathbf{C}, \mathbf{D})$, where $\mathbf{C} = (3,3) \in \overline{R}^2_+$ and $\mathbf{D} = (2,2,2,0) \in \overline{R}^4_+$, which is not uniform either:

$$\begin{array}{c} \\ 3 \\ 3 \end{array} \begin{array}{c} \begin{array}{cccc} 2 & 2 & 2 & 0 \end{array} \\ \begin{pmatrix} 2 & 0 & 1 & 0 \\ 0 & 2 & 1 & 0 \end{pmatrix} \end{array}$$

$$Y(\mathbf{C}, \mathbf{D})$$

Functional (1.9) from Chapter 1 for the matrix $Y(\mathbf{C}, \mathbf{D})$ does not satisfy (2.4): $\Phi_3\left(Y(\mathbf{C}, \mathbf{D})\right) = 9$, whereas the sum of the largest components of the vectors $\mathbf{C}$ and $\mathbf{D}$ is 5.

Lemma 2.8. *Suppose that a matrix $X(\mathbf{A}, \mathbf{B}) = (x_{ij})$ with nonnegative elements, where $\mathbf{A} = (a_1, \ldots, a_n) \in R^n_+$ and $\mathbf{B} = (b_1, \ldots, b_m) \in R^m_+$, is not uniform. Then it contains a submatrix $Y(\mathbf{C}, \mathbf{D}) = (x_{i_r j_t})$ $(1 \le r \le p$, $1 \le t \le q)$ with p rows and q columns (which may be identical with $Y(\mathbf{C}, \mathbf{D}) = X(\mathbf{A}, \mathbf{B})$), where $\mathbf{C} = \mathbf{C}(Y) = (c_{i_1}, \ldots, c_{i_p})$ and $\mathbf{D} = \mathbf{D}(Y) = (d_{j_1}, \ldots, d_{j_q})$, such that functional* (1.9) *from Chapter* 1 *for the matrix $Y(\mathbf{C}, \mathbf{D})$ satisfies the inequality $\Phi_3\left(Y(\mathbf{C}, \mathbf{D})\right) > \max\{c_{i_r} : 1 \le r \le p\} + \max\{d_{j_t} : 1 \le t \le q\}$.*

Proof. By Lemma 2.7, there exists at least one row or column in the matrix $X(\mathbf{A}, \mathbf{B})$ in which the uniformity is violated. For definiteness, let it be the row having the smallest number i_1: $x_{i_1 j_1} < x_{i_1 j_2}$ for $b_{j_1} \ge b_{j_2}$. The inequality $b_{j_1} \ge b_{j_2}$ implies that there exists an index i_2 such that $x_{i_2 j_1} < x_{i_2 j_2}$. Without loss of generality, suppose that $i_1 < i_2$. Consider a submatrix with two rows and two columns $Y(\mathbf{C}, \mathbf{D}) = (x_{i_r j_t})$ $(1 \le r, t \le 2)$. Then $\mathbf{C} = \mathbf{C}(Y) = (c_{i_1}, c_{i_2})$ and $\mathbf{D} = \mathbf{D}(Y) = (d_{j_1}, d_{j_2})$, where

$$c_{i_1} = x_{i_1 j_1} + x_{i_1 j_2}, \quad c_{i_2} = x_{i_2 j_1} + x_{i_2 j_2},$$

$$d_{i_1} = x_{i_1 j_1} + x_{i_2 j_1}, \quad d_{i_2} = x_{i_1 j_2} + x_{i_2 j_2}.$$

By virtue of the inequalities $x_{i_1j_1} < x_{i_1j_2}$ and $x_{i_2j_1} < x_{i_2j_2}$, functional (1.9) from Chapter 1 for the submatrix $Y(\mathbf{C},\mathbf{D})$ can be written as

$$\Phi_3\left(Y(\mathbf{C},\mathbf{D})\right) = \max_r x_{i_rj_1} + \max_r x_{i_rj_2} + \max_t x_{i_1j_t} + \max_t x_{i_2j_t} =$$
$$\sum_{t=1}^{2} \max_r x_{i_rj_t} + x_{i_1j_2} + x_{i_2j_1} = \Phi_2\left(Y(\mathbf{C},\mathbf{D})\right) + x_{i_1j_2} + x_{i_2j_1} \geq$$
$$\max\{c_{i_r} : 1 \leq r \leq 2\} + x_{i_1j_2} + x_{i_2j_1} >$$
$$\max\{c_{i_r} : 1 \leq r \leq 2\} + \max\{x_{i_1j_1} + x_{i_2j_1}, x_{i_1j_2} + x_{i_2j_2}\} =$$
$$\max\{c_{i_r} : 1 \leq r \leq 2\} + \max\{d_{j_t} : 1 \leq t \leq 2\},$$

which proves the lemma.

Now, we state the final result.

Theorem 2.4. *Suppose $X(\mathbf{A},\mathbf{B}) \in M(\mathbf{A},\mathbf{B})$, where $\mathbf{A} = (a_1,\dots,a_n) \in R_+^n$ and $\mathbf{B} = (b_1,\dots,b_m) \in R_+^m$, and $Y(\mathbf{C},\mathbf{D}) = (x_{i_rj_t})$ $(1 \leq r \leq p, 1 \leq t \leq q)$ is a submatrix of the matrix $X(\mathbf{A},\mathbf{B}) = (x_{ij})$, where $\mathbf{C} = (c_{i_1},\dots,c_{i_p})$ and $\mathbf{D} = (d_{j_1},\dots,d_{j_q})$. Functional* (1.9) *from Chapter* 1 *can be expressed as*

$$\Phi_3\left(Y(\mathbf{C},\mathbf{D})\right) = \max\left\{c_{i_r} : 1 \leq r \leq p\right\} + \max\left\{d_{j_t} : 1 \leq t \leq m\right\}$$

for any submatrix $Y(\mathbf{C},\mathbf{D})$ if and only if $X(\mathbf{A},\mathbf{B}) \in \widetilde{M}(\mathbf{A},\mathbf{B})$; that is, $X(\mathbf{A},\mathbf{B})$ is a uniform matrix.

2.5. Minimization by Means of Uniform Matrices

In what follows, we use uniform matrices to minimize functionals (1.12) and (1.13) form Chapter 1. Let us calculate the values of these functionals for a uniform matrix. Recall that $\mathfrak{M}(X)$ is the set of all submatrices of a matrix X, and the functionals defined by (1.12) and (1.13) in Chapter 1 have the form

$$\Phi_6\left(X(\mathbf{A},\mathbf{B})\right) = \sum_{Y \in \mathfrak{M}\left(X(\mathbf{A},\mathbf{B})\right)} \Phi_2(Y),$$

$$\Phi_7\left(X(\mathbf{A},\mathbf{B})\right) = \sum_{Y \in \mathfrak{M}\left(X(\mathbf{A},\mathbf{B})\right)} \Phi_3\left(X(\mathbf{A},\mathbf{B})\right).$$

Suppose $X(\mathbf{A},\mathbf{B}) \in \widetilde{M}(\mathbf{A},\mathbf{B})$, where $\mathbf{A} = (a_1,\dots,a_n) \in R_+^n$ and $\mathbf{B} = (b_1,\dots,b_m) \in R_+^m$, and $Y(\mathbf{C},\mathbf{D}) = (x_{i_rj_t})$ $(1 \leq r \leq p,\ 1 \leq t \leq q)$ is an arbitrary submatrix of $X(\mathbf{A},\mathbf{B}) = (x_{ij})$, with $\mathbf{C} = (c_{i_1},\dots,c_{i_p})$ and $\mathbf{D} = (d_{j_1},\dots,d_{j_q})$. By Lemma 2.5, $Y(\mathbf{C},\mathbf{D}) \in \widetilde{M}(\mathbf{A},\mathbf{B})$ (i.e., $Y(\mathbf{C},\mathbf{D})$ is a uniform matrix). Accordingly, (2.4) implies that formulas (2.8) are equivalent to the following expressions for functionals (1.8) and (1.9) from Chapter 1:

$$\Phi_2\left(Y(\mathbf{C},\mathbf{D})\right) = \sum_{t=1}^{q} x_{i_k j_t}, \quad \Phi_3\left(Y(\mathbf{C},\mathbf{D})\right) = \sum_{t=1}^{q} x_{i_k j_t} + \sum_{r=1}^{p} x_{i_r j_l}, \tag{2.9}$$

where $c_{i_k} = \max\{c_{i_r} : 1 \le r \le p\}$ and $d_{j_l} = \max\{d_{j_t} : 1 \le t \le q\}$.

If a matrix X_1 is constructed by permuting rows and columns in the matrix X, then it is clear that $\Phi_6(X_1) = \Phi_6(X)$ and $\Phi_7(X_1) = \Phi_7(X)$ (here, X may not be a uniform matrix). Consequently, $\Phi_6(\widetilde{X}(\widetilde{\mathbf{A}},\widetilde{\mathbf{B}})) = \Phi_6(X(\mathbf{A},\mathbf{B}))$ and $\Phi_7(\widetilde{X}(\widetilde{\mathbf{A}},\widetilde{\mathbf{B}})) = \Phi_7(X(\mathbf{A},\mathbf{B}))$, where the matrix $\widetilde{X}(\widetilde{\mathbf{A}},\widetilde{\mathbf{B}})$ is constructed by permuting the rows and columns of $X(\mathbf{A},\mathbf{B})$ so that the components of vectors $\widetilde{\mathbf{A}}$ and $\widetilde{\mathbf{B}}$ are arranged in nonascending order. Therefore, to simplify the calculation of functionals (1.12) and (1.13) from Chapter 1, we assume that $\mathbf{A} \in \overline{R}^n_+$ and $\mathbf{B} \in \overline{R}^m_+$ without loss of generality. The expressions in (2.9) are then rewritten as

$$\Phi_2\left(Y(\mathbf{C},\mathbf{D})\right) = \sum_{t=1}^{q} x_{i_1 j_t}, \quad \Phi_3\left(Y(\mathbf{C},\mathbf{D})\right) = \sum_{t=1}^{q} x_{i_1 j_t} + \sum_{r=1}^{p} x_{i_r j_1}, \tag{2.10}$$

because $c_{i_k} = \max\{c_{i_r} : 1 \le r \le p\}$ and $d_{j_l} = \max\{d_{j_t} : 1 \le t \le q\}$ for uniform submatrices under consideration. According to (2.10), to calculate functionals (1.8) and (1.9) from Chapter 1 for a uniform submatrix, it is sufficient to know only the elements of the first row and first column of the submatrix.

Consider functional (1.12) from Chapter 1 for a uniform submatrix $X(\mathbf{A},\mathbf{B})$, where $\mathbf{A} \in \overline{R}^n_+$ and $\mathbf{B} \in \overline{R}^m_+$. Represent the set of submatrices of X as

$$\mathfrak{M}\left(X(\mathbf{A},\mathbf{B})\right) = \sum_{k=1}^{n} \mathfrak{M}_k\left(X(\mathbf{A},\mathbf{B})\right), \tag{2.11}$$

where $\mathfrak{M}_k\left(X(\mathbf{A},\mathbf{B})\right)$ is the set of those submatrices of $X(\mathbf{A},\mathbf{B})$ whose first rows are constructed by rearranging the kth row of the matrix $X(\mathbf{A},\mathbf{B})$. Note that $\mathfrak{M}_n\left(X(\mathbf{A},\mathbf{B})\right)$ is the set of all submatrices from $\mathfrak{M}\left(X(\mathbf{A},\mathbf{B})\right)$ (which are row vectors) constructed by rearranging the last row of $X(\mathbf{A},\mathbf{B})$. It well known that the number of all nonempty sets of an n-element set is $2^n - 1$. Any submatrix of a matrix $X = (x_{ij})$ $(1 \le i \le n,\ 1 \le j \le m)$ is defined by the nonempty subsets of rows and columns. Therefore, the number of (nonempty) matrices of $X(\mathbf{A},\mathbf{B})$ is $\left|\mathfrak{M}\left(X(\mathbf{A},\mathbf{B})\right)\right| = (2^n-1)(2^m-1)$. Since every submatrix from $\mathfrak{M}_k\left(X(\mathbf{A},\mathbf{B})\right)$ contains a subrow of the kth row of $X(\mathbf{A},\mathbf{B})$, and the remaining rows of the submatrix are determined by the rows of $X(\mathbf{A},\mathbf{B})$ whose numbers are greater than k (the submatrix may be a row vector), the number of submatrices in $\mathfrak{M}_k\left(X(\mathbf{A},\mathbf{B})\right)$ is $\left|\mathfrak{M}_k\left(X(\mathbf{A},\mathbf{B})\right)\right| = (2^m - 1)2^{n-k}$.

Calculating the value of functional (1.12) from Chapter 1 for a uniform matrix $X(\mathbf{A},\mathbf{B})$ by (2.11), we obtain

$$\Phi_6\left(X(\mathbf{A},\mathbf{B})\right)=\sum_{k=1}^{n}\sum_{Y\in\mathfrak{M}_k(X(\mathbf{A},\mathbf{B}))}\Phi_2(Y).$$

Using the first expression in (2.10) and the fact that $i_1=k$ for every submatrix $Y\in\mathfrak{M}_k\left(X(\mathbf{A},\mathbf{B})\right)$, we rewrite functional (1.12) from Chapter 1 as

$$\Phi_6\left(X(\mathbf{A},\mathbf{B})\right)=\sum_{k=1}^{n}\sum_{Y\in\mathfrak{M}_k(X(\mathbf{A},\mathbf{B}))}\sum_{t=1}^{q}x_{kj_t}. \tag{2.12}$$

Consider the inner double sum in this formula:

$$\sum_{Y\in\mathfrak{M}_k(X(\mathbf{A},\mathbf{B}))}\Phi_2(Y)=\sum_{Y\in\mathfrak{M}_k(X(\mathbf{A},\mathbf{B}))}\sum_{t=1}^{q}x_{kj_t}. \tag{2.13}$$

Let x_{kj} be any element of the kth row of the matrix $X(\mathbf{A},\mathbf{B})$. Let us count the number of times that x_{kj} appears on the right-hand side of (2.13), which is equal to the number of submatrices from $\mathfrak{M}_k\left(X(\mathbf{A},\mathbf{B})\right)$ containing a certain element x_{kj}. This element is contained in 2^{m-1} different rows and is the first element of 2^{n-k} different columns of the submatrices under consideration. Note that x_{kj} can represent a row or a column. Therefore, the number of different submatrices from $\mathfrak{M}_k\left(X(\mathbf{A},\mathbf{B})\right)$ containing the element x_{kj} is $2^{m-1}2^{n-k}=2^{m+n-k-1}$, which is independent of the second index j. Thus, it follows from (2.13) that

$$\sum_{Y\in\mathfrak{M}_k(X(\mathbf{A},\mathbf{B}))}\Phi_2(Y)=2^{m-1}2^{n-k}\sum_{j=1}^{m}x_{kj}=2^{m-1}2^{n-k}a_k.$$

Formula (2.12) yields

$$\Phi_6\left(X(\mathbf{A},\mathbf{B})\right)=\sum_{k=1}^{n}2^{m-1}2^{n-k}a_k.$$

In summary, we have the following proposition.

Lemma 2.9. *Functional* (1.9) *from Chapter* 1 *has the same value for all uniform matrices* $X(\mathbf{A},\mathbf{B})$ *from* $\widetilde{M}(\mathbf{A},\mathbf{B})$, *where* $\mathbf{A}=(a_1,\ldots,a_n)\in R_+^n$ *and* $\mathbf{B}=(b_1,\ldots,b_m)\in R_+^m$. *If* $\widetilde{\mathbf{A}}=(\tilde{a}_1,\ldots,\tilde{a}_n)$ *is a vector constructed by arranging the components of* $\mathbf{A}$ *in nonascending order, then*

$$\Phi_6\left(X(\mathbf{A},\mathbf{B})\right)=2^{m-1}\sum_{k=1}^{n}2^{n-k}\tilde{a}_k. \tag{2.14}$$

Now, functional (1.13) from Chapter 1 is readily calculated for a uniform matrix. Since functionals (1.12) and (1.13) from Chapter 1 are related by the equation $\Phi_7\left(X(\mathbf{A},\mathbf{B})\right) = \Phi_6\left(X(\mathbf{A},\mathbf{B})\right) + \Phi_6\left(X(\mathbf{B},\mathbf{A})\right)$ for any matrix from $M(\mathbf{A},\mathbf{B})$, (2.14) leads to the following proposition.

Lemma 2.10. *Functional* (1.13) *from Chapter* 1 *has the same value for all uniform matrices from* $X(\mathbf{A},\mathbf{B})$ *from* $\widetilde{M}(\mathbf{A},\mathbf{B})$, *where* $\mathbf{A} = (a_1,\ldots,a_n) \in R_+^n$ *and* $\mathbf{B} = (b_1,\ldots,b_m) \in R_+^m$. *Moreover, if* $\widetilde{\mathbf{A}} = (\tilde{a}_1,\ldots,\tilde{a}_n)$ *and* $\widetilde{\mathbf{B}} = (\tilde{b}_1,\ldots,\tilde{b}_m)$ *are vectors constructed by arranging the components of* $\mathbf{A}$ *and* $\mathbf{B}$ *in nonascending order, then*

$$\Phi_7\left(X(\mathbf{A},\mathbf{B})\right) = 2^{m-1}\sum_{k=1}^{n} 2^{n-k}\tilde{a}_k + 2^{n-1}\sum_{k=1}^{m} 2^{m-k}\tilde{b}_k. \tag{2.15}$$

Consider functional (1.12) from Chapter 1 for an arbitrary matrix $X(\mathbf{A},\mathbf{B})$ from $M(\mathbf{A},\mathbf{B})$:

$$\Phi_6\left(X(\mathbf{A},\mathbf{B})\right) = \sum_{Y\in\mathfrak{M}\left(X(\mathbf{A},\mathbf{B})\right)} \Phi_2(Y).$$

For an arbitrary submatrix $Y(\mathbf{C},\mathbf{D}) \in \mathfrak{M}\left(X(\mathbf{A},\mathbf{B})\right)$, functional (1.8) from Chapter 1 can be expressed as (cf. (2.9) and (2.10))

$$\Phi_2\left(Y(\mathbf{C},\mathbf{D})\right) = \sum_{t=1}^{q} \max x_{i_r j_t} \geq \sum_{t=1}^{q} x_{i_k j_t}.$$

Since $\mathbf{A} \in \overline{R}_+^n$ and $\mathbf{B} \in \overline{R}_+^m$ (which does not restrict the generality of calculation of these functionals),

$$\Phi_2\left(Y(\mathbf{C},\mathbf{D})\right) \geq \sum_{t=1}^{q} x_{i_1 j_t}.$$

Therefore, functional (1.12) from Chapter 1 satisfies, instead of (2.12), the inequality

$$\Phi_6\left(X(\mathbf{A},\mathbf{B})\right) \geq \sum_{k=1}^{n} \sum_{Y\in\mathfrak{M}\left(X(\mathbf{A},\mathbf{B})\right)} \sum_{t=1}^{q} x_{k j_t}.$$

Accordingly, for an arbitrary matrix $X(\mathbf{A},\mathbf{B})$ from $M(\mathbf{A},\mathbf{B})$, we have the inequality

$$\Phi_6\left(X(\mathbf{A},\mathbf{B})\right) \geq 2^{m-1}\sum_{k=1}^{n} 2^{n-k} a_k, \tag{2.16}$$

which can be readily shown to be a strict inequality when the matrix $X(\mathbf{A},\mathbf{B})$ contains a column that is not uniform with respect to the vector $\mathbf{A}$.

Since functionals (1.12) and (1.13) from Chapter 1 are related by the equation $\Phi_7(X(\mathbf{A},\mathbf{B})) = \Phi_6(X(\mathbf{A},\mathbf{B})) + \Phi_6(X(\mathbf{B},\mathbf{A}))$,

$$\Phi_7(X(\mathbf{A},\mathbf{B})) \geq 2^{m-1}\sum_{k=1}^{n} 2^{n-k} a_k + 2^{n-1}\sum_{k=1}^{m} 2^{m-k} b_k, \qquad (2.17)$$

which is a strict inequality when $X(\mathbf{A},\mathbf{B}) \notin \widetilde{M}(\mathbf{A},\mathbf{B})$ by Lemma 2.8 or Theorem 2.4.

Combining Lemmas 2.9 and 2.10 with relations (2.14)–(2.17), we state the final result as a theorem about minimization of objective functions (1.12) and (1.13) from Chapter 1.

Theorem 2.5. *Suppose vectors* $\mathbf{A} = (a_1,\ldots,a_n) \in R_+^n$ *and* $\mathbf{B} = (b_1,\ldots,b_m) \in R_+^m$ *are characterized by equal sums of components. Any uniform matrix* $\overline{X}(\mathbf{A},\mathbf{B})$ *minimizes the functionals* $\Phi_6(X(\mathbf{A},\mathbf{B}))$ *and* $\Phi_7(X(\mathbf{A},\mathbf{B}))$:

$$\min_{X(\mathbf{A},\mathbf{B})\in M(\mathbf{A},\mathbf{B})} \Phi_6(X(\mathbf{A},\mathbf{B})) = \Phi_6(\overline{X}(\mathbf{A},\mathbf{B})) = 2^{m-1}\sum_{k=1}^{n} 2^{n-k}\tilde{a}_k,$$

$$\min_{X(\mathbf{A},\mathbf{B})\in M(\mathbf{A},\mathbf{B})} \Phi_7(X(\mathbf{A},\mathbf{B})) = \Phi_7(\overline{X}(\mathbf{A},\mathbf{B})) =$$

$$2^{m-1}\sum_{k=1}^{n} 2^{n-k}\tilde{a}_k + 2^{n-1}\sum_{k=1}^{m} 2^{m-k}\tilde{b}_k,$$

where the vectors $\widetilde{\mathbf{A}} = (\tilde{a}_1,\ldots,\tilde{a}_n)$ *and* $\widetilde{\mathbf{B}} = (\tilde{b}_1,\ldots,\tilde{b}_m)$ *are constructed by arranging the components of* $\mathbf{A}$ *and* $\mathbf{B}$ *in nonascending order. In particular, the hereditarily minimax matrix* $\overset{*}{X}(\mathbf{A},\mathbf{B})$ *also minimizes functionals* (1.12) *and* (1.13) *from Chapter* 1.

§3. Hereditarily Minimax Matrices in Transportation Problems with Minimax Criteria

In this section, we minimize the remaining minimax functionals (1.14)–(1.16) defined in Chapter 1. As a result, we obtain a unique solution given by a hereditarily minimax matrix.

3.1. The Largest Elements of Submatrices of the Transportation Matrix and Uniformity

An analysis of the values of functional (1.7) from Chapter 1 for submatrices shows that functional (1.14) from Chapter 1 is minimized by a certain uniform matrix.

Consider the functional

$$\Phi_8\left(X(\mathbf{A},\mathbf{B})\right) = \sum_{Y\in\mathfrak{M}(X(\mathbf{A},\mathbf{B}))} \Phi_1(Y) = \sum_{Y\in\mathfrak{M}(X(\mathbf{A},\mathbf{B}))} \max\{x_{i_r j_t} : (x_{i_r j_t}) = Y\}.$$

Its value can be readily calculated when $X(\mathbf{A},\mathbf{B})$ is a uniform matrix.

Lemma 3.1. *Suppose that $X = X(\mathbf{A},\mathbf{B}) = (x_{ij})$ is a uniform matrix, where $\mathbf{A} \in R_+^n$ and $\mathbf{B} \in R_+^m$, and the matrix $\widetilde{X} = (\tilde{x}_{ij})$ is constructed by arranging the rows and columns of X so that $\widetilde{X} \in M(\widetilde{\mathbf{A}},\widetilde{\mathbf{B}})$ ($\widetilde{\mathbf{A}} \in \overline{R}_+^n$, $\widetilde{\mathbf{B}} \in \overline{R}_+^m$). Then,*

$$\Phi_8\left(X(\mathbf{A},\mathbf{B})\right) = \Phi_8(\widetilde{X}) = \sum_{i=1}^{n}\sum_{j=1}^{m} 2^{n+m-i-j}\tilde{x}_{ij}. \tag{3.1}$$

Proof. It is obvious that $\Phi_8\left(X(\mathbf{A},\mathbf{B})\right) = \Phi_8(\widetilde{X})$. Since $\widetilde{X} = (\tilde{x}_{ij})$ is a uniform matrix, $\widetilde{\mathbf{A}} \in \overline{R}_+^n$, and $\widetilde{\mathbf{B}} \in \overline{R}_+^m$, $\tilde{x}_{ij} \geq \tilde{x}_{pq}$ for $i \leq p$ and $j \leq q$. Therefore, the largest element of any submatrix $\widetilde{Y} = (\tilde{x}_{i_r j_t})$ from $\mathfrak{M}(\widetilde{X})$ equals $\tilde{x}_{i_1 j_1}$: $\Phi_1(\widetilde{Y}) = \tilde{x}_{i_1 j_1}$. Accordingly, to calculate the functional $\Phi_8(\widetilde{X})$ (or $\Phi_8\left(X(\mathbf{A},\mathbf{B})\right)$) for arbitrary i and j ($1 \leq i \leq n$, $1 \leq j \leq m$), one must determine the number of submatrices from $\mathfrak{M}(\widetilde{X})$ whose first rows and columns correspond to the indices i and j.

Denote by $\mathfrak{M}_{ij}(\widetilde{X}) = \{\widetilde{Y} = (\tilde{x}_{i_r j_t}) : i_1 = i,\ j_1 = j\}$ the subset of submatrices from $\mathfrak{M}(\widetilde{X})$ for which i and j correspond to the first row and the first column, respectively. Any matrix from $\mathfrak{M}_{ij}(\widetilde{X})$ is defined by an arbitrary set of rows (which may be empty) with numbers greater than i and by an arbitrary set of columns (which may be empty) with numbers greater than j. Therefore, the number of submatrices in the set $\mathfrak{M}_{ij}(\widetilde{X})$ is $|\mathfrak{M}_{ij}(\widetilde{X})| = 2^{n-i}2^{m-j} = 2^{n+m-i-j}$. Since $\Phi_1(\widetilde{Y}) = \tilde{x}_{ij}$ for every $\widetilde{Y} \in \mathfrak{M}_{ij}(\widetilde{X})$, we have formula (3.1).

The following lemma states conditions for matrices from $M(\mathbf{A},\mathbf{B})$ under which the value of functional (1.14) from Chapter 1 can be reduced.

Lemma 3.2. *Suppose that a matrix $X = (x_{ij})$ from $M(\mathbf{A},\mathbf{B})$, where $\mathbf{A} \in R_+^n$ and $\mathbf{B} \in R_+^m$ ($n \geq 2$, $m \geq 2$), contains four elements x_{ku}, x_{kv}, x_{lu}, and*

x_{lv} *satisfying either of the following conditions:*

$$\begin{aligned}
&\text{(a)}\quad x_{ku} < x_{kv},\ x_{lu} > x_{lv} \text{ and} \\
&\qquad x_{kv} = \max\{x_{iv} : x_{iu} < x_{iv},\ 1 \le i \le n\}, \\
&\qquad x_{lu} = \max\{x_{iu} : x_{iu} > x_{iv},\ 1 \le i \le n\}; \\
&\text{(b)}\quad x_{ku} < x_{lu},\ x_{lv} > x_{lv} \text{ and} \\
&\qquad x_{lu} = \max\{x_{lj} : x_{kj} < x_{lj},\ 1 \le j \le m\}, \\
&\qquad x_{kv} = \max\{x_{kj} : x_{kj} > x_{lj},\ 1 \le j \le m\}.
\end{aligned} \tag{3.2}$$

When condition (a) *or* (b) *holds, define* $\alpha = \left(\min\{x_{kv} - x_{ku},\ x_{lu} - x_{lv}\}\right)/2$ *or* $\alpha = \left(\min\{x_{lu} - x_{ku},\ x_{kv} - x_{lv}\}\right)/2$, *respectively. Then, if the inequalities*

$$\begin{aligned}
&\beta < \min\{x_{ij} - x_{ku} : x_{ij} > x_{ku},\ 1 \le i \le n,\ 1 \le j \le m\}, \\
&\beta < \min\{x_{ij} - x_{lv} : x_{ij} > x_{lv},\ 1 \le i \le n,\ 1 \le j \le m\}, \\
&\beta < \min\{x_{kv} - x_{ij} : x_{kv} > x_{ij},\ 1 \le i \le n,\ 1 \le j \le m\}, \\
&\beta < \min\{x_{lu} - x_{ij} : x_{lu} > x_{ij},\ 1 \le i \le n,\ 1 \le j \le m\},
\end{aligned} \tag{3.3}$$

are satisfied for $\beta \in R$ *and* $0 < \beta < \alpha$, *then* $X = (x'_{ij}) \in M(\mathbf{A}, \mathbf{B})$, *where*

$$x'_{ij} = \begin{cases}
x_{ku} + \beta, & i = k, \quad j = u, \\
x_{kv} - \beta, & i = k, \quad j = v, \\
x_{lu} - \beta, & i = l, \quad j = u, \\
x_{lv} + \beta, & i = l, \quad j = v, \\
x_{ij}, & (i, j) \notin \{(k, u), (k, v), (l, u), (l, v)\},
\end{cases} \tag{3.4}$$

and $\Phi_8(X') < \Phi_8(X)$ *for functional* (1.14) *from Chapter* 1.

Proof. It is obvious that the sums of elements of the rows and columns of a matrix $X = (x_{ij})$ from $M(\mathbf{A}, \mathbf{B})$ are preserved under the transformation defined by (3.4), and conditions (3.3) guarantee that the elements of the matrix $X' = (x'_{ij})$ are nonnegative. Thus, $X' \in M(\mathbf{A}, \mathbf{B})$.

There exists a one-to-one correspondence $f : \mathfrak{M}(X) \to \mathfrak{M}(X')$ between submatrices from $\mathfrak{M}(X)$ and $\mathfrak{M}(X')$, where $f(Y) = Y'$, if the submatrices Y from $\mathfrak{M}(X)$ and Y' from $\mathfrak{M}(X')$ are defined by equal number of rows and columns: $Y = (x_{i_r j_t})$ and $Y' = (x'_{i_r j_t})$ ($1 \le i_r \le n$, $1 \le j_t \le m$, $1 \le r \le p$, $1 \le t \le q$).

The four elements singled out in the matrices X and X', with row numbers k and l and column numbers u and v, can be combined to construct (2^2-1)-by-$(2^2-1) = 9$ submatrices from either $\mathfrak{M}(X)$ or $\mathfrak{M}(X')$, with the dimensions

2-by-2, 2-by-1, 1-by-2, and 1-by-1. We introduce a special notation for these submatrices:

$$Y_1 = \begin{pmatrix} x_{ku} & x_{kv} \\ x_{lu} & x_{lv} \end{pmatrix}, \quad Y_1' = \begin{pmatrix} x'_{ku} & x'_{kv} \\ x'_{lu} & x'_{lv} \end{pmatrix},$$

$$Y_2 = (x_{ku}, x_{kv}), \quad Y_2' = (x'_{ku}, x'_{kv}), \quad Y_3 = (x_{lu}, x_{lv}), \quad Y_3' = (x'_{lu}, x'_{lv}),$$

$$Y_4 = \begin{pmatrix} x_{ku} \\ x_{lu} \end{pmatrix}, \quad Y_4' = \begin{pmatrix} x'_{ku} \\ x'_{lu} \end{pmatrix}, \quad Y_5 = \begin{pmatrix} x_{kv} \\ x_{lv} \end{pmatrix}, \quad Y_5' = \begin{pmatrix} x'_{kv} \\ x'_{lv} \end{pmatrix},$$

$$Y_6 = (x_{ku}), \quad Y_6' = (x'_{ku}), \quad Y_7 = (x_{kv}), \quad Y_7' = (x'_{kv}),$$

$$Y_8 = (x_{lu}), \quad Y_8' = (x'_{lu}), \quad Y_9 = (x_{lv}), \quad Y_9' = (x_{lv'}).$$

Compare the values of functional (1.7) from Chapter 1 for the submatrices $Y \in \mathfrak{M}(X)$ and $Y' \in \mathfrak{M}(X')$, where $f(Y) = Y'$, as depending on the relation between these submatrices and the submatrices Y_s and Y_s', where $1 \leq s \leq 9$.

If Y_1 is a submatrix of Y, and Y_1' is a submatrix of Y', it is clear that $\Phi_1(Y') \leq \Phi_1(Y)$ and $\Phi_1(Y_1') < \Phi_1(Y_1)$.

Suppose that Y_1 and Y_1' are not submatrices of Y and Y', but the submatrices Y_s and Y_s' are submatrices of Y and Y', respectively, for a certain s, $2 \leq s \leq 5$. Then, it is obvious that $\Phi_1(Y') \leq \Phi_1(Y)$ and $\Phi_1(Y_s') < \Phi_1(Y_s)$ for any s, $2 \leq s \leq 5$.

Consider the submatrices from $\mathfrak{M}(X)$ and $\mathfrak{M}(X')$ that contain only one of the singled-out elements of X and X'. Without loss of generality, assume that the four elements x_{ku}, x_{kv}, x_{lu}, and x_{lv} satisfy condition (a) of Lemma 3.2. Single out the submatrices from $\mathfrak{M}(X)$ and $\mathfrak{M}(X')$ that contain the 1-element submatrices Y_6, Y_7 and Y_6', Y_7' and do not contain the remaining 1-element submatrices Y_8, Y_9 and Y_8', Y_9'; and denote the corresponding subsets by $\mathfrak{M}(X;6)$, $\mathfrak{M}(X;7)$ and $\mathfrak{M}(X';6)$, $\mathfrak{M}(X';7)$, respectively.

Consider the one-to-one correspondences $\varphi_1 : \mathfrak{M}(X;6) \to \mathfrak{M}(X;7)$ and $\varphi_2 : \mathfrak{M}(X';6) \to \mathfrak{M}(X';7)$ between the matrix sets $M(X;6)$, $\mathfrak{M}(X;7)$ and $M(X';6)$, $\mathfrak{M}(X';7)$, respectively. We have $\varphi_1\left(Y(6)\right) = Y(7)$ and $\varphi_2\left(Y'(6)\right) = Y'(7)$ if and only if the submatrices $Y(6) \in \mathfrak{M}(X;6)$, $Y(7) \in \mathfrak{M}(X;7)$, $Y'(6) \in \mathfrak{M}(X';6)$, and $Y'(7) \in \mathfrak{M}(X';7)$ differ only by their uth and vth columns. Then the submatrices of $Y(6)$, $Y(7)$ and $Y'(6)$, $Y'(7)$ formed by removing the uth and vth columns, respectively, are identical.

Let $\varphi : \left\{\left(Y(6), Y(7)\right)\right\} \to \left\{\left(Y'(6), Y'(7)\right)\right\}$ be the one-to-one correspondence between the sets of submatrix pairs $\left\{\left(Y(6), Y(7)\right)\right\}$ and $\left\{\left(Y'(6), Y'(7)\right)\right\}$, where $\varphi_1\left(Y(6)\right) = Y(7)$ and $\varphi_2\left(Y'(6)\right) = Y'(7)$. We have $\varphi : \left(\left(Y(6), Y(7)\right)\right) = \left(Y'(6), Y'(7)\right)$ if and only if the submatrices $Y(6)$ and $Y'(6)$ differ only by elements contained in the kth row and uth column. This implies that the submatrices $Y(7)$ and $Y'(7)$ differ only by elements contained in the kth row and vth column.

Consider the sums of the functionals (1.7) from Chapter 1 calculated for the submatrices that constitute the pairs $\left(Y(6), Y(7)\right)$ and $\left(Y'(6), Y'(7)\right)$, where $\varphi\left(\left(Y(6), Y(7)\right)\right) = \left(Y'(6), Y'(7)\right)$:

$$\Phi_1\left(Y(6)\right) + \Phi_1\left(Y(7)\right) \text{ and } \Phi_1\left(Y'(6)\right) + \Phi_1\left(Y'(7)\right).$$

Suppose $\Phi_1\left(Y(6)\right) = \alpha_1 \geq x_{ku}$, $\Phi_1\left(Y(7)\right) = \alpha_2 \geq x_{kv}$, and $\Phi_1\left(Y(6)\right) + \Phi_1\left(Y(7)\right) = \alpha_1 + \alpha_2$. The cases $\alpha_1 > x_{ku}$ and $\alpha_1 = x_{ku}$ can be considered separately.

When $\alpha_1 > x_{ku}$, we have $\Phi_1\left(Y'(6)\right) = \alpha_1$, because β is sufficiently small by conditions (3.3). When $\alpha_2 > x_{kv}$, it is obvious that $\Phi_1\left(Y'(7)\right) = \alpha_2$. If $\alpha_2 = x_{kv}$, then it is clear that $\Phi_1\left(Y'(7)\right) \leq \alpha_2$. Accordingly, $\Phi_1\left(Y'(7)\right) = x_{kv}$ if the submatrix $Y'(7)$ contains at least two elements equal to x_{kv} and $\Phi_1\left(Y'(7)\right) = x_{kv} - \beta$ if $Y'(7)$ contains only one element equal to x_{kv}. As a result, we have $\Phi_1\left(Y'(6)\right) + \Phi_1\left(Y'(7)\right) \leq \alpha_1 + \alpha_2$ when $\alpha_1 > x_{ku}$.

When $\alpha_1 = x_{ku}$, it is obvious that $\Phi_1\left(Y'(6)\right) = x_{ku} + \beta$. Let us show that this entails $\alpha_2 = x_{kv}$. Suppose that $\alpha_2 > x_{kv}$. If the element α_2 of $Y(7)$ is not contained in the vth column, then $\alpha_1 = \alpha_2 > x_{ku}$, which contradicts the condition $\alpha_1 = x_{ku}$. If the element α_2 is contained in the vth column of the submatrix $Y(7)$, that is, $\alpha_2 = x_{k*v}$, then conditions (3.2) entail $x_{k*u} \geq \alpha_2 > x_{ku}$ and $\Phi_1\left(Y(6)\right) \alpha_1 > x_{ku}$, which results in a contradiction. Thus, $\alpha_1 = x_{ku}$ implies $\alpha_2 = x_{kv}$. Moreover, $Y(7)$ does not contain other elements equal to x_{kv}; otherwise, we would again come to a contradiction with the equation $\alpha_1 = x_{ku}$. Thus, conditions (3.3) imply that $\Phi_1\left(Y'(7)\right) = x_{kv} - \beta$ and $\Phi_1\left(Y'(6)\right) + \Phi_1\left(Y'(7)\right) = \alpha_1 + \alpha_2$.

Consequently,

$$\Phi_1\left(Y'(6)\right) + \Phi_1\left(Y'(7)\right) \leq \Phi_1\left(Y(6)\right) + \Phi_1\left(Y(7)\right),$$

for every pair of submatrices $\left(Y(6), Y(7)\right)$ and $\left(Y'(6), Y'(7)\right)$, where $\varphi((Y(6), Y(7))) = (Y'(6), Y'(7))$.

By a similar reasoning, it is established that $\Phi_1\left(Y'(8)\right) + \Phi_1\left(Y'(9)\right) \leq \Phi_1\left(Y(8)\right) + \Phi_1\left(Y(9)\right)$, where the submatrices contain 1-element submatrices Y_8', Y_9', Y_8, and Y_9 and do not contain 1-element submatrices Y_6', Y_7', Y_6, and Y_7.

Suppose that $f(Y) = Y'$, and Y_s and Y_s' $(1 \leq s \leq 9)$ are not submatrices of Y and Y'. Then, it is obvious that $Y = Y'$ and $\Phi_1(Y') = \Phi_1(Y)$.

The relations obtained for functional (1.7) from Chapter 1 of submatrices from $\mathfrak{M}(X)$ and $\mathfrak{M}(X')$ imply that

$$\Phi_8(X') = \sum_{Y' \in \mathfrak{M}(X')} \Phi_1(Y') < \Phi_8(X) = \sum_{Y \in \mathfrak{M}(X)} \Phi_1(Y),$$

which proves the lemma.

Now, we show that a matrix X with nonnegative elements that is not uniform contains four elements satisfying conditions (a) and (b) of Lemma 3.2.

Lemma 3.3. *Suppose that a matrix $X = (x_{ij})$ from $M(\mathbf{A}, \mathbf{B})$, where $\mathbf{A} = (a_1, \ldots, a_n) \in R_+^n$ and $\mathbf{B} = (b_1, \ldots, b_m) \in R_+^m$ $(n \geq 2,\ m \geq 2)$, is not uniform. Then, the matrix contains rows with numbers k and l and columns with numbers u and v such that the corresponding elements of X satisfy the inequalities $x_{ku} < x_{kv}$ and $x_{lu} > x_{lv}$ or $x_{ku} < x_{lu}$ and $x_{kv} > x_{lv}$.*

Proof. Since X is not a uniform matrix, we can apply Definition 1.1 from Chapter 1 to establish either (or both) of the following facts.

(a) There exist $u, v, k \in Z$ $(1 \leq u, v \leq m,\ u \neq v,\ 1 \leq k \leq n)$ for which either the inequality $b_u > b_v$ entails $x_{ku} < x_{kv}$ or the equality $b_u = b_v$ entails $x_{ku} \neq x_{kv}$. In the latter case, for definiteness, assume that $x_{ku} < x_{kv}$.

(b) There exist $k, l, u \in Z$ $(1 \leq k, l \leq n,\ k \neq l,\ 1 \leq u \leq m)$ for which either the inequality $a_k > a_l$ entails $x_{ku} < x_{lu}$ or the equality $a_k = a_l$ entails $x_{ku} \neq x_{lu}$, where it is assumed that $x_{ku} < x_{lu}$ for definiteness.

Without loss of generality, let (a) be the case. Since

$$b_j = \sum_{i=1}^{n} x_{ij}, \quad 1 \leq j \leq m,$$

relations $b_u > b_v$, $x_{ku} < x_{kv}$ or $b_u = b_v$, $x_{ku} < x_{kv}$ imply that there exists a row of the matrix X, with a number l, such that $x_{lu} > x_{lv}$, which completes the proof.

3.2. Minimization by Hereditarily Minimax Matrices

Now, we minimize the remaining functionals (1.14)–(1.16) from Chapter 1 by means of a hereditarily minimax matrix and give the values of the minimums.

It is clear that the minimization problem for objective function (1.14) from Chapter 1,

$$\Phi_8\left(X(\mathbf{A}, \mathbf{B})\right) \to \min, \quad X(\mathbf{A}, \mathbf{B}) \in M(\mathbf{A}, \mathbf{B})$$

has a solution. Combining Lemmas 3.2 and 3.3, we have the following proposition.

Lemma 3.4. *Let $\mathbf{A} \in R_+^n$ and $\mathbf{B} \in R_+^m$ be arbitrary vectors with equal sums of components. The minimization problem for functional* (1.14) *from*

Chapter 1 *on the matrix set* $M(\mathbf{A},\mathbf{B})$ *has a solution, and if*

$$\min_{X(\mathbf{A},\mathbf{B})\in M(\mathbf{A},\mathbf{B})} \Phi_8\left(X(\mathbf{A},\mathbf{B})\right) = \Phi_8\left(\overline{X}(\mathbf{A},\mathbf{B})\right),$$

then $\overline{X}(\mathbf{A},\mathbf{B})$ *is a uniform matrix:* $\overline{X}(\mathbf{A},\mathbf{B}) \in \widetilde{M}(\mathbf{A},\mathbf{B})$.

The matrix that minimizes functional (1.14) from Chapter 1 on the matrix set $M(\mathbf{A},\mathbf{B})$ was constructed in Algorithm 2.1.

Theorem 3.1. *Suppose vectors* $\mathbf{A} = (a_1,\ldots,a_n) \in R_+^n$ *and* $\mathbf{B} = (b_1,\ldots,b_m) \in R_+^m$ *satisfy the balance condition. The minimization problem for objective function* (1.14) *from Chapter* 1 *on the matrix set* $M(\mathbf{A},\mathbf{B})$ *has a unique solution defined by the hereditarily minimax matrix* $\overset{*}{X}(\mathbf{A},\mathbf{B}) = (\overset{*}{x}_{ij})$:

$$\min_{X(\mathbf{A},\mathbf{B})\in M(\mathbf{A},\mathbf{B})} \Phi_8\left(X(\mathbf{A},\mathbf{B})\right) = \Phi_8(\overset{*}{X}(\mathbf{A},\mathbf{B})).$$

Proof. When $n = 1$, the matrix set $M(\mathbf{A},\mathbf{B})$ contains only one matrix (row vector), which is obviously uniform and is a hereditarily minimax matrix; thus, the assertion of the theorem is proved in this special case.

To complete the proof, we apply mathematical induction on the number of rows in the matrix $X(\mathbf{A},\mathbf{B})$.

Suppose the theorem is true when the number of rows in $X(\mathbf{A},\mathbf{B})$ is not greater than n, and let $\mathbf{A} \in R_+^{n+1}$. Suppose also that

$$\min_{X(\mathbf{A},\mathbf{B})\in M(\mathbf{A},\mathbf{B})} \Phi_8\left(X(\mathbf{A},\mathbf{B})\right) = \Phi_8\left(\overline{X}(\mathbf{A},\mathbf{B})\right). \tag{3.5}$$

By Lemma 3.4, $\overline{X}(\mathbf{A},\mathbf{B}) = (\overline{x}_{ij})$ $(1 \le i \le n+1,\ 1 \le j \le m)$ is a uniform matrix.

Let us show that $\overline{X}(\mathbf{A},\mathbf{B})$ is a hereditarily minimax matrix. Since the value of functional (1.14) from Chapter 1 is invariant with respect to permutation of rows and columns of the matrix, we may consider $\mathbf{A} \in \overline{R}_+^{n+1}$ and $\mathbf{B} \in \overline{R}_+^m$ without loss of generality. Then, the uniformity of $\overline{X}(\mathbf{A},\mathbf{B})$ implies that $x_{ij} \ge x_{kl}$ if $i \le k$ and $j \le l$. Formula (3.1) gives the value of functional (1.14) from Chapter 1 for the matrix $\overline{X}(\mathbf{A},\mathbf{B})$.

Let us represent the set of all submatrices of the matrix $\overline{X} = \overline{X}(\mathbf{A},\mathbf{B})$ as $\mathfrak{M}(\overline{X}) = \mathfrak{M}'(\overline{X}) \cup \mathfrak{M}_{n+1}(\overline{X})$, where $\mathfrak{M}'(\overline{X})$ is the set of submatrices that do not contain elements from the last row of $\overline{X}(\mathbf{A},\mathbf{B})$, and $\mathfrak{M}_{n+1}(\overline{X})$ is the set of submatrices containing elements of the $(n+1)$th row. Then,

$$\Phi_8\left(\overline{X}(\mathbf{A},\mathbf{B})\right) = \sum_{\overline{Y}\in\mathfrak{M}(\overline{X})} \Phi_1(\overline{Y}) = \sum_{\overline{Y}\in\mathfrak{M}'(\overline{X})} \Phi_1(\overline{Y}) + \sum_{\overline{Y}\in\mathfrak{M}_{n+1}(\overline{X})} \Phi_1(\overline{Y}). \tag{3.6}$$

Any submatrix of a uniform matrix is a uniform matrix; moreover, if $\overline{Y} = (\overline{x}_{i_r j_t}) \in \mathfrak{M}(\overline{X})$, where $1 \leq r \leq p$ and $1 \leq t \leq q$, then $\Phi_1(\overline{Y}) = \max\{\overline{x}_{i_r j_t} : 1 \leq r \leq p,\ 1 \leq t \leq q\} = \overline{x}_{i_1 j_1}$. Suppose $\overline{Y} \in \mathfrak{M}_{n+1}(\overline{X})$; then, $i_p = n+1$. We represent the set of submatrices of the matrix $\overline{X}(\mathbf{A}, \mathbf{B})$ that contain elements of the $(n+1)$th row as

$$\mathfrak{M}_{n+1}(\overline{X}) = \mathfrak{M}'_{n+1}(\overline{X}) \cup \mathfrak{M}''_{n+1}(\overline{X}),$$

where every submatrix from $\mathfrak{M}'_{n+1}(\overline{X})$ has at least two rows and every submatrix in $\mathfrak{M}''_{n+1}(\overline{X})$ has only one row (the submatrices belonging to $\mathfrak{M}''_{n+1}(\overline{X})$ are row vectors consisting of the elements of the last row of $\overline{X} = \overline{X}(\mathbf{A}, \mathbf{B})$). There exists a one-to-one correspondence

$$f : \mathfrak{M}'_{n+1}(\overline{X}) \to \mathfrak{M}'(\overline{X}), \quad f(\overline{Y}') = \overline{Y}$$

between the sets $\mathfrak{M}'_{n+1}(\overline{X})$ and $\mathfrak{M}'(\overline{X})$ if and only if $\overline{Y}' = (\overline{x}_{i_r j_t})$ $(1 \leq r \leq p,\ 1 \leq t \leq q,\ i_p = n+1)$ and $\overline{Y} = (\overline{x}_{i_r j_t})$ $(1 \leq r \leq p-1,\ 1 \leq t \leq q)$; note that the submatrix $\overline{Y}$ is obtained by removing the last row from the matrix $\overline{Y}'$. It is obvious that if $f(\overline{Y}') = \overline{Y}$, then $\Phi_1(\overline{Y}) = \Phi_1(\overline{Y}') = \overline{x}_{i_1 j_1}$. Therefore, (3.6) entails

$$\Phi_8\left(\overline{X}(\mathbf{A}, \mathbf{B})\right) = 2 \sum_{\overline{Y} \in \mathfrak{M}'(\overline{X})} \Phi_1(\overline{Y}) + \sum_{\overline{Y} \in \mathfrak{M}''_{n+1}(\overline{X})} \Phi_1(\overline{Y}). \tag{3.7}$$

Let $\overline{X}^* = (\overline{x}_{ij})$ $(1 \leq i \leq n,\ 1 \leq j \leq m)$ be the matrix obtained by removing the last matrix from $\overline{X}$. It is clear that

$$\Phi_8(\overline{X}^*) = \sum_{\overline{Y}^* \in \mathfrak{M}(\overline{X}^*)} \Phi_1(\overline{Y}^*) = \sum_{\overline{Y} \in \mathfrak{M}'(\overline{X})} \Phi_1(\overline{Y}).$$

Now, we show that equality (3.5) implies that $\overline{X}^*$ is a hereditarily minimax matrix.

It can be readily shown for the matrix $\overline{X}^* \in M(\mathbf{A}', \mathbf{B}')$ that the components of $\mathbf{A}' = (a'_1, \ldots, a'_n)$ and $\mathbf{B}' = (b'_1, \ldots, b'_m)$ that satisfy the conditions

$$a'_i = a_i, \quad 1 \leq i \leq n, \quad b'_j = b_j - x_{n+1\, j}, \quad 1 \leq j \leq m,$$

are arranged in nonascending order; that is, $\mathbf{A}' \in \overline{R}^n_+$ and $\mathbf{B}' \in \overline{R}^m_+$.

Suppose that $\overline{X}^*$ is not a hereditarily minimax matrix and $\overset{*}{\overline{X}} = (\overset{*}{\overline{x}}_{ij})$ is a hereditarily minimax matrix from the set $M(\mathbf{A}', \mathbf{B}')$. Then, the latter is a uniform matrix by Lemma 2.4, and the induction hypothesis implies that

$$\Phi_8(\overset{*}{\overline{X}}) = \min_{X(\mathbf{A}', \mathbf{B}') \in M(\mathbf{A}', \mathbf{B}')} \Phi_8\left(X(\mathbf{A}', \mathbf{B}')\right) < \Phi_8(\overline{X}^*).$$

Let us construct a matrix $\dot{X} = (\dot{x}_{ij})$ by attaching the last row of $\overset{*}{\overline{X}}$ to $\overline{X}$ as its $(n+1)$th row:

$$\dot{x}_{ij} = \begin{cases} \overset{*}{\overline{x}}_{ij}, & 1 \le i \le n, \ 1 \le j \le m, \\ \overline{x}_{ij}, & i = n+1, \ 1 \le j \le m. \end{cases}$$

Consider the value of functional (1.14) from Chapter 1 calculated for the matrix $\dot{X}$, using the same notation as in (3.6) and (3.7):

$$\begin{aligned} \Phi_8(\dot{X}) = \sum_{\dot{Y} \in \mathfrak{M}'(\dot{X})} \Phi_1(\dot{Y}) + \sum_{\dot{Y} \in \mathfrak{M}_{n+1}(\dot{X})} \Phi_1(\dot{Y}) = \\ \sum_{\dot{Y} \in \mathfrak{M}'(\dot{X})} \Phi_1(\dot{Y}) + \sum_{\dot{Y} \in \mathfrak{M}'_{n+1}(\dot{X})} \Phi_1(\dot{Y}) + \sum_{\dot{Y} \in \mathfrak{M}''_{n+1}(\dot{X})} \Phi_1(\dot{Y}) \end{aligned} \tag{3.8}$$

It is clear that

$$\begin{aligned} \sum_{\dot{Y} \in \mathfrak{M}'(\dot{X})} \Phi_1(\dot{Y}) = \Phi_8(\overset{*}{\overline{X}}) < \Phi_8(\overline{X}^*) = \sum_{\overline{Y} \in \mathfrak{M}'(\overline{X})} \Phi_1(\overline{Y}), \\ \sum_{\dot{Y} \in \mathfrak{M}''_{n+1}(\dot{X})} \Phi_1(\dot{Y}) = \sum_{\overline{Y} \in \mathfrak{M}''_{n+1}(\overline{X})} \Phi_1(\overline{Y}). \end{aligned} \tag{3.9}$$

Let us compare the values of functional (1.7) from Chapter 1 calculated for the submatrices from $\mathfrak{M}'_{n+1}(\overline{X})$ and $\mathfrak{M}'_{n+1}(\dot{X})$ that have rows and columns with common numbers. Suppose

$$\overline{Y} = (\overline{x}_{i_r j_t}) \in \mathfrak{M}'_{n+1}(\overline{X}), \quad \dot{Y} = (\dot{x}_{i_r j_t}) \in \mathfrak{M}'_{n+1}(\dot{X}),$$

$$1 \le r \le p, \quad 1 \le t \le q, \quad i_p = n+1, \quad \dot{x}_{n+1\, j_t} = \overline{x}_{n+1\, j_t}.$$

Since uniformity is a hereditary property of transportation matrices, and a hereditarily minimax matrix is uniform,

$$\Phi_1(\overline{Y}) = \max\{\overline{x}_{i_r j_t} : 1 \le r \le p, \ 1 \le t \le q\} = \overline{x}_{i_1 j_1},$$

$$\Phi_1(\dot{Y}) = \max\{\dot{x}_{i_r j_t} : 1 \le r \le p, \ 1 \le t \le q\} = \max\{\dot{x}_{i_1 j_1}, \overline{x}_{n+1\, j_1}\}.$$

There exists a natural one-to-one correspondence between the submatrix sets $\mathfrak{M}'_{n+1}(\overline{X})$, $\mathfrak{M}(\overline{X}^*)$ and $\mathfrak{M}'_{n+1}(\dot{X})$, $\mathfrak{M}(\overset{*}{\overline{X}})$, defined by removing the last rows of the submatrices from $\mathfrak{M}'_{n+1}(\overline{X})$ and $\mathfrak{M}'_{n+1}(\dot{X})$. Therefore, it holds that

$$\sum_{\mathfrak{M}'_{n+1}(\overline{X})} \Phi_1(\overline{Y}) \ge \sum_{\mathfrak{M}'_{n+1}(\dot{X})} \Phi_1(\dot{Y}).$$

Since

$$\sum_{\overline{Y}\in\mathfrak{M}'_{n+1}(\overline{X})} \Phi_1(\overline{Y}) = \sum_{\overline{Y}\in\mathfrak{M}'(\overline{X})} \Phi_1(\overline{Y}) = \Phi_8(\overline{X}^*), \tag{3.10}$$

(3.8) and (3.9) can be combined to show that

$$\Phi_8(\dot{X}) < 2\Phi_8(\overline{X}^*) + \sum_{\overline{Y}\in\mathfrak{M}''_{n+1}(\overline{X})} \Phi_1(\overline{Y}). \tag{3.11}$$

Comparing (3.10), (3.11), and (3.7), we see that $\Phi_8(\dot{X}) < \Phi_8\left(X(\mathbf{A},\mathbf{B})\right)$, which is inconsistent with condition (3.5). The contradiction shows that the matrix $\overline{X}^*$ from the set $M(\mathbf{A}',\mathbf{B}')$ is a hereditarily minimax one.

Thus, by condition (3.5) and induction hypothesis, the submatrix $\overline{X}^*$ constructed by removing the last row from $\overline{X}(\mathbf{A},\mathbf{B})$ is a hereditarily minimax one. Now, we show that $\overline{X}(\mathbf{A},\mathbf{B})$ is also a hereditarily minimax matrix.

Consider a submatrix $\overline{X}^* = (\overline{x}_{ij})$, where $1 \le i \le n$ and $1 \le j \le m$. Being a hereditarily minimax matrix, $\overline{X}^*$ is a minimax one; that is, $\overline{X}^* \in M\left(\mathbf{A}',\mathbf{B}';c(\mathbf{A}',\mathbf{B}')\right)$, where $c(\mathbf{A}',\mathbf{B}') = \overline{x}_{11}$ is the minimax value for the matrix set $M(\mathbf{A}',\mathbf{B}')$. Therefore, $\overline{x}_{ij} \le c(\mathbf{A}',\mathbf{B}')$, where $1 \le i \le n$ and $1 \le j \le m$. By Lemma 1.1, there exists $p \in Z$ $(1 \le p \le n)$ such that $c(\mathbf{A}',\mathbf{B}')p \le b'_1$ and

$$\delta_p\left(\mathbf{A}',\mathbf{B}';c(\mathbf{A}',\mathbf{B}')\right) = \sum_{b'_j<c(\mathbf{A}',\mathbf{B}')p} b'_j + c(\mathbf{A}',\mathbf{B}')pl_p\left(\mathbf{B}';c(\mathbf{A}',\mathbf{B}')\right) - \sum_{i=1}^{k} a_i = 0,$$

where $a_i = a'_i$ $(1 \le i \le n)$.

Since $\overline{X}(\mathbf{A},\mathbf{B})$ is a uniform matrix, its last row is such that $\overline{x}_{n+1\,j} \le c(\mathbf{A}',\mathbf{B}')$ $(1 \le j \le m)$ and, therefore, $\overline{X}(\mathbf{A},\mathbf{B}) \in M\left(\mathbf{A},\mathbf{B};c(\mathbf{A}',\mathbf{B}')\right)$ and $\delta_k\left(\mathbf{A},\mathbf{B};c(\mathbf{A}',\mathbf{B}')\right) \ge 0$ for any $k \in Z$ $(1 \le k \le n+1)$. Recall that $b_j = b'_j + \overline{x}_{n+1\,j}$ for $1 \le j \le m$. Let us show that $\overline{X}(\mathbf{A},\mathbf{B})$ is a minimax matrix with the minimax value $c(\mathbf{A},\mathbf{B}) = c(\mathbf{A}',\mathbf{B}')$.

Suppose that $p < n$. We have two cases: $l_p\left(\mathbf{B}';c(\mathbf{A}',\mathbf{B}')\right) = m$ and $l_p\left(\mathbf{B}';c(\mathbf{A}',\mathbf{B}')\right) < m$.

In the former case, it is clear that

$$\delta_p\left(\mathbf{A},\mathbf{B};c(\mathbf{A}',\mathbf{B}')\right) = \delta_p\left(\mathbf{A}',\mathbf{B}';c(\mathbf{A}',\mathbf{B}')\right) = 0.$$

In the latter case, $\overline{x}_{ij} = 0$ for $p+1 \le i \le n$ and $l_p\left(\mathbf{B}';c(\mathbf{A}',\mathbf{B}')\right) + 1 \le j \le m$ by Theorem 3.2 from Chapter 1. Since $\overline{X}(\mathbf{A},\mathbf{B})$ is a uniform matrix, $\overline{x}_{n+1\,j} = 0$ for $l_p\left(\mathbf{B}';c(\mathbf{A}',\mathbf{B}')\right) + 1 \le j \le m$. Therefore, $l_p\left(\mathbf{B};c(\mathbf{A}',\mathbf{B}')\right) = l_p\left(\mathbf{B}';c(\mathbf{A}',\mathbf{B}')\right)$ and $b_j = b'_j$ if $l_p\left(\mathbf{B}';c(\mathbf{A}',\mathbf{B}')\right) + 1 \le j \le m$, and

$$\delta_p\left(\mathbf{A},\mathbf{B};c(\mathbf{A}',\mathbf{B}')\right) = \delta_p\left(\mathbf{A}',\mathbf{B}';c(\mathbf{A}',\mathbf{B}')\right) = 0.$$

Suppose that $p = n$. If $l_n(\mathbf{B}'; c(\mathbf{A}', \mathbf{B}')) = m$, then $l_n(\mathbf{B}; c(\mathbf{A}', \mathbf{B}')) = m$ a fortiori, and

$$\delta_n(\mathbf{A}, \mathbf{B}; c(\mathbf{A}', \mathbf{B}')) = \delta_n(\mathbf{A}', \mathbf{B}'; c(\mathbf{A}', \mathbf{B}')) = 0.$$

The inequality $l_n(\mathbf{B}'; c(\mathbf{A}', \mathbf{B}')) < m$ implies that $b'_j < c(\mathbf{A}', \mathbf{B}')n$ for $l_n(\mathbf{B}'; c(\mathbf{A}', \mathbf{B}')) + 1 \le j \le m$. Therefore, for every j that satisfies this condition, there exists an element $\overline{x}_{ij}$ of the matrix $\overline{X}^*$ such that $\overline{x}_{ij} < c(\mathbf{A}', \mathbf{B}')$. Since $X(\mathbf{A}, \mathbf{B})$ is a uniform matrix, $\overline{x}_{n+1\,j} < c(\mathbf{A}', \mathbf{B}')$ for $l_n(\mathbf{B}'; c(\mathbf{A}', \mathbf{B}')) + 1 \le j \le m$. Therefore, again, $l_n(\mathbf{B}; c(\mathbf{A}', \mathbf{B}')) = l_n(\mathbf{B}'; c(\mathbf{A}', \mathbf{B}'))$ and

$$\delta_n(\mathbf{A}, \mathbf{B}; c(\mathbf{A}', \mathbf{B}')) = \delta_n(\mathbf{A}', \mathbf{B}'; c(\mathbf{A}', \mathbf{B}')) = 0.$$

Thus, we have shown that $\overline{X}(\mathbf{A}, \mathbf{B})$ is a minimax matrix with minimax value $c(\mathbf{A}, \mathbf{B}) = c(\mathbf{A}', \mathbf{B}')$.

It remains to show that every submatrix $\overline{Y}$ of $\overline{X} = \overline{X}(\mathbf{A}, \mathbf{B})$ ($\overline{Y} \in \mathfrak{M}(\overline{X})$) is a minimax matrix. More specifically, since $\overline{X}^*$ is a hereditarily minimax submatrix of $\overline{X}(\mathbf{A}, \mathbf{B})$, it is sufficient to show that every submatrix $\overline{Y}$ having common elements with the last row of $\overline{X}(\mathbf{A}, \mathbf{B})$ is a minimax matrix.

Suppose that $\overline{Y} = (\overline{x}_{i_r j_t}) \in \mathfrak{M}_{n+1}(\overline{X})$, where $1 \le r \le p$, $1 \le t \le q$, and $i_p = n + 1$. When $p = 1$, $\overline{Y}$ is a row vector. Such a matrix is obviously a minimax one. When $p > 1$, $\overline{Y}$ is a uniform submatrix, and the submatrix $(\overline{x}_{i_r j_t})$ $(1 \le r \le p - 1,\ 1 \le t \le q)$, being a submatrix of the hereditarily minimax matrix $\overline{X}^*$, is also a minimax one. Therefore, by analogy with the cases of the uniform matrix $\overline{X}(\mathbf{A}, \mathbf{B})$ and its hereditarily minimax submatrix X^*, it can be shown that $\overline{Y}$ is a minimax matrix with the minimax value $\overline{x}_{i_1 j_1}$.

Consequently, $\overline{X}(\mathbf{A}, \mathbf{B})$ is a hereditarily minimax matrix, and $\overline{X}(\mathbf{A}, \mathbf{B}) = \overset{*}{X}(\mathbf{A}, \mathbf{B}) = (\overset{*}{x}_{ij})$. By Theorem 2.1, the hereditarily minimax matrix is unique, which completes the proof.

Remark 3.1. It should be stressed that the matrix $\overline{X}(\mathbf{A}, \mathbf{B})$ employed in the proof of Theorem 3.1 is a hereditarily minimax one by virtue of condition (3.5) alone, and not because $\overline{X}(\mathbf{A}, \mathbf{B})$ is a uniform matrix and $\overline{X}^*$ is its hereditarily minimax submatrix.

Example 3.1. These are matrices that can be readily shown to be hereditarily minimax ones:

$$\begin{pmatrix} 6 & 6 & 3 & 3 \\ 4 & 2 & 0 & 0 \end{pmatrix}, \quad \begin{pmatrix} 6 & 6 & 4 & 4 \\ 6 & 6 & 4 & 4 \end{pmatrix}.$$

Let us add a third row to each of these matrices so that the resulting matrices are uniform:

$$\begin{pmatrix} 6 & 6 & 3 & 3 \\ 4 & 2 & 0 & 0 \\ 2 & 1 & 0 & 0 \end{pmatrix}, \quad \begin{pmatrix} 6 & 6 & 4 & 4 \\ 6 & 6 & 4 & 4 \\ 4 & 4 & 2 & 2 \end{pmatrix}.$$

The first newly constructed matrix is a minimax one, but not a hereditarily minimax one, whereas the second one is not a minimax matrix. Now, we construct matrices that are defined, respectively, by the same vectors, but are hereditarily minimax ones:

$$\begin{pmatrix} 6 & 6 & 3 & 3 \\ 3 & 3 & 0 & 0 \\ 3 & 0 & 0 & 0 \end{pmatrix}, \quad \begin{pmatrix} 16/3 & 16/3 & 14/3 & 14/3 \\ 16/3 & 16/3 & 14/3 & 14/3 \\ 16/3 & 16/3 & 2/3 & 2/3 \end{pmatrix}.$$

Formula (3.1) gives the value of functional (1.14) from Chapter 1 for a uniform matrix. Since a hereditarily minimax matrix is uniform by Lemma 2.4, the minimal value of objective function (1.14) from Chapter 1 is determined by Lemma 3.1 and Theorem 3.1.

Theorem 3.2. *For arbitrary vectors* $\mathbf{A} \in R_+^n$ *and* $\mathbf{B} \in R_+^m$ *with equal sums of components, the minimal value of functional* (1.14) *from Chapter* 1 *on the matrix set* $M(\mathbf{A},\mathbf{B})$ *is expressed as*

$$\min_{X(\mathbf{A},\mathbf{B})\in M(\mathbf{A},\mathbf{B})} \Phi_8(X(\mathbf{A},\mathbf{B})) = \Phi_8(\overset{*}{X}(\mathbf{A},\mathbf{B})) = \sum_{i=1}^{n} \sum_{j=1}^{m} 2^{n+m-i-j} \tilde{\overset{*}{x}}_{ij}, \quad (3.12)$$

where the hereditarily minimax matrix $\tilde{\overset{*}{X}}(\tilde{\mathbf{A}},\tilde{\mathbf{B}}) = (\tilde{\overset{*}{x}}_{ij})$ *is constructed by permuting the rows and columns of the hereditarily minimax matrix* $\overset{*}{X}(\mathbf{A},\mathbf{B}) = (\overset{*}{x}_{ij})$, *where* $\tilde{\mathbf{A}} \in \overline{R}_+^n$ *and* $\tilde{\mathbf{B}} \in \overline{R}_+^m$.

Remark 3.2. The elements of the hereditarily minimax matrix $\tilde{\overset{*}{X}}(\tilde{\mathbf{A}},\tilde{\mathbf{B}}) = \tilde{\overset{*}{X}}(\tilde{\mathbf{A}},\tilde{\mathbf{B}})$ can be calculated by means of Algorithm 2.1.

When $|M(\mathbf{A},\mathbf{B};c(\mathbf{A},\mathbf{B}))| = 1$ (see Theorems 1.4, 1.5, and 1.6), the matrix $\tilde{\overset{*}{X}}(\tilde{\mathbf{A}},\tilde{\mathbf{B}})$ can be efficiently constructed either by applying Theorem 3.2 from Chapter 1, where $p \in Z$ is such that the function defined by (1.22) in Chapter 1 vanishes, or, equivalently, by applying formula (1.7).

To calculate the value of functional (1.14) from Chapter 1 for a uniform matrix, one can use relation (3.7). Suppose that $X(\mathbf{A},\mathbf{B}) = (x_{ij})$, where $\mathbf{A} \in R_+^n$ and $\mathbf{B} \in R_+^m$. Let us introduce a special notation for certain submatrices consisting of rows of the matrix $X(\mathbf{A},\mathbf{B})$: if $k_1, k_2 \in Z$ are such that $1 \leq k_1 \leq k_2 \leq n$, then $X(k_1,k_2) = (x_{ij})$ $(k_1 \leq i \leq k_2,\ 1 \leq j \leq m)$ is a submatrix consisting of the rows of $X(\mathbf{A}, \mathbf{B})$ with numbers from k_1 to k_2. Then $X(1, n) = X(\mathbf{A}, \mathbf{B})$. If $k \in Z$ is such that $1 \leq k \leq n$, then $X(k, k) = (x_{ij})$ $(1 \leq j \leq m)$ is a submatrix consisting of the kth row of $X(\mathbf{A}, \mathbf{B})$.

The value of functional (1.14) from Chapter 1 for a uniform matrix $X(\mathbf{A}, \mathbf{B})$, where $\mathbf{A} \in \overline{R}^n_+$ and $\mathbf{B} \in \overline{R}^m_+$, can be expressed in terms of the values of the functional for 1-row matrices. Now, we return to the proof of Theorem 3.1. Using the notation just introduced, we rewrite (3.7) as a recursion relation:

$$\Phi_8\left(X(1, n)\right) = 2\Phi_8\left(X(1, n-1)\right) + \Phi_8\left(X(n, n)\right).$$

Functional (1.14) from Chapter 1 for the matrix $X(1, n-1)$ $(1 \leq i \leq n-1, 1 \leq j \leq m)$ is then expressed as follows:

$$\Phi_8\left(X(1, n-1)\right) = 2\Phi_8\left(X(1, n-2)\right) + \Phi_8\left(X(n-1, n-1)\right),$$

$$\Phi_8\left(X(1, n)\right) = 2^2\Phi_8\left(X(1, n-2)\right) + 2\Phi_8\left(X(n-1, n-1)\right) + \Phi_8\left(X(n, n)\right).$$

This leads to the following final result.

Theorem 3.3. *Let* $\mathbf{A}$ *and* $\mathbf{B}$ *be vectors from* $\overline{R}^n_+$ *and* $\overline{R}^m_+$, *respectively, characterized by equal sums of components. If* $X(\mathbf{A}, \mathbf{B}) = X(1, n)$ *and* $X(\mathbf{B}, \mathbf{A}) = X(1, m)$ *are uniform matrices, then the corresponding values of functional* (1.14) *from Chapter* 1 *are expressed as*

$$\Phi_8\left(X(1, n)\right) = \sum_{k=0}^{n-1} 2^k \Phi_8\left(X(n-k, n-k)\right),$$

$$\Phi_8\left(X(1, m)\right) = \sum_{k=0}^{m-1} 2^k \Phi_8\left(X(m-k, m-k)\right).$$

In particular, these formulas apply to hereditarily minimax matrices $X(1, n)$ *and* $X(1, m)$.

Example 3.2. Let is minimize functional (1.14) from Chapter 1 on the matrix set $M(\mathbf{A}, \mathbf{B})$, where $\mathbf{A} = (13, 10, 3, 1)$ and $\mathbf{B} = (11, 9, 5, 2)$.

By Theorem 3.1, the problem

$$\Phi_8\left(X(\mathbf{A}, \mathbf{B})\right) \to \min, \quad X(\mathbf{A}, \mathbf{B}) \in M(\mathbf{A}, \mathbf{B})$$

has a unique solution, which is defined by a hereditarily minimax matrix:

$$\min_{X(\mathbf{A}, \mathbf{B}) \in M(\mathbf{A}, \mathbf{B})} \Phi_8(X(\mathbf{A}, \mathbf{B})) = \Phi_8(\overset{*}{X}(\mathbf{A}, \mathbf{B})).$$

We apply Algorithm 2.1 to calculate the elements of the hereditarily minimax matrix $\overset{*}{X}(\mathbf{A},\mathbf{B}) = (\overset{*}{x}_{ij})$ for the given vectors $\mathbf{A}$, $\mathbf{B} \in \overline{R}^4_+$. Algorithm 2.1 calculates the elements of certain submatrices of $\overset{*}{X}(\mathbf{A},\mathbf{B})$ by applying Algorithm 1.1.

Algorithm 1.1 is based on the system of relations (1.9), which involves two integer parameters, k and k_i. Since $\mathbf{A}$, $\mathbf{B} \in \overline{R}^4_+$, the components of these vectors are positive, and the components of $\mathbf{B}$ are mutually different, $1 \le k \le 4$ and $1 \le k_i = i \le 4$. Therefore, it is sufficient to consider, at most, 16 systems of relations (1.9) to calculate the minimax value $c(\mathbf{A},\mathbf{B})$. Among the solutions $c(k,i)$ to the equation in (1.9) that correspond to various admissible values of k and i and satisfy the inequalities in (1.9), we must determine the highest value of $c(k,i)$. Then, $c(\mathbf{A},\mathbf{B}) = c(k,i)$ by Theorem 1.7. In this example, $\max\{c(k,i) : 1 \le k, i \le 4\} = c(2,2) = 4$, and $c(\mathbf{A},\mathbf{B}) = 4$ is the minimax value. It is clear that $l_2(\mathbf{B};4) = 2$. Theorem 3.2 from Chapter 1 determines some elements of $\overset{*}{X}(\mathbf{A},\mathbf{B})$:

$$\overset{*}{x}_{ij} = \begin{cases} 4, & 1 \le i,\ j \le 2, \\ 0, & 3 \le i,\ j \le 4. \end{cases}$$

Then, the submatrix $(\overset{*}{x}_{ij})$ $(1 \le i \le 2,\ 3 \le j \le 4)$ is defined by the vectors $(5,2)$ and $(5,2)$, and the submatrix $(\overset{*}{x}_{ij})$ $(3 \le i \le 4,\ 1 \le j \le 2)$ is defined by the vectors $(3,1)$ and $(3,1)$. Obviously, $c\left((5,2),(5,2)\right) = 3$ and $c\left((3,1),(3,1)\right) = 2$. Accordingly, the remaining elements of $\overset{*}{X}(\mathbf{A},\mathbf{B})$ are $\overset{*}{x}_{13} = 3$, $\overset{*}{x}_{14} = \overset{*}{x}_{23} = 2$, $\overset{*}{x}_{24} = 0$, $\overset{*}{x}_{31} = 2$, $\overset{*}{x}_{32} = \overset{*}{x}_{41} = 1$, $\overset{*}{x}_{42} = 0$. The hereditarily minimax matrix $\overset{*}{X}(\mathbf{A},\mathbf{B})$ is thus constructed:

$$\overset{*}{X}(\mathbf{A},\mathbf{B}) = \begin{pmatrix} 4 & 4 & 3 & 2 \\ 4 & 4 & 2 & 0 \\ 2 & 1 & 0 & 0 \\ 1 & 0 & 0 & 0 \end{pmatrix}.$$

We calculate the value of objective function (1.14) from Chapter 1 for this matrix by formula (3.12):

$$\Phi_8(\overset{*}{X}(\mathbf{A},\mathbf{B})) = 2^6 \cdot 4 + 2^5 \cdot 4 + 2^4 \cdot 3 + 2^3 \cdot 2 + 2^5 \cdot 4 + 2^4 \cdot 4 + 2^3 \cdot 2 + 2^2 \cdot 0 + 2^4 \cdot 2 + 2^3 \cdot 1 + 2^2 \cdot 0 + 2 \cdot 0 + 2^3 \cdot 1 + 2^2 \cdot 0 + 2 \cdot 0 + 1 \cdot 0 = 708.$$

Thus, the solution to the minimization problem considered here is

$$\min_{X(\mathbf{A},\mathbf{B})\in M(\mathbf{A},\mathbf{B})} \Phi_8(X(\mathbf{A},\mathbf{B})) = \Phi_8(\overset{*}{X}(\mathbf{A},\mathbf{B})) = 708,$$

$$\overset{*}{X}(\mathbf{A},\mathbf{B}) = \begin{pmatrix} 4 & 4 & 3 & 2 \\ 4 & 4 & 2 & 0 \\ 2 & 1 & 0 & 0 \\ 1 & 0 & 0 & 0 \end{pmatrix}.$$

When 1-element matrices are ignored in the analysis of the minimization problem for functional (1.14) from Chapter 1, the hereditarily minimax matrix minimizes the functional as well. Suppose vectors $\mathbf{A}$ and $\mathbf{B}$ from R^n_+ and R^m_+, respectively, are characterized by equal sums of components and $\mathfrak{M}_0(X)$ is the set of non-1-element submatrices of a matrix $X = X(\mathbf{A},\mathbf{B}) \in M(\mathbf{A},\mathbf{B})$. The number of 1-element submatrices is nm; therefore, $|\mathfrak{M}_0(X)| = (2^n-1)(2^m-1) - nm$. It is obvious that the sum of all functionals (1.7) from Chapter 1 calculated for all 1-element submatrices of any matrix $X(\mathbf{A},\mathbf{B})$ is equal to the sum of components of $\mathbf{A}$ or $\mathbf{B}$. Therefore, the functional

$$\Phi_8'\left(X(\mathbf{A},\mathbf{B})\right) = \sum_{Y\in\mathfrak{M}_0(X)} \Phi_1(Y)$$

can be expressed as

$$\Phi_8'\left(X(\mathbf{A},\mathbf{B})\right) = \Phi_8\left(X(\mathbf{A},\mathbf{B})\right) - \sum_{i=1}^{n} a_i,$$

$$\min_{X(\mathbf{A},\mathbf{B})\in M(\mathbf{A},\mathbf{B})} \Phi_8'(X(\mathbf{A},\mathbf{B})) = \Phi_8'(\overset{*}{X}(\mathbf{A},\mathbf{B})),$$

and the hereditarily minimax matrix is the only solution to the problem.

Consider functionals (1.15) and (1.16) from Chapter 1:

$$\Phi_9\left(X(\mathbf{A},\mathbf{B})\right) = \sum_{Y\in\mathfrak{M}(X)} \Phi_1(Y) + \sum_{Y\in\mathfrak{M}(X)} \Phi_2(Y),$$

$$\Phi_{10}\left(X(\mathbf{A},\mathbf{B})\right) = \sum_{Y\in\mathfrak{M}(X)} \Phi_1(Y) + \sum_{Y\in\mathfrak{M}(X)} \Phi_3(Y),$$

where $X = X(\mathbf{A},\mathbf{B})$ is any matrix from $M(\mathbf{A},\mathbf{B})$. Recall that $(X(\mathbf{A},\mathbf{B}) = (x_{ij}))$

$$\Phi_2\left(X(\mathbf{A},\mathbf{B})\right) = \sum_{j=1}^{m} \max_i x_{ij},$$

$$\Phi_3\left(X(\mathbf{A},\mathbf{B})\right) = \sum_{j=1}^{m} \max_i x_{ij} + \sum_{i=1}^{n} \max_j x_{ij}.$$

Applying relations (1.17) from Chapter 1, we see that we are prepared to minimize objective functions (1.15) and (1.16) from Chapter 1 on the matrix set $M(\mathbf{A},\mathbf{B})$. The following theorem is a corollary to Theorems 2.5, 3.1, and 3.2.

Theorem 3.4. *Suppose that vectors* $\mathbf{A} = (a_1, \ldots, a_n) \in R^n_+$ *and* $\mathbf{B} = (b_1, \ldots, b_m) \in R^m_+$ *satisfy the balance condition. The minimization problems for objective functions* (1.15) *and* (1.16) *from Chapter* 1 *have unique solutions determined by the hereditarily minimax matrix* $\overset{*}{X}(\mathbf{A},\mathbf{B}) = (\overset{*}{x}_{ij})$:

$$\min_{X(\mathbf{A},\mathbf{B})\in M(\mathbf{A},\mathbf{B})} \Phi_9(X(\mathbf{A},\mathbf{B})) = \Phi_9(\overset{*}{X}(\mathbf{A},\mathbf{B})),$$

$$\min_{X(\mathbf{A},\mathbf{B})\in M(\mathbf{A},\mathbf{B})} \Phi_{10}(X(\mathbf{A},\mathbf{B})) = \Phi_{10}(\overset{*}{X}(\mathbf{A},\mathbf{B})).$$

Moreover, if $\overset{*}{\widetilde{X}}(\widetilde{\mathbf{A}},\widetilde{\mathbf{B}}) = (\overset{*}{\tilde{x}}_{ij})$ *is the hereditarily minimax matrix of a transportation polyhedron* $M(\widetilde{\mathbf{A}},\widetilde{\mathbf{B}})$, *where the vectors* $\widetilde{\mathbf{A}} = (\tilde{a}_1, \ldots, \tilde{a}_n)$ *and* $\widetilde{\mathbf{B}} = (\tilde{b}_1, \ldots, \tilde{b}_m)$ *are constructed by arranging the components of* $\mathbf{A}$ *and* $\mathbf{B}$ *in nonascending order, then*

$$\Phi_9(\overset{*}{X}(\mathbf{A},\mathbf{B})) = \Phi_9(\overset{*}{\widetilde{X}}(\widetilde{\mathbf{A}},\widetilde{\mathbf{B}})) = \sum_{i=1}^{n}\sum_{j=1}^{m} 2^{n+m-i-j}\overset{*}{\tilde{x}}_{ij} + 2^{m-1}\sum_{k=1}^{n} 2^{n-k}\tilde{a}_k,$$

$$\Phi_{10}(\overset{*}{X}(\mathbf{A},\mathbf{B})) = \Phi_{10}(\overset{*}{\widetilde{X}}(\widetilde{\mathbf{A}},\widetilde{\mathbf{B}})) =$$
$$\sum_{i=1}^{n}\sum_{j=1}^{m} 2^{n+m-i-j}\overset{*}{\tilde{x}}_{ij} + 2^{m-1}\sum_{k=1}^{n} 2^{n-k}\tilde{a}_k + 2^{n-1}\sum_{k=1}^{m} 2^{m-k}\tilde{b}_k.$$

In summary, in this chapter we have determined a matrix, called the hereditarily minimax one, that minimizes the functionals constructed in Chapter 1 for an arbitrary transportation polyhedron. We have shown that the hereditarily minimax matrix is the only matrix that minimizes all of these functionals.

References to Chapter 2

[1] Alekseev V. M., Tikhomirov V. M., and Fomin S. V., *Optimal'noe upravlenie* (Optimal Control), Moscow: Nauka, 1979.

[2] Astakhov N. D., Fedoseev A. V., and Chan Van Khung, *O primenenii uskorennykh algoritmov resheniya zadach transportnogo tipa* (On the Use of Accelerated Algorithms for Solving Transportation Problems), Moscow: Computing Center of Russ. Acad. Sci., 1994.

[3] Bellman R. *Dynamic Programming*, Princeton: Princeton Univ. Press, 1957.

[4] Gol'shtein E. G. and Yudin D. B., *Zadachi lineinogo programmirovania transportnogo tipa* (Transportation Problems in Linear Programming), Moscow: Nauka, 1969.

[5] Dem'yaninov V. F. and Malozemov V. N., *Vvedenie v minimaks* (Introduction to Minimax), Moscow: Nauka, 1972.

[6] Emelichev V. A. and Kovalev M. M., *Mnogogranniki, grafy, optimizatsiya* (Polyhedrons, Graphs, and Optimization), Moscow: Nauka, 1981.

[7] Emelichev V. A. and Kononenko A. M., On a Class of Transportation Polyhedrons, *Izv. Belarus. Akad. Nauk., Ser. Fiz. Mat.*, 1971, no. 3, pp. 21–23.

[8] Karmanov V. G., *Matematicheskoe programmirovanie* (Mathematical programming), Moscow: Nauka, 1986.

[9] Krasnoshchekov P. S. and Petrov A. A., *Printsipy postroeniya modelei* (Principles of Model Construction), Moscow: Mosk. Gos. Univ., 1983.

[10] Kuznetsov A.V., Sakovich V. A., and Kholod N. N., *Vysshaya matematika. Matematicheskoe programmirovanie* (Higher Mathematics: Mathematical programming), Minsk: Vysshaya Shkola, 1994.

[11] Mironov A. A., Properties of the Sets of Degrees of the Vertices of Generalized Graphs, *Dokl. Akad. Nauk*, 1992, vol. 324, no. 5, pp. 959–963.

[12] Mironov A. A., Generalized Graphs with Constraints for the Degrees of Vertices, *Dokl. Akad. Nauk*, 1993, vol. 336, no. 4, pp. 437–439.

[13] Mironov A. A. and Tsurkov V. I., Network Models with Fixed Parameters at the Connection Nodes I, *Izv. Akad. Nauk, Ser. Tekhn. Kibern.*, 1993, no. 4, pp. 212–223.

[14] Mironov A. A. and Tsurkov V. I., Network Models with Fixed Parameters at the Connection Nodes II, *Izv. Akad. Nauk, Ser. Tekhn. Kibern.*, 1993, no. 6, pp. 3–14.

[15] Mironov A. A. and Tsurkov V. I., A Class of Distributed Problems with Minimax Criteria, *Dokl. Akad. Nauk*, 1992, vol. 336, no. 1, pp. 35–38.

[16] Mironov A. A. and Tsurkov V. I., Transportation and Network Problems with Minimax Criteria, *Zh. Vychisl. Mat. Mat. Fiz.*, 1995, vol. 35, no. 1, pp. 24–45.

[17] Mironov A. A. and Tsurkov V. I., Transportation Problems with Minimax Criteria, *Avtom. Telemekh.*, 1995, no. 12, pp. 109–118.

[18] Mironov A. A. and Tsurkov V. I., Transportation Problems with Minimax Criteria, *Dokl. Akad. Nauk*, 1996, vol. 346, no. 2, pp. 1–4.

[19] *Sovremennoe sostoyanie teorii issledovaniya operatsii* (Modern Theory of Operations Research), Moiseev, N.N., Ed., Moscow: Nauka, 1979.

[20] Moder J. and Elmaghraby S. S., *Handbook of Operations Research*, New York: Van Nostrand, 1978. Translated under the title *Issledovanie operatsii. Metodologicheskie osnovy i matematicheskie metody,* vol. 1, Moscow: Mir, 1981.

[21] Mukhacheva E. A. and Rubinshtein G. Sh., *Matematicheskoe programmirovanie* (Mathematical Programming), Novosibirsk: Nauka, 1977.

[22] Panos M. Pardalos, *Minimax and Applications*, University of Minnesota and Institute of Applied Mathematics, Beijing: Academia Sinica, 1995.

[23] Pshenichnyi B. N., *Vypuklyi analiz i ekstremal'nye zadachi* (Convex Analysis and Extremum Value Problems), Moscow: Nauka, 1980.

[24] Rockafellar R. T., *Convex Analysis*, Princeton: Princeton Univ. Press, 1970. Translated under the title *Vypuklyi analiz,*, Moscow: Mir, 1973.

[25] Sukharev A. G., Timokhov, A.V., and Fedorov, V.V., *Kurs metodov optimizatsii* (Lectures on Optimization Methods), Moscow: Nauka, 1986.

[26] Trius E. B., *Zadachi matematicheskogo programmirovaniya transportnogo tipa* (Transportation Problems in Mathematical Programming), Moscow: Sovetskoe Radio, 1967.

[27] Tsurkov V. I., *Dinamicheskie zadachi bol'shoi razmernosti* (Dynamic Large-Scale Problems), Moscow: Nauka, 1988.

Chapter 3

Integer Transportation Matrices

In the statements of various specific problems, it is commonly required that the variables be integer [9, 26, 27]. For example, recall the knapsack problem or the traveling salesman problem [9, 10, 16, 22]. In such cases, discrete optimization methods are applied [8, 18, 22, 25]. The most popular of these are Gomory's truncation method and the branch-and-bound method [4, 5, 16, 25]. Examples of transportation problems with integer variables include the assignment problem and the goal distribution problems [3, 10, 11, 16, 23]. The discrete optimization methods mentioned above are also applied to transportation problems.

In this chapter, we analyze the construction of admissible integer solutions in truncated transportation polyhedrons. Our method is based on the characteristic functions and equations introduced in the preceding chapters, which demonstrates the wide applicability of these mathematical tools.

In Section 1, we formulate an algorithm for integer variables in the general case.

In Section 2, we specify the algorithm for Boolean variables. The solutions obtained are then used to construct the vertices of transportation polyhedrons.

§1. Integer Matrices with Common Bounds on Their Elements

In Chapter 1, we analyzed the construction of an admissible solution to the transportation problem in the case when all unknown elements are bounded by the same constant. We gave a criterion for the existence of a solution in the form of a set of inequalities. We formulated an algorithm for constructing such solutions. These results apply to integer matrices as well. The corresponding constructions based on characteristic functions are simpler and more graphic. In the final analysis, solutions are reduced to calculations of the rows of the desired matrices.

1.1. Auxiliary Constructions

Suppose that $\mathbf{A} = (a_1, \ldots, a_n) \in Z_+^n$, $\mathbf{B} = (b_1, \ldots, b_m) \in Z_+^m$, $c \in Z$, and $c \geq 0$ and let $M_Z(\mathbf{A}, \mathbf{B}; c)$ be the set of integer matrices from $M(\mathbf{A}, \mathbf{B}; c)$. The condition that the characteristic function defined by (1.22) in Chapter 1 (where the components of the first row vector are arranged in ascending order) is nonnegative is the criterion for the nonemptiness of $M_Z(\mathbf{A},\mathbf{B};c)$ as well (see Theorem 2.2 in Chapter 1). A certain matrix from $M_Z(\mathbf{A},\mathbf{B};c)$ can be constructed by applying Algorithm 2.1 from Chapter 1. Algorithm 2.2 from Chapter 1 can be used to construct any matrix from $M_Z(\mathbf{A},\mathbf{B};c)$. When integer matrices and vectors are considered, and the elements of matrices have a common upper bound, simpler and more graphic rules can be constructed for calculation of the elements of matrices from $M_Z(\mathbf{A},\mathbf{B};c)$.

To construct a matrix from $M_Z(\mathbf{A},\mathbf{B};c)$, we give the rules for determination of the rows of the desired matrix with elements not greater than a certain number and for changing the components of the second vector when the components of $\mathbf{A}\in Z_+^n$ are consecutively taken out.

We now give some definitions to be used as a basis for constructing a criterion for $M_Z(\mathbf{A},\mathbf{B};c)\neq\varnothing$ and an algorithm for calculating the elements of matrices from $M_Z(\mathbf{A},\mathbf{B};c)$.

Definition 1.1. Suppose $\mathbf{A}=(a_1,\dots,a_n)\in Z_+^n$, $\mathbf{B}=(b_1,\dots,b_m)\in Z_+^m$ and $p\in Z$, $1\le p\le n$. If there exists a pair of vectors $\mathbf{A}_1^{(1)}=(a_1^{(1)},\dots,a_n^{(1)})\in Z_+^n$ and $\mathbf{B}_1^{(1)}=(b_1^{(1)},\dots,b_m^{(1)})\in Z_+^m$ whose components are defined by the conditions

$$a_i^{(1)}=\begin{cases}a_i, & 1\le i\le n, \quad i\neq p,\\ 0, & i=p,\end{cases}$$
$$0\le b_j^{(1)}\le b_j,\quad 1\le j\le m, \tag{1.1}$$
$$\sum_{j=1}^{m}(b_j-b_j^{(1)})=a_p,$$

it is called the *reductive pair of vectors.* To emphasize the fact that the pth component of $\mathbf{A}$ is set equal to zero, the pair $\mathbf{A}^{(1)}=\mathbf{A}^{(1)}(p)$, $\mathbf{B}^{(1)}$ is reffered to as a (p)-*reductive vector pair.*

Definition 1.2. If there exists a reductive pair of vectors $\mathbf{A}$ and $\mathbf{B}$ corresponding to $\mathbf{A}^{(1)}$ and $\mathbf{B}^{(1)}$, then the transition from $\mathbf{A}$, $\mathbf{B}$ to $\mathbf{A}^{(1)}$, $\mathbf{B}^{(1)}$ is called a *reductive step.*

It is obvious that the difference between the sums of components of the vectors remains invariant when a reductive step is performed.

The idea behind Definitions 1.1 and 1.2 is that the pth row vector $(x_{p1},\dots,x_{pm})$ of a matrix $X(\mathbf{A},\mathbf{B})$ from $M_Z(\mathbf{A},\mathbf{B})$ is constructed by executing the reductive step that corresponds to the transition from $\mathbf{A}$ and $\mathbf{B}$ to the (p)-reductive pair of $\mathbf{A}^{(1)}(p)$ and $\mathbf{B}^{(1)}$ (provided that $\mathbf{A}$ and $\mathbf{B}$ satisfy the balance condition). If $X(\mathbf{A},\mathbf{B})$ is a matrix from $M_Z(\mathbf{A},\mathbf{B})$, then the corresponding reductive step is the transition to the matrix $X(\mathbf{A}^{(1)}(p), \mathbf{B}^{(1)})$ obtained by setting all elements of the pth row of $X(\mathbf{A}, \mathbf{B})$ equal to zero.

Definition 1.3. Suppose that $c \in Z$, $c \geq 0$, and the pairs of vectors $\mathbf{A}$, $\mathbf{B}$ and $\mathbf{A}^{(1)}(p) = \mathbf{A}^{(1)}$, $\mathbf{B}^{(1)}$ satisfy Definition 1.1. A reductive pair of vectors $\mathbf{A}^{(1)}$ and $\mathbf{B}^{(1)}$ is called a *c-reductive pair* if the components of $\mathbf{B}$ and $\mathbf{B}^{(1)}$ are such that

$$b_j - b_j^{(1)} \leq c, \quad 1 \leq j \leq m. \tag{1.2}$$

The vector pair $\mathbf{A}^{(1)}(p) = \mathbf{A}^{(1)}$, $\mathbf{B}^{(1)}$ corresponding to the index p of the component of $\mathbf{A}$ that is set equal to zero is called the *$(c;p)$-reductive pair.*

Note that the balance condition for the vectors is not violated under a reductive step.

Since every pair of vectors $\mathbf{A}$ and $\mathbf{B}$ with nonnegative integer components that satisfies the balance condition defines a nonempty set $M_Z(\mathbf{A}, \mathbf{B})$ of plan matrices, there exists a (p)-reductive pair of vectors $\mathbf{A}^{(1)}(p)$ and $\mathbf{B}^{(1)}$ that corresponds to the pair $\mathbf{A}$, $\mathbf{B}$ for any p $(1 \leq p \leq n)$. Transition to the $(c;p)$-reductive pair of vectors $\mathbf{A}^{(1)}(p)$ and $\mathbf{B}^{(1)}$ is equivalent to constructing the pth row $(x_{p1}, \ldots, x_{pm})$ of a matrix from $M_Z(\mathbf{A}, \mathbf{B})$ such that $x_{pj} \leq c$ for any j $(1 \leq j \leq m)$. The following example demonstrates that there exist pairs of $\mathbf{A}$ and $\mathbf{B}$ and a value of c such that the unified set of conditions (1.1) and (1.2) does not determine any reductive vector pair $\mathbf{A}^{(1)}$, $\mathbf{B}^{(1)}$.

Example 1.1. Suppose $n = m = 3$. Consider the vectors $\mathbf{A} = (a_1, a_2, a_3) = (7, 6, 4)$ and $\mathbf{B} = (b_1, b_2, b_3) = (6, 5, 6)$.

(a) Since $a_i \geq 3$ for any i $(1 \leq i \leq 3)$, there does not exist any $(1;p)$-reductive vector pair $\mathbf{A}^{(1)}(p)$, $\mathbf{B}^{(1)}$ for any $p \in Z$ $(1 \leq p \leq 3)$.

(b) It is obvious that $(7, 0, 4)$, $(4, 3, 4)$ and $(7, 6, 0)$, $(4, 4, 5)$ are $(2;2)$- and $(2;3)$-reductive vector pairs.

(c) Since $a_1 = 7 > 2m = 6$, there does not exist any $(2;1)$-reductive vector pair $\mathbf{A}^{(1)}(1)$, $\mathbf{B}^{(1)}$.

(d) Since $a_i \leq 3m = 9$ for any i $(1 \leq i \leq 3)$ and $b_j \geq 3$ for any j $(1 \leq j \leq 3)$, there exists a $(3;p)$-reductive pair of $\mathbf{A}^{(1)}(p)$ and $\mathbf{B}^{(1)}$ for any $p \in Z$ $(1 \leq p \leq 3)$. In particular, $\mathbf{A}^{(1)}(1) = (0, 6, 4)$ and $\mathbf{B}^{(1)} = (3, 3, 4)$ make up a $(3;1)$-reductive pair.

Note that when $\mathbf{A}^{(1)}(p)$ and $\mathbf{B}^{(1)}$ make up a $(c;p)$-reductive pair corresponding to the pair of $\mathbf{A}$ and $\mathbf{B}$, they also make up a $(d;p)$-reductive pair for the same $\mathbf{A}$ and $\mathbf{B}$ by Definition 1.3, with any $d \in Z$, $d \geq c$.

Definition 1.4. The pair of vectors $\mathbf{A} = (a_1, \ldots, a_n) \in Z_+^n$ and $\mathbf{B} \in Z_+^m$ is called a *c-reducible pair* ($c \in Z$, $c \geq 0$) if there exists a $(c;p)$-reductive vector pair $\mathbf{A}^{(1)}(p)$, $\mathbf{B}^{(1)}$ such that $p \in Z$ $(1 \leq p \leq n)$ and $a_p = \max\{a_i : 1 \leq i \leq n\}$.

Since the inequality $a_q \leq a_p$ holds for any q $(1 \leq q \leq n)$, Definition 1.4 implies that there exists a $(c;q)$-reductive pair $\mathbf{A}^{(1)}(q)$, $\mathbf{B}^{(1)}$ that corresponds to a c-reducible pair $\mathbf{A}$, $\mathbf{B}$ for any q $(1 \leq q \leq n)$.

Lemma 1.1. *If $M_Z(\mathbf{A}, \mathbf{B}; c) \neq \varnothing$, then the vectors $\mathbf{A}$ and $\mathbf{B}$ make up a c-reducible pair.*

Proof. Consider an arbitrary matrix $X(\mathbf{A}, \mathbf{B}) = (x_{ij})$ from $M_Z(\mathbf{A}, \mathbf{B}; c)$. Each of its elements is not greater than c. Setting the elements of its pth row equal to zero, we obtain a matrix $X(\mathbf{A}^{(1)}(p), \mathbf{B}^{(1)}) \in M_Z(\mathbf{A}^{(1)}(p), \mathbf{B}^{(1)}; c)$, where $\mathbf{A}^{(1)}(p)$ and $\mathbf{B}^{(1)}$ make up a $(c;p)$-reducible pair of vectors. Therefore, there exists a c-reductive pair of vectors that corresponds to the largest component of $\mathbf{A}$. Definition 1.4 implies that $\mathbf{A}$ and $\mathbf{B}$ make up a c-reducible pair.

Lemma 1.1 shows that the c-reducibility of $\mathbf{A}$ and $\mathbf{B}$ is a necessary condition for $M_Z(\mathbf{A}, \mathbf{B}; c) \neq \varnothing$. The following example shows that the c-reducibility condition is not sufficient for $M_Z(\mathbf{A}, \mathbf{B}; c)$ to be a nonempty set.

Example 1.2. The vectors $\mathbf{A} = (4, 2)$ and $\mathbf{B} = (3, 1, 1, 1)$ are characterized by equal sums of components and obviously make up a 1-reducible pair. Every matrix $X(\mathbf{A}, \mathbf{B})$ in $M_Z(\mathbf{A}, \mathbf{B}) \neq \varnothing$ has two rows and four columns, and the sum of components of the first column is 3. Consequently, there exists an element in the first column whose value is greater than 1. Therefore, $M_Z(\mathbf{A}, \mathbf{B}; 1) \neq \varnothing$.

Let $\mathbf{A}^{(1)}(p)$, $\mathbf{B}^{(1)}$ be a $(c;p)$-reductive vector pair that corresponds to $\mathbf{A} = (a_1, \ldots, a_n)$ and $\mathbf{B} = (b_1, \ldots, b_m)$. The vector $\mathbf{B}^{(1)}$ can be constructed step by step by decreasing a positive component by unity at each step:

$$\mathbf{B}_0 = \mathbf{B}, \mathbf{B}_1, \ldots, \mathbf{B}_k, \mathbf{B}_{k+1}, \ldots, \mathbf{B}_{a_p} = \mathbf{B}^{(1)}, \tag{1.3}$$

where each vector $\mathbf{B}_{k+1}$ (when $a_p > 0$) is derived from $\mathbf{B}_k$ by decreasing only one of its components by unity $(0 \leq k \leq a_p - 1)$. When $a_p = 0$, it is clear that $\mathbf{B} = \mathbf{B}_0 = \mathbf{B}^{(1)}$.

Consider the c-reducible pair of vectors $\mathbf{A} = (a_1, \ldots, a_n)$ and $\mathbf{B} = (b_1, \ldots, b_m)$. As we have already pointed out, Definition 1.4 implies that there exists a $(c;p)$-reductive vector pair $\mathbf{A}^{(1)}(p)$ and $\mathbf{B}^{(1)}$ for any $p \in Z$ $(1 \leq p \leq n)$. Let us use the sequence of vectors in (1.3) to construct, for an arbitrary $p \in Z$ $(1 \leq p \leq n)$, a special pair of vectors denoted by $\mathbf{A}^{(1)}(p)$ and $\mathbf{B}_*^{(1)}$ by the following rule. We introduce a new notation for the vector $\mathbf{B}$ and its components: $\mathbf{B} = (b_1, \ldots, b_m) = \mathbf{B}_0 = (\beta_1^{(0)}, \ldots, \beta_m^{(0)})$. We also introduce the following notation for vectors in sequence (1.3): $\mathbf{B}_k = (\beta_1^{(k)}, \ldots, \beta_m^{(k)})$. We rewrite sequence (1.3) as

$$\mathbf{B}_0 = (\beta_1^{(0)}, \ldots, \beta_m^{(0)}), \ldots, \mathbf{B}_k = (\beta_1^{(k)}, \ldots, \beta_m^{(k)}), \ldots, \\ \mathbf{B}_{a_p} = (\beta_1^{(a_p)}, \ldots, \beta_m^{(a_p)}) = \mathbf{B}_*^{(1)}. \tag{1.4}$$

The sequence in (1.4) is constructed so that only one component, $\beta_q^{(k)}$, of the vector $\mathbf{B}_k$ is decreased by unity when this vector is changed into $\mathbf{B}_{k+1}$, where $0 \leq k \leq a_p - 1$ (if $a_p \geq 1$). This component satisfies the condition

$$\beta_q^{(k)} = \max\left\{\beta_j^{(k)} : 1 \leq j \leq m,\ \beta_j^{(0)} - \beta_j^{(k)} < c\right\}. \tag{1.5}$$

In what follows (see Lemma 1.4), we show that if vectors $\mathbf{A}$ and $\mathbf{B}$ are c-reducible, then there exists a vector $\mathbf{B}_*^{(1)}$ for any $p \in Z$ $(1 \leq p \leq n)$, and $\mathbf{A}^{(1)}(p)$ and $\mathbf{B}_*^{(1)}$ make up a $(c;p)$-reductive pair of vectors.

Definition 1.5. Suppose that $\mathbf{A}$ and $\mathbf{B}$ make up a c-reducible pair of vectors. The vector pair $\mathbf{A}^{(1)}(p)$, $\mathbf{B}_*^{(1)}$ defined by conditions (1.3)–(1.5) is called a *principal* $(c;p)$*-reductive pair.*

Note that the principal $(c;p)$-reductive pair of vectors is not uniquely defined, because the component $\beta_q^{(k)}$ defined by (1.5) is not unique.

Let us analyze the concept of c-reducibility. We show that it guarantees the existence of a principal $(c;p)$-reductive pair $\mathbf{A}^{(1)}(p)$, $\mathbf{B}_*^{(1)}$ for any admissible value of p.

Lemma 1.2. *Vectors* $\mathbf{A} = (a_1, \ldots, a_n) \in Z_+^n$ *and* $\mathbf{A} = (a_1, \ldots, a_n) \in Z_+^n$ *make up a c-reducible pair if and only if*

$$\max\{a_i : 1 \leq i \leq n\} \leq \sum_{j=1}^{m} \min(b_j, c). \tag{1.6}$$

Proof.

Necessity. Suppose that $\mathbf{A}$ and $\mathbf{B}$ make up a c-reducible vector pair and $a_p = \max\{a_i : 1 \leq i \leq n\}$. By Definitions 1.1, 1.3, and 1.4, there exists a vector $\mathbf{B}^{(1)} = (b_1^{(1)}, \ldots, b_m^{(1)}) \in Z_+^m$ whose components satisfy conditions (1.1) and (1.2). Suppose that

$$a_p > \sum_{j=1}^{m} \min(b_j, c).$$

Then, (1.1) entails

$$\sum_{j=1}^{m} \min(b_j, c) < \sum_{j=1}^{m} (b_j - b_j^{(1)}).$$

Subtracting the right-hand side of this inequality from its left-hand side, we obtain

$$\sum_{j=1}^{m} \min(b_j, c) - \sum_{j=1}^{m} (b_j - b_j^{(1)}) =$$
$$\sum_{b_j<c} \min(b_j, c) + \sum_{b_j \geq c} \min(b_j, c) - \sum_{b_j<c} (b_j - b_j^{(1)}) - \sum_{b_j \geq c} (b_j - b_j^{(1)}) =$$
$$\sum_{b_j<c} b_j^{(1)} + \sum_{b_j \geq c} \left(c - (b_j - b_j^{(1)})\right) \geq 0.$$

This follows from (1.2) combined with the fact that the components of $\mathbf{B}^{(1)}$ are nonnegative. The resulting contradiction proves inequality (1.6) when the vectors $\mathbf{A}$ and $\mathbf{B}$ make up a c-reducible pair.

Sufficiency. Suppose inequality (1.6) is satisfied and $a_p = \max\{a_i : 1 \leq i \leq n\}$. For the vector $\mathbf{B}$ and number c, define $s_0(\mathbf{B}; c) = 0$ and

$$s_q(\mathbf{B}; c) = \sum_{j=1}^{q} \min(b_j, c),$$

$$k(\mathbf{B}; c) = \max\left\{q : q \in Z,\ 0 \leq q \leq m,\ s_q(\mathbf{B}; c) \leq a_p\right\},$$

where $q \in Z$ $(1 \leq q \leq m)$. Using $s_q(\mathbf{B}; c)$ and $k(\mathbf{B}; c)$, construct the vector $(x_{p1}, \ldots, x_{pm})$ whose components are defined by the conditions

$$x_{pj} = \begin{cases} \min(b_j, c), \quad j \leq k(\mathbf{B}; c) \text{ (if } k(\mathbf{B}; c) \geq 1), \\ a_p - s_{k(\mathbf{B};c)}(\mathbf{B}; c), \quad j = k(\mathbf{B}; c) + 1 \text{ (if } k(\mathbf{B}; c) < m), \\ 0, \quad j > k(\mathbf{B}; c) + 1 \text{ (if } k(\mathbf{B}; c) < m). \end{cases}$$

The components of $\mathbf{B} = (b_1, \ldots, b_m)$ and $(x_{p1}, \ldots, x_{pm})$ make up a new vector $\mathbf{B}^{(1)} = (b_1^{(1)}, \ldots, b_m^{(1)})$, where $b_j^{(1)} = b_j - x_{pj}$ $(1 \leq j \leq m)$. It is obvious that $\mathbf{B}^{(1)} \in Z_+^m$ and the vectors $\mathbf{B}$ and $\mathbf{B}^{(1)}$ and the components a_p of $\mathbf{A}$ satisfy conditions (1.1) and (1.2). Therefore, $\mathbf{A}^{(1)}(p)$ and $\mathbf{B}^{(1)}$ make up a $(c;p)$-reductive vector pair corresponding to $\mathbf{A}$ and $\mathbf{B}$; that is, the pair of vectors $\mathbf{A}$ and $\mathbf{B}$ is c-reducible, which proves the lemma.

Since inequality (1.6) holds if and only if the characteristic function defined by (1.22) in Chapter 1 is nonnegative for $k = 1$, Lemma 1.2 entails the following proposition.

Lemma 1.3. *The pair of vectors* $\mathbf{A} \in Z_+^n$ *and* $\mathbf{B} \in Z_+^m$ *is c-reducible if and only if* $\delta_1(\widetilde{\mathbf{A}}, \mathbf{B}; c) \geq 0$.

Proof. Combine definition (1.22) given in Chapter 1 with the identity

$$\sum_{j=1}^{m} b_j - \sum_{b_j \geq c} (b_j - c) = \sum_{j=1}^{m} \min(b_j, c).$$

Now, we show that the c-reducibility of a pair of vectors entails the existence of a principal $(c;p)$-reductive pair for any admissible p.

Lemma 1.4. *Vectors* $\mathbf{A} = (a_1, \ldots, a_n) \in Z_+^n$ *and* $\mathbf{B} = (b_1, \ldots, b_m) \in Z_+^m$ *make up a c-reducible pair if and only if there exists a principal* $(c;p)$*-reductive pair of vectors* $\mathbf{A}^{(1)}(p) = (a_1^{(1)}, \ldots, a_n^{(1)})$ *and* $\mathbf{B}_*^{(1)} = (b_1^{(1)}, \ldots, b_m^{(1)})$ *for any* p $(1 \leq p \leq n)$.

Proof.

Necessity. Suppose that the vector $\mathbf{B}_*^{(1)}$ defined by the sequence of vectors (1.4) in Definition 1.5 does not exist for a certain value of p, where $1 \leq p \leq n$. Then, there exists k $(0 \leq k \leq a_p - 1)$ such that the sequence of vectors in (1.4) includes $\mathbf{B}_k$, but does not include $\mathbf{B}_{k+1}$. The vector $\mathbf{B}_{k+1}$ does not exist because the compont $\beta_q^{(k)}$ cannot be determined. Therefore, the components of $\mathbf{B}_k = (\beta_1^{(k)}, \ldots, \beta_m^{(k)})$ are such that either $\beta_j^{(0)} - \beta_j^{(k)} = c$ if $\beta_j^{(0)} \geq c$ or $\beta_j^{(k)} = 0$ if $\beta_j^{(0)} < c$. Since each kth vector in (1.4) is obtained by decreasing a positive component by unity, we have

$$k = \sum_{j=1}^{m} (\beta_j^{(0)} - \beta_j^{(k)}) = \sum_{j=1}^{m} \min(\beta_j^{(0)}, c).$$

Recall that $\mathbf{B} = (b_1, \ldots, b_m) = \mathbf{B}_0 = (\beta_1^{(0)}, \ldots, \beta_m^{(0)})$. Consequently,

$$k = \sum_{j=1}^{m} \min(b_j, c).$$

Then, Lemma 1.2 entails

$$k = \sum_{j=1}^{m} \min(b_j, c) \geq \max\{a_i : 1 \leq i \leq n\} \geq a_p,$$

which contradicts the condition $k \leq a_p - 1$.

Sufficiency follows from Definition 1.4.

Definitions 1.4 and 1.5 combined with Lemma 1.4 imply that if there exists a $(c;p)$-reductive vector pair corresponding to $\mathbf{A} \in Z_+^n$, $\mathbf{B} \in Z_+^m$ when $a_p = \max\{a_i : 1 \le i \le n\}$, then there exists a principal $(c;q)$-reductive pair for any q $(1 \le q \le n)$.

The concepts of c-reducibility and $(c;p)$-reductiveness can be used to establish necessary and sufficient conditions for the nonemptiness of $M_Z(\mathbf{A},\mathbf{B};c)$ and to construct an algorithm for calculation of matrices from $M_Z(\mathbf{A},\mathbf{B};c)$.

Let $\mathbf{A} = (a_1,\dots,a_n) \in Z_+^n$ and $\mathbf{B} = (b_1,\dots,b_m) \in Z_+^m$ be arbitrary vectors, and define a sequence $p_1,\dots,p_k$ of indices (component numbers) of the vector $\mathbf{A}$, where $1 \le k \le n$ and $1 \le p_i \le n$ $(p_i \ne p_j,\ i \ne j)$. If the vector pair $\mathbf{A}$, $\mathbf{B}$ is c-reducibile, then we have a corresponding $(c;p_1)$-reductive vector pair $\mathbf{A}^{(1)} = \mathbf{A}^{(1)}(p_1) = (a_1^{(1)},\dots,a_n^{(1)})$, $\mathbf{B}^{(1)} = (b_1^{(1)},\dots,b_m^{(1)})$. The pair $\mathbf{A}^{(1)}$, $\mathbf{B}^{(1)}$ may also be c-reducible. Then, applying a reductive step to $\mathbf{A}^{(1)}$ and $\mathbf{B}^{(1)}$, we obtain a $(c;p_2)$-reductive pair of vectors $\mathbf{A}^{(2)} = \mathbf{A}^{(2)}(p_2) = (a_1^{(2)},\dots,a_n^{(2)})$ and $\mathbf{B}^{(2)} = (b_1^{(2)},\dots,b_m^{(2)})$.

The new pair $\mathbf{A}^{(2)}$, $\mathbf{B}^{(2)}$ is derived from $\mathbf{A}$ and $\mathbf{B}$ by applying successive reductive steps as follows:

$$\mathbf{A}^{(2)} = \mathbf{A}^{(2)}(p_2) = \left(\mathbf{A}^{(1)}(p_1)\right)^{(1)}(p_2) = \mathbf{A}^{(2)}(p_1,p_2), \quad \mathbf{B}^{(2)} = \left(\mathbf{B}^{(1)}\right)^{(1)}.$$

The pair $\mathbf{A}^{(2)} = \mathbf{A}^{(2)}(p_1,p_2)$, $\mathbf{B}^{(2)}$ is called the $(c;p_1,p_2)$-*reductive pair* (corresponding to $\mathbf{A}$ and $\mathbf{B}$). Suppose $k-1$ reductive steps can be applied to the pair of $\mathbf{A}$ and $\mathbf{B}$. Assume that the $(c;p_1,\dots,p_{k-1})$-reductive pair of vectors has been constructed for $\mathbf{A}$ and $\mathbf{B}$ as a result of $k-1$ reductive steps:

$$\begin{aligned}\mathbf{A}^{(k-1)} &= \mathbf{A}^{(k-1)}(p_1,\dots,p_{k-1}) = (a_1^{(k-1)},\dots,a_n^{(k-1)}),\\ \mathbf{B}^{(k-1)} &= (b_1^{(k-1)},\dots,b_m^{(k-1)}).\end{aligned}$$

If $\mathbf{A}^{(k-1)}$ and $\mathbf{B}^{(k-1)}$ make up a c-reducible pair of vectors, then a reductive step (i.e., the kth reductive step with respect to $\mathbf{A}$ and $\mathbf{B}$) applied to it will result in a $(c;p_k)$-reductive pair of vectors $\mathbf{A}^{(k)} = (a_1^{(k)},\dots,a_n^{(k)})$, $\mathbf{B}^{(k)} = (b_1^{(k)},\dots,b_m^{(k)})$, which may be described as follows:

$$\begin{aligned}&\mathbf{A}^{(k)} = \left(\mathbf{A}^{(k-1)}\right)^{(1)}(p_k) = \left(\mathbf{A}^{(k-1)}(p_1,\dots,p_{k-1})\right)^{(1)}(p_k) = \mathbf{A}^{(k)}(p_1,\dots,p_k),\\ &\mathbf{B}^{(k)} = \left(\mathbf{B}^{(k-1)}\right)^{(1)}.\end{aligned}$$

The pair $\mathbf{A}^{(k)}$, $\mathbf{B}^{(k)}$ is called a $(c;p_1,\dots,p_k)$-*reductive pair* of vectors if it is obtained by applying k reductive steps to $\mathbf{A}$ and $\mathbf{B}$.

If each kth reductive step above results in constructing a principal $(c; p_k)$-reductive pair of vectors (see Definition 1.5), we refer to this pair as the *principal* $(c; p_1, \ldots, p_k)$*-reductive pair* $\mathbf{A}^{(k)}$, $\mathbf{B}_*^{(k)}$ $\left(\mathbf{B}_*^{(k)} = \left(\mathbf{B}_*^{(k-1)}\right)_*^{(1)}\right)$ corresponding to $\mathbf{A}$ and $\mathbf{B}$.

Note that the reduction procedures based on Definitions 1.1–1.5 and Lemmas 1.2–1.4 do not rely on the balance conditions for the vectors under consideration. However, the conservation of the difference between the sums of components of these vectors was required at each reductive step.

Lemma 1.5. *Suppose vectors* $\mathbf{A} \in Z_+^n$ *and* $\mathbf{B} \in Z_+^m$ *satisfy the balance condition. If the pair* $\mathbf{A}$, $\mathbf{B}$ *admits every* k*th reductive step* $(1 \leq k \leq n)$ *resulting in the c-reductive pair* $(\mathbf{A}^{(k)}, \mathbf{B}^{(k)})$, *then the reductive pair of vectors* $\mathbf{A}^{(n)}$ *and* $\mathbf{B}^{(n)}$ *obtained at the* n*th reductive step consists of zero vectors.*

1.2. Reductive Criterion for the Existence of Transportation Matrices

The following theorem states necessary and sufficient conditions under which $M_Z(\mathbf{A}, \mathbf{B}; c) \neq \varnothing$, based on the concepts introduced by Definitions 1.1–1.5.

Theorem 1.1. *Suppose vectors* $\mathbf{A} = (a_1, \ldots, a_n) \in Z_+^n$ *and* $\mathbf{B} = (b_1, \ldots, b_n) \in Z_+^m$ $(n > 1)$ *are characterized by equal sums of components,* $c \in Z$ $(c \geq 0)$, *and the sequence* $p_1, \ldots, p_n$ *is obtained by an arbitrary permutation of the indices of components of* $\mathbf{A}$. *The matrix set* $M_Z(\mathbf{A}, \mathbf{B}; c)$ *is not empty if and only if* $\mathbf{A}$ *and* $\mathbf{B}$ *make up a c-reducible pair of vectors and so is every principal* $(c; p_1, \ldots, p_k)$*-reductive pair*

$$\mathbf{A}^{(k)} = \mathbf{A}^{(k)}(p_1, \ldots, p_k), \quad \mathbf{B}_*^{(k)}, \quad 1 \leq k \leq n-1,$$

obtained at the k*th reductive step for the pair of vectors* $\mathbf{A}$ *and* $\mathbf{B}$.

Proof.

Sufficiency. Suppose that $\mathbf{A}$ and $\mathbf{B}$ make up a c-reducible pair for which the sequence of principal $(c; p_1, \ldots, p_k)$-reductive pairs $\mathbf{A}^{(k)}$, $\mathbf{B}_*^{(k)}$ is constructed:

$$\mathbf{A}^{(k)} = \mathbf{A}^{(k)}(p_1, \ldots, p_k) = (a_1^{(k)}, \ldots, a_n^{(k)}), \quad 1 \leq k \leq n-1,$$

$$\mathbf{B}_*^{(k)} = (b_1^{(k)}, \ldots, b_m^{(k)}), \quad 1 \leq k \leq n-1,$$

where each pair is c-reducible.

We now introduce a new notation for the vector $\mathbf{B}$ and its components: $\mathbf{B} = (b_1, \ldots, b_m) = \mathbf{B}_*^{(0)} = (b_1^{(0)}, \ldots, b_m^{(0)})$. We construct a matrix $X = (x_{ij})$ from $M_Z(\mathbf{A}, \mathbf{B}; c)$, with elements calculated by using reductive pair of vectors $\mathbf{A}^{(k)}$ and $\mathbf{B}_*^{(k)}$. Each vector $\mathbf{B}_*^{(k)}$ defines a row of X with index p_k $(1 \leq k \leq$

$n-1$). We assume that the components of the row $(x_{p_k1}, \ldots, x_{p_km})$ are such that

$$x_{p_kj} = b_j^{(k-1)} - b_j^{(k)}, \quad 1 \le j \le m, \quad 1 \le k \le n-1.$$

The elements of these $n-1$ rows of X determine its last row $(x_{p_n1}, \ldots, x_{p_nm})$, whose index is p_n:

$$x_{p_nj} = b_j^{(0)} - \sum_{k=1}^{n-1} x_{p_kj}, \quad 1 \le j \le m.$$

For $\mathbf{A}^{(k)}$ and $\mathbf{B}_*^{(k)}$ making up a c-reductive pair of vectors, we have

$$0 \le x_{p_kj} \le c, \quad 1 \le j \le m, \quad 1 \le k \le n-1,$$

$$a_{p_k} = \sum_{j=1}^{m} x_{p_kj}, \quad 1 \le k \le n-1.$$

Since the pair $\mathbf{A}^{(n-1)}$, $\mathbf{B}_*^{(n-1)}$ is c-reducible, and the vectors $\mathbf{A}$ and $\mathbf{B}$ satisfy balance condition (1.1), the elements of the row of X whose index is p_n are such that

$$0 \le x_{p_nj} \le c, \quad 1 \le j \le m, \quad a_{p_n} = \sum_{j=1}^{m} x_{p_nj}.$$

It is clear that the following equality is true for any j $(1 \le j \le m)$:

$$\sum_{k=1}^{n} x_{p_kj} = b_j^{(0)} = b_j.$$

Consequently, the constructed matrix, $X = (x_{p_kj})$ $(1 \le k \le n, 1 \le j \le m)$, belongs to the set $M_Z(\mathbf{A}, \mathbf{B}; c)$; therefore, $M_Z(\mathbf{A}, \mathbf{B}; c)$ is not empty, which proves the sufficiency.

Necessity. For convenience, we assume that the components of $\mathbf{A}$ are arranged in nonascending order without loss of generality. Suppose that $M_Z(\mathbf{A}, \mathbf{B}; c) \neq \varnothing$, where $\mathbf{A} \in \overline{R}_+^n$, $\mathbf{B} \in R_+^m$, and the sequence $p_1, \ldots, p_n$ is obtained by an arbitrary permutation of the indices of components of $\mathbf{A}$. Let us show that the pair of vectors $\mathbf{A}$ and $\mathbf{B}$ is c-reducible and so is every principal $(c; p_1, \ldots, p_n)$-reductive pair $\mathbf{A}^{(k)}$, $\mathbf{B}_*^{(k)}$ $(1 \le k \le n-1)$.

We prove the necessity of the assumptions above by induction on the number of components of $\mathbf{A}$, using the characteristic function defined by (1.22) in Chapter 1, with $c \in Z$.

When $n = 1$, the vectors $\mathbf{A}$ and $\mathbf{B}$ obviously make up a c-reducible pair. Suppose that the condition $M_Z(\mathbf{A}, \mathbf{B}; c) \neq \varnothing$ entails the necessity of the assumptions of the theorem when $\mathbf{A} \in \overline{R}_+^{n-1}$, and let $\mathbf{A} \in \overline{R}_+^n$.

Because $M_Z(\mathbf{A},\mathbf{B};c) \neq \varnothing$, Lemma 1.1 (or Lemma 1.3) establishes that the pair of vectors $\mathbf{A}$ and $\mathbf{B}$ is c-reducible, and Lemma 1.4 entails the existence of a principal $(c;p_1)$-reductive pair of vectors $\mathbf{A}^{(1)}(p_1) = (a_1^{(1)},\ldots,a_n^{(1)})$, $\mathbf{B}_*^{(1)} = (b_1^{(1)},\ldots,b_m^{(1)})$ for $\mathbf{A}$ and $\mathbf{B}$. Let us show that $\delta_k(\widetilde{\mathbf{A}}^{(1)}(p_1),\mathbf{B}_*^{(1)};c) \geq 0$ for any admissible k. Then (by Theorem 2.2 from Chapter 1), $M_Z(\widetilde{\mathbf{A}}^{(1)}(p_1),\mathbf{B}_*^{(1)};c) \neq \varnothing$, and the induction hypothesis will be applied to prove the necessity of the assumptions of Theorem 1.1.

To do this, we denote any $(c;p)$-reductive pair of vectors corresponding to $\mathbf{A}$ and $\mathbf{B}$ by $\mathbf{A}^{(1)}(p)$, $\mathbf{B}' = (b'_1,\ldots,b'_m)$ to distinguish between the components of the second vectors of the principal $(c;p)$-reductive pair and an arbitrary $(c;p)$-reductive pair. We set

$$
\begin{aligned}
r(\mathbf{B},\mathbf{B}';t) &= \sum_{b_j \geq t}(b_j - t) - \sum_{b'_j \geq t}(b'_j - t), \\
r(\mathbf{B},\mathbf{B}_*^{(1)};t) &= \sum_{b_j \geq t}(b_j - t) - \sum_{b_j^{(1)} \geq t}(b_j^{(1)} - t),
\end{aligned}
\tag{1.7}
$$

where $t \in Z$ $(t \geq 0)$. The construction of the reductive pairs and the principal reductive pair, defined by (1.3) and (1.4), respectively, and by the condition for choosing the component to be decreased in the case of the principal reductive pair defined by (1.5), clearly shows that

$$
r(\mathbf{B},\mathbf{B}_*^{(1)};t) \geq r(\mathbf{B},\mathbf{B}';t), \tag{1.8}
$$

where $\mathbf{B}'$ is the second vector of an arbitrary $(c;p)$-reductive pair.

Suppose $X(\mathbf{A},\mathbf{B}) = (x_{ij}) \in M_Z(\mathbf{A},\mathbf{B};c)$. Then, the vectors $\mathbf{A}^{(1)}(p)$ and $\mathbf{B}' = (b'_1,\ldots,b'_m)$, where $b'_j = b_j - x_{p_1 j}$ $(1 \leq j \leq m)$, make up a $(c;p_1)$-reductive pair, and $M_Z(\mathbf{A}^{(1)}(p_1),\mathbf{B}';c) \neq \varnothing$, which is equivalent to the inequality $\delta_k(\widetilde{\mathbf{A}}^{(1)}(p_1),\mathbf{B}';c) \geq 0$ for every admissible k. Now, consider the function defined by (1.22) in Chapter 1 for the pairs $\mathbf{A}$, $\mathbf{B}$; $\widetilde{\mathbf{A}}^{(1)}(p_1)$, $\mathbf{B}'$; and $\widetilde{\mathbf{A}}^{(1)}(p_1)$, $\mathbf{B}_*^{(1)}$.

Since the characteristic function of the first two pairs is nonnegative, and

$$
\sum_{i=1}^{k} \tilde{a}_i = \begin{cases} \displaystyle\sum_{i=1}^{k} a_i, & k < p_1, \\ \displaystyle\sum_{i=1}^{k} a_i - a_{p_1}, & k \geq p_1, \end{cases}
$$

we have

$$
\delta_k(\mathbf{A},\mathbf{B};c) = \sum_{j=1}^{m} b_j - \sum_{b_j \geq ck}(b_j - ck) - \sum_{i=1}^{k} a_i \geq 0,
$$

$$\delta_k(\widetilde{\mathbf{A}}^{(1)}(p_1), \mathbf{B}'; c) = \sum_{j=1}^{m} b'_j - \sum_{b'_j \geq ck} (b'_j - ck) - \sum_{i=1}^{k} \tilde{a}_i \geq 0,$$

$$\delta_k(\widetilde{\mathbf{A}}^{(1)}(p_1), \mathbf{B}_*^{(1)}; c) = \sum_{j=1}^{m} b_j^{(1)} - \sum_{b_j^{(1)} \geq ck} (b_j^{(1)} - ck) - \sum_{i=1}^{k} \tilde{a}_i.$$

Let us examine the remainders $\delta_k(\mathbf{A}, \mathbf{B}; c) - \delta_k(\widetilde{\mathbf{A}}^{(1)}(p_1), \mathbf{B}'; c)$ and $\delta_k(\mathbf{A}, \mathbf{B}; c) - \delta_k(\widetilde{\mathbf{A}}^{(1)}(p_1), \mathbf{B}_*^{(1)}; c)$, using (1.7):

$$\delta_k(\mathbf{A}, \mathbf{B}; c) - \delta_k(\widetilde{\mathbf{A}}^{(1)}(p_1), \mathbf{B}'; c) = a_p - r(\mathbf{B}, \mathbf{B}'; ck) - \sum_{i=1}^{k} a_i + \sum_{i=1}^{k} \tilde{a}_i,$$

$$\delta_k(\mathbf{A}, \mathbf{B}; c) - \delta_k(\widetilde{\mathbf{A}}^{(1)}(p_1), \mathbf{B}_*^{(1)}; c) = a_p - r(\mathbf{B}, \mathbf{B}_*^{(1)}; ck) - \sum_{i=1}^{k} a_i + \sum_{i=1}^{k} \tilde{a}_i.$$

Inequality (1.8) implies that $r(\mathbf{B}, \mathbf{B}_*^{(1)}; ck) \geq r(\mathbf{B}, \mathbf{B}'; ck)$ and

$$\delta_k(\mathbf{A}, \mathbf{B}; c) - \delta_k(\widetilde{\mathbf{A}}^{(1)}(p_1), \mathbf{B}'; c) \geq \delta_k(\mathbf{A}, \mathbf{B}; c) - \delta_k(\widetilde{\mathbf{A}}^{(1)}(p_1), \mathbf{B}_*^{(1)}; c).$$

Consequently,

$$\delta_k(\widetilde{\mathbf{A}}^{(1)}(p_1), \mathbf{B}_*^{(1)}; c) \geq \delta_k(\widetilde{\mathbf{A}}^{(1)}(p_1), \mathbf{B}'; c) \geq 0,$$

and $M_Z(\mathbf{A}^{(1)}(p_1), \mathbf{B}_*^{(1)}; c) \neq \varnothing$ by Theorem 2.2 from Chapter 1. By Lemma 1.1, the principal $(c; p_1)$-reductive pair of vectors $\widetilde{\mathbf{A}}^{(1)}(p_1)$ and $\mathbf{B}_*^{(1)}$ is c-reducible.

Here, $\mathbf{A}^{(1)}(p_1) = (a_1^{(1)}, \ldots, a_n^{(1)}) \in R_+^n$ and $a_{p_1}^{(1)} = 0$, because the row with index p_1 contains only zeros in every matrix from $M_Z(\mathbf{A}^{(1)}(p_1), \mathbf{B}_*^{(1)}; c)$. Therefore, the induction hypothesis applies: each principal $(c; p_2, \ldots, p_k)$-reductive pair

$$\left(\widetilde{\mathbf{A}}^{(1)}(p_1)\right)^{(k-1)}(p_2, \ldots, p_k), \quad \left(\mathbf{B}_*^{(1)}\right)_*^{(k-1)}, \quad 2 \leq k \leq n-1,$$

corresponding to $\widetilde{\mathbf{A}}^{(1)}(p_1)$ and $\mathbf{B}_*^{(1)}$, where $n > 2$, is c-reducible. It is obvious that every principal $(c; p_2, \ldots, p_k)$-reductive pair of vectors that corresponds to $\mathbf{A}^{(1)}(p_1)$ and $\mathbf{B}_*^{(1)}$ is a $(c; p_1, \ldots, p_k)$-reductive pair corresponding to $\mathbf{A}$ and $\mathbf{B}$:

$$\left(\mathbf{A}^{(1)}(p_1)\right)^{(k-1)}(p_2, \ldots, p_k) = \mathbf{A}^{(k)}(p_1, \ldots, p_k), \quad \left(\mathbf{B}_*^{(1)}\right)_*^{(k-1)} = \mathbf{B}_*^{(k)}.$$

Thus, each principal $(c; p_1, \ldots, p_k)$-reductive pair of vectors $\mathbf{A}^{(k)}(p_1, \ldots, p_k)$ and $\mathbf{B}_*^{(k)}$ is c-reducible, which proves the necessity of the assumptions of Theorem 1.1.

Remark 1.1. Theorem 1.1 does not rely on the existence of a $(c; p_1, \ldots, p_k)$-reductive pair of vectors, which is guaranteed by the fact that the preceding principal reductive pair (or the starting pair) of vectors is c-reducible by Lemma 1.4.

Remark 1.2. In view of Lemma 1.5, it is not necessary to consider a $(c; p_1, \ldots, p_n)$-reductive vector pair in the context of Theorem 1.1.

Remark 1.3. The proof of the necessity of the assumptions of Theorem 1.1 is based on inequality (1.8), which may be interpreted as follows. If $M_Z(\mathbf{A}, \mathbf{B}; c) \neq \varnothing$, and $\mathbf{A}^{(1)}(p)$, $\mathbf{B}^{(1)}$ and $\mathbf{A}^{(1)}(p)$, $\mathbf{B}_*^{(1)}$ are, respectively, reductive and principal reductive pairs of vectors, then the condition $M_Z(\mathbf{A}^{(1)}(p), \mathbf{B}^{(1)}; c) \neq \varnothing$ implies that $M_Z(\mathbf{A}^{(1)}(p), \mathbf{B}_*^{(1)}; c) \neq \varnothing$ a fortiori.

1.3. Reduction Algorithms for Constructing Transportation Matrices

The following algorithm is constructed by invoking Theorem 1.1. Note that if $\mathbf{A}$ and $\mathbf{B}$ are integer, nonnegative vectors with equal sums of components, then $M_Z(\mathbf{A}, \mathbf{B}; 0) \neq \varnothing$ if and only if $\mathbf{A}$ and $\mathbf{B}$ are zero vectors.

Algorithm 1.1. Construction of an integer matrix with a fixed upper bound for all of its elements.

Let the vectors $\mathbf{A} = (a_1, \ldots, a_n) \in Z_+^n$ and $\mathbf{B} = (b_1, \ldots, b_m) \in Z_+^m$ be characterized by equal sums of components, $c \in Z$ $(c > 0)$, and $p_1, \ldots, p_n$ is an arbitrary sequence of different integers ranging from 1 to n.

Step 1. Set $p_0 = 0$ and $\mathbf{A}^{(0)}(p_0) = \mathbf{A}$, $\mathbf{B}_*^{(0)} = (b_1^{(0)}, \ldots, b_m^{(0)}) = \mathbf{B} = (b_1, \ldots, b_m)$.

Step 2. Set $k = 1$.

Step 3. If possible, construct the principal $(c; p_0, \ldots, p_k)$-reductive pair of vectors

$$\mathbf{A}^{(k)}(p_0, \ldots, p_k) = \left(\mathbf{A}^{(k-1)}(p_0, \ldots, p_{k-1})\right)^{(1)}(p_k),$$

$$\mathbf{B}_*^{(k)} = \left(\mathbf{B}_*^{(k-1)}\right)_*^{(1)} = (b_1^{(k)}, \ldots, b_m^{(k)}),$$

where $\mathbf{B}_*^{(k)}$ is derived from $\mathbf{B}_*^{(k-1)}$ by using the sequence defined by (1.4) and rule (1.5) for choosing the component to be decreased by unity.

Step 4. If the principal $(c; p_0, \ldots, p_k)$-reductive pair $\mathbf{A}^{(k)}(c; p_0, \ldots, p_k)$, $\mathbf{B}_*^{(k)}$ is constructed, then go to Step 5. If a principal $(c; p_0, \ldots, p_k)$-reductive pair $\mathbf{A}^{(k-1)}(c; p_0, \ldots, p_{k-1})$, $\mathbf{B}_*^{(k-1)}$ cannot be constructed, then go to Step 7.

Step 5. If $k \leq n - 1$, then increase the value of k by unity and go to Step 3. If $k = n$, then go to the next step.

Step 6. Define the elements of the desired matrix $X(\mathbf{A}, \mathbf{B}) = (x_{ij})$ from $M_Z(\mathbf{A}, \mathbf{B}; c)$ as

$$x_{p_k j} = b_j^{(k-1)} - b_j^{(k)}, \quad 1 \leq j \leq m, \quad 1 \leq k \leq n,$$

and stop.

Step 7. Set $M_Z(\mathbf{A}, \mathbf{B}; c) = \varnothing$.

Remark 1.4. In contrast to Theorem 1.1, Algorithm 1.1 does not rely on the concept of c-reducibility, because c-reducibility is equivalent to the existence of a c-reductive pair of vectors by Definition 1.4.

Algorithm 1.1 is better suited for calculating the elements of integer transportation matrices, as compared to Algorithm 2.1 described in Chapter 1. When combined with Algorithm 2.1 from Chapter 2, Algorithm 1.1 provides a rational method for constructing integer minimax matrices.

Example 1.3. Application of Algorithm 1.1. It can be readily checked that $\mathbf{A} = (18, 17, 15, 8, 6) \in \overline{Z}_+^5$ and $\mathbf{B} = (13, 13, 12, 12, 7, 5, 2) \in \overline{Z}_+^7$ satisfy the balance condition. Suppose $c = 3$ and $p_k = k$. We consecutively construct the principal $(3; 1, \ldots, k)$-reductive pairs of vectors corresponding to $\mathbf{A}$ and $\mathbf{B}$.

It is obvious that $\mathbf{A}^{(1)}(1) = (0, 17, 15, 8, 6)$, $\mathbf{A}^{(2)}(1, 2) = (0, 0, 15, 8, 6)$, $\mathbf{A}^{(3)}(1, 2, 3) = (0, 0, 0, 8, 6)$, $\mathbf{A}^{(4)}(1, 2, 3, 4) = (0, 0, 0, 0, 6)$, $\mathbf{A}^{(5)}(1, 2, 3, 4, 5) = (0, 0, 0, 0, 0)$.

Now, we construct the vectors $\mathbf{B}_*^{(k)}$ $(1 \leq k \leq 5)$ by using the sequence defined by (1.4) and applying rule (1.5) to determine the component to be decreased by unity. Here, we demonstrate in detail only the transition from $\mathbf{B}$ to $\mathbf{B}_*^{(1)}$:

$$\mathbf{B} = \mathbf{B}_0 = (13, 13, 12, 12, 7, 5, 2), \quad \mathbf{B}_1 = (13, 12, 12, 12, 7, 5, 2),$$

$$\mathbf{B}_2 = (12, 12, 12, 12, 7, 5, 2), \quad \mathbf{B}_3 = (12, 12, 12, 11, 7, 5, 2),$$

$$\mathbf{B}_4 = (12, 12, 11, 11, 7, 5, 2), \quad \mathbf{B}_5 = (12, 11, 11, 11, 7, 5, 2),$$

$$\mathbf{B}_6 = (11, 11, 11, 11, 7, 5, 2), \quad \mathbf{B}_7 = (11, 11, 11, 10, 7, 5, 2),$$

$$\mathbf{B}_8 = (11, 11, 10, 10, 7, 5, 2), \quad \mathbf{B}_9 = (11, 10, 10, 10, 7, 5, 2),$$

$$\mathbf{B}_{10} = (10, 10, 10, 10, 7, 5, 2), \quad \mathbf{B}_{11} = (10, 10, 10, 9, 7, 5, 2),$$

$$\mathbf{B}_{12} = (10, 10, 9, 9, 7, 5, 2), \quad \mathbf{B}_{13} = (10, 10, 9, 9, 6, 5, 2),$$
$$\mathbf{B}_{14} = (10, 10, 9, 9, 5, 5, 2), \quad \mathbf{B}_{15} = (10, 10, 9, 9, 5, 4, 2),$$
$$\mathbf{B}_{16} = (10, 10, 9, 9, 4, 4, 2), \quad \mathbf{B}_{17} = (10, 10, 9, 9, 4, 3, 2),$$
$$\mathbf{B}_*^{(1)} = \mathbf{B}_{18} = (10, 10, 9, 9, 4, 2, 2).$$

In constructing the remaining vectors $\mathbf{B}_*^{(k)}$ $(2 \le k \le 5)$, we write out the indices of the decreased components above each vector $\mathbf{B}_*^{(k-1)}$. Then, each component of $\mathbf{B}_*^{(k)}$ is equal to the difference between the corresponding component of $\mathbf{B}_*^{(k-1)}$ and the number of indices written out above the component:

$$\mathbf{B}_*^{(1)} = (\overset{\substack{10\\6\\2}}{10},\ \overset{\substack{9\\5\\1}}{10},\ \overset{\substack{12\\8\\4}}{9},\ \overset{\substack{11\\7\\3}}{9},\ \overset{\substack{17\\14\\13}}{4},\ \overset{16}{2},\ \overset{15}{2}), \quad \mathbf{B}_*^{(2)} = (\overset{\substack{10\\6\\2}}{7},\ \overset{\substack{9\\5\\1}}{7},\ \overset{\substack{12\\8\\4}}{6},\ \overset{\substack{11\\7\\3}}{6},\ \overset{15}{1},\ \overset{14}{1},\ \overset{13}{1}),$$

$$\mathbf{B}_*^{(3)} = (\overset{\substack{6\\2}}{4},\ \overset{\substack{5\\1}}{4},\ \overset{\substack{8\\4}}{3},\ \overset{\substack{7\\3}}{3},\ 0,\ 0,\ 0), \quad \mathbf{B}_*^{(4)} = (\overset{\substack{6\\2}}{2},\ \overset{\substack{5\\1}}{2},\ \overset{4}{1},\ \overset{3}{1},\ 0,\ 0,\ 0),$$

$$\mathbf{B}_*^{(5)} = (0,\ 0,\ 0,\ 0,\ 0,\ 0,\ 0).$$

Since all of the principal $(3; 1, \ldots, k)$-reductive pairs of vectors exist $(1 \le k \le 5)$, it holds that $M_Z(\mathbf{A}, \mathbf{B}; 3) \ne \varnothing$. Define the elements of a matrix $X(\mathbf{A}, \mathbf{B}) = (x_{ij})$ from $M_Z(\mathbf{A}, \mathbf{B}; 3)$ as

$$x_{ij} = b_j^{(i-1)} - b_j^{(i)}, \quad 1 \le i \le 5, \quad 1 \le j \le 7.$$

It can be readily checked that the element x_{ij} equals the number of indices written out above the component $b_j^{(i-1)}$ of $\mathbf{B}_*^{(i-1)}$. The matrix

$$X(\mathbf{A}, \mathbf{B}) = \begin{pmatrix} 3 & 3 & 3 & 3 & 3 & 3 & 0 \\ 3 & 3 & 3 & 3 & 3 & 1 & 1 \\ 3 & 3 & 3 & 3 & 1 & 1 & 1 \\ 2 & 2 & 2 & 2 & 0 & 0 & 0 \\ 2 & 2 & 1 & 1 & 0 & 0 & 0 \end{pmatrix}$$

belongs to the set $M_Z(\mathbf{A}, \mathbf{B}; 3)$. Since $a_1/m = 18/7 > 2$, it is an integer minimax matrix.

Example 1.4. Application of Algorithm 1.1. The sums of components of both $\mathbf{A} = (10, 9, 7, 3, 3, 2) \in \overline{Z}_+^6$ and $\mathbf{B} = (11, 9, 7, 3, 2, 2) \in \overline{Z}_+^6$ are equal to 34. Suppose $p_k = k$ $(1 \le k \le 6)$.

We successively construct the principal 2-reductive pairs of vectors $\mathbf{A}^{(k)}$, $\mathbf{B}_*^{(k)}$ $(k \geq 1)$ writing out the indices of the components of $\mathbf{B}_*^{(k-1)}$ $(\mathbf{B}_*^{(0)} = \mathbf{B})$ that are decreased by unity as prescribed by (1.4) and (1.5):

$$\mathbf{A} = (10,\ 9,\ 7,\ 3,\ 3,\ 2), \quad \mathbf{B}_*^{(0)} = (\overset{\substack{2\\1}}{11},\ \overset{\substack{4\\3}}{9},\ \overset{\substack{6\\5}}{7},\ \overset{\substack{10\\7}}{3},\ \overset{9}{2},\ \overset{8}{2}),$$

$$\mathbf{A}^{(1)}(1) = (0,\ 9,\ 7,\ 3,\ 3,\ 2), \quad \mathbf{B}_*^{(1)} = (\overset{\substack{2\\1}}{9},\ \overset{\substack{4\\3}}{7},\ \overset{\substack{6\\5}}{5},\ \overset{9}{1},\ \overset{8}{1},\ \overset{7}{1}),$$

$$\mathbf{A}^{(2)}(1,2) = (0,\ 0,\ 7,\ 3,\ 3,\ 2), \quad \mathbf{B}_*^{(2)} = (\overset{\substack{2\\1}}{7},\ \overset{\substack{4\\3}}{5},\ \overset{\substack{6\\5}}{3},\ 0,\ 0,\ 0),$$

$$\mathbf{A}^{(3)}(1,2,3) = (0,\ 0,\ 0,\ 3,\ 3,\ 2).$$

Since $a_3 = 7$, the total number of indices written out above the components of $\mathbf{B}_*^{(2)}$ is 7. However, the vector $\mathbf{B}_*^{(2)}$ does not have a seventh component. Therefore, there does not exist a $(2;1,2,3)$-reductive pair of vectors corresponding to the pair $\mathbf{A}^{(2)}(1,2)$, $\mathbf{B}_*^{(2)}$; that is, this pair is not 2-reducible. Thus, $M_Z(\mathbf{A},\mathbf{B};2) = \varnothing$ and $c_Z(\mathbf{A},\mathbf{B}) \geq 3$.

Let us show that $M_Z(\mathbf{A},\mathbf{B};3) \neq \varnothing$ and construct the matrix $X = (x_{ij})$ $(0 \leq x_{ij} \leq 3,\ 1 \leq i,j \leq 6)$. Starting from the pair $\mathbf{A}^{(2)}(2;1,2)$, $\mathbf{B}_*^{(2)}$, we construct the principal 3-reductive pairs:

$$\mathbf{A}^{(2)} = (0,\ 0,\ 7,\ 3,\ 3,\ 2), \quad \mathbf{B}_*^{(2)} = (\overset{\substack{4\\2\\1}}{7},\ \overset{\substack{7\\5\\3}}{5},\ \overset{6}{3},\ 0,\ 0,\ 0),$$

$$\mathbf{A}^{(3)} = (0,\ 0,\ 0,\ 3,\ 3,\ 2), \quad \mathbf{B}_*^{(3)} = (\overset{\substack{2\\1}}{4},\ 2,\ \overset{3}{2},\ 0,\ 0,\ 0),$$

$$\mathbf{A}^{(4)} = (0,\ 0,\ 0,\ 0,\ 3,\ 2), \quad \mathbf{B}_*^{(4)} = (\overset{2}{2},\ \overset{1}{2},\ \overset{3}{1},\ 0,\ 0,\ 0),$$

$$\mathbf{A}^{(5)} = (0,\ 0,\ 0,\ 0,\ 0,\ 2), \quad \mathbf{B}_*^{(5)} = (\overset{2}{1},\ \overset{1}{1},\ 0,\ 0,\ 0,\ 0),$$

$$\mathbf{A}^{(6)} = (0,\ 0,\ 0,\ 0,\ 0,\ 0), \quad \mathbf{B}_*^{(6)} = (0,\ 0,\ 0,\ 0,\ 0,\ 0).$$

Now, $M_Z(\mathbf{A},\mathbf{B};3) \neq \varnothing$ and the integer minimax is 3. We define the elements of the matrix $X(\mathbf{A},\mathbf{B}) = (x_{ij})$ as

$$x_{ij} = b_j^{(i-1)} - b_j^{(i)}, \quad 1 \leq i \leq 6, \quad 1 \leq j \leq 6.$$

This completes the construction of the integer minimax matrix

$$X(\mathbf{A},\mathbf{B}) = \begin{pmatrix} 2 & 2 & 2 & 2 & 1 & 1 \\ 2 & 2 & 2 & 1 & 1 & 1 \\ 3 & 3 & 1 & 0 & 0 & 0 \\ 2 & 0 & 1 & 0 & 0 & 0 \\ 1 & 1 & 1 & 0 & 0 & 0 \\ 1 & 1 & 0 & 0 & 0 & 0 \end{pmatrix} \in M_Z(\mathbf{A},\mathbf{B};3).$$

It can be readily checked that the characteristic function defined by (1.22) from Chapter 1 is $\delta_3(\mathbf{A},\mathbf{B};2) = -1 < 0$ when $c = 2$ and $k = 3$ for the starting vectors in this example. Accordingly, $M_Z(\mathbf{A},\mathbf{B};2) = \varnothing$ by Theorem 2.2 from Chapter 1. Furthermore, $\delta_k(\mathbf{A},\mathbf{B};3) \geq 0$ for any k $(1 \leq k \leq 6)$, and $M_Z(\mathbf{A},\mathbf{B};3) \neq \varnothing$ by Theorem 2.2 from Chapter 1. A matrix from the set $M_Z(\mathbf{A},\mathbf{B};c)$ can be constructed by combining Algorithm 1.1 with Theorem 2.2 from Chapter 1. Then, we either establish that $M_Z(\mathbf{A},\mathbf{B};c) = \varnothing$ or construct a matrix from $M_Z(\mathbf{A},\mathbf{B};c)$ if $M_Z(\mathbf{A},\mathbf{B};c) \neq \varnothing$.

Algorithm 1.2. Construction of a matrix from the set $M_Z(\mathbf{A},\mathbf{B};c)$ by applying Theorem 2.2 from Chapter 1.

Suppose that vectors $\mathbf{A} \in Z_+^n$ and $\mathbf{B} \in Z_+^m$ satisfy the balance condition, $c \in Z$ $(c \geq 0)$, and $p_1, \ldots, p_n$ is an arbitrary sequence of different integers ranging from 1 to n.

Step 1. Calculate the value of the characteristic function defined by (1.22) from Chapter 1 for $c \in Z$ and vectors $\widetilde{\mathbf{A}}$ and $\mathbf{B}$ for every $k \in Z$ satisfying the assumptions of Theorem 2.2 from Chapter 1:

$$k \in \Big\{k \in Z : k^*(\mathbf{B};n,c) \leq k \leq k^{**}(\mathbf{B};n,c) \quad \text{and}$$
$$\tilde{a}_k > \tilde{a}_{k+1} \text{ if } k^*(\mathbf{B};n,c) < k < k^{**}(\mathbf{B};n,c)\Big\} = K(\mathbf{B},\mathbf{A};n,c),$$

where

$$k^*(\mathbf{B};n,c) = \max\left\{k \in Z : 0 \leq k \leq n,\ \min\{b_j : 1 \leq j \leq m\} \geq ck\right\} + 1,$$

$$k^{**}(\mathbf{B};n,c) = \min\left\{k \in Z : 0 \leq k \leq n,\ \max\{b_j : 1 \leq j \leq m\} \leq ck\right\} - 1.$$

Step 2. Apply Theorem 2.2 from Chapter 1: if all values of function (1.22) are nonnegative for $k \in K(\mathbf{B},\mathbf{A};n,c)$ or $K(\mathbf{B},\mathbf{A};n,c) = \varnothing$, then go to the next step; otherwise, when the function has a negative value at $k \in K(\mathbf{B},\mathbf{A};n,c)$, go to Step 4.

Step 3. Since $M_Z(\mathbf{A},\mathbf{B};c) \neq \varnothing$, apply Algorithm 1.1 (except for Step 4) to construct the matrix $X(\mathbf{A},\mathbf{B})$ from $M_Z(\mathbf{A},\mathbf{B};c)$ and stop.

Step 4. Set $M_Z(\mathbf{A},\mathbf{B};c) = \varnothing$.

If $M_Z(\mathbf{A}, \mathbf{B}; c) \neq \varnothing$, then Theorem 1.1 cannot be applied to determine all matrices in $M_Z(\mathbf{A}, \mathbf{B}; c)$. The impossibility of constructing all of the different matrices by means of Algorithms 1.1 and 1.2 is dictated by the stringent rule (1.5) for choosing the component β_q to be decreased by unity in constructing principal c-reductive pairs of vectors by applying the sequence of vectors in (1.4). For any $c \in Z$ $(c > 0)$ there exist vectors $\mathbf{A}$ and $\mathbf{B}$ with matrix $X(\mathbf{A}, \mathbf{B}) \in M_Z(\mathbf{A}, \mathbf{B}; c) \neq \varnothing$ that cannot be constructed by these algorithms. Thus, not all of the matrices are determined by Theorem 1.1 and, therefore, constructed by Algorithms 1.1 and 1.2. This is illustrated by the following example.

Example 1.5. A matrix that cannot be determined by applying Theorem 1.1. The vectors $\mathbf{A} = (4, 4, 4, 4) \in \overline{Z}_+^4$ and $\mathbf{B} = (3, 3, 3, 3, 1, 1, 1, 1) \in \overline{Z}_+^8$ are characterized by equal sums of components. Set $c = 1$.

The matrices constructed in this case,

$$X_1(\mathbf{A}, \mathbf{B}) = \begin{pmatrix} 1 & 1 & 1 & 0 & 1 & 0 & 0 & 0 \\ 1 & 1 & 0 & 1 & 0 & 1 & 0 & 0 \\ 1 & 0 & 1 & 1 & 0 & 0 & 1 & 0 \\ 0 & 1 & 1 & 1 & 0 & 0 & 0 & 1 \end{pmatrix}, \quad X_2(\mathbf{A}, \mathbf{B}) = \begin{pmatrix} 1 & 1 & 1 & 1 & 0 & 0 & 0 & 0 \\ 1 & 1 & 1 & 1 & 0 & 0 & 0 & 0 \\ 0 & 0 & 1 & 1 & 1 & 1 & 0 & 0 \\ 1 & 1 & 0 & 0 & 0 & 0 & 1 & 1 \end{pmatrix},$$

belong to the set $M_Z(\mathbf{A}, \mathbf{B}; 1)$. However, the first one cannot be constructed by means of Algorithm 1.1. Indeed, recall the definition of sequence (1.4) and rule (1.5) for choosing the component to decreased by unity. It is obvious that every matrix constructed by Algorithm 1.1 must contain at least two rows in which the first four elements are ones and the last four elements are zeros. The second matrix can be constructed by a method involving the principal 1-reductive pairs of vectors, that is, by Algorithm 1.1.

Let c be any positive integer, $\mathbf{A}' = c\mathbf{A} = (4c, 4c, 4c, 4c)$, and $\mathbf{B}' = c\mathbf{B} = (3c, 3c, 3c, 3c, c, c, c, c)$. Then, the matrices $X_1'(\mathbf{A}', \mathbf{B}') = cX_1(\mathbf{A}, \mathbf{B})$ and $X_2'(\mathbf{A}', \mathbf{B}') = cX_2(\mathbf{A}, \mathbf{B})$ belong to the set $M_Z(\mathbf{A}', \mathbf{B}'; c)$. It is obvious for the reasons explained above that the matrix $X_1'(\mathbf{A}', \mathbf{B}')$ cannot be constructed by means of Algorithm 1.1 and $X_2'(\mathbf{A}', \mathbf{B}')$ can be constructed by the algorithm.

The following Theorem characterizes every matrix from $M_Z(\mathbf{A}, \mathbf{B}; c)$ in the case when $M_Z(\mathbf{A}, \mathbf{B}; c) \neq \varnothing$.

Theorem 1.2. *Suppose that vectors $\mathbf{A} \in Z_+^n$ and $\mathbf{B} \in Z_+^m$ satisfy the balance condition and $c \in Z$ $(c \geq 0)$. Define $\mathbf{A}^{(0)} = \mathbf{A}$ and $\mathbf{B}^{(0)} = \mathbf{B}$.*

(a) *The matrix set $M_Z(\mathbf{A}, \mathbf{B}; c)$ is not empty if and only if there exists a sequence of $(c; 1, \ldots, k)$-reductive pairs of vectors*

$$\mathbf{A}^{(k)} = \mathbf{A}^{(k)}(1, \ldots, k), \quad \mathbf{B}^{(k)}, \quad 1 \leq k \leq n, \tag{1.9}$$

corresponding to $\mathbf{A}$ and $\mathbf{B}$.

(b) *Assume that a* $(c;1,\dots,k-1)$*-reductive pair of vectors* $\mathbf{A}^{(k-1)}$ *and* $\mathbf{B}^{(k-1)}$, *where* $1 \le k \le n$, *is constructed. There exists a* $(c;1,\dots,k)$*-reductive pair* $\mathbf{A}^{(k)}$, $\mathbf{B}^{(k)}$ *such that the vector* $\mathbf{B}^{(k-1)} - \mathbf{B}^{(k)}$ *is the* k*th row of a certain matrix from* $M_Z(\mathbf{A},\mathbf{B};c)$ *if and only if* $\delta_p(\widetilde{\mathbf{A}}^{(k)},\mathbf{B}^{(k)};c) \ge 0$ *for any* p $(1 \le p \le n)$.

(c) *If* $M_Z(\mathbf{A},\mathbf{B};c) \ne \varnothing$, *then every matrix from* $M_Z(\mathbf{A},\mathbf{B};c)$ *is uniquely determined by the corresponding sequence* (1.9) *of* $(c;1,\dots,k)$*-reductive pairs of vectors, and vice versa.*

Proof is based on Theorem 2.2 from Chapter 1 and Theorem 1.1.

The following Algorithm is constructed on the basis of Theorem 1.2.

Algorithm 1.3. Construction of a general matrix from the set $M_Z(\mathbf{A},\mathbf{B};c)$.

Suppose $M_Z(\mathbf{A},\mathbf{B};c) \ne \varnothing$, where $\mathbf{A}=(a_1,\dots,a_n)\in Z_+^n$, $\mathbf{B}=(b_1,\dots,b_m)\in Z_+^m$, and $c \ge 1$.

Step 1. Define $\mathbf{B}^{(0)} = (b_1^{(0)},\dots,b_m^{(0)}) = \mathbf{B} = (b_1,\dots,b_m)$ and $\mathbf{A}^{(0)} = \mathbf{A}$.

Step 2. Set $i=1$.

Step 3. Define $\mathbf{A}^{(i)} = \mathbf{A}^{(i)}(1,\dots,i) = (a_1^{(i)},\dots,a_n^{(i)})$, where

$$a_p^{(i)} = \begin{cases} 0, & 1 \le p \le i, \\ a_p, & i < p \le n, \end{cases}$$

and $\widetilde{\mathbf{A}}^{(i)}(1,\dots,i) = (\tilde{a}_1^{(i)},\dots,\tilde{a}_n^{(i)})$, where $\tilde{a}_p^{(i)} = a_p^{(i)} = 0$ for $1 \le p \le i$ and the components $\tilde{a}_p^{(i)}$ with $i < p \le n$ are obtained by rearranging components $a_p^{(i)}$ $(i < p \le n)$ of $\mathbf{A}^{(i)}(1,\dots,i)$ in nonascending order. (The present notation is employed only in Algorithm 1.3.)

Step 4. Find any integer solution $x_{i1},\dots,x_{im}$ to the set of relations

$$\begin{cases} 0 \le x_{ij} \le \min(b_j^{(i-1)},c), \quad 1 \le j \le m, \\ \displaystyle\sum_{j=1}^{m} x_{ij} = a_i, \\ \displaystyle\sum_{j=1}^{m}(b_j^{(i-1)} - x_{ij}) - \sum_{b_j^{(i-1)}-x_{ij}\ge ck}(b_j^{(i-1)} - x_{ij} - ck) - \sum_{p=i+1}^{i+k}\tilde{a}_p^{(i)} \ge 0, \quad 1 \le k \le n-i. \end{cases} \tag{1.10}$$

Step 5. Define the ith row of the desired matrix $X(\mathbf{A},\mathbf{B}) \in M_Z(\mathbf{A},\mathbf{B};c)$ as the solution to (1.10).

Step 6. Construct the vector $\mathbf{B}^{(i)} = (b_1^{(i)}, \ldots, b_m^{(i)})$ with $b_j^{(i)} = b_j^{(i-1)} - x_{ij}$ $(1 \le j \le m)$.

Step 7. If $i \le n-1$, then increase i by unity and go to Step 3. If $i = n$, then go to the next step.

Step 8. Define $X(\mathbf{A}, \mathbf{B}) = (x_{ij})$, where $1 \le i \le n$ and $1 \le j \le m$.

Since $M_Z(\mathbf{A}^{(0)}, \mathbf{B}^{(0)}; c) \neq \varnothing$, it is obvious that (1.10) has a solution for $i = 1$, and the last relation in (1.10) is equivalent to the inequality $\delta_k\left((\tilde{a}_{i+1}^{(i)}, \ldots, \tilde{a}_n^{(i)}), \mathbf{B}^{(i)}; c\right) \ge 0$ $(1 \le k \le n-i)$. Consequently, $M_Z(\mathbf{A}^{(i)}, \mathbf{B}^{(i)}; c) \neq \varnothing$, which validates the algorithm.

The rows of every matrix $X(\mathbf{A}, \mathbf{B})$ from $M_Z(\mathbf{A}, \mathbf{B}; c)$ are consecutive solutions to (1.10). Therefore, every matrix from $M_Z(\mathbf{A}, \mathbf{B}; c)$ can be constructed by means of Algorithm 1.3.

Both the assumption of proposition (b) in Theorem 1.2 and Step 4 in Algorithm 1.3 can be simplified. This is done in the next section for $c = 1$.

§2. Matrices Consisting of Zeros and Units

The results obtained in the preceding section are specified here for boolean variables. We formulate necessary and sufficient conditions for the existence of admissible solutions to the transportation problem in the case when the unknown elements are zeros and ones. The combinatorial constructions proposed here can be used to find all admissible matrices.

For this problem, necessary and sufficient conditions were independently obtained by D. Gale [2, 6] and H.J. Ryser [19]. Their results are equivalent to those presented here. In [24], L.R. Ford and D. Fulkerson used the conditions formulated by Gale and Ryser to construct an algorithm that finds certain admissible solutions. Note also the results presented in [20, 21], which establish certain combinatorial properties of the matrices discussed here.

2.1. Existence Criterion and Characteristic Equations

Suppose that vectors $\mathbf{A} = (a_1, \ldots, a_n) \in Z_+^n$ and $\mathbf{B} = (b_1, \ldots, b_m) \in Z_+^m$ satisfy the balance condition. The characteristic function of these vectors and $c = 1$, defined by (1.22) Chapter 1, has the form

$$\delta_k(\mathbf{A}, \mathbf{B}; 1) = \sum_{j=1}^{m} b_j - \sum_{b_j \ge k} (b_j - k) - \sum_{i=1}^{k} a_i, \quad 1 \le k \le n.$$

By Lemma 2.5 from Chapter 1 and Remark 2.1 from Chapter 1, the fact that the function is nonnegative when the components of the first vector are arranged in nonascending order dictates the conditions under which the matrix set $M_Z(\mathbf{A},\mathbf{B};1)$ is not empty.

Lemma 2.1. *If vectors* $\mathbf{A}=(a_1,\dots,a_n)\in Z_+^n$ *and* $\mathbf{B}=(b_1,\dots,b_m)\in Z_+^m$ *have equal sums of components, there exists a matrix* $X(\mathbf{A},\mathbf{B})$ *of ones and zeros if and only if*

$$\delta_k(\widetilde{\mathbf{A}},\mathbf{B};1)\geq 0,\quad 1\leq k\leq n, \tag{2.1}$$

where $\widetilde{\mathbf{A}}=(\tilde{a}_1,\dots,\tilde{a}_n)$ *is a vector derived from* $\mathbf{A}$ *by arranging its components in nonascending order.*

Note that the inequalities in (2.1) are not necessarily independent in the general case. Therefore, we can apply Theorem 2.2 from Chapter 1, with $M_Z(\mathbf{A},\mathbf{B};1)\neq\varnothing$ and all redundant inequalities eliminated from (2.1), to verify that $c=1$.

In the following theorem, which is a corollary to Theorem 3.1 from Chapter 1, the vectors $\mathbf{A}$ and $\mathbf{B}$ determine the zero and unit entries in every matrix from $M_Z(\mathbf{A},\mathbf{B};1)\neq\varnothing$.

Theorem 2.1. *Suppose* $M_Z(\mathbf{A},\mathbf{B};1)\neq\varnothing$, *where* $\mathbf{A}\in Z_+^n$ *and* $\mathbf{B}\in Z_+^m$.

(a) *For any matrix* $X(\mathbf{A},\mathbf{B})=(x_{ij})$ *from* $M_Z(\mathbf{A},\mathbf{B};1)$ *it holds that*

$$\delta_k(\mathbf{A},\mathbf{B};1)=kl_k(\mathbf{B};1)-\sum_{i=1}^{k}\sum_{b_i\geq k}x_{ij}+\sum_{i>k}\sum_{b_i<k}x_{ij} \tag{2.2}$$

for any k $(1\leq k\leq n)$.

(b) *If* $\mathbf{B}\in\overline{Z}_+^m$, *then every matrix* $X(\mathbf{A},\mathbf{B})=(x_{ij})$ *from* $M_Z(\mathbf{A},\mathbf{B};1)$ *is such that*

$$\delta_k(\mathbf{A},\mathbf{B};1)=kl_k(\mathbf{B};1)-\sum_{i=1}^{k}\sum_{j\leq l_k(\mathbf{B};1)}x_{ij}+\sum_{i>k}\sum_{j>l_k(\mathbf{B};1)}x_{ij} \tag{2.3}$$

for any k $(1\leq k\leq n)$.

Definition 2.1. Equations (2.2) and (2.3) in integer variables x_{ij}, where $x_{ij}=0$ or $x_{ij}=1$, are called the *characteristic equations* corresponding to matrices from the set $M_Z(\mathbf{A},\mathbf{B};1)$.

The characteristic equations (2.2) and (2.3) that correspond to equation (3.1) from Chapter 1, with $c=1$, define the general properties of matrices from $M_Z(\mathbf{A},\mathbf{B};1)$. It is obvious that every matrix $X(\mathbf{A},\mathbf{B})=(x_{ij})\in$

$M_Z(\mathbf{A}, \mathbf{B}; 1)$ is such that

$$kl_k(\mathbf{B}; 1) - \sum_{i=1}^{k} \sum_{b_i \geq k} x_{ij} \geq 0, \quad \sum_{i>k} \sum_{b_i<k} x_{ij} \geq 0$$

for any $k \in Z$ $(1 \leq k \leq n)$. Suppose that $\overline{X} = (x_{ij})$, where $1 \leq i \leq k$ and $b_j \geq k$, and $\overline{\overline{X}} = (x_{ij})$, where $i > k$ and $b_j < k$, are submatrices of $X(\mathbf{A}, \mathbf{B})$. The inequalities above imply that the distribution of zeros and ones in the submatrices $\overline{X}$ and $\overline{\overline{X}}$ depends on the values of the left-hand sides of (2.2) and (2.3): the less are these values, the more specifically these submatrices are defined. Since $\delta_k(\tilde{\mathbf{A}}, \mathbf{B}; 1) \leq \delta_k(\mathbf{A}, \mathbf{B}; 1)$, it is reasonable to arrange the components of the first vector in nonascending order when analyzing matrices and matrix sets by means of characteristic equations.

Example 2.1. Application of the characteristic equations in calculating the number of matrices in the set $M_Z(\mathbf{A}, \mathbf{B}; 1)$. Consider the vectors $\mathbf{A} = (4, 4, 4, 4, 3, 3) \in \overline{Z}_+^6$ and $\mathbf{B} = (5, 5, 5, 5, 1, 1) \in \overline{Z}_+^6$. For each vector, the sum of its components is 22; $\delta_k(\mathbf{A}, \mathbf{B}; 1) = 2$ $(1 \leq k \leq 4)$; $\delta_5(\mathbf{A}, \mathbf{B}; 1) = 3$; and $\delta_6(\mathbf{A}, \mathbf{B}; 1) = 0$. Therefore, $M_Z(\mathbf{A}, \mathbf{B}; 1) \neq \varnothing$.

Apply characteristic equation (2.3) with $k = 4$. Here, $l_4(\mathbf{B}; 1) = 4$, and every matrix $X(\mathbf{A}, \mathbf{B}) = (x_{ij}) \in M_Z(\mathbf{A}, \mathbf{B}; 1)$ is such that

$$2 = 4 \cdot 4 - \sum_{i=1}^{4} \sum_{j=1}^{4} x_{ij} + \sum_{i=5}^{6} \sum_{j=5}^{6} x_{ij}.$$

Rewrite this equation using the same notation as in Example 3.3 from Chapter 1 as follows:

$$16 - s(\overline{X}) + s(\overline{\overline{X}}) = 2, \tag{2.4}$$

where $\overline{X} = (x_{ij})$, where $1 \leq i, j \leq 4$, and $\overline{\overline{X}} = (x_{ij})$, where $5 \leq i, j \leq 6$. Recall Theorem 3.4 from Chapter 1, which establishes that there exists a matrix from $M_Z(\mathbf{A}, \mathbf{B}; 1)$ corresponding to each solution of equation (2.4). Let us enumerate all of the solutions: $s(\overline{X}) = 16$, $s(\overline{\overline{X}}) = 2$; $s(\overline{X}) = 15$, $s(\overline{\overline{X}}) = 1$; $s(\overline{X}) = 14$, $s(\overline{\overline{X}}) = 0$. There exists a matrix subset corresponding to each solution, which contains only zeros and ones:

$$M_1(\mathbf{A},\mathbf{B};1) = \Big\{X = X(\mathbf{A},\mathbf{B}) : X(\mathbf{A},\mathbf{B}) \in M_Z(\mathbf{A},\mathbf{B};1),\ s(\overline{X}) = 16,\ s(\overline{\overline{X}}) = 2\Big\},$$

$$M_2(\mathbf{A},\mathbf{B};1) = \Big\{X = X(\mathbf{A},\mathbf{B}) : X(\mathbf{A},\mathbf{B}) \in M_Z(\mathbf{A},\mathbf{B};1),\ s(\overline{X}) = 15,\ s(\overline{\overline{X}}) = 1\Big\},$$

$$M_3(\mathbf{A},\mathbf{B};1) = \Big\{X = X(\mathbf{A},\mathbf{B}) : X(\mathbf{A},\mathbf{B}) \in M_Z(\mathbf{A},\mathbf{B};1),\ s(\overline{X}) = 14,\ s(\overline{\overline{X}}) = 0\Big\}.$$

Now, we calculate the number of matrices in each subset thus determined. We also construct some of the matrices belonging to these subsets, which is a simple task. In doing so, we write out the sums of components of rows and columns on the left of the rows and above the columns, respectively.

Consider the following arbitrary matrix $X_1(\mathbf{A},\mathbf{B})$ in $M_1(\mathbf{A},\mathbf{B};1)$:

$$\begin{matrix} & 5 & 5 & 5 & 5 & 1 & 1 \\ 4 & 1 & 1 & 1 & 1 & 0 & 0 \\ 4 & 1 & 1 & 1 & 1 & 0 & 0 \\ 4 & 1 & 1 & 1 & 1 & 0 & 0 \\ 4 & 1 & 1 & 1 & 1 & 0 & 0 \\ 3 & 0 & 1 & 0 & 1 & 0 & 1 \\ 3 & 1 & 0 & 1 & 0 & 1 & 0 \end{matrix} = X_1(\mathbf{A},\mathbf{B}) \in M_1(\mathbf{A},\mathbf{B};1).$$

Each matrix $X = (x_{ij}) \in M_1(\mathbf{A},\mathbf{B};1)$ contains a uniquely defined submatrix $\overline{X}$ (whose elements are ones), whereas the submatrix $\overline{\overline{X}}$ can be defined as one containing two ones on either diagonal. It is clear that the submatrix $\overline{X}' = (x_{ij})$ $(1 \le i \le 4,\ 5 \le j \le 6)$ contains only zeros and, therefore, is uniquely defined. The submatrix $\overline{\overline{X}}' = (x_{ij})$ $(5 \le i \le 6,\ 1 \le j \le 4)$ contains four ones (each column contains one, and each row contains two). Therefore, there exist six different submatrices $\overline{\overline{X}}'$. Thus, the number of matrices in the subset under consideration is $\big|M_1(\mathbf{A},\mathbf{B};1)\big| = 2 \cdot 6 = 12$.

Next, we consider the following arbitrary matrix in subset $M_2(\mathbf{A},\mathbf{B};1)$:

$$\begin{matrix} & 5 & 5 & 5 & 5 & 1 & 1 \\ 4 & 1 & 1 & 1 & 1 & 0 & 0 \\ 4 & 1 & 1 & 1 & 1 & 0 & 0 \\ 4 & 1 & 1 & 1 & 1 & 0 & 0 \\ 4 & 1 & 1 & 1 & 0 & 1 & 0 \\ 3 & 0 & 1 & 1 & 1 & 0 & 0 \\ 3 & 1 & 0 & 0 & 1 & 0 & 1 \end{matrix} = X_2(\mathbf{A},\mathbf{B}) \in M_2(\mathbf{A},\mathbf{B};1).$$

The only zero in the submatrix $\overline{X}$ can occupy one of 16 entries; the single only unit element in $\overline{\overline{X}}$, one of four entries. The submatrix $\overline{X}'$ contains a single unit element. It belongs to the row that contains the zero in $\overline{X}$ and to the column in $\overline{\overline{X}}$ that does not contain ones. Therefore, the submatrix $\overline{X}'$ is uniquely defined. In $\overline{\overline{X}}'$, the two unit elements obviously belong to the column that contains the zero of $\overline{X}$. It is clear that the remaining three elements of $\overline{\overline{X}}'$ can be located in three different combinations. Thus, we have $\left|M_2(\mathbf{A},\mathbf{B};1)\right| = 16\cdot 4\cdot 3 = 192$ in this case.

Lastly, we consider the matrix subset $M_3(\mathbf{A},\mathbf{B};1)$. The submatrix $\overline{\overline{X}}$ of each matrix $X = (x_{ij}) \in M_3(\mathbf{A},\mathbf{B};1)$ contains only zeros and is thus uniquely defined. The submatrix $\overline{X}$ of every matrix from $M_3(\mathbf{A},\mathbf{B};1)$ contains two zeros. According to the arrangement of the zero elements, we divide $M_3(\mathbf{A},\mathbf{B};1)$ in two subsets. It is clear that each column of a submatrix $\overline{X}$ of every matrix $X \in M_Z(\mathbf{A},\mathbf{B};1)$ contains not less than three unit elements. Denote by $M_3'(\mathbf{A},\mathbf{B};1)$ the set of matrix from $M_3(\mathbf{A},\mathbf{B};1)$ for which the submatrix $\overline{X}$ contain two zeros in the same row and by $M_3''(\mathbf{A},\mathbf{B};1)$ the set of matrices from $M_3(\mathbf{A},\mathbf{B};1)$ whose zeros belong to different rows (and different columns). Then, $M_3(\mathbf{A},\mathbf{B};1) = M_3'(\mathbf{A},\mathbf{B};1) \cup M_3''(\mathbf{A},\mathbf{B};1)$.

Consider now the following matrix belonging to the matrix subset $M_3'(\mathbf{A},\mathbf{B};1)$:

$$\begin{array}{c} \\ 4\\4\\4\\4\\3\\3 \end{array}\!\!\begin{array}{c} \begin{array}{cccccc} 5&5&5&5&1&1 \end{array} \\ \begin{pmatrix} 0&0&1&1&1&1\\ 1&1&1&1&0&0\\ 1&1&1&1&0&0\\ 1&1&1&1&0&0\\ 1&1&1&0&0&0\\ 1&1&0&1&0&0 \end{pmatrix} \end{array} = X_3'(\mathbf{A},\mathbf{B}) \in M_3'(\mathbf{A},\mathbf{B};1).$$

The submatrix $\overline{X}$ of every $X \in M_3'(\mathbf{A},\mathbf{B};1)$ has four rows, and the two zeros can be arranged in six different combinations in each row. Thus, there exist 24 different matrices $\overline{X}$. The submatrix $\overline{X}'$ contains two ones, which belong to the row containing two zeros of $\overline{X}$. Therefore, the submatrix $\overline{X}'$ is uniquely defined for every submatrix $\overline{X}$. The four ones in $\overline{\overline{X}}'$ belong to the two columns that contain the zeros of $\overline{X}$, whereas the remaining two ones belong to different rows (and different columns). Consequently, there exist two different submatrices $\overline{\overline{X}}'$ (for a given $\overline{X}$). Thus, $\left|M_3'(\mathbf{A},\mathbf{B};1)\right| = 24\cdot 2 =$ 48.

Finally, consider the following arbitrary matrix belonging to the set $M_3''(\mathbf{A},\mathbf{B};1)$:

$$\begin{array}{c} \\ 4 \\ 4 \\ 4 \\ 4 \\ 3 \\ 3 \end{array}\begin{array}{c} \begin{matrix} 5 & 5 & 5 & 5 & 1 & 1 \end{matrix} \\ \begin{pmatrix} 1 & 1 & 1 & 1 & 0 & 0 \\ 1 & 1 & 1 & 1 & 0 & 0 \\ 1 & 1 & 0 & 1 & 1 & 0 \\ 1 & 1 & 1 & 0 & 0 & 1 \\ 1 & 0 & 1 & 1 & 0 & 0 \\ 0 & 1 & 1 & 1 & 0 & 0 \end{pmatrix} \end{array} = X_3''(\mathbf{A},\mathbf{B}) \in M_3''(\mathbf{A},\mathbf{B};1).$$

There now exist $4^2 \cdot 3^2/2 = 72$ variants for arranging the two zeros of $\overline{X}$. It is obvious that each submatrix $\overline{X}$ defines two matrices $\overline{X}'$ and two matrices $\overline{\overline{X}}'$. Therefore, $|M_3''(\mathbf{A},\mathbf{B};1)| = 72 \cdot 2 \cdot 2 = 288$, and

$$|M_3(\mathbf{A},\mathbf{B};1)| = |M_3'(\mathbf{A},\mathbf{B};1)| + |M_3''(\mathbf{A},\mathbf{B};1)| = 48 + 288 = 336.$$

Thus, the total number of matrices in $M_Z(\mathbf{A},\mathbf{B};1)$ is

$$|M_Z(\mathbf{A},\mathbf{B};1)| = |M_1(\mathbf{A},\mathbf{B};1)| + |M_2(\mathbf{A},\mathbf{B};1)| + |M_3(\mathbf{A},\mathbf{B};1)| = 12 + 192 + 336 = 540.$$

Let $M_Z(\mathbf{A},\mathbf{B};1)$ be a nonempty set of minimax matrices; that is, $c(\mathbf{A},\mathbf{B}) = c_Z(\mathbf{A},\mathbf{B}) = 1$. Then, the corresponding characteristic equations (2.2) and (2.3) uniquely define the elements of the diagonal submatrices of every matrix in $M_Z(\mathbf{A},\mathbf{B};1)$ (and in $M(\mathbf{A},\mathbf{B};1)$). As a corollary to Theorem 2.1, we state the following proposition, which can be viewed as a special case implied by Theorem 3.2 from Chapter 1.

Theorem 2.2. *Suppose that* $M_Z(\mathbf{A},\mathbf{B};1) \neq \varnothing$, *where* $\mathbf{A} = (a_1,\dots,a_n) \in Z_+^n$, $\mathbf{B} = (b_1,\dots,b_m) \in Z_+^m$, $\max\{b_j : 1 \le j \le m\} \ge 1$, *and* k $(1 \le k \le n)$ *is such an integer that* $\max\{b_j : 1 \le j \le m\} \ge k$ *and* $\delta_k(\widetilde{\mathbf{A}},\mathbf{B};1) = 0$. *Then, the following is true.*

(a) *If* $a_p = \max\{a_i : 1 \le i \le n\}$ *and* $b_q = \max\{b_j : 1 \le j \le m\}$, *then* $x_{pq} = 1$ *for every matrix* $X(\mathbf{A},\mathbf{B}) = (x_{ij})$ *from* $M_Z(\mathbf{A},\mathbf{B};1)$.

(b) *For every matrix* $X(\mathbf{A},\mathbf{B}) = (x_{ij})$ *from* $M_Z(\mathbf{A},\mathbf{B};1)$, *where* $\mathbf{A} \in \overline{Z}_+^n$,

$$x_{ij} = \begin{cases} 1, & i \le k, \quad b_j \ge k, \\ 0, & k < i \le n, \quad b_j < k. \end{cases}$$

(c) *If* $\mathbf{A} \in \overline{Z}_+^n$ *and* $\mathbf{B} \in \overline{Z}_+^m$, *then the elements of any* $X(\mathbf{A},\mathbf{B}) = (x_{ij}) \in M_Z(\mathbf{A},\mathbf{B};1)$ *are as follows:*

$$x_{ij} = \begin{cases} 1, & 1 \le i \le k, \quad 1 \le j \le l_k(\mathbf{B};1), \\ 0, & k < i \le n, \quad l_k(\mathbf{B};1) < j \le m. \end{cases} \tag{2.5}$$

Remark 2.1. Every proposition in Theorem 2.2 applies to the matrices from $M(\mathbf{A},\mathbf{B};1)$ as well.

Theorem 2.2 demonstrates that if $M_Z(\mathbf{A},\mathbf{B};1)$ is a set of minimax matrices, then it is equivalent to two matrix subsets, each containing less rows and columns.

Example 2.2. In the case of $\mathbf{A} = (8,8,8,7,4,4,2) \in \overline{Z}^7_+$ and $\mathbf{B} = (6,6,6,6,5,5,3,2,2) \in \overline{Z}^9_+$, it can be readily shown that $M_Z(\mathbf{A},\mathbf{B};1) \neq \varnothing$. Applying formula (2.5) from Theorem 2.2, with $k = 4$, we see that every matrix $X(\mathbf{A},\mathbf{B}) = (x_{ij})$ from $M_Z(\mathbf{A},\mathbf{B};1)$ has the form

$$X(\mathbf{A},\mathbf{B}) = \begin{pmatrix} 1 & 1 & 1 & 1 & 1 & 1 & x_{17} & x_{18} & x_{19} \\ 1 & 1 & 1 & 1 & 1 & 1 & x_{27} & x_{28} & x_{29} \\ 1 & 1 & 1 & 1 & 1 & 1 & x_{37} & x_{38} & x_{39} \\ 1 & 1 & 1 & 1 & 1 & 1 & x_{47} & x_{48} & x_{49} \\ x_{51} & x_{52} & x_{53} & x_{54} & x_{55} & x_{56} & 0 & 0 & 0 \\ x_{61} & x_{62} & x_{63} & x_{64} & x_{65} & x_{66} & 0 & 0 & 0 \\ x_{71} & x_{72} & x_{73} & x_{74} & x_{75} & x_{76} & 0 & 0 & 0 \end{pmatrix},$$

where

$$\begin{pmatrix} x_{17} & x_{18} & x_{19} \\ x_{27} & x_{28} & x_{29} \\ x_{37} & x_{38} & x_{39} \\ x_{47} & x_{48} & x_{49} \end{pmatrix} \in M_Z\left((2,2,2,1),(3,2,2);1\right),$$

$$\begin{pmatrix} x_{51} & x_{52} & x_{53} & x_{54} & x_{55} & x_{56} \\ x_{61} & x_{62} & x_{63} & x_{64} & x_{65} & x_{66} \\ x_{71} & x_{72} & x_{73} & x_{74} & x_{75} & x_{76} \end{pmatrix} \in M_Z\left((4,4,2),(2,2,2,2,1,1);1\right).$$

Note that a similar representation applies to the matrices from $M(\mathbf{A},\mathbf{B};1)$ (see Remark 2.1).

Now, we seek an upper estimate for the number of matrices in the set $M_Z(\mathbf{A},\mathbf{B};1)$:

$$\left|M_Z(\mathbf{A},\mathbf{B};1)\right| = \left|M_Z\left((2,2,2,1),(3,2,2);1\right)\right| \times \left|M_Z\left((4,4,2),(2,2,2,2,1,1);1\right)\right|.$$

Obviously, the number of n-by-m matrices in question that contain exactly p unit elements is C^p_{nm}. Consequently, $\left|M_Z(\mathbf{A},\mathbf{B};1)\right| \le C^{2+2+2+1}_{4\cdot 3} \cdot C^{4+4+2}_{3\cdot 5} = C^7_{12} \cdot C^{10}_{15}$ and $\left|M_Z(\mathbf{A},\mathbf{B};1)\right| \le 2378376$. The matrices considered in this example are such that this estimate can be improved. It is clear that

$$\left|M_Z\left((2,2,2,1),(3,2,2);1\right)\right| \le 3 \cdot C^6_9 = 252,$$

$$\left|M_Z\left((4,4,2),(2,2,2,2,1,1);1\right)\right| \le 3^2 \cdot C^8_{12} = 4455.$$

Obviously, the weak inequalities above can be changed to the respective strict ones. Thus, $|M_Z(\mathbf{A},\mathbf{B};1)| < 1122660$. Note that this is a rough estimate. It can be substantially improved by applying characteristic equation (2.3) to the matrix sets $M_Z((2,2,2,1),(3,2,2);1)$ and $M_Z((4,4,2),(2,2,2,2,1,1);1)$. In what follows, we calculate the exact number of matrices in the set $M_Z(\mathbf{A},\mathbf{B};1)$ from Example 2.2 by applying reductive transformations.

2.2. Reduction Algorithms

The condition $M_Z(\mathbf{A},\mathbf{B};1) \neq \varnothing$ can be verified by invoking Theorem 1.1, with $c = 1$. A matrix from this set can be constructed by means of Algorithm 1.1, also with $c = 1$.

Definition 2.2. The pair of vectors $\mathbf{A} \in Z_+^n$ and $\mathbf{B} \in Z_+^m$ is said to be *reducible* if $\mathbf{A}$ and $\mathbf{B}$ make up a 1-reducible pair of vectors (see Definition 1.4).

The reduction theorem and algorithm stated below can be viewed as special cases of Theorem 1.1 and Algorithm 1.1, respectively. The hypotheses and transformations involved in Theorem 2.3 and Algorithm 2.1 are identical with the (0, 1)-matrix rule formulated in [24].

Theorem 2.3. *Suppose that vectors* $\mathbf{A} \in \overline{Z}_+^n$ *and* $\mathbf{B} \in Z_+^m$ $(n > 1)$ *satisfy the balance condition. The matrix set* $M_Z(\mathbf{A},\mathbf{B};1)$ *is not empty if and only if* $\mathbf{A}$ *and* $\mathbf{B}$ *make up a reducible pair of vectors and so is every principal* $(1;1,\ldots,k)$*-reductive pair of vectors*

$$\mathbf{A}^{(k)} = \mathbf{A}^{(k)}(1,\ldots,k), \quad \mathbf{B}_*^{(k)}, \quad 1 \leq k \leq n-1$$

constructed at the kth reductive step for the pair of vectors $\mathbf{A}$ *and* $\mathbf{B}$.

Algorithm 2.1. Construction of a matrix containing only zeros and units.

Suppose that the vectors $\mathbf{A} = (a_1,\ldots,a_n) \in \overline{Z}_+^n$ and $\mathbf{B} = (b_1,\ldots,b_m) \in Z_+^m$ are characterized by equal sums of components.

Step 1. Set $\mathbf{A}^{(0)}(0) = \mathbf{A}$ and $\mathbf{B}_*^{(0)} = (b_1^{(0)},\ldots,b_m^{(0)}) = \mathbf{B} = (b_1,\ldots,b_m)$.

Step 2. Set $k = 1$.

Step 3. When possible, construct the principal $(1;1,\ldots,k) = (1;0,1,\ldots,k)$-reductive pair of vectors

$$\mathbf{A}^{(k)}(0,1,\ldots,k) = \left(\mathbf{A}^{(k-1)}(0,1,\ldots,k-1)\right)^{(1)}(k),$$

$$\mathbf{B}_*^{(k)} = \left(\mathbf{B}_*^{(k-1)}\right)_*^{(1)} = (b_1^{(k)},\ldots,b_m^{(k)}),$$

where the vector $\mathbf{B}_*^{(k)}$ is derived from $\mathbf{B}_*^{(k-1)}$ by applying the sequence of vectors defined by (1.4) and rule (1.5) for selecting the component to be decreased by unity for $c = 1$. Here, (1.5) should be written as

$$\beta_q^{(k)} = \max\left\{\beta_j^{(k)} : 1 \le j \le m,\ \beta_j^{(0)} - \beta_j^{(k)} < 1\right\} = \max\left\{\beta_j^{(k)} : 1 \le j \le m,\ \beta_j^{(0)} = \beta_j^{(k)}\right\}. \tag{2.6}$$

Step 4. If the principal $(1;0,1,\ldots,k)$-reductive pair $\mathbf{A}^{(k)}(0,1,\ldots,k)$, $\mathbf{B}_*^{(k)}$ is constructed, then go to the next step. If the principal $(1;0,1,\ldots,k)$-reductive pair corresponding to $\mathbf{A}^{(k-1)}(0,1,\ldots,k-1)$ and $\mathbf{B}_*^{(k-1)}$ cannot be constructed, go to Step 7.

Step 5. If $k \le n-1$, increase the value of k by unity and go to Step 3. If $k = n$, then go to the next step.

Step 6. Define the elements of the desired matrix $X(\mathbf{A},\mathbf{B}) = (x_{ij})$ from $M_Z(\mathbf{A},\mathbf{B};1)$ as

$$x_{kj} = b_j^{(k-1)} - b_j^{(k)}, \quad 1 \le j \le m, \quad 1 \le k \le n,$$

and stop.

Step 7. Set $M_Z(\mathbf{A},\mathbf{B};1) = \varnothing$.

For convenience, this algorithm can be unified with Lemma 2.1.

Algorithm 2.2. Construction of a matrix from $M_Z(\mathbf{A},\mathbf{B};1)$ by Theorem 2.2 from Chapter 1, with $c = 1$, or by Lemma 2.1.

Suppose that vectors $\mathbf{A} \in \overline{Z}_+^n$ and $\mathbf{B} \in Z_+^m$ satisfy the balance condition.

Step 1. Calculate the value of the characteristic function defined by (1.22) in Chapter 1, with $c = 1$, for every $k \in Z$ that satisfies the assumptions of Theorem 2.2 from Chapter 1:

$$\delta_k(\mathbf{A},\mathbf{B};1) = \sum_{j=1}^{m} b_j - \sum_{b_j \ge k} (b_j - k) - \sum_{i=1}^{k} a_i, \quad 1 \le k \le n,$$

where $k \in K(\mathbf{B},\mathbf{A};n,1)$ (see Algorithm 1.2).

Step 2. If either $\delta_k(\mathbf{A},\mathbf{B};1) \ge 0$ for every $k \in K(\mathbf{B},\mathbf{A};n,1)$ or $K(\mathbf{B},\mathbf{A};n,1) = \varnothing$, go to the next step. If the value of $\delta_k(\mathbf{A},\mathbf{B};1)$ is negative for some $k \in K(\mathbf{B},\mathbf{A};n,1)$, go to Step 4.

Step 3. Construct a matrix $X(\mathbf{A},\mathbf{B})$ from $M_Z(\mathbf{A},\mathbf{B};1)$ by means of Algorithm 2.1, with the exception of Step 4.

Step 4. Set $M_Z(\mathbf{A},\mathbf{B};1) = \varnothing$.

Example 2.3. Application of Algorithm 2.2. Consider the vectors

$$\mathbf{A} = (8,8,8,7,3,3) \in \overline{Z}_+^6$$

and

$$\mathbf{B} = (5,5,5,5,4,4,3,3,1,1,1) \in \overline{Z}_+^{11}.$$

The sum of components of either vector is 37, and the balance condition is satisfied by the pair.

We have the following values of $\delta_k(\mathbf{A},\mathbf{B};1)$ for $1 \le k \le 6$: $\delta_1(\mathbf{A},\mathbf{B};1) = 3$, $\delta_2(\mathbf{A},\mathbf{B};1) = 3$, $\delta_3(\mathbf{A},\mathbf{B};1) = 3$, $\delta_4(\mathbf{A},\mathbf{B};1) = 2$, $\delta_5(\mathbf{A},\mathbf{B};1) = 3$, and $\delta_6(\mathbf{A},\mathbf{B};1) = 0$. Therefore, $M_Z(\mathbf{A},\mathbf{B};1) \neq \varnothing$. Note that the analysis could be restricted to calculating the characteristic function for $k \in K(\mathbf{B},\mathbf{A};6,1) = \{2,3,4\}$.

Now, we construct a matrix $X(\mathbf{A},\mathbf{B}) = (x_{ij}) \in M_Z(\mathbf{A},\mathbf{B};1)$ by means of Algorithm 2.1. Each reductive step uniquely defined a row of the matrix. Accordingly, we construct the principal $(1;1,\ldots,k)$-reductive pairs of vectors $\mathbf{A}^{(k)}$ and $\mathbf{B}_*^{(k)}$ for $1 \le k \le 6$ and calculate the elements of the corresponding row of the desired matrix simultaneously. To make the calculations more graphic, we mark each component of the second vectors that is decreased by unity by a dot:

$$\mathbf{A} = (8,8,8,7,3,3), \quad \mathbf{B} = (\dot{5},\dot{5},\dot{5},\dot{5},\dot{4},\dot{4},\dot{3},\dot{3},1,1,1);$$

$$\mathbf{A}^{(1)} = (0,8,8,7,3,3), \quad \mathbf{B}_*^{(1)} = (\dot{4},\dot{4},\dot{4},\dot{4},\dot{3},\dot{3},\dot{2},\dot{2},1,1,1),$$

$$x_{1j} = 1, \quad 1 \le j \le 8, \quad x_{1j} = 0, \quad 9 \le j \le 11;$$

$$\mathbf{A}^{(2)} = (0,0,8,7,3,3), \quad \mathbf{B}_*^{(2)} = (\dot{3},\dot{3},\dot{3},\dot{3},\dot{2},\dot{2},1,1,1,\dot{1},\dot{1}),$$

$$x_{2j} = 1, \quad 1 \le j \le 8, \quad x_{2j} = 0, \quad 9 \le j \le 11;$$

$$\mathbf{A}^{(3)} = (0,0,0,7,3,3), \quad \mathbf{B}_*^{(3)} = (\dot{2},\dot{2},\dot{2},\dot{2},1,1,\dot{1},\dot{1},\dot{1},0,0),$$

$$x_{3j} = 1, \quad 1 \le j \le 6, \quad 10 \le j \le 11, \quad x_{3j} = 0, \quad 7 \le j \le 9;$$

$$\mathbf{A}^{(4)} = (0,0,0,0,3,3), \quad \mathbf{B}_*^{(4)} = (1,1,1,\dot{1},\dot{1},\dot{1},0,0,0,0,0),$$

$$x_{4j} = 1, \quad 1 \le j \le 4, \quad 7 \le j \le 9, \quad x_{4j} = 0, \quad 5 \le j \le 6, \quad 10 \le j \le 11;$$

$$\mathbf{A}^{(5)} = (0,0,0,0,0,3), \quad \mathbf{B}_*^{(5)} = (\dot{1},\dot{1},\dot{1},0,0,0,0,0,0,0,0),$$

$$x_{5j} = 1, \quad 4 \le j \le 6, \quad x_{5j} = 0, \quad 1 \le j \le 3, \quad 7 \le j \le 11;$$

$$\mathbf{A}^{(6)} = (0,0,0,0,0,0), \quad \mathbf{B}_*^{(6)} = (0,0,0,0,0,0,0,0,0,0,0),$$

$$x_{6j} = 1, \quad 1 \le j \le 3, \quad x_{6j} = 0, \quad 4 \le j \le 11.$$

The desired matrix is constructed:

$$X(\mathbf{A},\mathbf{B}) = \begin{pmatrix} 1&1&1&1&1&1&1&1&0&0&0 \\ 1&1&1&1&1&1&1&1&0&0&0 \\ 1&1&1&1&1&1&0&0&0&1&1 \\ 1&1&1&1&0&0&1&1&1&0&0 \\ 0&0&0&1&1&1&0&0&0&0&0 \\ 1&1&1&0&0&0&0&0&0&0&0 \end{pmatrix} \in M_Z(\mathbf{A},\mathbf{B};1).$$

In Chapter 4, the components of most vectors to be analyzed are arranged in nonascending order. This explains the use of the first vector $\mathbf{A} \in \overline{Z}_+^n$ in Theorem 2.3 and Algorithms 2.1 and 2.2 above. Note that a matrix from $M_Z(\mathbf{A}, \mathbf{B}; 1) \neq \varnothing$ can be constructed by means of Algorithms 1.2 and 1.3, with $c = 1$. Theorem 1.2 (with $c = 1$) also applies; therefore, every matrix from $M_Z(\mathbf{A}, \mathbf{B}; 1)$ can be constructed by reduction.

Let us derive a formula that expresses the number of matrices in $M_Z(\mathbf{A}, \mathbf{B}; 1)$ in terms of the numbers of matrices in the sets $M_Z(\mathbf{A}^{(1)}, \mathbf{B}^{(1)}; 1)$ corresponding to different reductive pairs $\mathbf{A}^{(1)}$ and $\mathbf{B}^{(1)}$. Suppose that vectors $\mathbf{A} = (a_1, \ldots, a_n) \in Z_+^n$ and $\mathbf{B} = (b_1, \ldots, b_m) \in Z_+^m$ are characterized by equal sums of components and $n \geq 2$. Let $\beta_1, \ldots, \beta_p$ be all of the different positive values of the components of $\mathbf{B}$, where $p \geq 1$, and $m_t = \big|\{j : b_j = \beta_t,\ 1 \leq j \leq m\}\big|$ is the number of components of $\mathbf{B}$ equal to β_t, where $1 \leq t \leq p$.

Each reductive pair $\mathbf{A}^{(1)}$, $\mathbf{B}^{(1)}$ corresponds to an integer-valued solution of the equation $a_1 = x_1 + \cdots + x_p$, with $0 \leq x_t \leq m_t$ and $1 \leq t \leq p$, where x_t is the number of components of $\mathbf{B}$ equal to β_t and decreased by unity at the reductive step. We refer to this reductive pair as $\mathbf{A}^{(1)}$, $\mathbf{B}^{(1)}(x_1, \ldots, x_p)$. It is obvious that

$$\big|M_Z(\mathbf{A}, \mathbf{B}; 1)\big| = \sum_{\substack{x_1, \ldots, x_p \\ x_1 + \cdots + x_p = a_1 \\ x_t \in Z,\ 0 \leq x_t \leq m_t,\ 1 \leq t \leq p}} C_{m_1}^{x_1} \ldots C_{m_p}^{x_p} \left|M_Z\left(\mathbf{A}^{(1)}, \mathbf{B}^{(1)}(x_1, \ldots, x_p); 1\right)\right|. \quad (2.7)$$

Thus, we have expression (2.7) for the number of n-row integer matrices of a transportation polyhedron in terms of the number of $n-1$-row integer matrices of the transportation polyhedron, which could be used to construct an algorithm that calculates the exact value of $\big|M_Z(\mathbf{A}, \mathbf{B}; 1)\big|$. However, the algorithm thus constructed would involve too many operations, which makes it inefficient as applied to practical problems. Therefore, we propose here the following mixed procedure for estimating the number of matrices in $M_Z(\mathbf{A}, \mathbf{B}; 1)$. In the initial steps, expression (2.7) is used, and then the method of characteristic equations is applied.

Example 2.4. Application of (2.7). Let us calculate the numbers of integer matrices in the following sets:

(a) $M_Z\big((8,7), (2,2,2,2,2,1,1,1,1,1); 1\big)$,

(b) $M_Z\big((3,3,3,3), (2,2,2,2,2,2); 1\big)$,

(c) $M_Z\big((5,5,5,5,5,5), (5,5,5,5,5,5); 1\big)$.

(a) Here, the components of the second vector of any reductive pair of vectors that defines a nonempty matrix set must be not greater than unity. Therefore,

$$\left|M_Z\left((8,7),(2,2,2,2,2,1,1,1,1,1);1\right)\right| = C_5^5C_5^3\left|M_Z\left((7),(1,1,1,1,1,1,1,0,0,0);1\right)\right| = C_5^5C_5^3 = 10.$$

(b) $$\left|M_Z\left((3,3,3,3),(2,2,2,2,2,2);1\right)\right| =$$
$$C_6^3\left|M_Z\left((3,3,3),(2,2,2,1,1,1);1\right)\right| =$$
$$20\Big(C_3^0C_3^3\left|M_Z\left((3,3),(2,2,2,0,0,0);1\right)\right| +$$
$$C_3^1C_3^2\left|M_Z\left((3,3),(2,2,1,1,0,0);1\right)\right| +$$
$$C_3^2C_3^1\left|M_Z\left((3,3),(2,1,1,1,1,0);1\right)\right| +$$
$$C_3^3C_3^0\left|M_Z\left((3,3),(1,1,1,1,1,1);1\right)\right|\Big) =$$
$$20\Big(1+9\Big(C_2^1C_2^2\left|M_Z\left((3),(2,1,0,0,0,0);1\right)\right| +$$
$$C_2^2C_2^1\left|M_Z\left((3),(1,1,1,0,0,0);1\right)\right|\Big)+$$
$$9\Big(C_1^0C_4^3\left|M_Z\left((3),(2,1,0,0,0,0);1\right)\right| +$$
$$C_1^1C_4^2\left|M_Z\left((3),(1,1,1,0,0,0);1\right)\right|\Big)+$$
$$C_6^3\left|M_Z\left((3),(1,1,1,0,0,0);1\right)\right|\Big) =$$
$$20\left(1+9(0+2)+9(0+6)+20\right) = 1860.$$

(c) It is clear that

$$\left|M_Z\left((5,5,5,5,5,5),(5,5,5,5,5,5);1\right)\right| = \left|M_Z\left((1,1,1,1,1,1),(1,1,1,1,1,1);1\right)\right|.$$

Accordingly,

$$\left|M_Z\left((5,5,5,5,5,5),(5,5,5,5,5,5);1\right)\right| = 6! = 720.$$

Example 2.5. (Extension of Example 2.2.) Let us calculate the number of matrices in the sets $M_Z((2,2,2,1),(3,2,2);1)$ and $M_Z((4,4,2),(2,2,2,2,1,1);1)$ by reduction with respect to both vectors.

$$\left|M_Z\left((2,2,2,1),(3,2,2);1\right)\right| = \left|M_Z\left((2,2,2),(2,2,2);1\right)\right| +$$
$$2\left|M_Z\left((2,2,2),(3,2,1);1\right)\right| = 3\left|M_Z\left((2,2),(2,1,1);1\right)\right| +$$
$$2\left|M_Z\left((1,1,1),(2,1);1\right)\right| = 3\left|M_Z\left((1,1),(1,1);1\right)\right| +$$
$$6\left|M_Z\left((1),(1);1\right)\right| = 3\cdot 2\left|M_Z\left((1),(1);1\right)\right| + 6 = 6+6 = 12,$$

$$\left|M_Z\left((4,4,2),(2,2,2,2,1,1);1\right)\right| = \left|M_Z\left((4,2),(1,1,1,1,1,1);1\right)\right| + \\ 8\left|M_Z\left((4,2),(2,1,1,1,1);1\right)\right| + 6\left|M_Z\left((4,2),(2,2,1,1);1\right)\right| = \\ 15\left|M_Z\left((2),(1,1);1\right)\right| + 8\Big(4\left|M_Z\left((2),(1,1);1\right)\right| + \left|M_Z\left((2),(2);1\right)\right|\Big) + \\ 6\left|M_Z\left((2),(1,1);1\right)\right| = 15 + 8(4+0) + 6 = 53.$$

Consequently,

$$\left|M_Z(\mathbf{A},\mathbf{B};1)\right| = \left|M_Z\left((2,2,2,1),(3,2,2);1\right)\right| \times \\ \left|M_Z\left((4,4,2),(2,2,2,2,1,1);1\right)\right| = 12\cdot 53 = 636.$$

In item (b) of Example 2.4, some reductive pairs make up empty matrix sets. Here, we formulate conditions under which only the 1-reductive pairs of vectors that correspond to transportation matrices containing only zeros and ones can be constructed.

Suppose that $M_Z(\mathbf{A},\mathbf{B};1) \neq \varnothing$, where the components of the first vector are arranged in nonascending order, and $\mathbf{A}^{(1)} = \mathbf{A}^{(1)}(1)$, $\mathbf{B}^{(1)}$ is a 1-reductive pair of vectors. If $M_Z(\mathbf{A}^{(1)},\mathbf{B}^{(1)};1) \neq \varnothing$, then $\mathbf{B} - \mathbf{B}^{(1)}$ is the first row of a matrix from $M_Z(\mathbf{A},\mathbf{B};1)$. The following assertions from theorems stated above apply to the 1-reductive pair of vectors $\mathbf{A}^{(1)}$ and $\mathbf{B}^{(1)}$.

(a) Theorem 2.3 states sufficient conditions for $M_Z(\mathbf{A}^{(1)},\mathbf{B}^{(1)};1)$ to be a nonempty set. If $M_Z(\mathbf{A},\mathbf{B};1) \neq \varnothing$ and $\mathbf{B}^{(1)} = \mathbf{B}_*^{(1)}$, where $\mathbf{B}_*^{(1)}$ is the vector constructed by applying the sequence defined by (1.4) and rule (2.6), then the matrix set $M_Z(\mathbf{A}^{(1)},\mathbf{B}^{(1)};1)$ is not empty.

(b) Proposition (b) of Theorem 1.2 (with $c = 1$) states a necessary and sufficient condition for $M_Z(\mathbf{A}^{(1)},\mathbf{B}^{(1)};1)$ to be a nonempty set: the characteristic function $\delta_k(\mathbf{A}^{(1)},\mathbf{B}^{(1)};1)$ must be nonnegative.

Now, we use the characteristic functions and equations to simplify proposition (b) of Theorem 1.3 in the case of $c = 1$.

Suppose that vectors $\mathbf{A} = (a_1,\ldots,a_n) \in \overline{Z}_+^n$ and $\mathbf{B} = (b_1,\ldots,b_m) \in \overline{Z}_+^m$ are characterized by equal sums of components. Let $a_1 \leq l_1(\mathbf{B};1)$ and $\mathbf{A}^{(1)} = \mathbf{A}^{(1)}(1)$, $\mathbf{B}^{(1)}$ be a 1-reductive pair of vectors, which must exist by Definitions 1.1 and 1.3. Set $(x_{11},\ldots,x_{1m}) = \mathbf{B} - \mathbf{B}^{(1)}$. It is obvious that this vector contains only zeros and units and the sum of its components is a_1. For $(x_{11},\ldots,x_{1m})$ and $k \in Z$ $(1 \leq k \leq n)$, denote by x_k and y_k the numbers of unit components of $(x_{11},\ldots,x_{1m})$ in the cases $j \leq l_k(\mathbf{B};1)$ and $j > l_k(\mathbf{B};1)$, respectively:

$$x_k = \left|\{j : j \leq l_k(\mathbf{B};1),\ x_{1j} = 1\}\right|, \quad y_k = \left|\{j : j > l_k(\mathbf{B};1),\ x_{1j} = 1\}\right|$$

Note the following obvious properties of the integers x_k and y_k $(1 \le k \le n)$: $x_1 = a_1$ and $y_1 = 0$,

$$\begin{cases} x_k + y_k = a_1, \\ 0 \le x_k \le \min\left\{a_1, l_k(\mathbf{B};1)\right\}, \\ 0 \le y_k \le \min\left\{a_1, l_1(\mathbf{B};1) - l_k(\mathbf{B};1)\right\}; \end{cases}$$

if $n \ge 2$, then for $1 \le k \le n-1$ we have

$$\begin{cases} x_k \ge x_{k+1}, \\ x_k - x_{k+1} \le l_k(\mathbf{B};1) - l_{k+1}(\mathbf{B};1), \end{cases}$$

where $x_k = x_{k+1}$ when $l_k(\mathbf{B};1) = l_{k+1}(\mathbf{B};1)$.

Let us establish the conditions that must be satisfied by the numbers x_k $(1 \le k \le n)$ for $M_Z(\mathbf{A}^{(1)}, \mathbf{B}^{(1)}; 1)$ to be a nonempty set.

Definition 2.3. If $M_Z(\mathbf{A}^{(1)}, \mathbf{B}^{(1)}; 1) \ne \varnothing$, then a 1-reductive pair of vectors $\mathbf{A}^{(1)}$, $\mathbf{B}^{(1)}$ is referred to as *admissible.*

Note that if $\mathbf{A}^{(1)}$ and $\mathbf{B}^{(1)}$ make up an admissible pair, then $M_Z(\mathbf{A}, \mathbf{B}; 1) \ne \varnothing$.

Consider a matrix $X(\mathbf{A}, \mathbf{B}) = (x_{ij})$ from $M_Z(\mathbf{A}, \mathbf{B}; 1)$. Expression (2.3) implies that

$$\delta_k(\mathbf{A}, \mathbf{B}; 1) \ge k l_k(\mathbf{B}; 1) - \sum_{i=1}^{k} \sum_{j \le l_k(\mathbf{B};1)} x_{ij}, \quad 1 \le k \le n-1.$$

The right-hand side of this inequality is the number of zeros in (x_{ij}), where $1 \le i \le k$ and $1 \le j \le l_k(\mathbf{B};1)$. Since $l_k(\mathbf{B};1) - x_k$ is the number of zeros in the first row of the submatrix, then $\delta_k(\mathbf{A}, \mathbf{B}; 1) \ge l_k(\mathbf{B};1) - x_k$. Consequently, inequality $x_k \ge l_k(\mathbf{B};1) - \delta_k(\mathbf{A}, \mathbf{B}; 1)$, for any k $(1 \le k \le n-1)$, is a necessary condition for the 1-reductive pair $\mathbf{A}^{(1)}$, $\mathbf{B}^{(1)}$ to be admissible. Let us show that this is also a sufficient condition for $M_Z(\mathbf{A}^{(1)}, \mathbf{B}^{(1)}; 1)$ to be a nonempty set.

Suppose that $a_1 \le l_1(\mathbf{B};1)$ and the 1-reductive pair of vectors $\mathbf{A}^{(1)} = \mathbf{A}^{(1)}(1)$ and $\mathbf{B}^{(1)}$ constructed for $\mathbf{A}$ and $\mathbf{B}$ corresponds to the vector $(x_{11}, \ldots, x_{1m})$. Consider the difference between the characteristic functions $\delta_{k-1}(\mathbf{A}^{(1)}, \mathbf{B}^{(1)}; 1)$ and $\delta_k(\mathbf{A}, \mathbf{B}; 1)$:

$$\delta_k(\mathbf{A}, \mathbf{B}; 1) = \sum_{j=1}^{m} b_j - \sum_{b_j \ge k} (b_j - k) - \sum_{i=1}^{k} a_i, \quad 1 \le k \le n;$$

$$\delta_{k-1}(\mathbf{A}^{(1)}, \mathbf{B}^{(1)}; 1) = \sum_{j=1}^{m} b_j^{(1)} - \sum_{b_j^{(1)} \geq k-1} (b_j^{(1)} - k + 1) - \sum_{i=2}^{k} a_i, \quad 2 \leq k \leq n;$$

$$\delta_{k-1}(\mathbf{A}^{(1)}, \mathbf{B}^{(1)}; 1) - \delta_k(\mathbf{A}, \mathbf{B}; 1) = - \sum_{b_j^{(1)} \geq k-1} (b_j^{(1)} - k + 1) + \sum_{b_j \geq k} (b_j - k) =$$

$$\sum_{\substack{b_j \geq k \\ b_j = b_j^{(1)} + 1}} (b_j - k) - \sum_{\substack{b_j^{(1)} \geq k-1 \\ b_j^{(1)} = b_j - 1}} (b_j^{(1)} - k + 1) + \sum_{\substack{b_j \geq k \\ b_j = b_j^{(1)}}} (b_j - k) - \sum_{\substack{b_j^{(1)} \geq k-1 \\ b_j^{(1)} = b_j}} (b_j^{(1)} - k + 1),$$

$$2 \leq k \leq n.$$

It can be easily shown that

$$\sum_{\substack{b_j \geq k \\ b_j = b_j^{(1)} + 1}} (b_j - k) - \sum_{\substack{b_j^{(1)} \geq k-1 \\ b_j^{(1)} = b_j - 1}} (b_j^{(1)} - k + 1) = 0,$$

$$\sum_{\substack{b_j \geq k \\ b_j = b_j^{(1)}}} (b_j - k) - \sum_{\substack{b_j^{(1)} \geq k-1 \\ b_j^{(1)} = b_j}} (b_j^{(1)} - k + 1) = \sum_{\substack{b_j \geq k \\ b_j = b_j^{(1)}}} (b_j - k) - \sum_{\substack{b_j^{(1)} \geq k \\ b_j^{(1)} = b_j}} (b_j^{(1)} - k + 1) =$$

$$-\left|\{j : b_j = b_j^{(1)},\ b_j \geq k\}\right| = -\left|\{j : j \leq l_k(\mathbf{B}; 1), x_{1j} = 0\}\right| =$$

$$-\left(l_k(\mathbf{B}; 1) - x_k\right) = x_k - l_k(\mathbf{B}; 1).$$

Accordingly,

$$\delta_{k-1}(\mathbf{A}^{(1)}, \mathbf{B}^{(1)}; 1) - \delta_k(\mathbf{A}, \mathbf{B}; 1) = x_k - l_k(\mathbf{B}; 1),$$

$$\delta_{k-1}(\mathbf{A}^{(1)}, \mathbf{B}^{(1)}; 1) = \delta_k(\mathbf{A}, \mathbf{B}; 1) + x_k - l_k(\mathbf{B}; 1), \quad 2 \leq k \leq n. \tag{2.8}$$

Thus, if $x_k \geq l_k(\mathbf{B}; 1) - \delta_k(\mathbf{A}, \mathbf{B}; 1)$ for any k $(2 \leq k \leq n)$, then $\delta_{k-1}(\mathbf{A}^{(1)}, \mathbf{B}^{(1)}; 1) \geq 0$, where $2 \leq k \leq n$. Therefore, $M_Z(\mathbf{A}^{(1)}, \mathbf{B}^{(1)}; 1) \neq \varnothing$ and $\mathbf{A}^{(1)}$, $\mathbf{B}^{(1)}$ is an admissible pair of vectors. This completes the proof of the following proposition, which states the criterion for a 1-reductive pair of vectors to be admissible.

Theorem 2.4. *Suppose that vectors* $\mathbf{A} = (a_1, \ldots, a_n) \in \overline{Z}_+^n$ *and* $\mathbf{B} = (b_1, \ldots, b_m) \in \overline{Z}_+^m$, *where* $a_1 \leq l_1(\mathbf{B}; 1)$, *are characterized by equal sums of components. Then, the following assertions concerning the* 1*-reductive pair of vectors* $\mathbf{A}^{(1)} = \mathbf{A}^{(1)}(1)$ *and* $\mathbf{B}^{(1)}$, *vector* $(x_{11}, \ldots, x_{1m}) = \mathbf{B} - \mathbf{B}^{(1)}$, *and numbers* $x_k = \left|\{j : j \leq l_k(\mathbf{B}; 1),\ x_{1j} = 1\}\right|$ $(1 \leq k \leq n)$ *are true.*

(a) *The vectors* $\mathbf{A}^{(1)}$ *and* $\mathbf{B}^{(1)}$ *make up an admissible pair if and only if* $x_k \geq l_k(\mathbf{B}; 1) - \delta_k(\mathbf{A}, \mathbf{B}; 1)$ *for any* $k \in Z$ $(1 \leq k \leq n)$.

(b) *A matrix* $Y(\mathbf{A}, \mathbf{B}) = (y_{ij})$ *from* $M_Z(\mathbf{A}, \mathbf{B}; 1)$ *such that* $(y_{11}, \ldots, y_{1m}) = (x_{11}, \ldots, x_{1m})$ *exists if and only if* $x_k \geq l_k(\mathbf{B}; 1) - \delta_k(\mathbf{A}, \mathbf{B}; 1)$ *for any* $k \in Z$ $(1 \leq k \leq n)$.

Remark 2.2. When $\delta_{k-1}(\mathbf{A}^{(1)}, \mathbf{B}^{(1)}; 1) \geq 0$ $(2 \leq k \leq n)$, the set of inequalities (2.8) proves the necessity of the assumptions of Theorem 2.4.

Remark 2.3. The assumption that the components of $\mathbf{B}$ are arranged in nonascending order may be added to the statement of Theorem 2.4. Then, the vector $(x_{11}, \ldots, x_{1m}) = \mathbf{B} - \mathbf{B}^{(1)}$ is such that $x_k = \left|\{j : b_j \geq k,\ x_{1j} = 1\}\right|$ $(1 \leq k \leq n)$.

Now, we formulate an algorithm that can be used to construct any matrix from $M_Z(\mathbf{A}, \mathbf{B}; c) \neq \varnothing$ on the basis on Theorem 2.4 and Remark 2.3.

Algorithm 2.3. Application of Lemma 2.1 in constructing a matrix from the set $M_Z(\mathbf{A}, \mathbf{B}; 1)$.

Suppose that vectors $\mathbf{A} = (a_1, \ldots, a_n) \in \overline{Z}_+^n$ and $\mathbf{B} = (b_1, \ldots, b_m) \in \overline{Z}_+^m$ are characterized by equal sums of components.

Step 1. Calculate the value of the characteristic function $\delta_k(\mathbf{A}, \mathbf{B}; 1)$ for every $k \in Z$ $(1 \leq k \leq n)$.

Step 2. Apply Lemma 2.1. If $\delta_k(\mathbf{A}, \mathbf{B}; 1) \geq 0$ for every k $(1 \leq k \leq n)$, then go to the next step. If there exist k such that $\delta_k(\mathbf{A}, \mathbf{B}; 1) < 0$, then go to Step 9.

Step 3. Set $\mathbf{A}^{(0)} = \mathbf{A}$ and $\mathbf{B}^{(0)} = (b_1^{(0)}, \ldots, b_m^{(0)}) = \mathbf{B}$.

Step 4. Set $p = 1$.

Step 5. Calculate the value of

$$\alpha_k^{(p)} = l_k(\mathbf{B}^{(p-1)}; 1) - \delta_k(\mathbf{A}^{(p-1)}, \mathbf{B}^{(p-1)}; 1)$$

for every k $(1 \leq k \leq n - p + 1)$.

Step 6. Construct a vector $(x_{p1}, \ldots, x_{pm})$ with zero and unit elements that satisfies the following conditions:

$$\sum_{j=1}^{m} x_{pj} = a_p, \quad \alpha_k^{(p)} \leq x_k^{(p)} \leq l_k(\mathbf{B}^{(p-1)}; 1), \quad 1 \leq k \leq n - p + 1,$$

where $x_k^{(p)} = \left|\{j : b_j^{(p-1)} \geq k,\ x_{pj} = 1\}\right|$, $1 \leq k \leq n - p + 1$.

Step 7. Define the pth row of the matrix $X(\mathbf{A}, \mathbf{B})$ from $M_Z(\mathbf{A}, \mathbf{B}; 1)$ as the vector $(x_{p1}, \ldots, x_{pm})$.

Step 8. If $p \leq n - 1$, then set

$$\mathbf{B}^{(p)} = (b_1^{(p)}, \ldots, b_m^{(p)}) = \mathbf{B}^{(p-1)} - (x_{p1}, \ldots, x_{pm}),$$

$$\mathbf{A}^{(p)} = \mathbf{A}^{(p)}(0, 1, \ldots, p) = (a_{p+1}, \ldots, a_n);$$

increase the value of p by unity; and go to Step 5. If $p = n$, then stop.

Step 9. Set $M_Z(\mathbf{A}, \mathbf{B}; 1) = \varnothing$.

Example 2.6. Consider the following vectors with equal sums of components: $\mathbf{A} = (6,6,6,2) \in \overline{Z}^4_+$ and $\mathbf{B} = (3,3,3,3,2,2,1,1,1,1) \in \overline{Z}^{10}_+$. Here, $\delta_1(\mathbf{A}, \mathbf{B}; 1) = 4$, $\delta_2(\mathbf{A}, \mathbf{B}; 1) = 4$, $\delta_3(\mathbf{A}, \mathbf{B}; 1) = 2$, and $\delta_4(\mathbf{A}, \mathbf{B}; 1) = 0$. Accordingly, $M_Z(\mathbf{A}, \mathbf{B}; 1) \neq \varnothing$.

For $\mathbf{A}$ and $\mathbf{B}$, calculate the numbers of all 1-reductive and all admissible pairs of vectors by taking out the first component of $\mathbf{A}$ and construct all of the first rows of the matrices belonging to $M_Z(\mathbf{A}, \mathbf{B}; 1)$.

It can be readily shown that the total number of 1-reductive pairs of the form $\mathbf{A}^{(1)}(1)$, $\mathbf{B}^{(1)}$ is $C^6_{10} = 210$.

Now, calculate the number of admissible vectors by applying Theorem 2.4 (or Steps 4 and 5 of Algorithm 2.3). Since $l_2(\mathbf{B}; 1) = 6$, $l_3(\mathbf{B}; 1) = 4$, and $l_4(\mathbf{B}; 1) = 0$, the number $\alpha_k = l_k(\mathbf{B}; 1) - \delta_k(\mathbf{A}, \mathbf{B}; 1)$ are $\alpha_2 = 2$, $\alpha_3 = 2$, and $\alpha_4 = 0$ for $2 \leq k \leq 4$. Therefore, the following inequalities apply to every vector $(x_{11}, \ldots, x_{1\,10})$ identical with the first row of a matrix from $M_Z(\mathbf{A}, \mathbf{B}; 1)$ and to the numbers $x_k = \left|\{j : j \leq l_k(\mathbf{B}; 1),\ x_{1j} = 1\}\right|$, where $k = 2$ or $k = 3$: $2 \leq x_2 \leq 6$, $2 \leq x_3 \leq 4$.

Let us write out all solutions of the set of inequalities

$$\begin{cases} 2 \leq x_2 \leq 6, \\ 2 \leq x_3 \leq 4, \\ x_2 \geq x_3, \\ x_2 - x_3 \leq l_2(\mathbf{B}; 1) - l_3(\mathbf{B}; 1) = 2. \end{cases}$$

For every solution (x_2, x_3), we construct the corresponding vector $(x_{11}, \ldots, x_{1\,10})$, identical with a row of a matrix from $M_Z(\mathbf{A}, \mathbf{B}; 1)$, marking off the components with the indices 4, 5 and 6, 7 for the sake of graphic representation:

$$(2,2), \quad (1,1,0,0|0,0|1,1,1,1);$$

$$(3,2), \quad (1,1,0,0|1,0|1,1,1,0);$$

$$(3,3), \quad (1,1,1,0|0,0|1,1,1,0);$$

$$(4,2), \quad (1,1,0,0|1,1|1,1,0,0);$$

$$(4,3), \quad (1,1,1,0|1,0|1,1,0,0);$$

$$(4,4), \quad (1,1,1,1|0,0|1,1,0,0);$$

$$(5,3), \quad (1,1,1,0|1,1|1,0,0,0);$$

$$(5,4), \quad (1,1,1,1|1,0|1,0,0,0);$$

$$(6,4), \quad (1,1,1,1|1,1|0,0,0,0).$$

Thus, we have found all of the nine solutions to the set of inequalities and constructed nine vectors that are identical with the first rows of some matrices from $M_Z(\mathbf{A},\mathbf{B};1)$. We have marked off the components of each vector by parentheses and bars to divide them in three subvectors. It is obvious that the set of the first rows of the set of matrices under consideration is the set of vectors constructed by rearranging the components of the subvectors. These rearrangements are mutually independent. Consequently, the number of admissible pairs of vectors is

$$C_4^2 + C_4^2C_2^1C_4^3 + C_4^3C_4^3 + C_4^2C_4^2 + C_4^3C_2^1C_4^2 + C_4^2 + C_4^3C_4^1 + C_2^1C_4^1 + 1 = 185.$$

Remark 2.4. Suppose that vectors $\mathbf{A} \in \overline{Z}_+^n$ and $\mathbf{B} \in \overline{Z}_+^m$ satisfying the balance condition are such that $a_1 \le l_1(\mathbf{B};1)$. If the vectors $\mathbf{A}^{(1)}(1)$ and $\mathbf{B}^{(1)} = (b_1^{(1)},\dots,b_m^{(1)})$, where

$$b_j^{(1)} = \begin{cases} b_j - 1, & l_1(\mathbf{B};1) - a_1 + 1 \le j \le l_1(\mathbf{B};1), \\ b_j, & j \le l_1(\mathbf{B};1) - a_1, \end{cases}$$

make up an admissible 1-reductive pair, then every 1-reductive pair is admissible.

Now, we formulate assumptions that are analogous to those of Theorem 2.4 and provide an admissibility criterion for the $(1;p)$-reductive pair of vectors.

Suppose that vectors $\mathbf{A} = (a_1,\dots,a_n) \in \overline{Z}_+^n$ and $\mathbf{B} = (b_1,\dots,b_m) \in \overline{Z}_+^m$, where $a_1 \le l_1(\mathbf{B};1)$, are characterized by equal sums of components. Consider a 1-reductive pair of vectors $\mathbf{A}^{(1)}(p)$ and $\mathbf{B}^{(1)}$ obtained by taking out the pth component of $\mathbf{A}$. The vector $(x_{p1},\dots,x_{pm}) = \mathbf{B} - \mathbf{B}^{(1)}$ contains only zeros and ones, and the sum of its components is a_p. For this vector and $k \in Z$ $(1 \le k \le n)$, denote by $x_k^{(p)}$ the number of ones in the vector $(x_{p1},\dots,x_{pm})$ for $j \le l_k(\mathbf{B};1)$:

$$x_k^{(p)} = \left|\{j : j \le l_k(\mathbf{B};1),\ x_{pj} = 1\}\right|.$$

It is obvious that the integers $x_k^{(p)}$ $(1 \le k \le n)$ have the following properties: $x_1^{(p)} = a_p$ and $0 \le x_k^{(p)} \le \min\{a_p, l_k(\mathbf{B};1)\}$; if $n \ge 2$, then

$$\begin{cases} x_k^{(p)} \ge x_{k+1}^{(p)}, \\ x_k^{(p)} - x_{k+1}^{(p)} \le l_k(\mathbf{B};1) - l_{k+1}(\mathbf{B};1), \end{cases}$$

and $x_k^{(p)} = x_{k+1}^{(p)}$ when $l_k(\mathbf{B};1) = l_{k+1}(\mathbf{B};1)$.

The following proposition extends Theorem 2.4 and states the necessary and sufficient conditions to be satisfied by the numbers $x_k^{(p)}$ $(1 \le k \le n)$ for $M_Z\left(\mathbf{A}^{(1)}(p), \mathbf{B}^{(1)}; 1\right) \ne \varnothing$ to be a nonempty set.

Theorem 2.5. *Suppose that vectors* $\mathbf{A} = (a_1, \ldots, a_n) \in \overline{Z}_+^n$ *and* $\mathbf{B} = (b_1, \ldots, b_m) \in \overline{Z}_+^m$, *where* $a_1 \le l_1(\mathbf{B}; 1)$, *satisfy the balance condition. The following assertions concerning the* $(1;p)$*-reductive pair of vectors* $\mathbf{A}^{(1)}(p)$ *and* $\mathbf{B}^{(1)}$ $(1 \le p \le n)$, *vector* $(x_{p1}, \ldots, x_{pm}) = \mathbf{B} - \mathbf{B}^{(1)}$, *and numbers* $x_k^{(p)} = \left|\{j : j \le l_k(\mathbf{B}; 1),\ x_{pj} = 1\}\right|$ $(1 \le k \le n)$ *are true:*

(a) *The vectors* $\mathbf{A}^{(1)}(p)$ *and* $\mathbf{B}^{(1)}$ *make up an admissible pair if and only if it holds, for any* $k \in Z$ $(1 \le k \le n)$, *that*

$$\begin{aligned} x_{k+1}^{(p)} &\ge a_p - \delta_k(\mathbf{A}, \mathbf{B}; 1), \quad k < p, \\ x_k^{(p)} &\ge l_k(\mathbf{B}; 1) - \delta_k(\mathbf{A}, \mathbf{B}; 1), \quad p \le k \le n. \end{aligned} \tag{2.9}$$

(b) *A matrix* $Y(\mathbf{A}, \mathbf{B}) = (y_{ij})$ *from* $M_Z(\mathbf{A}, \mathbf{B}; 1)$ *such that* $(y_{p1}, \ldots, y_{pm}) = (x_{p1}, \ldots, x_{pm})$ *exists if and only if* (2.9) *is true for every* $k \in Z$ $(1 \le k \le n)$.

Proof. Consider the remainder $\delta_k(\mathbf{A}^{(1)}(p), \mathbf{B}^{(1)}; 1) - \delta_k(\mathbf{A}, \mathbf{B}; 1)$ for $k < p$ $(p > 1)$:

$$\delta_k(\mathbf{A}, \mathbf{B}; 1) = \sum_{j=1}^{m} b_j - \sum_{b_j \ge k} (b_j - k) + \sum_{i=1}^{k} a_i,$$

$$\delta_k(\mathbf{A}^{(1)}(p), \mathbf{B}^{(1)}; 1) = \sum_{j=1}^{m} b_j^{(1)} - \sum_{b_j^{(1)} \ge k} (b_j^{(1)} - k) + \sum_{i=1}^{k} a_i,$$

$$\delta_k(\mathbf{A}^{(1)}(p), \mathbf{B}^{(1)}; 1) - \delta_k(\mathbf{A}, \mathbf{B}; 1) = -a_p - \sum_{b_j^{(1)} \ge k} (b_j^{(1)} - k) + \sum_{b_j \ge k} (b_j - k) =$$

$$\sum_{\substack{b_j \ge k \\ b_j = b_j^{(1)}}} (b_j - k) - \sum_{\substack{b_j^{(1)} \ge k \\ b_j^{(1)} = b_j}} (b_j^{(1)} - k) + \sum_{\substack{b_j \ge k \\ b_j = b_j^{(1)} + 1}} (b_j - k) - \sum_{\substack{b_j^{(1)} \ge k \\ b_j^{(1)} = b_j - 1}} (b_j^{(1)} - k) - a_p.$$

Since

$$\sum_{\substack{b_j \ge k \\ b_j = b_j^{(1)}}} (b_j - k) - \sum_{\substack{b_j^{(1)} \ge k \\ b_j^{(1)} = b_j}} (b_j^{(1)} - k) = 0,$$

then

$$\delta_k(\mathbf{A}^{(1)}(p), \mathbf{B}^{(1)}; 1) - \delta_k(\mathbf{A}, \mathbf{B}; 1) = \sum_{\substack{b_j \geq k \\ b_j = b_j^{(1)} + 1}} (b_j - k) - \sum_{\substack{b_j^{(1)} \geq k \\ b_j^{(1)} = b_j - 1}} (b_j^{(1)} - k) - a_p =$$

$$\sum_{\substack{b_j \geq k+1 \\ b_j = b_j^{(1)} + 1}} (b_j - k) - \sum_{\substack{b_j^{(1)} \geq k \\ b_j^{(1)} = b_j - 1}} (b_j^{(1)} - k) - a_p =$$

$$\left|\{j : b_j \geq k+1,\ b_j^{(1)} = b_j - 1\}\right| - a_p = \left|\{j : b_j \geq k+1,\ x_{pj} = 1\}\right| - a_p =$$

$$\left|\{j : j \leq l_{k+1}(\mathbf{B}; 1),\ x_{pj} = 1\}\right| - a_p = x_{k+1}^{(p)} - a_p.$$

Therefore,

$$\delta_k(\mathbf{A}^{(1)}(p), \mathbf{B}^{(1)}; 1) = \delta_k(\mathbf{A}, \mathbf{B}; 1) + x_{k+1}^{(p)} - a_p, \quad k \leq p. \tag{2.10}$$

Suppose $p \leq k \leq n$. Then, the difference between the characteristic functions $\delta_{k-1}(\mathbf{A}^{(1)}(p), \mathbf{B}^{(1)}; 1)$ and $\delta_k(\mathbf{A}, \mathbf{B}; 1)$ is calculated as follows (see the proof of Theorem 2.4):

$$\delta_{k-1}(\mathbf{A}^{(1)}(p), \mathbf{B}^{(1)}; 1) - \delta_k(\mathbf{A}, \mathbf{B}; 1) = -\sum_{b_j^{(1)} \geq k-1} (b_j^{(1)} - k + 1) + \sum_{b_j \geq k} (b_j - k) =$$

$$\sum_{\substack{b_j \geq k \\ b_j = b_j^{(1)}}} (b_j - k) - \sum_{\substack{b_j^{(1)} \geq k \\ b_j^{(1)} = b_j}} (b_j^{(1)} - k + 1) = -\left|\{j : b_j = b_j^{(1)},\ b_j \geq k\}\right| =$$

$$-\left|\{j : j \leq l_k(\mathbf{B}; 1),\ x_{pj} = 0\}\right| = x_k^{(p)} - l_k(\mathbf{B}; 1).$$

Thus, we then have

$$\delta_{k-1}(\mathbf{A}^{(1)}(p), \mathbf{B}^{(1)}; 1) = \delta_k(\mathbf{A}, \mathbf{B}; 1) + x_k^{(p)} - l_k(\mathbf{B}; 1), \quad p \leq k \leq n. \tag{2.11}$$

Combining (2.10) with (2.11), we see that $\delta_k(\mathbf{A}^{(1)}(p), \mathbf{B}^{(1)}; 1) \geq 0$ for every k $(1 \leq k \leq n-1)$ if and only if the numbers $x_k^{(p)}$ $(1 \leq k \leq n)$ satisfy the inequalities in (2.9), which completes the proof.

The following example illustrates the distinction between the admissibility conditions for the $(1;p)$-reductive pairs of vectors that correspond to different values of p.

Example 2.7. The vectors $\mathbf{A} = (5, 5, 3, 1) \in \overline{Z}_+^4$ and $\mathbf{B} = (3, 3, 2, 2, 2, 1, 1) \in \overline{Z}_+^7$ have equal sums of components. Calculating the values of $\delta_k(\mathbf{A}, \mathbf{B}; 1)$ and $l_k(\mathbf{B}; 1)$ $(1 \leq k \leq 4)$, we obtain $\delta_1(\mathbf{A}, \mathbf{B}; 1) = 2$, $l_1(\mathbf{B}; 1) = 7$, $\delta_2(\mathbf{A}, \mathbf{B}; 1) = 2$,

$l_2(\mathbf{B};1) = 5$, $\delta_3(\mathbf{A},\mathbf{B};1) = 1$, $l_3(\mathbf{B};1) = 2$, $\delta_4(\mathbf{A},\mathbf{B};1) = 0$, and $l_4(\mathbf{B};1) = 0$. Therefore, $M_Z(\mathbf{A},\mathbf{B};1) \neq \varnothing$.

It can be readily shown that the numbers of $(1;p)$-reductive vector pairs that correspond to $p \in \{1,2,3,4\}$ are, respectively, $C_7^5 = 21$, $C_7^5 = 21$, $C_7^3 = 35$, $C_7^1 = 7$.

Let us apply Theorem 2.5 to the case of each $p \in \{1,2,3,4\}$. To do this, we calculate the values of $\alpha_k = l_k(\mathbf{B};1) - \delta_k(\mathbf{A},\mathbf{B};1)$ for $1 \le k \le 4$ and $\beta_k^{(p)} = a_p - \delta_k(\mathbf{A},\mathbf{B};1)$ for $2 \le p \le 4$, $1 \le k < p$. We obtain

$$\alpha_1 = 5, \quad \alpha_2 = 3, \quad \alpha_3 = 1, \quad \alpha_4 = 0;$$

$$\beta_1^{(2)} = 3, \quad \beta_1^{(3)} = 1, \quad \beta_2^{(3)} = 1, \quad \beta_1^{(4)} = -1, \quad \beta_2^{(4)} = -1, \quad \beta_3^{(4)} = 0.$$

We analyze each of the cases $p \in \{1,2,3,4\}$ separately, following a procedure similar to that used in Example 2.6.

(a) Suppose $p = 1$. The set of inequalities (2.9) is as follows ($x_1^{(1)} = 5$, $x_4^{(1)} = 0$):

$$\begin{cases} 3 \le x_2^{(1)} \le 5, \\ 1 \le x_3^{(1)} \le 2, \\ x_2^{(1)} \ge x_3^{(1)}, \\ x_2^{(1)} - x_3^{(1)} \le l_2(\mathbf{B};1) - l_3(\mathbf{B};1) = 3. \end{cases}$$

We write out all solutions and, for each of the solutions $(x_2^{(1)}), (x_3^{(1)})$, construct a vector $(x_{11}, \ldots, x_{17})$ that is identical with the first row of a matrix from $M_Z(\mathbf{A},\mathbf{B};1)$:

$$(3,1), \quad (1,0|1,1,0|1,1);$$

$$(3,2), \quad (1,1|1,0,0|1,1);$$

$$(4,1), \quad (1,0|1,1,1|1,0);$$

$$(4,2), \quad (1,1|1,1,0|1,0);$$

$$(5,2), \quad (1,1|1,1,1|0,0).$$

The number of admissible $(1;1)$-reductive pairs of vectors is

$$C_2^1 C_3^2 + C_3^1 + C_2^1 C_2^1 + C_3^2 C_2^1 + 1 = 20.$$

In this case, only one of the $(1;1)$-reductive pairs is not admissible and corresponds to the vector $(0,0,1,1,1,1,1)$, namely

$$(0,5,3,1), (3,3,1,1,1,0,0) = (3,3,2,2,2,1,1) - (0,0,1,1,1,1,1).$$

(b) If $p = 2$, then the resulting $(1;2)$-reductive pairs of vectors are the same as in item (a). In this case, the set of inequalities is

$$\begin{cases} 3 \le x_2^{(2)} \le 5, \\ 1 \le x_3^{(2)} \le 2, \\ x_2^{(2)} \ge x_3^{(2)}, \\ x_2^{(2)} - x_3^{(2)} \le 3. \end{cases}$$

(c) Suppose $p = 3$. Then $(x_2^{(3)} \le \min\{a_3, l_2(\mathbf{B};1)\})$,

$$\begin{cases} 1 \le x_2^{(3)} \le 3, \\ 1 \le x_3^{(3)} \le 2, \\ x_2^{(3)} \ge x_3^{(3)}, \\ x_2^{(3)} - x_3^{(3)} \le 3. \end{cases}$$

We write out the solutions and the corresponding vectors:

$$(1,1), \quad (1,0|0,0,0|1,1);$$

$$(2,1), \quad (1,0|1,0,0|1,0);$$

$$(2,2), \quad (1,1|0,0,0|1,0);$$

$$(3,1), \quad (1,0|1,1,0|0,0);$$

$$(3,2), \quad (1,1|1,0,0|0,0).$$

The total number of admissible $(1;3)$-reductive pairs of vectors is

$$C_2^1 + C_2^1 C_3^1 C_2^1 + C_2^1 + C_2^1 C_3^2 + C_3^1 = 25.$$

The set of the third rows of matrices from $M_Z(\mathbf{A},\mathbf{B};1)$ is the set of vectors obtained by rearranging components within the three groups in all of the five constructed vectors.

(d) Suppose $p = 4$. Then $(x_2^{(4)} \ge \max(0, \beta_1(4)),\ x_3^{(4)} \ge \max(0, \beta_2(4)))$,

$$\begin{cases} 0 \le x_2^{(4)} \le 1, \\ 0 \le x_3^{(4)} \le 1, \\ x_2^{(4)} \ge x_3^{(4)}, \\ x_2^{(4)} - x_3^{(4)} \le 3. \end{cases}$$

The corresponding solutions and vectors are

$$(0,0), \quad (0,0|0,0,0|1,0);$$

$$(1,0), \quad (0,0|1,0,0|0,0);$$

$$(1,1), \quad (1,0|0,0,0|0,0).$$

Here, every $(1;4)$-reductive pair of vectors is admissible.

The second inequality in (2.9) is used as a basis for analyzing the admissibility of $(1;1)$-reductive pairs of vectors (see Theorem 2.4). In Example 2.7, the first and second inequalities in (2.9) are equivalent when $p = 1$ and $p = 2$ in the first and second inequality, respectively. The following example provides a more explicit illustration of the use of the first unequality in (2.9).

Example 2.8. (Application of Theorem 2.5 in constructing all $(1;p)$-reductive pairs of vectors.)

Consider the vectors

$$\mathbf{A} = (7,7,7,7,5,4,2,2,1) \in \overline{Z}^9_+ \text{ and } \mathbf{B} = (8,8,6,6,5,5,2,2,1) \in \overline{Z}^9_+,$$

with the sums of components equal to 42. For $\delta_k(\mathbf{A},\mathbf{B};1)$ $(1 \le k \le 8)$, we have $\delta_1(\mathbf{A},\mathbf{B};1) = 2$, $\delta_2(\mathbf{A},\mathbf{B};1) = 2$, $\delta_3(\mathbf{A},\mathbf{B};1) = 1$, $\delta_4(\mathbf{A},\mathbf{B};1) = 0$, $\delta_5(\mathbf{A},\mathbf{B};1) = 1$, $\delta_6(\mathbf{A},\mathbf{B};1) = 1$, $\delta_7(\mathbf{A},\mathbf{B};1) = 1$, $\delta_8(\mathbf{A},\mathbf{B};1) = 1$, and $\delta_9(\mathbf{A},\mathbf{B};1) = 0$. Since the characteristic function is nonnegative, $M_Z(\mathbf{A},\mathbf{B};1) \ne \varnothing$.

Suppose $p = 5$. Using (2.9), construct the set $\mathcal{V}$ of 9-component vectors that are the fifth rows of matrices from $Y(\mathbf{A},\mathbf{B}) \in M_Z(\mathbf{A},\mathbf{B};1)$:

$$\mathcal{V} = \Big\{(x_{51},\dots,x_{59}) : \exists Y(\mathbf{A},\mathbf{B}) = (y_{ij}) \in M_Z(\mathbf{A},\mathbf{B};1), \\ (y_{51},\dots,y_{59}) = (x_{51},\dots,x_{59})\Big\}.$$

For $l_k(\mathbf{B};1)$, we have the numbers

$$l_1(\mathbf{B};1) = 9, \quad l_2(\mathbf{B};1) = 7, \quad l_3(\mathbf{B};1) = l_4(\mathbf{B};1) = l_5(\mathbf{B};1) = 6,$$

$$l_6(\mathbf{B};1) = 4, \quad l_7(\mathbf{B};1) = l_8(\mathbf{B};1) = 2, \quad l_9(\mathbf{B};1) = 0.$$

Let $\mathbf{A}^{(1)}(5)$ and $\mathbf{B}^{(1)}$ make up a $(1;5)$-reductive pair and $(x_{51},\dots,x_{59}) = \mathbf{B} - \mathbf{B}^{(1)}$. Here, $x_1^{(5)} = a_5 = 5$ and the fact that $x_k^{(p)} \le \min\{a_p, l_k(\mathbf{B};1)\}$ entails the following inequalities for the vector $(x_{51},\dots,x_{59})$ and integers $x_k^{(5)}$ $(2 \le k \le 9)$:

$$x_2^{(5)} \le 5, \quad x_3^{(5)} \le 5, \quad x_4^{(5)} \le 5,$$

$$x_5^{(5)} \le 5, \quad x_6^{(5)} \le 4, \quad x_7^{(5)} \le 2, \quad x_8^{(5)} \le 2, \quad x_9^{(5)} = 0.$$

Note also that the equalities $l_k(\mathbf{B};1) = 6$ $(3 \le k \le 5)$ and $l_k(\mathbf{B};1) = 2$ $(7 \le k \le 8)$ entail $x_3^{(5)} = x_4^{(5)} = x_5^{(5)}$ and $x_7^{(5)} = x_8^{(5)}$. Therefore,

$$5 \ge x_2^{(5)} \ge x_3^{(5)} = x_4^{(5)} = x_5^{(5)} \ge x_6^{(5)} \ge x_7^{(5)} = x_8^{(5)} \ge 0,$$

where $x_8^{(5)} \le 2$. Using the inequality $x_k^{(p)} - x_{k+1}^{(p)} \le l_k(\mathbf{B};1) - l_{k+1}(\mathbf{B};1)$ for $k \in \{1,2,5,6\}$, we obtain

$$x_1^{(5)} - x_2^{(5)} \le 2, \quad x_2^{(5)} - x_3^{(5)} \le 1, \quad x_5^{(5)} - x_6^{(5)} \le 2, \quad x_6^{(5)} - x_7^{(5)} \le 2.$$

Since $x_1^{(5)} = 5$, we have $x_2^{(5)} \ge 3$. Note that all relations obtained above for $x_k^{(5)}$ $(1 \le k \le 9)$ apply to every vector $\mathbf{B} - \mathbf{B}^{(1)}$ (i.e., for every $(1;5)$-reductive pair of vectors).

Now, we calculate the right-hand sides of inequalities (2.9) for the given pair $\mathbf{A}$, $\mathbf{B}$ for $p = 5$. If $k < 5$, then $\beta_k = a_5 - \delta_k(\mathbf{A},\mathbf{B};1) = 5 - \delta_k(\mathbf{A},\mathbf{B};1)$:

$$\beta_1 = 3, \quad \beta_2 = 3, \quad \beta_3 = 4, \quad \beta_4 = 5.$$

If $k \ge 5$, then $\alpha_k = l_k(\mathbf{B};1) - \delta_k(\mathbf{A},\mathbf{B};1)$:

$$\alpha_5 = 5, \quad \alpha_6 = 3, \quad \alpha_7 = 1, \quad \alpha_8 = 1, \quad \alpha_9 = 0.$$

Accordingly, inequalities in (2.9) now are

$$x_2^{(5)} \ge 3, \quad x_3^{(5)} \ge 3, \quad x_4^{(5)} \ge 4, \quad x_5^{(5)} \ge 5, \quad x_6^{(5)} \ge 3, \quad x_7^{(5)} \ge 1, \quad x_8^{(5)} \ge 1.$$

Note here that the inequalities in (2.9) for $k = 4$ and $k = 5$ are identical.

Using the relations constructed for the numbers $x_k^{(5)}$ $(1 \le k \le 9)$ by analogy with those obtained both for an arbitrary $(1;5)$-reductive pair and for an admissible pair of vectors, we get

$$\begin{cases} 3 \le x_2^{(5)} \le 5, \\ 3 \le x_3^{(5)} = x_4^{(5)} = x_5^{(5)} \le 5, \\ 4 \le x_4^{(5)} \le 5, \\ 5 \le x_5^{(5)} \le 5, \\ 3 \le x_6^{(5)} \le 4, \\ 1 \le x_7^{(5)} = x_8^{(5)} \le 2, \\ x_2^{(5)} \ge x_3^{(5)}. \end{cases}$$

Hence, $x_2^{(5)} = x_3^{(5)} = x_4^{(5)} = x_5^{(5)} = 5$ and

$$\begin{cases} 3 \le x_6^{(5)} \le 4, \\ 1 \le x_7^{(5)} = x_8^{(5)} \le 2. \end{cases}$$

Consequently, all admissible pairs of vectors $\mathbf{A}^{(1)}(5)$ and $\mathbf{B}^{(1)}$ and all of the fifth rows of matrices from $M_Z(\mathbf{A},\mathbf{B};1)$ are identical with the solutions of the set of inequalities ($x_7^{(5)} = x_8^{(5)}$):

$$\begin{cases} x_5^{(5)} = 5, \\ 3 \leq x_6^{(5)} \leq 4, \\ 1 \leq x_7^{(5)} \leq 2. \end{cases}$$

Now, we write out the solutions $(x_5^{(5)}, x_6^{(5)}, x_7^{(5)})$ of these inequalities, taking into account the inequality $x_6^{(5)} - x_7^{(5)} \leq 2$. For each solution, we can construct a corresponding vector, which contains five ones and four zeros and is identical with the fifth row of a matrix from $M_Z(\mathbf{A},\mathbf{B};1)$:

$$(5,3,1), \quad (1,0,1,1,1,1,0,0,0);$$

$$(5,3,2), \quad (1,1,1,0,1,1,0,0,0);$$

$$(5,4,2), \quad (1,1,1,1,1,0,0,0,0).$$

Rearranging its components with indices 1 and 2, 3 and 4, and 5 and 6, we obtain all of the fifth rows of matrices from $M_Z(\mathbf{A},\mathbf{B};1)$:

$$\begin{aligned} \mathcal{V} = \Big\{ &(1,0,1,1,1,1,0,0,0), (0,1,1,1,1,1,0,0,0), \\ &(1,1,1,0,1,1,0,0,0), (1,1,0,1,1,1,0,0,0), \\ &(1,1,1,1,1,0,0,0,0), (1,1,1,1,0,1,0,0,0) \Big\}. \end{aligned}$$

The set of all admissible $(1;5)$-reductive pairs of vectors is described by the formula

$$\begin{aligned} &\big\{\mathbf{A}^{(1)}(5), \mathbf{B}^{(1)} \text{ make up an admissible pair}\big\} = \\ &\qquad\qquad \big\{\mathbf{A}^{(1)}(5), \mathbf{B} - (x_{51}, \ldots, x_{59}) : (x_{51}, \ldots, x_{59}) \in \mathcal{V}\big\}. \end{aligned}$$

The basic results presented in this chapter were published in [12–15].

Finally, we note that every integer transportation matrix is the adjacency matrix of a bichromatic multigraph [1, 7, 17]. Therefore, the results of this chapter can be used in constructing bichromatic multigraphs with constraints on the multiplicity of their edges.

References to Chapter 3

[1] Busacker R. G. and Saaty T. L., *Finite Graphs and Networks: An Introduction with Applications*, New York: McGraw-Hill, 1965. Translated under the title *Konechnye grafy i seti*, Moscow: Nauka, 1974.

[2] Gale D., A Theorem on Flow in Networks, *Pacific J. Math.*, 1957, vol. 7, pp. 1073–1082.

[3] Gol'shtein E. G. and Yudin D. B., *Zadachi lineinogo programmirovaniya transportnogo tipa* (Transportation Problems in Linear Programming), Moscow: Nauka, 1969.

[4] Gomory R. E., An Algorithm for Integer Solutions for Linear Programs, in *Recent Advances in Mathematical Programming*, New York: McGraw-Hill, 1963, pp. 269–302.

[5] Dakin R. J., A Tree Search Algorithm for Mixed Integer Programming Problems, *Computer J.*, 1965, vol. 8, pp. 250–255.

[6] Emelichev V. A., Kovalev M. M., and Kravtsov M. K., *Mnogogranniki, grafy, optimizatsiya* (Polyhedrons, Graphs, and Optimization), Moscow: Nauka, 1981.

[7] Zykov A. A., *Teoriya konechnykh grafov* (Theory of Finite Graphs), Novosibirsk: Nauka, 1969.

[8] Zhuravlev Yu. I., On a Class of Algorithms over Finite Sets, *Dokl. Akad. Nauk SSSR*, 1963, vol. 151, no. 5, pp. 1025–1028.

[9] Kuznetsov O. P. and Adel'son-Vel'skii G. M., *Diskretnaya matematika dlya inzhenerov* (Discrete Mathematics for Engineers), Moscow: Energiya, 1980.

[10] Kuznetsov A.V., Sakovich V. A., and Kholod N. N., *Vysshaya matematika. Matematicheskoe programmirovanie* (Higher Mathematics: Mathematical programming), Minsk: Vysshaya Shkola, 1994.

[11] Leont'ev V. K., On the Number of Optimal Plans in Assignment Problems, *Zh. Probl. Kibern.*, 1978, no. 34, pp. 36–43.

[12] Mironov A. A. and Tsurkov V. I., A Class of Distributed Problems with Minimax Criteria, *Dokl. Akad. Nauk*, 1992, vol. 336, no. 1, pp. 35–38.

[13] Mironov A. A. and Tsurkov V. I., Transportation and Network Problems with Minimax Criteria, *Zh. Vychisl. Mat. Mat. Fiz.*, 1995, vol. 35, no. 1, pp. 24–45.

[14] Mironov A. A. and Tsurkov V. I., Transportation Problems with Minimax Criteria, *Avtom. Telemekh.*, 1995, no. 12, pp. 109–118.

[15] Mironov A. A. and Tsurkov V. I., Transportation Problems with Minimax Criteria, *Dokl. Akad. Nauk*, 1996, vol. 346, no. 2, pp. 1–4.

[16] Moder J. and Elmaghraby S. S., *Handbook of Operations Research*, New York: Van Nostrand, 1978. Translated under the title *Issledovanie operatsii. Metodologicheskie osnovy i matematicheskie metody*, vol. 1, Moscow: Mir, 1981.

[17] *Algoritmy i programmy resheniya zadach na grafakh i setyakh* (Algorithms and Programs for Solving Problems on Graphs and Networks), Nechepurenko, M.I., Ed., Novosibirsk: Nauka, 1990.

[18] Propoi A. I., *Elementy teorii optimal'nykh diskretnykh protsessov* (&Elements of the Theory of Optimal Discrete Processes), Moscow: Nauka, 1973.

[19] Ryser H. J., Combinatorial Properties of Matrices of Zeros and Ones, *Can. J. Math.*, 1957, vol. 9, pp. 371–377.

[20] Ryser H. J., The Term Rank of a Matrix, *Can. J. Math.*, 1958, vol. 10, pp. 57–65.

[21] Ryser H. J., Traces of Matrices of Zeros and Ones, *Can. J. Math.*, 1960, vol. 12, pp. 463–476.

[22] Saaty T., *Optimization in Integers and Related Extremal Problems*, New York: McGraw-Hill, 1970. Translated under the title *Tselochislennye metody optimizatsii i svyazannye s nimi ekstremal'nye problemy,* Moscow: Mir, 1973.

[23] Trius E. B., *Zadachi matematicheskogo programmirovaniya transportnogo tipa* (Transportation Problems in Mathematical Programming), Moscow: Sovetskoe Radio, 1967.

[24] Ford L. R. and Fulkerson D., *Flows in Networks*, Princeton: Princeton Univ. Press, 1962. Translated under the title *Potoki v setyakh,* Moscow: Mir, 1966.

[25] Hu T., *Integer Programming and Network Flows*, Reading (USA): Addison-Wesley, 1969. Translated under the title *Tselochislennoe programmirovanie i potoki v setyakh,* Moscow: Mir, 1974.

[26] *Diskretnaya matematika i matematicheskie voprosy kibernetiki* (Discrete Mathematics and Mathematical Topics in Cybernetics), Yablonskii S. V. and Lupanov O. B., Eds., Moscow: Nauka, 1974, vol. 1.

[27] Yablonskii S. V., *Vvedenie v diskretnuyu matematiku* (An Introduction to Discrete Mathematics), Moscow: Nauka, 1979.

Chapter 4

Extremal Vector Pairs and Matrices

In this chapter, we consider the integer pairs of vectors that define the classical transportation polyhedron [9]. Feasible solutions of closed transportation problems [7, 27, 36] are taken here from the class of matrices of zeros and ones. Such matrices are used in integer programming [25, 27, 33, 40], in transportation problems [7, 11, 36], in the theory of communication networks [14, 17, 31], and in the theory of automata [16, 24]. For example, in graph theory, such a matrix may play the role of an incidence matrix, adjacency matrix, or a matrix of cycles and cuts [2, 3, 10, 26, 38].

In Section 1, we introduce the concepts of extremal pair of vectors and extremal matrix. We consider pairs of vectors whose components are arranged in nonascending order. If the corresponding polyhedron contains only one admissible matrix of zeros and ones, then the pair is said to be extremal. The binary matrix is also said to be extremal.

The importance of the concept thus introduced is explained as follows. It can be shown that all minimax solutions to problems of the types considered in Chapter 1 that correspond to arbitrary extremal pairs of vectors are given by extremal matrices. Therefore, the matrix is a generically minimax, uniform one. Moreover, the concepts of extremal, uniform, and generically minimax matrix are equivalent as applied to binary matrices.

We give various necessary and sufficient conditions of extremality. We calculate the numbers of extremal pairs of vectors and extremal matrices of given dimensions and describe some methods of their construction. In the next chapter, we invoke the well-known theorem about extreme points [6, 8] to establish some essential geometric properties of sets of extremal vector pairs and matrices.

In Section 2, we define partial order relations, two algebraic binary operations, and metrics for extremal vector pairs and matrices of equal dimensions. We show that these sets are isomorphic distributive lattices and isometric spaces.

In Section 3, we develop some algebraic and metric constructions. We construct special matrices, which are called the kernel and the hull. The kernel is defined as the intersection of all matrices of zeros and ones that belong to a given transportation polyhedron. The hull is defined as the union of all of these matrices. We show that both kernels and hulls are extremal. The kernel and hull define bounds for every matrix of the type under consideration and are the best extremal approximations of these matrices. This method is extended to minimax matrices.

The construction and analysis of kernels and hulls are based on the characteristic functions and equations introduced in Chapter 1. We formulate necessary and sufficient conditions for an extremal matrix to be the kernel or hull of a nontrivial matrix set.

Extremal vector pairs and matrices play a key role in the construction of transportation polyhedrons and in optimization problems analyzed in Chapter 5.

§1. Extremality

Any n-by-m matrix of zeros and ones may be treated as a distribution of n elements over m sets. These matrices play an important role in combinatorics [34, 39]. We refer here to comprehensive presentations of the theory of matrices of zeros and ones with prescribed sums of elements of their rows and columns [5, 28–30, 37].

In this section, we specify vector pairs such that each pair defines a unique transportation matrix of zeros and ones. These vector pairs and matrices are called extremal. We show that extremal matrices are hereditarity minimax and, therefore, uniform matrices. Vice versa, every uniform matrix of zeros and ones is an extremal one (when the components of the vectors under consideration are arranged in nonascending order). We formulate various extremality criteria for the matrices and vector pairs. We calculate the numbers of extremal pairs for given n and m.

1.1. Extremality Criteria for Vector Pairs and Matrices

In the constructions developed below, we use the concept of reducibility of a vector pair introduced in Section 2 of Chapter 3.

Let $l(0; \mathbf{B}) = \left|\{j,\ 1 \le j \le m,\ b_j = 0\}\right|$ be the number of zero components of a vector $\mathbf{B} = (b_1, \ldots, b_m) \in Z_+^m$.

Definition 1.1. The pair of vectors $\mathbf{A} = (a_1, \ldots, a_n) \in Z_+^n$ and $\mathbf{B} = (b_1, \ldots, b_m) \in Z_+^m$ is said to be *strictly reducible* if $\max\{a_i : 1 \le i \le n\} = m - l(0; \mathbf{B})$.

Combined with Definitions 1.4 and 2.2 from Chapter 3, this definition implies that every strictly reducible pair of vectors is reducible.

Lemma 1.1. *Suppose* $\mathbf{A} = (a_1, \ldots, a_n) \in Z_+^n$ *and* $\mathbf{B} = (b_1, \ldots, b_m) \in Z_+^m$.

(a) *If* $p \in Z$ $(1 \le p \le n)$ *and* $a_p = \max\{a_i : 1 \le i \le n\}$, *then the pair of vectors* $\mathbf{A}$ *and* $\mathbf{B}$ *is strictly reducible if and only if there exists a unique* $(1; p)$*-reductive pair of vectors* $\mathbf{A}^{(1)}(p)$ *and* $\mathbf{B}^{(1)}$.

(b) *Vectors* **A** *and* **B** *make up a strictly reducible pair if and only if*

$$\max\{a_i : 1 \le i \le n\} = \sum_{j=1}^{m} \min(b_j, 1). \tag{1.1}$$

Proof.

(a) If **A** and **B** make up a strictly reducible pair, then, by Definitions 1.4 and 2.2 from Chapter 3, there exists a $(1;p)$-reductive pair, and the pair is unique because $a_p = m - l(0;\mathbf{B})$.

If there exists a $(1;p)$-reductive pair of vectors $\mathbf{A}^{(1)}(p)$ and $\mathbf{B}^{(1)}$, then $a_p \le m - l(0;\mathbf{B})$; moreover, if the pair is unique, $a_p = m - l(0;\mathbf{B})$. Since $a_p = \max\{a_i : 1 \le i \le n\}$, **A** and **B** make up a strictly reducible pair of vectors.

(b) This proposition is proved by combining the equality

$$\sum_{j=1}^{m} \min(b_j, 1) = m - l(0;\mathbf{B})$$

with Definition 1.1.

The following proposition explicates the concept of strict reducibility.

Lemma 1.2. *Suppose* $\mathbf{A} = (a_1, \dots, a_n) \in Z_+^n$ *and* $\mathbf{B} = (b_1, \dots, b_m) \in Z_+^m$.

(a) *If* **A** *and* **B** *make up a strictly reducible pair of vectors and* $a_p = \max\{a_i : 1 \le i \le n\}$, *then the (unique)* $(1;p)$*-reductive pair of vectors* $\mathbf{A}^{(1)}(p)$ *and* $\mathbf{B}^{(1)}$ *is the principal* $(1;p)$*-reductive pair (i.e.,* $\mathbf{B}^{(1)} = \mathbf{B}_*^{(1)}$*).*

(b) *The vectors* **A** *and* **B** *are strictly reducible if and only if*

$$\delta_1(\widetilde{\mathbf{A}}, \mathbf{B}; 1) = \sum_{j=1}^{m} \min(b_j, 1) - \tilde{a}_1 = 0.$$

The lemma is readily proved by combining Definition 1.5 from Chapter 3 with Lemma 1.1.

By virtue of its strict reducibility, the pair of vectors **A** and **B** uniquely defines a certain row of any matrix from the nonempty set $M_Z(\mathbf{A}, \mathbf{B}; 1)$.

Lemma 1.3. *Suppose that* $\mathbf{A} = (a_1, \dots, a_n) \in Z_+^n$, $\mathbf{B} = (b_1, \dots, b_m) \in Z_+^m$, $M_Z(\mathbf{A}, \mathbf{B}; 1) \ne \varnothing$, and $a_p = \max\{a_i : 1 \le i \le n\}$. *The vectors* **A** *and* **B** *make up a strictly reducible pair if and only if the pth row of every matrix* $X(\mathbf{A}, \mathbf{B}) = (x_{ij})$ *from* $M_Z(\mathbf{A}, \mathbf{B}; 1)$ *is completely defined and*

$$x_{pj} = \begin{cases} 1, & 1 \le j \le m, \quad b_j \ge 1, \\ 0, & 1 \le i \le m, \quad b_j = 0. \end{cases}$$

The lemma is proved by the sequence of equalities

$$\sum_{j=1}^{m} x_{pj} = a_p = \max\{a_i : 1 \le i \le n\} = \sum_{j=1}^{m} \min(b_j, 1) = m - l(\mathbf{B}; 0).$$

Consider $M_Z(\mathbf{A}, \mathbf{B}; 1) \neq \varnothing$, where both $\mathbf{A}$ and $\mathbf{B}$ are nonzero vectors. Every matrix from $M_Z(\mathbf{A}, \mathbf{B}; 1)$ is integer and is a minimax matrix for the set $M_Z(\mathbf{A}, \mathbf{B})$, with the integer minimax value $c_Z(\mathbf{A}, \mathbf{B}) = 1$. In the general case, minimax values satisfy the inequality $c(\mathbf{A}, \mathbf{B}) \le c_Z(\mathbf{A}, \mathbf{B})$, which becomes an equality when $\mathbf{A}$ and $\mathbf{B}$ make up a strictly reducible pair by Lemma 1.2.

Lemma 1.4. *Suppose that* $M_Z(\mathbf{A}, \mathbf{B}; 1) \neq \varnothing$, *and* $\mathbf{A}$ *and* $\mathbf{B}$ *make up a strictly reducible pair of vectors. Then,* $c(\mathbf{A}, \mathbf{B}) = c_Z(\mathbf{A}, \mathbf{B})$, *and if* $\mathbf{A}$ *and* $\mathbf{B}$ *are nonzero vectors, then* $c(\mathbf{A}, \mathbf{B}) = c_Z(\mathbf{A}, \mathbf{B}) = 1$.

When a pair of vectors is strictly reducible and the components of the first vector are arranged in nonascending order, the values of the characteristic function defined by (1.22) in Chapter 1 are invariant for $c = 1$ under the transition to the $(1;1)$-reductive pair of vectors.

We now introduce the following notation associated with the vector $\mathbf{A} = (a_1, \dots, a_n) \in Z_+^n$. We denote by $\widetilde{\mathbf{A}} = (\tilde{a}_1, \dots, \tilde{a}_n)$ the vector constructed by arranging the components of $\mathbf{A}$ in nonascending order, by $\mathbf{A}^{(1)}(p) = (a_1^{(1)}, \dots, a_n^{(1)})$ the vector obtained by setting the pth component of $\mathbf{A}$ equal to zero ($a_i^{(1)} = a_i$, $1 \le i \le n$, $i \neq p$; $a_p^{(1)} = 0$), by $\widetilde{\mathbf{A}}^{(1)}(p) = (\tilde{a}_1^{(1)}, \dots, \tilde{a}_n^{(1)})$ the vector constructed by arranging the components of $\mathbf{A}^{(1)}(p)$ in nonascending order, by $\mathbf{A}'(p) = (a_1', \dots, a_{n-1}')$ the vector obtained by taking out the pth component of $\mathbf{A}$ (if $n > 1$, then $a_i' = a_i$ for $1 \le i < p$ and $a_i' = a_{i+1}$ for $p \le i \le n-1$), and by $\widetilde{\mathbf{A}}'(p) = (\tilde{a}_1', \dots, \tilde{a}_{n-1}')$ the vector constructed by arranging the components of $\mathbf{A}'(p)$ in nonascending order.

Lemma 1.5. *Suppose that vectors* $\mathbf{A} \in Z_+^n$ *and* $\mathbf{B} \in Z_+^m$, *where* $n > 1$, *make up a strictly reducible pair or, equivalently,* $\delta_1(\widetilde{\mathbf{A}}, \mathbf{B}; 1) = 0$ (*by Lemma* 1.2). *Then,* $\delta_{k+1}(\widetilde{\mathbf{A}}, \mathbf{B}; 1) = \delta_k(\widetilde{\mathbf{A}}'(p), \mathbf{B}^{(1)}; 1)$ *for any* $k \in Z$ $(1 \le k \le n-1)$, *where* p *is the index of the largest component of* $\mathbf{A}$, *and* $\mathbf{B}^{(1)}$ *is the second vector of the* $(1;p)$*-reductive pair* $\mathbf{A}^{(1)}(p)$, $\mathbf{B}^{(1)}$.

Note that the assumptions of the lemma imply that the $(1;p)$-reductive pair of vectors $\mathbf{A}^{(1)}(p)$ and $\mathbf{B}^{(1)}$ is unique and is the principal one; that is, $\mathbf{B}^{(1)} = \mathbf{B}_*^{(1)}$ (by Lemma 1.2).

Let us prove Lemma 1.5 for $\mathbf{A} \in \overline{Z}_+^n$, in which case its statement can be simplified without loss of generality.

Lemma 1.5′. *If the pair of vectors* $\mathbf{A} = (a_1, \ldots, a_n) \in \overline{Z}_+^n$ *and* $\mathbf{B} = (b_1, \ldots, b_m) \in Z_+^m$, *where* $n > 1$, *is strictly reducible, then* $\delta_{k+1}(\mathbf{A}, \mathbf{B}; 1) = \delta_k(\mathbf{A}'(1), \mathbf{B}^{(1)}; 1)$ *for any* k $(1 \leq k \leq n-1)$, *where* $\mathbf{B}^{(1)}$ *is the second vector in the* $(1;1)$*-reductive pair of vectors* $\mathbf{A}^{(1)}(p)$ *and* $\mathbf{B}^{(1)}$, *with* $\mathbf{B}^{(1)} = \mathbf{B}_*^{(1)}$.

Proof. Consider the values of function (1.22) from Chapter 1 for the pairs $\mathbf{A}$, $\mathbf{B}$ and $\mathbf{A}'(1) = (a_1', \ldots, a_{n-1}')$, $\mathbf{B}^{(1)} = (b_1^{(1)}, \ldots, b_m^{(1)})$ with $c = 1$:

$$\delta_{k+1}(\mathbf{A}, \mathbf{B}; 1) = \sum_{j=1}^{m} b_j - \sum_{b_j \geq k+1} (b_j - k - 1) - \sum_{i=1}^{k+1} a_i,$$

$$\delta_k(\mathbf{A}'(1), \mathbf{B}^{(1)}; 1) = \sum_{j=1}^{m} b_j^{(1)} - \sum_{b_j^{(1)} \geq k} (b_j^{(1)} - k) - \sum_{i=1}^{k} a_i',$$

where $1 \leq k \leq n-1$. Since

$$\sum_{i=1}^{k} a_i' = \sum_{i=2}^{k+1} a_i, \quad \sum_{j=1}^{m} b_j^{(1)} = \sum_{j=1}^{m} b_j - a_1,$$

the difference between consecutive values of function (1.22) from Chapter 1 is

$$\delta_{k+1}(\mathbf{A}, \mathbf{B}; 1) - \delta_k(\mathbf{A}'(1), \mathbf{B}^{(1)}; 1) = \sum_{b_j^{(1)} \geq k} (b_j^{(1)} - k) - \sum_{b_j \geq k+1} (b_j - k - 1). \quad (1.2)$$

We obtained (1.2) without relying on the strict reducibility. This result applies to a reducible pair of vectors. The strict reducibility of the pair of vectors $\mathbf{A}$ and $\mathbf{B}$ implies that $a_1 = \max\{a_i : 1 \leq i \leq n\} = m - l(0; \mathbf{B})$. Therefore, $b_j^{(1)} = b_j - 1$ $(1 \leq j \leq m,\ b_j \geq 1)$ and $b_j^{(1)} = b_j = 0$ $(1 \leq j \leq m,\ b_j < 1)$. Thus, the right-hand side of (1.2) is zero and $\delta_{k+1}(\mathbf{A}, \mathbf{B}; 1) = \delta_k(\mathbf{A}'(1), \mathbf{B}^{(1)}; 1)$, which proves Lemmas 1.5′ and 1.5.

The definition that follows should be considered in combination with Theorem 2.3 and Algorithm 2.1 from Chapter 3.

Definition 1.2. Suppose $\mathbf{A} \in \overline{Z}_+^n$ and $\mathbf{B} \in \overline{Z}_+^m$. The pair of vectors $\mathbf{A}$ and $\mathbf{B}$ is said to be *extremal* if $\mathbf{A}$ and $\mathbf{B}$ make up a strictly reducible pair and the $(1; 1, \ldots, k)$-reductive pair

$$\mathbf{A}^{(k)} = \mathbf{A}^{(k)}(1, \ldots, k), \quad \mathbf{B}^{(k)}, \quad 1 \leq k \leq n, \quad (1.3)$$

derived from the pair of vectors $\mathbf{A}$ and $\mathbf{B}$ at the kth reductive step is strictly reducible.

This definition is well substantiated by Definition 1.1 and Lemmas 1.1 and 1.2. In particular, the lemmas imply that every $(1;1,\ldots,k)$-reductive pair (1.3) is a principal one and $\mathbf{B}^{(k)} = \mathbf{B}_*^{(k)}$. Note that the $(1;1,\ldots,n)$-reductive pair $\mathbf{A}^{(n)}$, $\mathbf{B}^{(n)} = \mathbf{B}_*^{(n)}$ contains only zero elements, because $\mathbf{A}^{(n)}$ is a zero vector by construction and the fact that $\mathbf{A}^{(n)}$ and $\mathbf{B}^{(n)}$ make up a strictly reducible pair implies that the $\mathbf{B}^{(n)}$ is a zero vector as well.

Since a reductive step leaves the difference between the sums of components of the two vectors invariant and the pair $\mathbf{A}^{(n)}$, $\mathbf{B}^{(n)}$ derived from the extremal pair $\mathbf{A} \in \overline{Z}_+^n$, $\mathbf{B} \in \overline{Z}_+^m$ consists of zero vectors, we have the following proposition.

Lemma 1.6. *Every extremal pair of vectors satisfies the balance condition.*

As a corollary to Definition 1.2 combined with Theorem 2.2 from Chapter 3 and Lemma 1.6, we have the following proposition.

Lemma 1.7. *For any extremal pair of vectors* $\mathbf{A}$ *and* $\mathbf{B}$, *there exists an integer matrix* $X(\mathbf{A},\mathbf{B}) \in M_Z(\mathbf{A},\mathbf{B})$ *whose elements are zeros and ones.*

Consider $\mathbf{A} = (a_1,\ldots,a_n) \in R_+^n$ and $\mathbf{B} = (b_1,\ldots,b_m) \in R_+^m$. For the pair of vectors $\mathbf{A}$ and $\mathbf{B}$ and $c = 1$, the characteristic function defined by (1.22) in Chapter 1 is

$$\delta_k(\mathbf{A},\mathbf{B};1) = \sum_{j=1}^{m} b_j - \sum_{b_j \geq k} (b_j - k) - \sum_{i=1}^{k} a_i, \quad 1 \leq k \leq n. \tag{1.4}$$

The extremality of a vector pair can be determined by invoking the characteristic function given by (1.4). Lemmas 1.2 and 1.5′ combined entail the following proposition.

Lemma 1.8. *Vectors* $\mathbf{A} \in \overline{Z}_+^n$ *and* $\mathbf{B} \in \overline{Z}_+^m$ *make up an extremal pair of vectors if and only if* $\delta_k(\mathbf{A},\mathbf{B};1) = 0$ *for any* $k \in Z$ $(1 \leq k \leq n)$.

By Lemma 1.7, Definitions 1.2 and 1.3 are equivalent.

Definition 1.3. The pair of vectors $\mathbf{A} \in \overline{Z}_+^n$ and $\mathbf{B} \in \overline{Z}_+^m$ is said to be *extremal* if $\delta_k(\mathbf{A},\mathbf{B};1) = 0$ for any $k \in Z$ $(1 \leq k \leq n)$.

For an extremal pair of vectors and any admissible $k \in Z$, the characteristic equation (2.3) from Chapter 3 has a unique solution given by (2.5) in Chapter 3. Now we are prepared to formulate an extremality criterion. The proposition that follows is a corollary to Theorem 1.3 from Chapter 2.

Theorem 1.1. *Suppose that vectors* $\mathbf{A} \in \overline{Z}_+^n$ *and* $\mathbf{B} \in \overline{Z}_+^m$ *satisfy the balance condition. The matrix set* $M_Z(\mathbf{A},\mathbf{B};1)$ *contains only one matrix if and only if* $\mathbf{A}$ *and* $\mathbf{B}$ *make up an extremal pair of vectors. Moreover, when the* *components of* **A** *and* **B** *are integer, the matrix set* $M(\mathbf{A},\mathbf{B};1)$ *also contains only one matrix if and only if* **A** *and* **B** *make up an extremal pair of vectors.*

Theorem 1.1 for the set $M_Z(\mathbf{A},\mathbf{B};1)$ is proved by combining Lemmas 1.2, 1.3, and 1.5.

Since any extremal pair of vectors **A** and **B** is strictly reducible, the only matrix in $M_Z(\mathbf{A},\mathbf{B};1)$ is the minimax matrix with respect to the sets $M_Z(\mathbf{A},\mathbf{B})$ and $M(\mathbf{A},\mathbf{B})$ (by Lemma 1.4).

Using expressions (1.7) formulated in Theorem 1.4 of Chapter 2, we can construct a (unique) matrix $X(\mathbf{A},\mathbf{B})$ of zeros and ones for an extremal pair of vectors **A** and **B**. Using the properties of extremal pairs of vectors, we now formulate the rules for calculating the elements of $X(\mathbf{A},\mathbf{B})$. Note that, by Definition 1.2 or 1.3, the extremality of a pair of vectors can be viewed as a property independent of a particular matrix.

In the theorem stated below, the extremality of a vector pair is interpreted in terms of rules for calculating the elements of a matrix of zeros and ones.

Theorem 1.2. *Suppose that* $\mathbf{A} = (a_1,\ldots,a_n) \in \overline{Z}^n_+$, $\mathbf{B} = (b_1,\ldots,b_m) \in \overline{Z}^m_+$; *and* $X(\mathbf{A},\mathbf{B}) = (x_{ij}) \in M_Z(\mathbf{A},\mathbf{B};1)$ *is a nonempty set.*

(a) *The vectors* **A** *and* **B** *make up an extremal pair if and only if the equality* $x_{ij} = 1$ *is equivalent to the inequality* $l_i(\mathbf{B};1) \geq j$ *for any* i *and* j $(1 \leq i \leq n,\ 1 \leq j \leq m)$.

(b) *The vectors* **A** *and* **B** *make up an extremal pair if and only if the equality* $x_{ij} = 0$ *is equivalent to the inequality* $l_i(\mathbf{B};1) < j$ *for any* i *and* j $(1 \leq i \leq n,\ 1 \leq j \leq m)$.

Proof. Recall that $l_i(\mathbf{B};1) = \left|\{j : 1 \leq j \leq m,\ b_j \geq i\}\right|$ (see (1.25) in Chapter 1). Since $\mathbf{B} \in \overline{Z}^m_+$,

$$l_i(\mathbf{B};1) = \begin{cases} \max\{j : 1 \leq j \leq m,\ b_j \geq i\}, & b_1 \geq i, \\ 0, \quad b_1 < i. \end{cases}$$

We prove propositions (a) and (b) simultaneously by induction on the number of components of **A**.

If $n = 1$, then $M_Z(\mathbf{A},\mathbf{B})$ contains only one matrix, and $M_Z(\mathbf{A},\mathbf{B};1) = M_Z(\mathbf{A},\mathbf{B})$. Consequently, the only matrix $X(\mathbf{A},\mathbf{B}) = (x_{1j})$ $(1 \leq j \leq m)$ and the vector $\mathbf{B} = (b_1,\ldots,b_m) \in \overline{Z}^m_+$ have the following elements:

$$x_{1j} = b_j = 1, \quad 1 \leq j \leq l_1(\mathbf{B};1), \quad x_{1j} = b_j = 0, \quad l_1(\mathbf{B};1) < j \leq m.$$

It can also be readily shown that $\delta_1(\mathbf{A},\mathbf{B};1) = 0$ and the vectors **A** and **B** make up an extremal pair. Thus, propositions (a) and (b) are obviously true in this case.

Assume that Theorem 1.2 is also true if the number of components of $\mathbf{A}$ is not greater than $n-1$, and suppose that $\mathbf{A} \in \overline{Z}_+^n$.

Consider an arbitrary extremal pair of vectors $\mathbf{A} = (a_1, \ldots, a_n) \in \overline{Z}_+^n$ and $\mathbf{B} = (b_1, \ldots, b_m) \in \overline{Z}_+^m$. The pair $\mathbf{A}$, $\mathbf{B}$ is strictly reducible, and the first row of the matrix $X(\mathbf{A}, \mathbf{B}) = (x_{ij})$ has the following elements (by Lemma 1.3): $x_{1j} = 1$ $(1 \leq j \leq m,\ b_j \geq 1)$ and $x_{1j} = 0$ $(1 \leq j \leq m,\ b_j = 0)$. Since $\mathbf{B} \in \overline{Z}_+^m$, $x_{1j} = 1$ for $1 \leq j \leq l_1(\mathbf{B}; 1)$ and $x_{1j} = 0$ for $l_1(\mathbf{B}; 1) < j \leq m$. Thus, the extremality of the pair of vectors $\mathbf{A}$ and $\mathbf{B}$ suggests the rule for calculating the elements of the first row of $X(\mathbf{A}, \mathbf{B})$ formulated in the statement of Theorem 1.2.

Taking out the first component of $\mathbf{A}$, we obtain the pair of vectors $\mathbf{A}'(1) = (a'_1, \ldots, a'_{n-1}) \in \overline{Z}_+^n$ and $\mathbf{B}^{(1)} = (b_1^{(1)}, \ldots, b_m^{(1)})$. Since $\mathbf{A} \in \overline{Z}_+^n$ and $\mathbf{B} \in \overline{Z}_+^m$ make up a strictly reducible pair of vectors, it holds that $\mathbf{A}'(1) \in \overline{Z}_+^{n-1}$ and $\mathbf{B}^{(1)} \in \overline{Z}_+^m$. The vector $\mathbf{B}^{(1)}$ is uniquely defined (see Lemma 1.1), and its components can be expressed in terms of the components of $\mathbf{B}$ and the elements of the first row of $X(\mathbf{A}, \mathbf{B}) = (x_{ij})$ $(1 \leq i \leq n,\ 1 \leq j \leq m)$: $b_j^{(1)} = b_j - x_{1j}$ $(1 \leq j \leq m)$.

The vectors $\mathbf{A}'(1)$ and $\mathbf{B}^{(1)}$ make up an extremal pair by Lemma 1.5; taking out the first row of $X(\mathbf{A}, \mathbf{B})$, we obtain its submatrix $X'\left(\mathbf{A}'(1), \mathbf{B}^{(1)}\right) = (x'_{ij})$ $(1 \leq i \leq n-1,\ 1 \leq j \leq m)$, where $x_{i+1\,j} = x'_{ij}$ $(1 \leq i \leq n-1,\ 1 \leq j \leq m)$. By the induction hypothesis, the elements of the matrix $X'\left(\mathbf{A}'(1), \mathbf{B}^{(1)}\right)$ are as follows: $x'_{ij} = 1$ if and only if $j \leq l_i(\mathbf{B}^{(1)}; 1)$, and $x'_{ij} = 0$ if and only if $j > l_i(\mathbf{B}^{(1)}; 1)$. Since $\mathbf{A}$ and $\mathbf{B}$ make up a strictly reducible pair of vectors, it holds that $a_1 = m - l(0; \mathbf{B}) = l_1(\mathbf{B}; 1)$ and $b_j^{(1)} = b_j - 1$ for $j \leq l_1(\mathbf{B}; 1)$, and $b_j^{(1)} = b_j = 0$ for $j > l_1(\mathbf{B}; 1)$. Therefore, $l_{i+1}(\mathbf{B}; 1) = l_i(\mathbf{B}^{(1)}; 1)$ for $1 \leq i \leq n-1$. Consequently, $x_{i+1\,j} = x'_{ij} = 1$ if $1 \leq i \leq n-1$ and $j \leq l_{i+1}(\mathbf{B}; 1)$, and $x_{i+1\,j} = x'_{ij} = 0$ if $1 \leq i \leq n-1$ and $j > l_{i+1}(\mathbf{B}; 1)$.

Since the rule for calculating the elements of the first row of $X(\mathbf{A}, \mathbf{B})$ has been constructed, an arbitrary element x_{ij} of the matrix $X(\mathbf{A}, \mathbf{B})$ is as follows: $x_{ij} = 1$ if $1 \leq i \leq n$ and $j \leq l_i(\mathbf{B}; 1)$, and $x_{ij} = 0$ if $1 \leq i \leq n$ and $j > l_i(\mathbf{B}; 1)$.

Now, we reverse the proof. Assume that the elements of a matrix $X(\mathbf{A}, \mathbf{B}) = (x_{ij}) \in M_Z(\mathbf{A}, \mathbf{B}; 1)$, where $\mathbf{A} \in \overline{Z}_+^n$ and $\mathbf{B} \in \overline{Z}_+^m$, are defined as follows: $x_{ij} = 1$ if and only if $1 \leq j \leq l_i(\mathbf{B}; 1)$ and $1 \leq i \leq n$; equivalently, $x_{ij} = 0$ if and only if $l_i(\mathbf{B}; 1) < j \leq m$ and $1 \leq i \leq n$. Let us show that $\mathbf{A}$ and $\mathbf{B}$ make up an extremal pair of vectors.

For the elements of the first row of $X(\mathbf{A}, \mathbf{B})$, the assumption above implies that $a_1 = l_1(\mathbf{B}; 1) = m - l(0; \mathbf{B})$. Therefore, $\mathbf{A}$ and $\mathbf{B}$ make up a strictly reducible pair. Consider now the vectors $\mathbf{A}'(1)$ and $\mathbf{B}^{(1)}$, where $\mathbf{B}^{(1)}$ is uniquely defined (by Lemma 1.1). Taking out the first row of $X(\mathbf{A},\mathbf{B})$, we obtain the submatrix $X'\left(\mathbf{A}'(1),\mathbf{B}^{(1)}\right) = (x'_{ij})$, where $1 \le i \le n-1$ and $1 \le j \le m$. It is obvious that $x'_{ij} = x_{i+1\,j}$ $(1 \le i \le n-1,\ 1 \le j \le m)$, and the components of the vector $\mathbf{B}^{(1)} = (b_1^{(1)}, \dots, b_m^{(1)})$ are calculated as $b_j^{(1)} = b_j - x_{1j}$ $(1 \le j \le m)$. As applied to the elements of the first row, these relations yield $b_j^{(1)} = b_j - 1$ for $1 \le j \le l_1(\mathbf{B};1)$ and $b_j^{(1)} = b_j$ for $l_1(\mathbf{B};1) < j \le m$. Accordingly, $l_i(\mathbf{B}^{(1)};1) = l_{i+1}(\mathbf{B};1)$ for $1 \le i \le n-1$, and the elements of $X'\left(\mathbf{A}'(1),\mathbf{B}^{(1)}\right)$ are as follows: $x'_{ij} = x_{i+1\,j} = 1$ for $1 \le j \le l_i(\mathbf{B}^{(1)};1)$ and $1 \le i \le n-1$, and $x'_{ij} = x_{i+1\,j} = 0$ for $l_i(\mathbf{B}^{(1)};1) < j \le m$ and $1 \le i \le n-1$.

Since $\mathbf{A}'(1) \in \overline{Z}_+^{n-1}$ and $\mathbf{B}^{(1)} \in \overline{Z}_+^m$, $\mathbf{A}'(1)$ and $\mathbf{B}^{(1)}$ make up an extremal vector pair by the induction hypothesis. By invoking Definition 1.2, it can be readily shown that the extremality of the pair of vectors $\mathbf{A}^{(1)}(1)$ and $\mathbf{B}^{(1)}$ and the strict reducibility of $\mathbf{A}$ and $\mathbf{B}$ entail the extremality of the pair of $\mathbf{A}$ and $\mathbf{B}$, which proves the theorem.

Remark 1.1. In the proof of Theorem 1.2, it was noted that arbitrary vectors $\mathbf{A}$ and $\mathbf{B}$, where $\mathbf{A}$ is a single-component vector and $\mathbf{B} \in \overline{Z}_+^m$, make up an extremal pair of vectors provided that the set $M_Z(\mathbf{A},\mathbf{B};1)$ is not empty.

Definition 1.4. When $\mathbf{A}$ and $\mathbf{B}$ make up an extremal pair of vectors, the single matrix $X(\mathbf{A},\mathbf{B})$ in $M_Z(\mathbf{A},\mathbf{B};1)$ (and in $M(\mathbf{A},\mathbf{B};1)$) is said to be *extremal.*

Now, we give an explicit formula for the elements of extremal matrices, based on a simplified statement of Theorem 1.2.

Theorem 1.2′. *Suppose that* $\mathbf{A} \in \overline{Z}_+^n$, $\mathbf{B} \in \overline{Z}_+^m$, $X(\mathbf{A},\mathbf{B}) = (x_{ij}) \in M_Z(\mathbf{A},\mathbf{B};1)$, *and the set* $M_Z(\mathbf{A},\mathbf{B};1)$ *is not empty. The vectors* $\mathbf{A}$ *and* $\mathbf{B}$ *make up an extremal pair if and only if the elements of* $X(\mathbf{A},\mathbf{B})$ *are defined as*

$$x_{ij} = \begin{cases} 1, & 1 \le i \le n, \quad 1 \le j \le l_i(\mathbf{B};1), \\ 0, & 1 \le i \le n, \quad l_i(\mathbf{B};1) < j \le m. \end{cases} \tag{1.5}$$

Example 1.1. Construction of an extremal matrix. One can readily check that the vectors

$$\mathbf{A} = (11,11,10,9,9,5,5,3) \in \overline{Z}_+^8, \quad \mathbf{B} = (8,8,8,7,7,5,5,5,5,3,2,0) \in \overline{Z}_+^{12}$$

make up an extremal pair by Definitions 1.2 and 1.3. To construct the extremal matrix $X(\mathbf{A},\mathbf{B}) = (x_{ij})$ $(1 \le i \le 8,\ 1 \le j \le 12)$, calculate the values of $l_i(\mathbf{B};1)$ for $1 \le i \le 8$. Recall that $l_i(\mathbf{B};1) = \max\{j : 1 \le i \le n,\ b_j \ge i\}$ if $b_1 \ge i$. Therefore, it is obvious that $l_1(\mathbf{B};1) = l_2(\mathbf{B};1) = 11$, $l_3(\mathbf{B};1) = 10$,

$l_4(\mathbf{B};1) = l_5(\mathbf{B};1) = 9$, $l_6(\mathbf{B};1) = l_7(\mathbf{B};1) = 5$, and $l_8(\mathbf{B};1) = 3$. Formula (1.5) implies that the first $l_i(\mathbf{B};1)$ elements of the ith row of the matrix $X(\mathbf{A},\mathbf{B})$ are ones, and the remaining $m - l_i(\mathbf{B};1)$ elements are zeros. Thus,

$$X(\mathbf{A},\mathbf{B}) = \begin{pmatrix} 1&1&1&1&1&1&1&1&1&1&1&0 \\ 1&1&1&1&1&1&1&1&1&1&1&0 \\ 1&1&1&1&1&1&1&1&1&1&0&0 \\ 1&1&1&1&1&1&1&1&1&0&0&0 \\ 1&1&1&1&1&1&1&1&1&0&0&0 \\ 1&1&1&1&1&0&0&0&0&0&0&0 \\ 1&1&1&1&1&0&0&0&0&0&0&0 \\ 1&1&1&0&0&0&0&0&0&0&0&0 \end{pmatrix}$$

is the extremal matrix.

We now give an alternative definition of the extremal matrix, which does not rely on the extremality of a pair of vectors.

Definition 1.5. Matrix $X = (x_{ij})$, where $1 \le i \le n$ and $1 \le j \le m$, is said to be *extremal* if each of its elements is zero or one and the following is true for all i and j $(1 \le i \le n,\ 1 \le j \le m)$: the equality $x_{ij} = 1$ implies that $x_{pq} = 1$ for any p and q $(1 \le p \le i,\ 1 \le q \le j)$, and $x_{ij} = 0$ implies that $x_{pq} = 0$ for any p and q $(i < p \le n,\ j < q \le m)$.

Definitions 1.2 and 1.3 are equivalent by Theorem 1.2 or 1.2′.

Definition 1.5 demonstrates that the extremality of a matrix is its intrinsic property, independent of pairs of vectors).

Consider an arbitrary matrix $X(\mathbf{A},\mathbf{B})$ and its transpose $X^{\mathrm{T}} = X^{\mathrm{T}}(\mathbf{A},\mathbf{B})$. By Definition 1.5, the extremality of a matrix is preserved under transposition. Therefore, we have the following theorem.

Theorem 1.3. *The matrices* X *and* X^{T} *and vector pairs* $\mathbf{A}$, $\mathbf{B}$ *and* $\mathbf{B}$, $\mathbf{A}$ *are either simultaneously extremal or simultaneously nonextremal.*

The components of vectors that make up an extremal pair are related by the restrictive formulas in (1.5). As a corollary to Theorems 1.2 and 1.3, we have a proposition that characterizes the extremal vector pairs in terms of components of vector pairs.

Theorem 1.4. *Suppose* $\mathbf{A} = (a_1, \dots, a_n) \in \overline{Z}_+^n$ *and* $\mathbf{B} = (b_1, \dots, b_m) \in \overline{Z}_+^m$. *The vectors* $\mathbf{A}$ *and* $\mathbf{B}$ *make up an extremal pair if and only if either of the following conditions is satisfied:*

(a) $a_i = l_i(\mathbf{B};1), \quad 1 \le i \le n;$ (1.6)

(b) $b_j = l_j(\mathbf{A};1), \quad 1 \le j \le m.$ (1.7)

By Theorem 1.4, either vector of an extremal pair determines the other vector of the pair up to its zero components.

Theorem 1.5.

(a) *For any vector* $\mathbf{A} = (a_1, \ldots, a_n)$ *from* $\overline{Z}_+^n$ *and any* $m \in Z$ $(m \geq a_1)$, *there exists a unique vector* $\mathbf{B} \in \overline{Z}_+^m$ *such that* $\mathbf{A}$ *and* $\mathbf{B}$ *make up an extremal pair.*

(b) *For any vector* $\mathbf{B} = (b_1, \ldots, b_m)$ *from* $\overline{Z}_+^m$ *and any* $n \in Z$ $(n \geq b_1)$, *there exists a unique vector* $\mathbf{A} \in \overline{Z}_+^n$ *such that* $\mathbf{A}$ *and* $\mathbf{B}$ *make up an extremal pair.*

Propositions (a) and (b) of Theorem 1.5 follow from relations (1.6) and (1.7), respectively.

Example 1.2. Given

$$\mathbf{A}_1 = (7,6,6,4,4,3,2,2) \in \overline{Z}_+^8 \quad \text{and } \mathbf{B}_2 = (8,7,6,5,4,3,2,1) \in \overline{Z}_+^8,$$

let us construct vectors $\mathbf{B}_1$ and $\mathbf{A}_2$ with nonzero components such that $\mathbf{A}_1$, $\mathbf{B}_1$ and $\mathbf{A}_2$, $\mathbf{B}_2$ make up extremal pairs.

We use (1.6) and (1.7) to calculate the components of $\mathbf{B}_1$ and $\mathbf{A}_2$, respectively. Since $\mathbf{B}_1$ and $\mathbf{A}_2$ have nonzero components, $\mathbf{B}_1 \in \overline{Z}_+^7$ and $\mathbf{A}_2 \in \overline{Z}_+^8$. Accordingly,

$$\mathbf{B}_1 = \Big(l_1(\mathbf{A}_1;1), l_2(\mathbf{A}_1;1), l_3(\mathbf{A}_1;1), l_4(\mathbf{A}_1;1), l_5(\mathbf{A}_1;1), l_6(\mathbf{A}_1;1), l_7(\mathbf{A}_1;1)\Big) = (8,8,6,5,3,3,1),$$

$$\mathbf{A}_2 = \Big(l_1(\mathbf{B}_2;1), l_2(\mathbf{B}_2;1), l_3(\mathbf{B}_2;1), l_4(\mathbf{B}_2;1), l_5(\mathbf{B}_2;1), l_6(\mathbf{B}_2;1), l_7(\mathbf{B}_2;1), l_8(\mathbf{B}_2;1)\Big) = (8,7,6,5,4,3,2,1).$$

By Theorem 1.4, the vector pairs $\mathbf{A}_1$, $\mathbf{B}_1$ and $\mathbf{A}_2$, $\mathbf{B}_2$ are extremal.

By Definition 1.5, matrices of zeros and matrices of ones are the simplest examples of extremal matrices. The corresponding extremal vector pairs are

$$\underbrace{(0,\ldots,0)}_{n}, \underbrace{(0,\ldots,0)}_{m}; \quad \underbrace{(m,\ldots,m)}_{n}, \underbrace{(n,\ldots,n)}_{m}.$$

Comparing Definition 1.5 with Definition 1.1 from Chapter 1, we see that Definition 1.5 applies to uniform matrices of zeros and ones. Therefore, a matrix from $M_Z(\mathbf{A},\mathbf{B};1)$ is extremal if and only if it is uniform ($\mathbf{A} \in \overline{Z}_+^n$, $\mathbf{B} \in \overline{Z}_+^m$).

Theorem 1.6. *Suppose that* $\mathbf{A} \in \overline{Z}_+^n$, $\mathbf{B} \in \overline{Z}_+^m$, $X(\mathbf{A},\mathbf{B}) \in M_Z(\mathbf{A},\mathbf{B};1)$, *and the set* $M_Z(\mathbf{A},\mathbf{B};1)$ *is not empty. The matrix* $X(\mathbf{A},\mathbf{B})$ *(or the pair of vectors* $\mathbf{A}$ *and* $\mathbf{B}$*) is extremal if and only if* $X(\mathbf{A},\mathbf{B})$ *is a uniform matrix* $(X(\mathbf{A},\mathbf{B}) \in \widetilde{M}_Z(\mathbf{A},\mathbf{B}))$.

We note here that if $\mathbf{A} \in \overline{Z}_+^n$, $\mathbf{B} \in \overline{Z}_+^m$, and $X(\mathbf{A},\mathbf{B})$ is a uniform matrix from the set $M_Z(\mathbf{A},\mathbf{B};1)$ that is not empty, then $|M_Z(\mathbf{A},\mathbf{B};1)| = 1$ and $|M(\mathbf{A},\mathbf{B};1)| = 1$.

1.2. Calculation of the Number of Extremal Matrices and Vector Pairs of Equal Dimensions

Henceforth, we generally parenthesize vector pairs. To distinguish extremal vector pairs from arbitrary ones, we denote the first vector of an extremal pair by $\overline{\mathbf{E}} = (e_1, \ldots, e_n)$ and the second vector by $\overline{\mathbf{F}} = (f_1, \ldots, f_m)$; thus, $(\overline{\mathbf{E}}, \overline{\mathbf{F}})$ is an arbitrary extremal pair of vectors. We also introduce special notation for the sets of extremal vector pairs and matrices:

$$\overline{E}^{n,m} = \left\{(\overline{\mathbf{E}}, \overline{\mathbf{F}}) : \overline{\mathbf{E}} \in \overline{Z}_+^n,\ \overline{\mathbf{F}} \in \overline{Z}_+^m,\ (\overline{\mathbf{E}}, \overline{\mathbf{F}}) \text{ is the extremal vector pair}\right\}$$

denotes the set of all extremal pairs of n- and m-dimensional vectors, respectively;

$$\overline{E}_{n,m} = \left\{X(\overline{\mathbf{E}}, \overline{\mathbf{F}}) = (x_{ij}) : (\overline{\mathbf{E}}, \overline{\mathbf{F}}) \in \overline{E}^{n,m},\ X(\overline{\mathbf{E}}, \overline{\mathbf{F}}) \in M_Z(\overline{\mathbf{E}}, \overline{\mathbf{F}};1)\right\}$$

is the set of all extremal n-by-m matrices.

Let us now count the extremal vector pairs and matrices in the sets $\overline{E}^{n,m}$ and $\overline{E}_{n,m}$, using the following property of binomial coefficients:

$$\sum_{i=0}^{q-1} C_{p+i}^p = C_{p+q}^{p+1}. \tag{1.8}$$

Relation (1.8) (offered as an exercise in some texts on discrete mathematics and combinatorics [4, 32]) can be readily proved by induction on q.

For $q = 1$ and $q = 2$, we have, respectively,

$$C_p^p = C_{p+1}^{p+1} = 1, \quad C_p^p + C_{p+1}^p = C_{p+2}^{p+1} = p+2.$$

Assuming that (1.8) is true for i not greater than $q-2$ and applying this hypothesis inductively, we obtain

$$\sum_{i=0}^{q-1} C_{p+i}^p = \sum_{i=0}^{q-2} C_{p+i}^p + C_{p+q-1}^p = C_{p+q-1}^{p+1} + C_{p+q-1}^p =$$

$$\frac{(p+q-1)!}{(p+1)!(q-2)!} + \frac{(p+q-1)!}{p!(q-1)!} = \frac{(p+q-1)!}{p!(q-2)!} \cdot \left(\frac{1}{p+1} + \frac{1}{q-1}\right) =$$

$$\frac{(p+q-1)!}{p!(q-2)!} \cdot \frac{p+q}{(p+1)(q-1)} = \frac{(p+q)!}{(p+1)!(q-1)!} = C_{p+q}^{p+1},$$

which proves relation (1.8).

Theorem 1.7. *The number of elements in both $\overline{E}^{n,m}$ and $\overline{E}_{n,m}$ is C_{n+m}^n.*

Proof. Any extremal vector pair is uniquely defined by its first vector (by Theorem 1.5); moreover, the vector $\overline{\mathbf{E}} = (e_1, \ldots, e_n)$ in an arbitrary pair $(\overline{\mathbf{E}}, \overline{\mathbf{F}}) \in \overline{E}^{n,m}$ must be such that $\overline{\mathbf{E}} \in \overline{Z}_+^n$ and $e_1 \le m$. Therefore, the theorem can be proved by calculating the number of the integer vectors with n nonnegative components arranged in nonascending order whose largest components are not greater than m. We denote this number by

$$P_n(m) = \left|\{\overline{\mathbf{E}} = (e_1, \ldots, e_n) : \overline{\mathbf{E}} \in \overline{Z}_+^n,\ e_1 \le m\}\right|.$$

It is clear that $P_n(m) = |\overline{E}^{n,m}| = |\overline{E}_{n,m}|$. Let us construct a recursive formula relating $P_n(m)$ to $P_{n-1}(t)$, where $t \le m$.

The set of vectors of the type in question can be represented as

$$\left\{\overline{\mathbf{E}} = (e_1, \ldots, e_n) : \overline{\mathbf{E}} \in \overline{Z}_+^n,\ e_1 \le m\right\} = \bigcup_{t=0}^{m} \left\{\overline{\mathbf{E}} = (e_1, \ldots, e_n) : \overline{\mathbf{E}} \in \overline{Z}_+^n,\ e_1 = t\right\}.$$

Taking out the first component of every vector in any set on the right-hand side, we obtain

$$\left|\left\{\overline{\mathbf{E}} = (e_1, e_2, \ldots, e_n) : \overline{\mathbf{E}} \in \overline{Z}_+^n,\ e_1 \le m\right\}\right| = \sum_{t=0}^{m} \left|\left\{(e_2, \ldots, e_n) : \overline{\mathbf{E}} \in \overline{Z}_+^{n-1},\ e_1 = t,\ e_2 \le t\right\}\right|.$$

Consequently, it holds that

$$P_n(m) = \sum_{t=0}^{m} P_{n-1}(t). \tag{1.9}$$

We prove that $P_n(m) = C_{m+n}^n$ by induction on n. When $n = 1$, the only component of the vector may be any $t \in Z$ such that $0 \le t \le m$. Therefore, $P_1(m) = m + 1 = C_{1+m}^1$. Assuming that $P_{n-1}(m) = C_{n-1+m}^{n-1}$ and applying this hypothesis combined with (1.9) inductively, we write

$$P_n(m) = \sum_{t=0}^{m} C_{n-1+t}^{n-1}.$$

Using the property of binomial coefficients expressed by (1.8), we obtain

$$P_n(m) = C_{n+m}^n = |\overline{E}^{n,m}| = |\overline{E}_{n,m}|,$$

which proves the theorem.

Definition 1.6. The sets of extremal vector pairs $\overline{E}^{n,m}$ and extremal matrices $\overline{E}_{n,m}$ are called the *fundamental systems.*

The idea behind Definition 1.6 is explained below. We introduce a special notation for the subsystems of fundamental systems.

Let $(\overline{\mathbf{E}}, \overline{\mathbf{F}}) = ((e_1, \ldots, e_n), (f_1, \ldots, f_m))$ be an arbitrary extremal vector pair from $\overline{E}^{n,m}$. Denote by

$$\overline{E}^{n^+,m} = \left\{(\overline{\mathbf{E}}, \overline{\mathbf{F}}) : (\overline{\mathbf{E}}, \overline{\mathbf{F}}) \in \overline{E}^{n,m},\ e_n \geq 1\right\},$$

$$\overline{E}^{n,m^+} = \left\{(\overline{\mathbf{E}}, \overline{\mathbf{F}}) : (\overline{\mathbf{E}}, \overline{\mathbf{F}}) \in \overline{E}^{n,m},\ e_1 = m\right\},$$

$$\overline{E}^{n^+,m^+} = \left\{(\overline{\mathbf{E}}, \overline{\mathbf{F}}) : (\overline{\mathbf{E}}, \overline{\mathbf{F}}) \in \overline{E}^{n^+,m} \cap \overline{E}^{n,m^+}\right\}$$

the sets of extremal vector pairs such that the first vector, the second one, and both vectors have only nonzero components, respectively. Also, denote by

$$\overline{E}_{n^+,m} = \left\{X(\overline{\mathbf{E}}, \overline{\mathbf{F}}) : (\overline{\mathbf{E}}, \overline{\mathbf{F}}) \in \overline{E}^{n^+,m},\ X(\overline{\mathbf{E}}, \overline{\mathbf{F}}) \in M_Z(\overline{\mathbf{E}}, \overline{\mathbf{F}}; 1)\right\},$$

$$\overline{E}_{n,m^+} = \left\{X(\overline{\mathbf{E}}, \overline{\mathbf{F}}) : (\overline{\mathbf{E}}, \overline{\mathbf{F}}) \in \overline{E}^{n,m^+},\ X(\overline{\mathbf{E}}, \overline{\mathbf{F}}) \in M_Z(\overline{\mathbf{E}}, \overline{\mathbf{F}}; 1)\right\},$$

$$\overline{E}_{n^+,m^+} = \left\{X(\overline{\mathbf{E}}, \overline{\mathbf{F}}) : (\overline{\mathbf{E}}, \overline{\mathbf{F}}) \in \overline{E}^{n^+,m^+} X(\overline{\mathbf{E}}, \overline{\mathbf{F}}) \in M_Z(\overline{\mathbf{E}}, \overline{\mathbf{F}}; 1)\right\}$$

the sets of extremal matrices that do not have zero rows, columns, and rows and columns, respectively.

Let us now count the elements of the fundamental subsystems thus defined, noting that $|\overline{E}^{n^+,m}| = |\overline{E}_{n^+,m}|$, $|\overline{E}^{n^+,m}| = |\overline{E}_{n^+,m}|$, and $|\overline{E}^{n^+,m^+}| = |\overline{E}_{n^+,m^+}|$.

If $\mathbf{A} \in R^n$ and $\mathbf{B} \in R^m$, then, by analogy with matrices, the vector pair $(\mathbf{A}, \mathbf{B})$ is referred to as an n-by-m *pair*. We transform the fundamental subsystems $\overline{E}^{n^+,m}$, $\overline{E}^{n,m^+}$, and $\overline{E}^{n^+,m^+}$ of n-by-m extremal vector pairs into fundamental systems of n-by-$(m-1)$, $(n-1)$-by-m, and $(n-1)$-by-$(m-1)$ vector pairs.

Here, we use the following notation for an extremal vector pair $(\overline{\mathbf{E}}, \overline{\mathbf{F}})$, where $\overline{\mathbf{E}} = (e_1, \ldots, e_n)$ $(e_1 \leq m, e_n \geq 0)$. We denote by $(\overline{\mathbf{E}}', \overline{\mathbf{F}}')$, $(\overline{\mathbf{E}}'', \overline{\mathbf{F}}'')$, and $(\overline{\mathbf{E}}''', \overline{\mathbf{F}}''')$ such extremal vector pairs that the components of the respective first vectors $\overline{\mathbf{E}}' = (e_1', \ldots, e_n')$, $\overline{\mathbf{E}}'' = (e_1'', \ldots, e_{n-1}'')$, and $\overline{\mathbf{E}}''' = (e_1''', \ldots, e_{n-1}''')$ satisfy the relations

$$e_i' = e_i - 1, \quad 1 \leq i \leq n, \text{if } e_n \geq 1;$$

$$e_i'' = e_{i+1}, \quad 1 \leq i \leq n-1, \text{ if } e_1 = m;$$

$$e_i''' = e_{i+1} - 1, \quad 1 \le i \le n-1, \text{ if } e_n \ge 1 \text{ and } e_1 = m.$$

Applying Theorem 1.7 to calculate the numbers of elements of the fundamental subsystems, we obtain

$$\left|\overline{E}^{n^+,m}\right| = \left|\left\{(\overline{\mathbf{E}},\overline{\mathbf{F}}) \in \overline{E}^{n,m} : e_1 \le m,\ e_n \ge 1\right\}\right| =$$

$$\left|\left\{(\overline{\mathbf{E}}',\overline{\mathbf{F}}') \in \overline{E}^{n,m-1} : e_1' \le m-1,\ e_n' \ge 0\right\}\right| =$$

$$\left|\left\{(\overline{\mathbf{E}}',\overline{\mathbf{F}}') \in \overline{E}^{n,m-1} : e_1' \le m-1\right\}\right| = P_n(m-1) = C^n_{n+m-1};$$

$$\left|\overline{E}^{n,m^+}\right| = \left|\left\{(\overline{\mathbf{E}},\overline{\mathbf{F}}) \in \overline{E}^{n,m} : e_1 = m,\ e_n \ge 0\right\}\right| =$$

$$\left|\left\{(\overline{\mathbf{E}}'',\overline{\mathbf{F}}'') \in \overline{E}^{n-1,m} : e_1'' \le m,\ e_{n-1}'' \ge 0\right\}\right| =$$

$$\left|\left\{(\overline{\mathbf{E}}'',\overline{\mathbf{F}}'') \in \overline{E}^{n-1,m} : e_1'' \le m\right\}\right| = P_{n-1}(m) = C^{n-1}_{n+m-1};$$

$$\left|\overline{E}^{n^+,m^+}\right| = \left|\left\{(\overline{\mathbf{E}},\overline{\mathbf{F}}) \in \overline{E}^{n,m} : e_1 = m,\ e_n \ge 1\right\}\right| =$$

$$\left|\left\{(\overline{\mathbf{E}}''',\overline{\mathbf{F}}''') \in \overline{E}^{n-1,m-1} : e_1''' \le m-1,\ e_{n-1}''' \ge 0\right\}\right| =$$

$$\left|\left\{(\overline{\mathbf{E}}''',\overline{\mathbf{F}}''') \in \overline{E}^{n-1,m-1} : e_1''' \le m-1\right\}\right| = P_{n-1}(m-1) = C^{n-1}_{n+m-2}.$$

Thus, we have proved the following theorem.

Theorem 1.8. *The numbers of elements of fundamental subsystems of vector pairs and matrices are*

$$|\overline{E}^{n^+,m}| = |\overline{E}_{n^+,m}| = C^n_{n+m-1}, \quad |\overline{E}^{n,m^+}| = |\overline{E}_{n,m^+}| = C^{n-1}_{n+m-1},$$

$$|\overline{E}^{n^+,m^+}| = |\overline{E}_{n^+,m^+}| = C^{n-1}_{n+m-2}.$$

1.3. Perfect Vector Pairs and Perfect Matrices

We now generalize the concept of extremality of vector pairs and matrices.

Definition 1.7. Vectors **A** and **B** are said to make up a *perfect pair* if the vector pair $\widetilde{\mathbf{A}}$, $\widetilde{\mathbf{B}}$ derived from **A** and **B** by arranging their components in nonascending order is an extremal one.

Definition 1.8. Matrix X is called *perfect* if X is a uniform matrix whose elements are either zeros or ones.

Combined with Definitions 1.5 and 1.8, Theorem 1.6 implies that any perfect matrix can be transformed into an extremal one by permuting its rows and columns; moreover, if $\mathbf{A}$ and $\mathbf{B}$ make up a perfect vector pair and $X(\mathbf{A},\mathbf{B})$ is a perfect matrix ($X(\mathbf{A},\mathbf{B}) \in M_Z(\mathbf{A},\mathbf{B};1)$), then the matrix $\widetilde{X}(\widetilde{\mathbf{A}},\widetilde{\mathbf{B}})$ obtained by permuting the rows and columns of $X(\mathbf{A},\mathbf{B})$ is an extremal one. Note also that Definitions 1.7 and 1.8 imply that an extremal vector pair or matrix is perfect.

It is clear that every perfect vector pair is strictly reducible. Therefore, a perfect matrix can be constructed by applying Definition 1.2, provided that the largest component (i.e., the component that determines the strict reducibility) of the vector is found at each reductive step. One may also apply Lemma 1.3. However, the construction and analysis of a perfect matrix are simplified when its rows and columns are permuted so as to obtain an extremal matrix and then the original matrix is reconstructed by reversing the permutation. It is obvious that all properties of extremal matrices that are invariant under the permutation of rows and columns are retained by perfect matrices. For example, perfect matrices are also minimax ones, and the extremal matrices in Theorems 1.1, 1.2′, 1.3, and 1.6 can be replaced by perfect matrices.

Theorem 1.9. *Suppose that vectors $\mathbf{A} \in Z_+^n$ and $\mathbf{B} \in Z_+^m$ are characterized by equal sums of components. The set $M_Z(\mathbf{A},\mathbf{B};1)$ contains only one matrix if and only if $\mathbf{A}$ and $\mathbf{B}$ make up a perfect pair of vectors. Moreover, when $\mathbf{A}$ and $\mathbf{B}$ have integer components, the set $M(\mathbf{A},\mathbf{B};1)$ contains only one matrix if and only if $\mathbf{A}$ and $\mathbf{B}$ make up a perfect vector pair.*

Theorem 1.10. *Suppose that $\mathbf{A} \in Z_+^n$, $\mathbf{B} \in Z_+^m$, $X(\mathbf{A},\mathbf{B}) = (x_{ij}) \in M_Z(\mathbf{A},\mathbf{B};1)$, and the set $M_Z(\mathbf{A},\mathbf{B};1)$ is not empty. The vectors $\mathbf{A}$ and $\mathbf{B}$ make up a perfect pair if and only if the elements of the matrix $\widetilde{X}(\widetilde{\mathbf{A}},\widetilde{\mathbf{B}}) = (\tilde{x}_{ij})$ constructed by permuting the rows of $X(\mathbf{A},\mathbf{B})$ are defined as*

$$\tilde{x}_{ij} = \begin{cases} 1, & 1 \le i \le n, \quad b_j \ge i, \\ 0, & 1 \le i \le n, \quad b_j < i. \end{cases}$$

Theorem 1.11. *The matrices X and X^{T} and vector pairs $\mathbf{A}$, $\mathbf{B}$ and $\mathbf{B}$, $\mathbf{A}$ are either simultaneously perfect or simultaneously imperfect (X^{T} is the transpose of X).*

Theorem 1.12. *Suppose that $\mathbf{A} \in Z_+^n$, $\mathbf{B} \in Z_+^m$, $X(\mathbf{A},\mathbf{B}) \in M_Z(\mathbf{A},\mathbf{B};1)$, and the set $M_Z(\mathbf{A},\mathbf{B};1)$ is not empty. The matrix $X(\mathbf{A},\mathbf{B})$ and the vector pair $\mathbf{A}$, $\mathbf{B}$ are perfect if and only if $X(\mathbf{A},\mathbf{B})$ is a uniform matrix.*

Some properties of extremal and perfect matrices admit simpler definitions in the case of a perfect matrix. This is illustrated by Theorem 1.13. In what follows, we show that perfect matrices are hereditarity minimax matrices. An analogous proposition is true for extremal matrices.

A matrix of ones is called a *complete*, or 1-*complete*, matrix. Suppose that $X = (x_{ij})$ $(1 \le i \le n,\ 1 \le j \le m)$ is an arbitrary matrix of zeros and ones. The matrix $\overline{X}' = (\overline{x}'_{ij}) = (1 - x_{ij})$ $(1 \le i \le n,\ 1 \le j \le m)$ is called the *complement*, or 1-*complement*, to the complete matrix for X. If X is an n-by-m complete matrix, then $X = X(\mathbf{A}, \mathbf{B}) \in M_Z(\mathbf{A}, \mathbf{B}; 1)$, where the vectors

$$\mathbf{A} = \underbrace{(m, \ldots, m)}_{n}, \quad \mathbf{B} = \underbrace{(n, \ldots, n)}_{m}$$

make up an extremal pair.

Definition 1.9. n-by-m matrices $X = (x_{ij})$ and $\overline{X}' = (\overline{x}'_{ij})$ are said to be *mutually complementary* if x_{ij}, $\overline{x}'_{ij} \ge 0$, and $x_{ij} + \overline{x}'_{ij} = 1$ for $1 \le i \le n$ and $1 \le j \le m$.

Theorem 1.13. *Let $X = (x_{ij})$ $(1 \le i \le n,\ 1 \le j \le m)$ be a matrix of zeros and ones. Then, X is a perfect matrix if and only if its complement $\overline{X}'$ to the complete matrix is perfect.*

Proof. If a matrix consists of zeros and ones, then it is uniform if and only if it is perfect (by Theorem 1.12). Therefore, we must prove the simultaneous uniformity or nonuniformity of X and $\overline{X}'$. Suppose $X = (x_{ij}) = X(\mathbf{A}, \mathbf{B})$, where $\mathbf{A} = (a_1, \ldots, a_n)$, $\mathbf{B} = (b_1, \ldots, b_m)$,

$$\sum_{j=1}^{m} x_{ij} = a_i, \quad \sum_{i=1}^{n} = b_j, \quad 1 \le i \le n, \quad 1 \le j \le m,$$

and

$$\overline{X}' = (\overline{x}'_{ij}) = \overline{X}'(\overline{\mathbf{A}}', \overline{\mathbf{B}}'),$$

where

$$\overline{\mathbf{A}}' = (\overline{a}'_1, \ldots, \overline{a}'_n), \quad \overline{\mathbf{B}}' = (\overline{b}'_1, \ldots, \overline{b}'_m),$$

$$\sum_{j=1}^{m} \overline{x}'_{ij} = \overline{a}'_i, \quad \sum_{i=1}^{n} \overline{x}'_{ij} = \overline{b}'_j, \quad 1 \le i \le n, \quad 1 \le j \le m.$$

Since $\overline{x}'_{ij} = 1 - x_{ij}$, we have $\overline{a}'_i = m - a_i$ and $\overline{b}'_j = n - b_j$, where $1 \le i \le n$ and $1 \le j \le m$. Hence, for the components of $\mathbf{A}$, $\overline{\mathbf{A}}'$, $\mathbf{B}$, and $\overline{\mathbf{B}}'$, the conditions $a_i \ge a_p$, $a_i = a_p$, $b_j \ge b_q$, and $b_j = b_q$ are equivalent to $\overline{a}'_i \le \overline{a}'_p$, $\overline{a}'_i = \overline{a}'_p$, $\overline{b}'_j \le \overline{b}'_q$, and $\overline{b}'_j = \overline{b}'_q$, respectively. Similarly, the conditions $x_{ij} \le x_{pq}$ and $x_{ij} = x_{pq}$ for the elements of X are equivalent to the conditions $\overline{x}'_{ij} \le \overline{x}'_{pq}$ and $\overline{x}'_{ij} = \overline{x}'_{pq}$ for the elements of $\overline{X}'$. The equivalence of the respective conditions for the components of $\mathbf{A}$, $\mathbf{B}$, $\overline{\mathbf{A}}'$, and $\overline{\mathbf{B}}'$ and for the elements of X and $\overline{X}'$ implies that the matrices X and $\overline{X}'$ are simultaneously uniform or nonuniform, which proves the theorem.

The perfectness of a matrix is a generic property of transportation matrices.

Theorem 1.14. *If X is a perfect matrix, then its every submatrix is also a perfect matrix.*

The proof can be based either on the fact that uniformity is a hereditary property (recall Lemma 2.5 from Chapter 2) or on Theorem 1.12.

We now show that a matrix is perfect if and only if it is a hereditarity minimax matrix.

Theorem 1.15. *Let X be a matrix of zeros and ones. Then, it is a hereditarity minimax matrix if and only if it is a perfect (in particular, extremal) matrix.*

Proof.

Necessity. If X is a hereditarity minimax matrix, then X is a uniform matrix by Lemma 2.4 from Chapter 2. By Theorem 1.12, the matrix X is then perfect.

Sufficiency. If X is a perfect matrix, then it is a minimax matrix. By Theorem 1.14, every submatrix of X is then a perfect and, therefore, minimax matrix, which completes the proof.

To distinguish perfect vector pairs from ordinary ones, we denote the first vector of a perfect pair by $\mathbf{E} = (e_1, \ldots, e_n)$ and its second vector by $\mathbf{F} = (f_1, \ldots, f_m)$. Let $(\overline{\mathbf{E}}, \overline{\mathbf{F}})$ be an arbitrary perfect pair of vectors. We introduce the following notation for sets of perfect vector pairs and matrices:

$$E^{n,m} = \Big\{(\mathbf{E}, \mathbf{F}) : \mathbf{E} \in Z_+^n,\ \mathbf{F} \in Z_+^m,\ (\mathbf{E}, \mathbf{F}) \text{ is the perfect vector pair}\Big\}$$

denotes the set of all perfect n-by-m vector pairs, and

$$E_{n,m} = \Big\{X(\mathbf{E}, \mathbf{F}) : (\mathbf{E}, \mathbf{F}) \in E^{n,m},\ X(\mathbf{E}, \mathbf{F}) \in M_Z(\mathbf{A}, \mathbf{B}; 1)\Big\}$$

denotes the set of all perfect n-by-m matrices. We also distinguish perfect and extremal matrices from arbitrary matrices. We denote arbitrary perfect and extremal matrices by $H = (h_{ij})$ and $\overline{H} = (h_{ij})$, respectively. If these matrices are defined by vector pairs $(\mathbf{E}, \mathbf{F}) \in E^{n,m}$ and $(\overline{\mathbf{E}}, \overline{\mathbf{E}}) \in \overline{E}^{n,m}$, respectively, then $H = H(\mathbf{E}, \mathbf{F})$ and $\overline{H} = \overline{H}(\overline{\mathbf{E}}, \overline{\mathbf{F}})$. Accordingly,

$$E_{n,m} = \Big\{H(\mathbf{E}, \mathbf{F}) : (\mathbf{E}, \mathbf{F}) \in E^{n,m}\Big\}$$

and

$$\overline{E}_{n,m} = \Big\{\overline{H}(\overline{\mathbf{E}}, \overline{\mathbf{F}}) : (\overline{\mathbf{E}}, \overline{\mathbf{F}}) \in \overline{E}^{n,m}\Big\}.$$

Let $\mathbf{A}$ and X be an arbitrary vector and arbitrary matrix, respectively. By $\widehat{\mathbf{A}}$ and $\widehat{X}$, we denote any vector constructed by permuting the components of $\mathbf{A}$ and any matrix constructed by permuting the rows and columns of X, respectively. By Definitions 1.7 and 1.8, if a vector pair $(\overline{\mathbf{E}}, \overline{\mathbf{F}})$ and matrix $\overline{H}$ are extremal, then $(\widehat{\overline{\mathbf{E}}}, \widehat{\overline{\mathbf{F}}})$ and $\widehat{\overline{H}}$ are perfect vector pair and matrix, respectively. Using this notation, we define the sets of perfect vector pairs and matrices as follows:

$$E^{n,m} = \left\{(\widehat{\overline{\mathbf{E}}}, \widehat{\overline{\mathbf{F}}}) : (\overline{\mathbf{E}}, \overline{\mathbf{F}}) \in \overline{E}^{n,m}\right\}, \quad E_{n,m} = \left\{\widehat{\overline{H}} : \overline{H} \in \overline{E}_{n,m}\right\}.$$

As corollaries to Theorem 1.5 and Definition 1.7, we have the following propositions concerning perfect pairs of vectors.

Theorem 1.16. *Suppose that* $\mathbf{A} = (a_1, \dots, a_n) \in Z_+^n$, $\mathbf{B} = (b_1, \dots, b_m) \in Z_+^m$, *and vectors* $\widetilde{\mathbf{A}} = (\tilde{a}_1, \dots, \tilde{a}_n)$ *and* $\widetilde{\mathbf{B}} = (\tilde{b}_1, \dots, \tilde{b}_m)$ *are constructed by arranging the components of* $\mathbf{A}$ *and* $\mathbf{B}$ *in nonascending order. Then,* $\mathbf{A}$ *and* $\mathbf{B}$ *make up a perfect pair of vectors if and only if either of the following conditions is satisfied:*

(a) $\tilde{a}_i = l_i(\widetilde{\mathbf{B}}; 1) = l_i(\mathbf{B}; 1), \quad 1 \le i \le n,$

(b) $\tilde{b}_j = l_j(\widetilde{\mathbf{A}}; 1) = l_j(\mathbf{A}; 1), \quad 1 \le j \le m.$

Theorem 1.17.

(a) *For any vector* $\mathbf{E} = (e_1, \dots, e_n) \in Z_+^n$ *and any* $m \in Z$ *such that* $m \ge \max\{e_i : 1 \le i \le n\}$, *there exists a vector* $\mathbf{F} \in Z_+^m$ *such that* $(\mathbf{E}, \mathbf{F})$ *is a perfect vector pair and every vector pair* $(\widehat{\mathbf{E}}, \widehat{\mathbf{F}})$ *is perfect.*

(b) *For any vector* $\mathbf{F} = (f_1, \dots, f_m) \in Z_+^m$ *and any* $n \in Z$ *such that* $n \ge \max\{f_j : 1 \le j \le m\}$, *there exists a vector* $\mathbf{E} \in Z_+^n$ *such that* $(\mathbf{E}, \mathbf{F})$ *is a perfect vector pair and every vector pair* $(\widehat{\mathbf{E}}, \widehat{\mathbf{F}})$ *is perfect.*

Consider the minimization of functionals (1.7) and (1.14) from Chapter 1 on the matrix sets defined by perfect (or extremal) pairs of vectors. Each element of a perfect matrix is zero or one. Therefore, if $(\mathbf{E}, \mathbf{F}) \in E^{n,m}$, then

$$\Phi_1\left(H(\mathbf{E}, \mathbf{F})\right) = \max_{i,j} h_{ij} = \begin{cases} 1 \text{ if } \mathbf{E} \text{ is a nonzero vector,} \\ 0 \text{ if } \mathbf{E} \text{ is a zero vector,} \end{cases}$$

where $H(\mathbf{E}, \mathbf{F}) = (h_{ij}) \in E_{n,m}$. Since any perfect matrix is a minimax matrix,

$$\min_{X(\mathbf{E},\mathbf{F}) \in M(\mathbf{E},\mathbf{F})} \Phi_1\left(X(\mathbf{E}, \mathbf{F})\right) = \Phi_1\left(H(\mathbf{E}, \mathbf{F})\right) = 1$$

for any perfect vector pair $(\mathbf{E}, \mathbf{F})$, where $\mathbf{E}$ is a nonzero vector. By Theorem 1.15, any perfect (or extremal) matrix is a hereditarity minimax matrix.

Therefore, formula (3.12) from Chapter 2, which determines the minimum of objective function (1.14) from Chapter 1, simplifies when the vectors make up a perfect pair. If $(\mathbf{E},\mathbf{F}) \in E^{n,m}$, then $\overset{*}{X}(\mathbf{E},\mathbf{F}) = (\overset{*}{x}_{ij}) = H(\mathbf{E},\mathbf{F}) = (h_{ij}) \in E_{n,m}$, and the perfect matrix $H(\mathbf{E},\mathbf{F})$ minimizes functional (1.14) from Chapter 1 by Theorem 3.2 from Chapter 2 and Theorem 1.16:

$$\min_{X(\mathbf{E},\mathbf{F})\in M(\mathbf{E},\mathbf{F})} \Phi_8\left(X(\mathbf{E},\mathbf{F})\right) = \Phi_8\left(H(\mathbf{E},\mathbf{F})\right) = \Phi_8(H(\widetilde{\mathbf{E}},\widetilde{\mathbf{F}})) = \sum_{i=1}^{n}\sum_{j=1}^{\tilde{e}_i} 2^{n+m-i-j},$$

where $(\widetilde{\mathbf{E}} = (\tilde{e}_1,\ldots,\tilde{e}_n), \widetilde{\mathbf{F}} = (\tilde{f}_1,\ldots,\tilde{f}_m))$ is an extremal vector pair constructed by arranging the components of the vector pair $(\mathbf{E},\mathbf{F})$ in nonascending order.

§2. Algebraic and Metric Properties of Extremal Vector Pairs and Extremal Matrices

In this section, we make use of the algebraic concept of distributive lattice [12, 15, 35]. Recall that the lattice is defined as a partially ordered set in which every two-element subset has exact upper and lower bounds. We consider extremal pairs of vectors and extremal matrices of equal dimensions. We introduce order relations and two binary operations on these sets. These operations define exact upper and lower bounds for every two elements. We show that fundamental systems of vector pairs $\overline{E}^{n,m}$ and matrices $\overline{E}_{n,m}$ are isomorphic distributive lattices.

We construct metric functions [1, 13] on the sets of extremal vector pairs and extremal matrices of equal dimensions. The fundamental systems of vector pairs and matrices, $\overline{E}^{n,m}$ and $\overline{E}_{n,m}$, are isometric as spaces with these metrics. We show that the metrics are consistent with the partial order relations on $\overline{E}^{n,m}$ and $\overline{E}_{n,m}$ and quantify the difference between every two elements of the sets. The algebraic and metric constructions developed here are utilized in the next section.

2.1. Operations on Vector Pairs and Matrices

Consider arbitrary vectors $\mathbf{A} = (a_1,\ldots,a_n)$ and $\mathbf{B} = (b_1,\ldots,b_n)$ from R^n. We say that $\mathbf{A} \le \mathbf{B}$ if $a_i \le b_i$ for any i $(1 \le i \le n)$, and $\mathbf{A} < \mathbf{B}$ if $\mathbf{A} \le \mathbf{B}$ and there exists $p \in Z$ $(1 \le p \le n)$ such that $a_p < b_p$. Suppose that the vector pairs $(\mathbf{A},\mathbf{B})$ and $(\mathbf{C},\mathbf{D})$ have equal dimensions: $\mathbf{A},\ \mathbf{C} \in R^n$ and $\mathbf{B},\ \mathbf{D} \in R^m$. The equality of vector pairs, $(\mathbf{A},\mathbf{B}) = (\mathbf{C},\mathbf{D})$, is defined by the equality of the corresponding vectors. We say that $(\mathbf{A},\mathbf{B}) \le (\mathbf{C},\mathbf{D})$ if $\mathbf{A} \le \mathbf{C}$

and $\mathbf{B} \leq \mathbf{D}$, and $(\mathbf{A},\mathbf{B}) < (\mathbf{C},\mathbf{D})$ if $(\mathbf{A},\mathbf{B}) \leq (\mathbf{C},\mathbf{D})$ and either inequality $\mathbf{A} < \mathbf{C}$ or $\mathbf{B} < \mathbf{D}$ is valid.

Consider arbitrary n-by-m matrices $X = (x_{ij})$ and $Y = (y_{ij})$ $(1 \leq i \leq n$, $1 \leq j \leq m)$. We say that $X \subseteq Y$ if $x_{ij} \leq y_{ij}$ for all i and j $(1 \leq i \leq n$, $1 \leq j \leq m)$, and $X \subset Y$ if $X \subseteq Y$ and there exist p, $q \in Z$ $(1 \leq p \leq n$, $1 \leq q \leq m)$ such that $x_{pq} < y_{pq}$.

One can easily check that the axioms that define a partially ordered set apply to vector pairs and matrices of equal dimensions:

$(\mathbf{A},\mathbf{B}) \leq (\mathbf{A},\mathbf{B})$ and $X \subseteq X$ for any vector pair and matrix, respectively;

if $(\mathbf{A},\mathbf{B}) \leq (\mathbf{C},\mathbf{D})$, $(\mathbf{C},\mathbf{D}) \leq (\mathbf{K},\mathbf{L})$ and $X \subseteq Y$, $Y \subseteq Z$, then $(\mathbf{A},\mathbf{B}) \leq (\mathbf{K},\mathbf{L})$ and $X \subseteq Z$;

if $(\mathbf{A},\mathbf{B}) \leq (\mathbf{C},\mathbf{D})$, $(\mathbf{C},\mathbf{D}) \leq (\mathbf{A},\mathbf{B})$ and $X \subseteq Y$, $Y \subseteq X$, then $(\mathbf{A},\mathbf{B}) = (\mathbf{C},\mathbf{D})$ and $X = Y$.

It is obvious that the sets of extremal vector pairs and extremal matrices with order relations "$\leq$" and "$\subseteq$" between their elements, $\overline{E}^{n,m}$ and $\overline{E}_{n,m}$, are partially ordered sets. We denote the (extremal) matrices of zeros and those of ones by $\overline{H}(0)$ and $\overline{H}(1)$, respectively. It is clear that any pair $(\overline{\mathbf{E}},\overline{\mathbf{F}}) \in \overline{E}^{n,m}$ and any matrix $\overline{H} \in \overline{E}_{n,m}$ satisfy, respectively, the relations

$$\Big((\underbrace{0,\ldots,0}_{n}),(\underbrace{0,\ldots,0}_{m})\Big) \leq (\overline{\mathbf{E}},\overline{\mathbf{F}}) \leq \Big((\underbrace{m,\ldots,m}_{n}),(\underbrace{n,\ldots,n}_{m})\Big),$$

and

$$\overline{H}(0) \subseteq \overline{H} \subseteq \overline{H}(1),$$

where $\overline{H}(0)$, $\overline{H}(1) \in \overline{E}_{n,m}$.

In the general case, the order relation "$\leq$" (or "$<$") for vector pairs should be checked separately for the first and second vectors. In the case of extremal vector pairs, it is sufficient to check the order relation either for the first vectors or for the second vectors only.

Lemma 2.1. *Suppose $(\overline{\mathbf{E}}_1,\overline{\mathbf{F}}_1)$, $(\overline{\mathbf{E}}_2,\overline{\mathbf{F}}_2) \in \overline{E}^{n,m}$. For the relation $(\overline{\mathbf{E}}_1,\overline{\mathbf{F}}_1) \leq (\overline{\mathbf{E}}_2,\overline{\mathbf{F}}_2)$ to be valid, it is sufficient that either $\overline{\mathbf{E}}_1 \leq \overline{\mathbf{E}}_2$ or $\overline{\mathbf{F}}_1 \leq \overline{\mathbf{F}}_2$. The order relation "$\leq$" can be replaced by either "$<$" or "$=$".*

Proof. For the relation "$=$", the assertion of the lemma follows from Theorems 1.4 and 1.5.

Consider the order relation "$\leq$". Let $\overline{\mathbf{E}}_1 = (e'_1,\ldots,e'_n) \leq \overline{\mathbf{E}}_2 = (e''_1,\ldots,e''_n)$. Then, $e'_i \leq e''_i$ for all i $(1 \leq i \leq n)$, and $l_j(\overline{\mathbf{E}}_1;1) \leq l_j(\overline{\mathbf{E}}_2;1)$ for $1 \leq j \leq m$. The equality in (1.7) implies that the components of $\overline{\mathbf{F}}_1 = (f'_1,\ldots,f'_m)$ and $\overline{\mathbf{F}}_2 = (f''_1,\ldots,f''_m)$ are such that $f'_j \leq f''_j$ for any j $(1 \leq j \leq m)$. Hence, $\overline{\mathbf{F}}_1 \leq \overline{\mathbf{F}}_2$ and $(\overline{\mathbf{E}}_1,\overline{\mathbf{F}}_1) \leq (\overline{\mathbf{E}}_2,\overline{\mathbf{F}}_2)$.

If $\overline{\mathbf{E}}_1 < \overline{\mathbf{E}}_2$, then $\overline{\mathbf{E}}_1 \leq \overline{\mathbf{E}}_2$ and there exists $p \in Z$ $(1 \leq p \leq n)$ such that $e'_p < e''_p$. It is obvious that $1 \leq e''_p \leq m$ and (1.7) entails $f'_{e''_p} < f''_{e''_p}$ because $l_{e''_p}(\overline{\mathbf{E}}_1; 1) \leq l_{e''_p}(\overline{\mathbf{E}}_2; 1)$. Hence, $\overline{\mathbf{F}}_1 < \overline{\mathbf{F}}_2$ and $(\overline{\mathbf{E}}_1, \overline{\mathbf{F}}_1) < (\overline{\mathbf{E}}_2, \overline{\mathbf{F}}_2)$.

The assertion of the lemma is proved by analogy for the second vectors, $\overline{\mathbf{F}}_1$ and $\overline{\mathbf{F}}_2$, by invoking the equality in (1.6).

The order relation "$\leq$" (or "$<$" and "$=$") introduced for extremal vector pairs and the order relation "$\subseteq$" (or "$\subset$" and "$=$") introduced for extremal matrices are mutually consistent.

Lemma 2.2. *Suppose that* $(\overline{\mathbf{E}}_1, \overline{\mathbf{F}}_1)$, $(\overline{\mathbf{E}}_2, \overline{\mathbf{F}}_2) \in \overline{E}^{n,m}$ *and* $\overline{H}_1$, $\overline{H}_2 \in \overline{E}_{n,m}$; *moreover,* $\overline{H}_1 = H_1(\overline{\mathbf{E}}_1, \overline{\mathbf{F}}_1)$ *and* $\overline{H}_2 = H_2(\overline{\mathbf{E}}_2, \overline{\mathbf{F}}_2)$. *Then, the relations* $(\overline{\mathbf{E}}_1, \overline{\mathbf{F}}_1) \leq (\overline{\mathbf{E}}_2, \overline{\mathbf{F}}_2)$, $(\overline{\mathbf{E}}_1, \overline{\mathbf{F}}_1) < (\overline{\mathbf{E}}_2, \overline{\mathbf{F}}_2)$, *and* $(\overline{\mathbf{E}}_1, \overline{\mathbf{F}}_1) = (\overline{\mathbf{E}}_2, \overline{\mathbf{F}}_2)$ *for the extremal vector pairs are equivalent to the relations* $\overline{H}_1 \subseteq \overline{H}_2$, $\overline{H}_1 \subset \overline{H}_2$, *and* $\overline{H}_1 = \overline{H}_2$, *respectively.*

Proof. As concerns the relation "$=$", the assertion of the lemma follows from Theorem 1.1, 1.2, or 1.2′.

Let $\overline{H}_1 = (h'_{ij})$ and $\overline{H}_2 = (h''_{ij})$, where $1 \leq i \leq n$ and $1 \leq j \leq m$. As a corollary to Lemma 2.1 and Theorem 1.4, we have the following sequences of equivalent relations:

$$(\overline{\mathbf{E}}_1, \overline{\mathbf{F}}_1) \leq (\overline{\mathbf{E}}_2, \overline{\mathbf{F}}_2); \quad \overline{\mathbf{E}}_1 = (e'_1, \ldots, e'_n) \leq \overline{\mathbf{E}}_2 = (e''_1, \ldots, e''_n),$$
$$e'_i \leq e''_i, \quad 1 \leq i \leq n; \quad l_i(\overline{\mathbf{F}}_1; 1) \leq l_i(\overline{\mathbf{F}}_2; 1), \quad 1 \leq i \leq n;$$

and

$$(\overline{\mathbf{E}}_1, \overline{\mathbf{F}}_1) < (\overline{\mathbf{E}}_2, \overline{\mathbf{F}}_2); \quad \overline{\mathbf{E}}_1 < \overline{\mathbf{E}}_2, \quad e'_i \leq e''_i, \quad 1 \leq i \leq n, \quad e'_p < e''_p;$$
$$l_i(\overline{\mathbf{F}}_1; 1) \leq l_i(\overline{\mathbf{F}}_2; 1), \quad 1 \leq i \leq n, \quad l_p(\overline{\mathbf{F}}_1; 1) < l_p(\overline{\mathbf{F}}_2; 1)$$

for a certain $p \in Z$ $(1 \leq p \leq n)$. Consider an arbitrary ith row of the matrix $\overline{H}_1$ and the corresponding row of $\overline{H}_2$. Applying (1.5) to the elements of these rows, we obtain

$$h'_{ij} = \begin{cases} 1, & 1 \leq j \leq l_i(\overline{\mathbf{F}}_1; 1), \\ 0, & l_i(\overline{\mathbf{F}}_1; 1) < j \leq m, \end{cases} \qquad h''_{ij} = \begin{cases} 1, & 1 \leq j \leq l_i(\overline{\mathbf{F}}_2; 1), \\ 0, & l_i(\overline{\mathbf{F}}_1; 1) < j \leq m. \end{cases}$$

Since the inequality $l_i(\overline{\mathbf{F}}_1; 1) \leq l_i(\overline{\mathbf{F}}_2; 1)$ holds for both sequences of relations, the elements of the ith rows of $\overline{H}_1$ and $\overline{H}_2$ are as follows:

$$h'_{ij} = h''_{ij} = 1, \quad 1 \leq j \leq l_i(\overline{\mathbf{F}}_1; 1);$$
$$h'_{ij} = 0, \quad h''_{ij} = 1, \quad l_i(\overline{\mathbf{F}}_1; 1) < j \leq l_i(\overline{\mathbf{F}}_2; 1);$$
$$h'_{ij} = h''_{ij} = 0, \quad l_i(\overline{\mathbf{F}}_2; 1) < j \leq m.$$

Hence, $h'_{ij} \le h''_{ij}$ for all i and j $(1 \le i \le n,\ 1 \le j \le m)$, and $\overline{H}_1 \subseteq \overline{H}_2$. For the second sequence of relations, it holds that $l_p(\overline{\mathbf{F}}_1; 1) < l_p(\overline{\mathbf{F}}_2; 1)$. Therefore, the set of indices $\{j : l_p(\overline{\mathbf{F}}_1; 1) < j \le l_p(\overline{\mathbf{F}}_2; 1)\}$ is not empty, and $h'_{pj} = 0$, $h''_{pj} = 1$ if $l_p(\overline{\mathbf{F}}_1; 1) < j \le l_p(\overline{\mathbf{F}}_2; 1)$. Thus, $h'_{pj} < h''_{pj}$ and $\overline{H}_1 \subset \overline{H}_2$ for this sequence.

If $\overline{H}_1 \subseteq \overline{H}_2$, then $l_i(\overline{\mathbf{F}}_1; 1) \le l_i(\overline{\mathbf{F}}_2; 1)$ for any i $(1 \le i \le n)$ by Theorem 1.2′, and $e'_i \le e''_i$ for any i $(1 \le i \le n)$ by Theorem 1.4. Therefore, $\overline{\mathbf{E}}_1 \le \overline{\mathbf{E}}_2$ and $(\overline{\mathbf{E}}_1, \overline{\mathbf{F}}_1) \le (\overline{\mathbf{E}}_2, \overline{\mathbf{F}}_2)$. If $\overline{H}_1 \subset \overline{H}_2$, then $l_p(\overline{\mathbf{F}}_1; 1) < l_p(\overline{\mathbf{F}}_2; 1)$ and $e'_p < e''_p$ for a certain p $(1 \le p \le n)$ by Theorems 1.2′ and 1.4. Thus, $\overline{\mathbf{E}}_1 < \overline{\mathbf{E}}_2$ and $(\overline{\mathbf{E}}_1, \overline{\mathbf{F}}_1) < (\overline{\mathbf{E}}_2, \overline{\mathbf{F}}_2)$, which completes the proof.

As a corollary to the lemma, we have the following proposition.

Lemma 2.3. *The sets of extremal vector pairs and matrices,* $\overline{E}^{n,m}$ *and* $\overline{E}_{n,m}$, *are isomorphic as partially ordered with the order relations "≤" and "⊆" and the isomorphism is defined as the mapping* $\varphi : \overline{E}^{n,m} \to \overline{E}_{n,m}$, *where* $\varphi\left((\overline{\mathbf{E}}, \overline{\mathbf{F}})\right) = \overline{H}(\overline{\mathbf{E}}, \overline{\mathbf{F}})$.

For vectors of equal dimensions, we introduce the operations of union, "∨", and intersection, "∧". Suppose $\mathbf{A} = (a_1, \ldots, a_n)$, $\mathbf{B} = (b_1, \ldots, b_n) \in R^n$. Define $\mathbf{A} \vee \mathbf{B} = \mathbf{C}' = (c'_1, \ldots, c'_n)$ and $\mathbf{A} \wedge \mathbf{B} = \mathbf{C}'' = (c''_1, \ldots, c''_n)$ as the vectors from R^n whose components satisfy the conditions

$$c'_i = \max(a_i, b_i), \quad c''_i = \min(a_i, b_i), \quad 1 \le i \le n.$$

It is obvious that these operations on vectors are such that $\mathbf{A} \wedge \mathbf{B} \le \mathbf{A} \vee \mathbf{B}$ and the following algebraic relations are valid:

$$\left.\begin{aligned}\mathbf{A} \vee \mathbf{B} &= \mathbf{B} \vee \mathbf{A},\\ \mathbf{A} \wedge \mathbf{B} &= \mathbf{B} \wedge \mathbf{A};\end{aligned}\right\} \quad \text{(commutative law)}$$

$$\left.\begin{aligned}\mathbf{A} \vee (\mathbf{B} \vee \mathbf{C}) &= (\mathbf{A} \vee \mathbf{B}) \vee \mathbf{C},\\ \mathbf{A} \wedge (\mathbf{B} \wedge \mathbf{C}) &= (\mathbf{A} \wedge \mathbf{B}) \wedge \mathbf{C};\end{aligned}\right\} \quad \text{(associative law)}$$

$$\left.\begin{aligned}\mathbf{A} \vee (\mathbf{A} \wedge \mathbf{B}) &= \mathbf{A},\\ \mathbf{A} \wedge (\mathbf{A} \vee \mathbf{B}) &= \mathbf{A}.\end{aligned}\right\} \quad \text{(absorptive law)}$$

It is also easy to verify the following distributive laws:

$$(\mathbf{A} \vee \mathbf{B}) \wedge \mathbf{C} = (\mathbf{A} \wedge \mathbf{C}) \vee (\mathbf{B} \wedge \mathbf{C}),$$
$$(\mathbf{A} \wedge \mathbf{B}) \vee \mathbf{C} = (\mathbf{A} \vee \mathbf{C}) \wedge (\mathbf{B} \vee \mathbf{C}).$$

We now extend the operations on vectors to vector pairs and introduce the union "∪" and intersection "∩" of matrices of equal dimensions. Consider

n-by-m vector pairs $(\mathbf{A}, \mathbf{B})$ and $(\mathbf{C}, \mathbf{D})$ and matrices X and Y. The union and intersection of the pairs $(\mathbf{A}, \mathbf{B})$ and $(\mathbf{C}, \mathbf{D})$ are the vector pairs $(\mathbf{A}, \mathbf{B}) \vee (\mathbf{C}, \mathbf{D})$ and $(\mathbf{A}, \mathbf{B}) \wedge (\mathbf{C}, \mathbf{D})$ defined, respectively, as

$$(\mathbf{A}, \mathbf{B}) \vee (\mathbf{C}, \mathbf{D}) = (\mathbf{A} \vee \mathbf{C}, \mathbf{B} \vee \mathbf{D}),$$
$$(\mathbf{A}, \mathbf{B}) \wedge (\mathbf{C}, \mathbf{D}) = (\mathbf{A} \wedge \mathbf{C}, \mathbf{B} \wedge \mathbf{D}).$$

The union and intersection of matrices $X = (x_{ij})$ and $Y = (y_{ij})$ are the matrices $X \cup Y$ and $X \cap Y$ defined, respectively, as

$$X \cup Y = \left(\max(x_{ij}, y_{ij})\right), \quad X \cap Y = \left(\min(x_{ij}, y_{ij})\right), \quad 1 \le i \le n, \quad 1 \le j \le m.$$

It is obvious that $(\mathbf{A}, \mathbf{B}) \wedge (\mathbf{C}, \mathbf{D}) \le (\mathbf{A}, \mathbf{B}) \vee (\mathbf{C}, \mathbf{D})$ and $X \cap Y \le X \cup Y$. It is also easy to check that the commutative, associative, absorptive, and distributive laws apply to arbitrary vector pairs and matrices of equal dimensions.

Consider the set $\overline{E}^{n,m}$ of extremal vector pairs, with the operations "$\vee$" and "$\wedge$" on pairs, and the set $\overline{E}_{n,m}$ of extremal matrices, with operations "$\cup$" and "$\cap$" on matrices. First, we demonstrate that the sets $\overline{E}^{n,m}$ and $\overline{E}_{n,m}$ are closed under union and intersection.

Lemma 2.4. *If* $(\overline{\mathbf{E}}_1, \overline{\mathbf{F}}_1)$, $(\overline{\mathbf{E}}_2, \overline{\mathbf{F}}_2) \in \overline{E}^{n,m}$, *then* $(\overline{\mathbf{E}}_1, \overline{\mathbf{F}}_1) \vee (\overline{\mathbf{E}}_2, \overline{\mathbf{F}}_2)$, $(\overline{\mathbf{E}}_1, \overline{\mathbf{F}}_1) \wedge (\overline{\mathbf{E}}_2, \overline{\mathbf{F}}_2) \in \overline{E}^{n,m}$.

Proof. For the vectors mentioned in the statement of the lemma, we have $\overline{\mathbf{E}}_1 = (e'_1, \ldots, e'_n)$, $\overline{\mathbf{F}}_1 = (f'_1, \ldots, f'_m)$, $\overline{\mathbf{E}}_2 = (e''_1, \ldots, e''_n)$, $\overline{\mathbf{F}}_2 = (f''_1, \ldots, f''_m)$, $(\overline{\mathbf{E}}_1, \overline{\mathbf{F}}_1) \vee (\overline{\mathbf{E}}_2, \overline{\mathbf{F}}_2) = (\mathbf{A}', \mathbf{B}')$, and $(\overline{\mathbf{E}}_1, \overline{\mathbf{F}}_1) \wedge (\overline{\mathbf{E}}_2, \overline{\mathbf{F}}_2) = (\mathbf{A}'', \mathbf{B}'')$, where $\mathbf{A}' = (a'_1, \ldots, a'_n)$, $\mathbf{B}' = (b'_1, \ldots, b'_m)$, $\mathbf{A}'' = (a''_1, \ldots, a''_n)$, $\mathbf{B}'' = (b''_1, \ldots, b''_m)$, and

$$a'_i = \max(e'_i, e''_i), \quad a''_i = \min(e'_i, e''_i), \quad 1 \le i \le n,$$
$$b'_j = \max(f'_j, f''_j), \quad b''_j = \min(f'_j, f''_j), \quad 1 \le j \le m.$$

It is clear that $\mathbf{A}'$, $\mathbf{A}'' \in \overline{Z}^n_+$ and $\mathbf{B}'$, $\mathbf{B}'' \in \overline{Z}^m_+$. Let us show that (1.6) holds for both $(\mathbf{A}', \mathbf{B}')$ and $(\mathbf{A}'', \mathbf{B}'')$. The right-hand side of (1.6) is expressed for $\overline{\mathbf{F}}_1$ and $\overline{\mathbf{F}}_2$ as

$$l_i(\overline{\mathbf{F}}_1; 1) = \begin{cases} \max\{j : f'_j \ge i\}, & f'_1 \ge i \\ 0, & f'_1 < i, \end{cases}$$

$$l_i(\overline{\mathbf{F}}_2; 1) = \begin{cases} \max\{j : f''_j \ge i\}, & f''_1 \ge i \\ 0, & f''_1 < i. \end{cases}$$

Since $(\overline{\mathbf{E}}_1, \overline{\mathbf{F}}_1)$ and $(\overline{\mathbf{E}}_2, \overline{\mathbf{F}}_2)$ are extremal vector pairs, (1.6) applies to the components of the vectors making up these pairs. Therefore, we can write the following relations for the components of $\mathbf{A}'$, $\mathbf{B}'$, $\mathbf{A}''$, and $\mathbf{B}''$:

$$a_i' = \max(e_i', e_i'') = \max\left(l_i(\overline{\mathbf{F}}_1; 1), l_i(\overline{\mathbf{F}}_2; 1)\right) =$$

$$\max\left\{\begin{pmatrix} \max\{j : f_j' \geq i\}, & f_1' \geq i \\ 0, \quad f_1' < i & \end{pmatrix}, \begin{pmatrix} \max\{j : f_j'' \geq i\}, & f_1'' \geq i \\ 0, \quad f_1'' < i & \end{pmatrix}\right\} =$$

$$\begin{cases} \max\left\{j : \max(f_j', f_j'') \geq i\right\}, & \max(f_1', f_1'') \geq i, \\ 0, \quad \max(f_1', f_1'') < i & \end{cases} =$$

$$\begin{cases} \max\{j : b_j' \geq i\}, & b_1' \geq i, \\ 0, \quad b_1' < i & \end{cases} = l_i(\mathbf{B}'; 1), \quad 1 \leq i \leq n;$$

$$a_i'' = \min(e_i', e_i'') = \min\left(l_i(\overline{\mathbf{F}}_1; 1), l_i(\overline{\mathbf{F}}_2; 1)\right) =$$

$$\min\left\{\begin{pmatrix} \max\{j : f_j' \geq i\}, & f_1' \geq i \\ 0, \quad f_1' < i & \end{pmatrix}, \begin{pmatrix} \max\{j : f_j'' \geq i\}, & f_1'' \geq i \\ 0, \quad f_1'' < i & \end{pmatrix}\right\} =$$

$$\begin{cases} \max\left\{j : \min(f_j', f_j'') \geq i\right\}, & \min(f_1', f_1'') \geq i, \\ 0, \quad \min(f_1', f_1'') < i & \end{cases} =$$

$$\begin{cases} \max\{j : b_j'' \geq i\}, & b_1'' \geq i, \\ 0, \quad b_1'' < i & \end{cases} = l_i(\mathbf{B}''; 1), \quad 1 \leq i \leq n.$$

Thus, $a_i' = l_i(\mathbf{B}'; 1)$ and $a_i'' = l_i(\mathbf{B}''; 1)$ for any i $(1 \leq i \leq n)$. By Theorem 1.4, $(\mathbf{A}', \mathbf{B}')$ and $(\mathbf{A}'', \mathbf{B}'')$ make up extremal vector pairs, and $(\overline{\mathbf{E}}_1, \overline{\mathbf{F}}_1) \vee (\overline{\mathbf{E}}_2, \overline{\mathbf{F}}_2)$, $(\overline{\mathbf{E}}_1, \overline{\mathbf{F}}_1) \wedge (\overline{\mathbf{E}}_2, \overline{\mathbf{F}}_2) \in \overline{E}^{n,m}$.

Lemma 2.5. *If $\overline{H}_1$, $\overline{H}_2 \in \overline{E}_{n,m}$, then $\overline{H}_1 \cup \overline{H}_2$, $\overline{H}_1 \cap \overline{H}_2 \in \overline{E}_{n,m}$.*

Proof. Let $\overline{H}_1 = (h_{ij}')$ and $\overline{H}_2 = (h_{ij}'')$, where $1 \leq i \leq n$ and $1 \leq j \leq m$. Define

$$\overline{H}_1 \cap \overline{H}_2 = \left(\max(h_{ij}', h_{ij}'')\right) = X' = (x_{ij}')$$

and

$$\overline{H}_1 \cup \overline{H}_2 = \left(\min(h_{ij}', h_{ij}'')\right) = X'' = (x_{ij}''),$$

where $1 \leq i \leq n$ and $1 \leq j \leq m$. It is obvious that both X' and X'' are matrices of zeros and ones. We now apply Definition 1.5 to the matrices X' and X''. Let x_{ij}' and x_{ij}'' be arbitrary elements of these matrices. Since $\overline{H}_1$ and $\overline{H}_2$ are extremal matrices, the conditions of Definition 1.5 apply to their elements. Thus, the elements x_{ij}' and x_{ij}'' satisfy the following conditions:

if $x_{ij}' = \max(h_{ij}', h_{ij}'') = 1$, then $x_{pq}' = \max(h_{pq}', h_{pq}'') = 1$ for $p \leq i$ and $q \leq j$;

if $x'_{ij} = \max(h'_{ij}, h''_{ij}) = 0$, then $x'_{pq} = \max(h'_{pq}, h''_{pq}) = 0$ for $p \geq i$ and $q \geq j$;

if $x''_{ij} = \min(h'_{ij}, h''_{ij}) = 1$, then $x''_{pq} = \min(h'_{pq}, h''_{pq}) = 1$ for $p \leq i$ and $q \leq j$;

and if $x''_{ij} = \min(h'_{ij}, h''_{ij}) = 0$, then $x''_{pq} = \min(h'_{pq}, h''_{pq}) = 0$ for $p \geq i$ and $q \geq j$.

By Definition 1.5, X' and X'' are extremal matrices. Therefore, $\overline{H}_1 \cup \overline{H}_2$, $\overline{H}_1 \cap \overline{H}_2 \in \overline{E}_{n,m}$.

2.2. Isomorphic Distributive Lattices of Fundamental Systems of Vector Pairs and Matrices

Since the sets $\overline{E}^{n,m}$ and $\overline{E}_{n,m}$ are closed under union and intersection, the operations "$\vee$", "$\wedge$", "$\cup$", and "$\cap$" on these sets obey the commutative, associative, absorptive, and distributive laws:

$$(\overline{\mathbf{E}}_1, \overline{\mathbf{F}}_1) \vee (\overline{\mathbf{E}}_2, \overline{\mathbf{F}}_2) = (\overline{\mathbf{E}}_2, \overline{\mathbf{F}}_2) \vee (\overline{\mathbf{E}}_1, \overline{\mathbf{F}}_1),$$

$$(\overline{\mathbf{E}}_1, \overline{\mathbf{F}}_1) \wedge (\overline{\mathbf{E}}_2, \overline{\mathbf{F}}_2) = (\overline{\mathbf{E}}_2, \overline{\mathbf{F}}_2) \wedge (\overline{\mathbf{E}}_1, \overline{\mathbf{F}}_1),$$

$$\overline{H}_1 \cup \overline{H}_2 = \overline{H}_2 \cup \overline{H}_1, \quad \overline{H}_1 \cap \overline{H}_2 = \overline{H}_2 \cap \overline{H}_1,$$

$$(\overline{\mathbf{E}}_1, \overline{\mathbf{F}}_1) \vee ((\overline{\mathbf{E}}_2, \overline{\mathbf{F}}_2) \vee (\overline{\mathbf{E}}_3, \overline{\mathbf{F}}_3)) = ((\overline{\mathbf{E}}_1, \overline{\mathbf{F}}_1) \vee (\overline{\mathbf{E}}_2, \overline{\mathbf{F}}_2)) \vee (\overline{\mathbf{E}}_3, \overline{\mathbf{F}}_3),$$

$$(\overline{\mathbf{E}}_1, \overline{\mathbf{F}}_1) \wedge ((\overline{\mathbf{E}}_2, \overline{\mathbf{F}}_2) \wedge (\overline{\mathbf{E}}_3, \overline{\mathbf{F}}_3)) = ((\overline{\mathbf{E}}_1, \overline{\mathbf{F}}_1) \wedge (\overline{\mathbf{E}}_2, \overline{\mathbf{F}}_2)) \wedge (\overline{\mathbf{E}}_3, \overline{\mathbf{F}}_3),$$

$$\overline{H}_1 \cup (\overline{H}_2 \cup \overline{H}_3) = (\overline{H}_1 \cup \overline{H}_2) \cup \overline{H}_3, \quad \overline{H}_1 \cap (\overline{H}_2 \cap \overline{H}_3) = (\overline{H}_1 \cap \overline{H}_2) \cap \overline{H}_3,$$

$$(\overline{\mathbf{E}}_1, \overline{\mathbf{F}}_1) \vee ((\overline{\mathbf{E}}_1, \overline{\mathbf{F}}_1) \wedge (\overline{\mathbf{E}}_2, \overline{\mathbf{F}}_2)) = (\overline{\mathbf{E}}_1, \overline{\mathbf{F}}_1), \quad \overline{H}_1 \cup (\overline{H}_1 \cap \overline{H}_2) = \overline{H}_1,$$

$$(\overline{\mathbf{E}}_1, \overline{\mathbf{F}}_1) \wedge ((\overline{\mathbf{E}}_1, \overline{\mathbf{F}}_1) \vee (\overline{\mathbf{E}}_2, \overline{\mathbf{F}}_2)) = (\overline{\mathbf{E}}_1, \overline{\mathbf{F}}_1), \quad \overline{H}_1 \cap (\overline{H}_1 \cup \overline{H}_2) = \overline{H}_1,$$

$$((\overline{\mathbf{E}}_1, \overline{\mathbf{F}}_1) \vee (\overline{\mathbf{E}}_2, \overline{\mathbf{F}}_2)) \wedge (\overline{\mathbf{E}}_3, \overline{\mathbf{F}}_3) = ((\overline{\mathbf{E}}_1, \overline{\mathbf{F}}_1) \wedge (\overline{\mathbf{E}}_3, \overline{\mathbf{F}}_3)) \vee ((\overline{\mathbf{E}}_2, \overline{\mathbf{F}}_2) \wedge (\overline{\mathbf{E}}_3, \overline{\mathbf{F}}_3)),$$

$$((\overline{\mathbf{E}}_1, \overline{\mathbf{F}}_1) \wedge (\overline{\mathbf{E}}_2, \overline{\mathbf{F}}_2)) \vee (\overline{\mathbf{E}}_3, \overline{\mathbf{F}}_3) = ((\overline{\mathbf{E}}_1, \overline{\mathbf{F}}_1) \vee (\overline{\mathbf{E}}_3, \overline{\mathbf{F}}_3)) \wedge ((\overline{\mathbf{E}}_2, \overline{\mathbf{F}}_2) \vee (\overline{\mathbf{E}}_3, \overline{\mathbf{F}}_3)),$$

$$(\overline{H}_1 \cup \overline{H}_2) \cap \overline{H}_3 = (\overline{H}_1 \cap \overline{H}_3) \cup (\overline{H}_2 \cap \overline{H}_3),$$

$$(\overline{H}_1 \cap \overline{H}_2) \cup \overline{H}_3 = (\overline{H}_1 \cup \overline{H}_3) \cap (\overline{H}_2 \cup \overline{H}_3),$$

where $(\overline{\mathbf{E}}_1, \overline{\mathbf{F}}_1)$, $(\overline{\mathbf{E}}_2, \overline{\mathbf{F}}_2)$, $(\overline{\mathbf{E}}_3, \overline{\mathbf{F}}_3) \in \overline{E}^{n,m}$; and $\overline{H}_1$, $\overline{H}_2$, $\overline{H}_3 \in \overline{E}_{n,m}$.

The union "$\vee$" and intersection "$\wedge$" of extremal vector pairs from $\overline{E}^{n,m}$, are naturally related to the union "$\cup$" and intersection "$\cap$" of extremal matrices from $\overline{E}_{n,m}$.

Lemma 2.6. *For arbitrary extremal vector pairs* $(\overline{\mathbf{E}}_1, \overline{\mathbf{F}}_1)$ *and* $(\overline{\mathbf{E}}_2, \overline{\mathbf{F}}_2)$ *from* $\overline{E}^{n,m}$ *and the corresponding extremal matrices* $\overline{H}_1(\overline{\mathbf{E}}_1, \overline{\mathbf{F}}_1)$ *and*

$\overline{H}_2(\overline{\mathbf{E}}_2, \overline{\mathbf{F}}_2)$ *from* $\overline{E}_{n,m}$ *it holds that*

$$\overline{H}_1(\overline{\mathbf{E}}_1, \overline{\mathbf{F}}_1) \cup \overline{H}_2(\overline{\mathbf{E}}_2, \overline{\mathbf{F}}_2) = \overline{H}'\left((\overline{\mathbf{E}}_1, \overline{\mathbf{F}}_1) \vee (\overline{\mathbf{E}}_2, \overline{\mathbf{F}}_2)\right),$$
$$\overline{H}_1(\overline{\mathbf{E}}_1, \overline{\mathbf{F}}_1) \cap \overline{H}_2(\overline{\mathbf{E}}_2, \overline{\mathbf{F}}_2) = \overline{H}''\left((\overline{\mathbf{E}}_1, \overline{\mathbf{F}}_1) \wedge (\overline{\mathbf{E}}_2, \overline{\mathbf{F}}_2)\right).$$

Proof. The unions and intersections of vector pairs and matrices are extremal by Lemmas 2.4 and 2.5. Let

$$\overline{H}_1(\overline{\mathbf{E}}_1, \overline{\mathbf{F}}_1) \cup \overline{H}_2(\overline{\mathbf{E}}_2, \overline{\mathbf{F}}_2) = (h'_{ij}),$$
$$\overline{H}'\left((\overline{\mathbf{E}}_1, \overline{\mathbf{F}}_1) \vee (\overline{\mathbf{E}}_2, \overline{\mathbf{F}}_2)\right) = (\tilde{h}'_{ij}), \quad 1 \le i \le n, \quad 1 \le j \le m,$$
$$\overline{H}_1(\overline{\mathbf{E}}_1, \overline{\mathbf{F}}_1) \cap \overline{H}_2(\overline{\mathbf{E}}_2, \overline{\mathbf{F}}_2) = (h''_{ij}),$$
$$\overline{H}''\left((\overline{\mathbf{E}}_1, \overline{\mathbf{F}}_1) \wedge (\overline{\mathbf{E}}_2, \overline{\mathbf{F}}_2)\right) = (\tilde{h}''_{ij}), \quad 1 \le i \le n, \quad 1 \le j \le m.$$

Applying relation (1.5) from Theorem 1.2′ to the elements of these matrices, we obtain

$$h'_{ij} = \begin{cases} 1, & 1 \le i \le n, \quad 1 \le j \le \max\left(l_i(\overline{\mathbf{F}}_1; 1), l_i(\overline{\mathbf{F}}_2; 1)\right), \\ 0, & 1 \le i \le n, \quad \max\left(l_i(\overline{\mathbf{F}}_1; 1), l_i(\overline{\mathbf{F}}_2; 1)\right) < j \le m, \end{cases}$$

$$\tilde{h}'_{ij} = \begin{cases} 1, & 1 \le i \le n, \quad 1 \le j \le l_i(\overline{\mathbf{F}}_1 \vee \overline{\mathbf{F}}_2; 1), \\ 0, & 1 \le i \le n, \quad l_i(\overline{\mathbf{F}}_1 \vee \overline{\mathbf{F}}_2; 1) < j \le m, \end{cases}$$

$$h''_{ij} = \begin{cases} 1, & 1 \le i \le n, \quad 1 \le j \le \min\left(l_i(\overline{\mathbf{F}}_1; 1), l_i(\overline{\mathbf{F}}_2; 1)\right), \\ 0, & 1 \le i \le n, \quad \min\left(l_i(\overline{\mathbf{F}}_1; 1), l_i(\overline{\mathbf{F}}_2; 1)\right) < j \le m, \end{cases}$$

$$\tilde{h}''_{ij} = \begin{cases} 1, & 1 \le i \le n, \quad 1 \le j \le l_i(\overline{\mathbf{F}}_1 \wedge \overline{\mathbf{F}}_2; 1), \\ 0, & 1 \le i \le n, \quad l_i(\overline{\mathbf{F}}_1 \wedge \overline{\mathbf{F}}_2; 1) < j \le m. \end{cases}$$

In proving Lemma 2.4, we showed that

$$\max\left(l_i(\overline{\mathbf{F}}_1; 1), l_i(\overline{\mathbf{F}}_2; 1)\right) = l_i(\overline{\mathbf{F}}_1 \vee \overline{\mathbf{F}}_2; 1),$$
$$\min\left(l_i(\overline{\mathbf{F}}_1; 1), l_i(\overline{\mathbf{F}}_2; 1)\right) = l_i(\overline{\mathbf{F}}_1 \wedge \overline{\mathbf{F}}_2; 1).$$

Hence, $h'_{ij} = \tilde{h}'_{ij}$ and $h''_{ij} = \tilde{h}''_{ij}$ for any i and j $(1 \le i \le n,\ 1 \le j \le m)$, which proves the lemma.

Lemmas 2.2–2.6 define isomorphic distributive lattices on the fundamental systems of vector pairs and matrices, $\overline{E}^{n,m}$ and $\overline{E}_{n,m}$.

Theorem 2.1. *The partially ordered sets* $\overline{E}^{n,m}$ *and* $\overline{E}_{n,m}$ *of extremal vector pairs and extremal matrices on which union and intersection* ("$\vee$",

"$\wedge$" and "$\cup$", "$\cap$", respectively) are isomorphic distributive lattices, with isomorphism defined as $\varphi : \overline{E}^{n,m} \to \overline{E}_{n,m}$, where $\varphi\left((\overline{\mathbf{E}}, \overline{\mathbf{F}})\right) = \overline{H}(\overline{\mathbf{E}}, \overline{\mathbf{F}})$.

2.3. Metrics on Sets of Extremal Vector Pairs and Matrices

For arbitrary vectors $\mathbf{A} = (a_1, \dots, a_n) \in R^n$ and $\mathbf{B} = (b_1, \dots, b_n) \in R^n$, define

$$\rho(\mathbf{A}, \mathbf{B}) = \sum_{i=1}^{n} |a_i - b_i|.$$

Let $(\mathbf{A}, \mathbf{B})$, $(\mathbf{C}, \mathbf{D})$ and X, Y be arbitrary n-by-m vector pairs and matrices, respectively. For $\mathbf{A} = (a_1, \dots, a_n)$, $\mathbf{C} = (c_1, \dots, c_n) \in R^n$, $\mathbf{B} = (b_1, \dots, b_m)$, $\mathbf{D} = (d_1, \dots, d_m) \in R^m$, and $X = (x_{ij})$, $Y = (y_{ij})$ $(1 \le i \le n$, $1 \le j \le m)$, define, respectively,

$$\begin{aligned} \rho\left((\mathbf{A}, \mathbf{B}), (\mathbf{C}, \mathbf{D})\right) &= \left(\rho(\mathbf{A}, \mathbf{C}) + \rho(\mathbf{B}, \mathbf{D})\right)/2, \\ r(X, Y) &= \sum_{i=1}^{n} \sum_{j=1}^{m} |x_{ij} - y_{ij}|. \end{aligned} \tag{2.1}$$

Consider the functions $\rho\left((\mathbf{A}, \mathbf{B}), (\mathbf{C}, \mathbf{D})\right)$ and $r(X, Y)$ as defined on the fundamental systems of vector pairs and matrices, $\overline{E}^{n,m}$ and $\overline{E}_{n,m}$.

The function $\rho\left((\overline{\mathbf{E}}_1, \overline{\mathbf{F}}_1), (\overline{\mathbf{E}}_2, \overline{\mathbf{F}}_2)\right)$ of extremal vector pairs is completely defined by their first or second vectors. This is established by the following proposition.

Lemma 2.7. *For extremal vector pairs* $(\overline{\mathbf{E}}_1, \overline{\mathbf{F}}_1)$, $(\overline{\mathbf{E}}_2, \overline{\mathbf{F}}_2) \in \overline{E}^{n,m}$, *it holds that*

$$\rho\left((\overline{\mathbf{E}}_1, \overline{\mathbf{F}}_1), (\overline{\mathbf{E}}_2, \overline{\mathbf{F}}_2)\right) = \rho(\overline{\mathbf{E}}_1, \overline{\mathbf{E}}_2) = \rho(\overline{\mathbf{F}}_1, \overline{\mathbf{F}}_2).$$

Proof. Let $\overline{\mathbf{E}}_1 = (e'_1, \dots, e'_n)$, $\overline{\mathbf{F}}_1 = (f'_1, \dots, f'_m)$, $\overline{\mathbf{E}}_2 = (e''_1, \dots, e''_n)$, and $\overline{\mathbf{F}}_2 = (f''_1, \dots, f''_m)$. By Lemma 2.4, $(\overline{\mathbf{E}}_1, \overline{\mathbf{F}}_1) \vee (\overline{\mathbf{E}}_2, \overline{\mathbf{F}}_2) = (\overline{\mathbf{E}}_3, \overline{\mathbf{F}}_3)$, $(\overline{\mathbf{E}}_1, \overline{\mathbf{F}}_1) \wedge (\overline{\mathbf{E}}_2, \overline{\mathbf{F}}_2) = (\overline{\mathbf{E}}_4, \overline{\mathbf{F}}_4) \in \overline{E}^{n,m}$. We henceforth refer to the vectors thus constructed as $\overline{\mathbf{E}}_3 = (e'''_1, \dots, e'''_n)$, $\overline{\mathbf{F}}_3 = (f'''_1, \dots, f'''_m)$, $\overline{\mathbf{E}}_4 = (e^{IV}_1, \dots, e^{IV}_n)$, and $\overline{\mathbf{F}}_4 = (f^{IV}_1, \dots, f^{IV}_m)$. Invoking the definitions of the operations "$\vee$" and "$\wedge$", we express the components of $\overline{\mathbf{E}}_3$, $\overline{\mathbf{F}}_3$, $\overline{\mathbf{E}}_4$, and $\overline{\mathbf{F}}_4$ in terms of the components of $\overline{\mathbf{E}}_1$, $\overline{\mathbf{F}}_1$, $\overline{\mathbf{E}}_2$, and $\overline{\mathbf{F}}_2$ as follows:

$$e'''_i = \max(e'_i, e''_i), \quad e^{IV}_i = \min(e'_i, e''_i), \quad 1 \le i \le n,$$

$$f'''_j = \max(f'_j, f''_j), \quad f^{IV}_j = \min(f'_j, f''_j), \quad 1 \le j \le m.$$

Consider the function $\rho(\mathbf{A}, \mathbf{B})$ of vectors mentioned in the statement of the lemma:

$$\rho(\overline{\mathbf{E}}_1, \overline{\mathbf{E}}_2) = \sum_{i=1}^{n} |e_i' - e_i''| = \sum_{i=1}^{n} \left(\max(e_i', e_i'') - \min(e_i', e_i'')\right) =$$

$$\sum_{i=1}^{n} \max(e_i', e_i'') - \sum_{i=1}^{n} \min(e_i', e_i'') = \sum_{i=1}^{n} e_i''' - \sum_{i=1}^{n} e_i^{IV} = \rho(\overline{\mathbf{E}}_3, \overline{\mathbf{E}}_4),$$

$$\rho(\overline{\mathbf{F}}_1, \overline{\mathbf{F}}_2) = \sum_{j=1}^{m} |f_j' - f_j''| = \sum_{j=1}^{m} \left(\max(f_j', f_j'') - \min(f_j', f_j'')\right) =$$

$$\sum_{j=1}^{m} \max(f_j', f_j'') - \sum_{j=1}^{m} \min(f_j', f_j'') = \sum_{j=1}^{m} f_j''' - \sum_{j=1}^{m} f_j^{IV} = \rho(\overline{\mathbf{F}}_3, \overline{\mathbf{F}}_4),$$

because it is obvious that $\overline{\mathbf{E}}_4 \leq \overline{\mathbf{E}}_3$ and $\overline{\mathbf{F}}_4 \leq \overline{\mathbf{F}}_3$. By Lemma 1.6,

$$\sum_{i=1}^{n} e_i''' = \sum_{j=1}^{m} f_j''', \quad \sum_{i=1}^{n} e_i^{IV} = \sum_{j=1}^{m} f_j^{IV};$$

therefor $\rho(\overline{\mathbf{E}}_1, \overline{\mathbf{E}}_2) = \rho(\overline{\mathbf{E}}_3, \overline{\mathbf{E}}_4) = \rho(\overline{\mathbf{F}}_3, \overline{\mathbf{F}}_4) = \rho(\overline{\mathbf{F}}_1, \overline{\mathbf{F}}_2)$. The assertion of the lemma follows from the equality $\rho(\overline{\mathbf{E}}_1, \overline{\mathbf{E}}_2) = \rho(\overline{\mathbf{F}}_1, \overline{\mathbf{F}}_2)$.

It is clear that the functions defined by (2.1) satisfy the following relations:

(a) $\rho\left((\mathbf{A}, \mathbf{B}), (\mathbf{C}, \mathbf{D})\right) = 0$ and $r(X, Y) = 0$ if and only if $(\mathbf{A}, \mathbf{B}) = (\mathbf{C}, \mathbf{D})$ and $X = Y$, respectively (the identity axiom);

(b) $\rho\left((\mathbf{A}, \mathbf{B}), (\mathbf{C}, \mathbf{D})\right) = \rho\left((\mathbf{C}, \mathbf{D}), (\mathbf{A}, \mathbf{B})\right)$, $r(X, Y) = r(Y, X)$ (the symmetry axiom);

(c) for every three vector pairs $(\mathbf{A}_1, \mathbf{B}_1)$, $(\mathbf{A}_2, \mathbf{B}_2)$, and $(\mathbf{A}_3, \mathbf{B}_3)$ and three matrices X_1, X_2, and X_3, having equal dimensions,

$$\rho\left((\mathbf{A}_1, \mathbf{B}_1), (\mathbf{A}_2, \mathbf{B}_2)\right) + \rho\left((\mathbf{A}_2, \mathbf{B}_2), (\mathbf{A}_3, \mathbf{B}_3)\right) \geq \rho\left((\mathbf{A}_1, \mathbf{B}_1), (\mathbf{A}_3, \mathbf{B}_3)\right),$$

$r(X_1, X_2) + r(X_2, X_3) \geq r(X_1, X_3)$ (the triangle axiom).

Therefore, the functions defined by (2.1) are metrics on the sets of vector pairs and matrices having equal dimensions. Accordingly, $\rho\left((\overline{\mathbf{E}}_1, \overline{\mathbf{F}}_1), (\overline{\mathbf{E}}_2, \overline{\mathbf{F}}_2)\right)$ and $r(\overline{H}_1, \overline{H}_2)$ define metric spaces on the sets of extremal vector pairs and extremal matrices, $\overline{E}^{n,m}$ and $\overline{E}_{n,m}$.

Lemma 2.8. *For every vector pair* $(\overline{\mathbf{E}}_1, \overline{\mathbf{F}}_1)$, $(\overline{\mathbf{E}}_2, \overline{\mathbf{F}}_2)$ *from* $\overline{E}^{n,m}$, *it holds that*

$$\rho\left((\overline{\mathbf{E}}_1, \overline{\mathbf{F}}_1), (\overline{\mathbf{E}}_2, \overline{\mathbf{F}}_2)\right) = r\left(\overline{H}_1(\overline{\mathbf{E}}_1, \overline{\mathbf{F}}_1), \overline{H}_2(\overline{\mathbf{E}}_2, \overline{\mathbf{F}}_2)\right).$$

Proof. We use here the notation for vectors introduced in the proof of Lemma 2.7. Let

$$\overline{H}_1 = \overline{H}_1(\overline{\mathbf{E}}_1, \overline{\mathbf{F}}_1) = (h'_{ij}), \quad \overline{H}_2 = \overline{H}_2(\overline{\mathbf{E}}_2, \overline{\mathbf{F}}_2) = (h''_{ij}),$$

$$\overline{H}_3 = \overline{H}_1 \cup \overline{H}_2 = (h'''_{ij}), \quad \overline{H}_4 = \overline{H}_1 \cap \overline{H}_2 = (h^{IV}_{ij}),$$

where $1 \leq i \leq n$ and $1 \leq j \leq m$.

Applying Lemma 2.7 to the first metric in (2.1), we obtain

$$\rho\left((\overline{\mathbf{E}}_1, \overline{\mathbf{F}}_1), (\overline{\mathbf{E}}_2, \overline{\mathbf{F}}_2)\right) = \rho\left((\overline{\mathbf{E}}_1, \overline{\mathbf{F}}_1) \vee (\overline{\mathbf{E}}_2, \overline{\mathbf{F}}_2), (\overline{\mathbf{E}}_1, \overline{\mathbf{F}}_1) \wedge (\overline{\mathbf{E}}_2, \overline{\mathbf{F}}_2)\right) =$$

$$\rho\left((\overline{\mathbf{E}}_3, \overline{\mathbf{F}}_3), (\overline{\mathbf{E}}_4, \overline{\mathbf{F}}_4)\right) = \rho(\overline{\mathbf{E}}_3, \overline{\mathbf{E}}_4) = \sum_{i=1}^{n} e'''_i - \sum_{i=1}^{n} e^{IV}_i.$$

It is clear that the second metric in (2.1) can be expressed as

$$r(\overline{H}_1, \overline{H}_2) = r(\overline{H}_1 \cup \overline{H}_2, \overline{H}_1 \cap \overline{H}_2) = r(\overline{H}_3, \overline{H}_4) =$$

$$\sum_{i=1}^{n} \sum_{j=1}^{m} (h'''_{ij} - h^{IV}_{ij}) = \sum_{j=1}^{m} \sum_{i=1}^{n} (h'''_{ij} - h^{IV}_{ij}).$$

Since

$$\sum_{j=1}^{m} (h'''_{ij} - h^{IV}_{ij}) = \sum_{j=1}^{m} h'''_{ij} - \sum_{j=1}^{m} h^{IV}_{ij} = e'''_i - e^{IV}_i,$$

we write

$$\sum_{i=1}^{n} \sum_{j=1}^{m} (h'''_{ij} - h^{IV}_{ij}) = \sum_{i=1}^{n} e'''_i - \sum_{i=1}^{n} e^{IV}_i$$

and

$$\rho\left((\overline{\mathbf{E}}_1, \overline{\mathbf{F}}_1), (\overline{\mathbf{E}}_2, \overline{\mathbf{F}}_2)\right) = r\left(\overline{H}_1(\overline{\mathbf{E}}_1, \overline{\mathbf{F}}_1), \overline{H}_2(\overline{\mathbf{E}}_2, \overline{\mathbf{F}}_2)\right).$$

2.4. Isometry of the Fundamental Systems of Vector Pairs and Matrices

As a corollary to Lemma 2.8, we have the following proposition.

Theorem 2.2. *The metric spaces $\overline{E}^{n,m}$ and $\overline{E}_{n,m}$ with metrics defined by (2.1) are isometric, and the isometry is defined by the mapping $\varphi : \overline{E}^{n,m} \to \overline{E}_{n,m}$, where $\varphi(\overline{\mathbf{E}}, \overline{\mathbf{F}}) = \overline{H}(\overline{\mathbf{E}}, \overline{\mathbf{F}})$.*

We now enumerate some properties of $\overline{E}^{n,m}$ and $\overline{E}_{n,m}$, treating them as partially ordered sets, lattices, and metric spaces.

Theorem 2.3.

(a) *The number of elements of both* $\overline{E}^{n,m}$ *and* $\overline{E}_{n,m}$ *is* $|\overline{E}^{n,m}| = |\overline{E}_{n,m}| = C^n_{n+m}$.

(b) *The diameter of both* $\overline{E}^{n,m}$ *and* $\overline{E}_{n,m}$ *is equal to* n-by-m.

(c) *For every two extremal vector pairs* $(\overline{\mathbf{E}}_1, \overline{\mathbf{F}}_1)$ *and* $(\overline{\mathbf{E}}_2, \overline{\mathbf{F}}_2)$ *from* $\overline{E}^{n,m}$ *and every two extremal matrices* $\overline{H}_1$ *and* $\overline{H}_2$ *from* $\overline{E}_{n,m}$, *it holds that*

$$\rho\left((\overline{\mathbf{E}}_1, \overline{\mathbf{F}}_1), (\overline{\mathbf{E}}_2, \overline{\mathbf{F}}_2)\right) = \rho\left((\overline{\mathbf{E}}_1, \overline{\mathbf{F}}_1) \vee (\overline{\mathbf{E}}_2, \overline{\mathbf{F}}_2), (\overline{\mathbf{E}}_1, \overline{\mathbf{F}}_1) \wedge (\overline{\mathbf{E}}_2, \overline{\mathbf{F}}_2)\right),$$

$$r(\overline{H}_1, \overline{H}_2) = r(\overline{H}_1 \cup \overline{H}_2, \overline{H}_1 \cap \overline{H}_2).$$

(d) *For every three elements from* $\overline{E}^{n,m}$ *and* $\overline{E}_{n,m}$ *satisfying the order relations* $(\overline{\mathbf{E}}_1, \overline{\mathbf{F}}_1) \leq (\overline{\mathbf{E}}_2, \overline{\mathbf{F}}_2) \leq (\overline{\mathbf{E}}_3, \overline{\mathbf{F}}_3)$ *and* $\overline{H}_1 \subseteq \overline{H}_2 \subseteq \overline{H}_3$, *respectively, the triangle inequality (the triangle axiom) reduces to a strict equality:*

$$\rho\left((\overline{\mathbf{E}}_1, \overline{\mathbf{F}}_1), (\overline{\mathbf{E}}_2, \overline{\mathbf{F}}_2)\right) + \rho\left((\overline{\mathbf{E}}_2, \overline{\mathbf{F}}_2), (\overline{\mathbf{E}}_3, \overline{\mathbf{F}}_3)\right) = \rho\left((\overline{\mathbf{E}}_1, \overline{\mathbf{F}}_1), (\overline{\mathbf{E}}_3, \overline{\mathbf{F}}_3)\right),$$

$$r(\overline{H}_1, \overline{H}_2) + r(\overline{H}_2, \overline{H}_3) = r(\overline{H}_1, \overline{H}_3).$$

(e) *Suppose* $(\overline{\mathbf{E}}_1, \overline{\mathbf{F}}_1)$, $(\overline{\mathbf{E}}_2, \overline{\mathbf{F}}_2) \in \overline{E}^{n,m}$ *and* $\overline{H}_1$, $\overline{H}_2 \in \overline{E}_{n,m}$. *If the metrics defined by* (2.1) *are such that* $\rho\left((\overline{\mathbf{E}}_1, \overline{\mathbf{F}}_1), (\overline{\mathbf{E}}_2, \overline{\mathbf{F}}_2)\right) = 1$ *and* $r(\overline{H}_1, \overline{H}_2) = 1$, *then the vector pairs satisfy the relation* $(\overline{\mathbf{E}}_1, \overline{\mathbf{F}}_1) < (\overline{\mathbf{E}}_2, \overline{\mathbf{F}}_2)$ *or* $(\overline{\mathbf{E}}_2, \overline{\mathbf{F}}_2) < (\overline{\mathbf{E}}_1, \overline{\mathbf{F}}_1)$, *and the matrices satisfy the relation* $\overline{H}_1 \subset \overline{H}_2$ *or* $\overline{H}_2 \subset \overline{H}_1$.

(f) *Let* $(\overline{\mathbf{E}}', \overline{\mathbf{F}}')$, $(\overline{\mathbf{E}}'', \overline{\mathbf{F}}'')$ *and* $\overline{H}'$, $\overline{H}''$ *be arbitrary elements of* $\overline{E}^{n,m}$ *and* $\overline{E}_{n,m}$, *respectively, such that* $(\overline{\mathbf{E}}', \overline{\mathbf{F}}') < (\overline{\mathbf{E}}'', \overline{\mathbf{F}}'')$ *and* $\overline{H}' \subset \overline{H}''$. *Then, the metrics defined by* (2.1) *determine the differences between these extremal vector pairs and matrices: if*

$$\rho\left((\overline{\mathbf{E}}', \overline{\mathbf{F}}'), (\overline{\mathbf{E}}'', \overline{\mathbf{F}}'')\right) = r(\overline{H}', \overline{H}'') = k \geq 2 \quad (k \in Z),$$

then for

$$\overline{\mathbf{E}}' = \overline{\mathbf{E}}_0, \quad \overline{\mathbf{E}}'' = \overline{\mathbf{E}}_k, \quad \overline{\mathbf{F}}' = \overline{\mathbf{F}}_0, \quad \overline{\mathbf{F}}'' = \overline{\mathbf{F}}_k, \quad \overline{H}' = \overline{H}_0, \quad \overline{H}'' = \overline{H}_k$$

there exist elements $(\overline{\mathbf{E}}_i, \overline{\mathbf{F}}_i)$ *and* $\overline{H}_i$ $(1 \leq i \leq k-1)$ *of* $\overline{E}^{n,m}$ *and* $\overline{E}_{n,m}$, *respectively, such that*

$$\rho\left((\overline{\mathbf{E}}_{i-1}, \overline{\mathbf{F}}_{i-1}), (\overline{\mathbf{E}}_i, \overline{\mathbf{F}}_i)\right) = 1 \ \text{ and } \ r(\overline{H}_{i-1}, \overline{H}_i) = 1, \quad 1 \leq i \leq k.$$

(g) *For arbitrary vector pair and matrix*

$$(\overline{\mathbf{E}},\overline{\mathbf{F}}) = \big((e_1,\dots,e_n),(f_1,\dots,f_m)\big) \in \overline{E}^{n,m} \quad \textit{and} \quad \overline{H} = h_{ij} \in \overline{E}_{n,m},$$

define $(\overline{\mathbf{E}}^1,\overline{\mathbf{F}}^1) = \big((m-e_1,\dots,m-e_n),(n-f_1,\dots,n-f_m)\big)$; $(\tilde{\overline{\mathbf{E}}}^1,\tilde{\overline{\mathbf{F}}}^1) = \big((m-e_n,\dots,m-e_1),(n-f_m,\dots,n-f_1)\big)$; $\overline{\overline{H}}^1 = (h^1_{ij}) = (1-h_{ij})$, *where* $1 \le i \le n$ *and* $1 \le j \le m$ ($\overline{\overline{H}}^1$ *is the complement of* $\overline{H}$ *to the complete matrix*); *and* $\tilde{\overline{H}}^1$ *as the matrix obtained by permuting the rows and columns of* $\overline{\overline{H}}^1$ *so as to arrange the sums of elements of rows and columns in nonascending order. Then, the following is true.*

(1) $(\overline{\mathbf{E}}^1,\overline{\mathbf{F}}^1)$ *and* $\overline{\overline{H}}^1$ *are perfect vector pair and perfect matrix, respectively* ($(\overline{\mathbf{E}}^1,\overline{\mathbf{F}}^1) \in E^{n,m}$, $\overline{\overline{H}}^1 \in E_{n,m}$).

(2) $(\tilde{\overline{\mathbf{E}}}^1,\tilde{\overline{\mathbf{F}}}^1)$ *and* $\tilde{\overline{H}}^1$ *are extremal vector pair and extremal matrix, respectively* ($(\tilde{\overline{\mathbf{E}}}^1,\tilde{\overline{\mathbf{F}}}^1) \in \overline{E}^{n,m}$, $\tilde{\overline{H}}^1 \in \overline{E}_{n,m}$).

(3) *If* $\overline{H} = \overline{H}(\overline{\mathbf{E}},\overline{\mathbf{F}})$, *then* $\tilde{\overline{H}}^1 = \tilde{\overline{H}}^1(\tilde{\overline{\mathbf{E}}}^1,\tilde{\overline{\mathbf{F}}}^1)$.

(4) *The mappings* $\alpha : \overline{E}^{n,m} \to \overline{E}^{n,m}$ *and* $\beta : \overline{E}_{n,m} \to \overline{E}_{n,m}$, *where* $\alpha(\overline{\mathbf{E}},\overline{\mathbf{F}}) = (\tilde{\overline{\mathbf{E}}}^1,\tilde{\overline{\mathbf{F}}}^1)$ *and* $\beta(\overline{H}) = \tilde{\overline{H}}^1$, *are isomorphisms of the distributive lattices* $\overline{E}^{n,m}$ *and* $\overline{E}_{n,m}$ *onto themselves (automorphisms) and isometries of the metric spaces* $\overline{E}^{n,m}$ *and* $\overline{E}_{n,m}$ *onto themselves.*

(5) *The mappings* α *and* β *have the following properties:*

$$\textit{if } (\overline{\mathbf{E}}_1,\overline{\mathbf{F}}_1) \le (\overline{\mathbf{E}}_2,\overline{\mathbf{F}}_2) \textit{ and } \overline{H}_1 \subseteq \overline{H}_2,$$
$$\textit{then } \alpha(\overline{\mathbf{E}}_2,\overline{\mathbf{F}}_2) \le \alpha(\overline{\mathbf{E}}_1,\overline{\mathbf{F}}_1) \textit{ and } \beta(\overline{H}_2) \subseteq \beta(\overline{H}_1);$$

the unions and intersections of the mappings satisfy the relations

$$\alpha\big((\overline{\mathbf{E}}_1,\overline{\mathbf{F}}_1) \vee (\overline{\mathbf{E}}_2,\overline{\mathbf{F}}_2)\big) = \alpha(\overline{\mathbf{E}}_1,\overline{\mathbf{F}}_1) \wedge \alpha(\overline{\mathbf{E}}_2,\overline{\mathbf{F}}_2),$$
$$\alpha\big((\overline{\mathbf{E}}_1,\overline{\mathbf{F}}_1) \wedge (\overline{\mathbf{E}}_2,\overline{\mathbf{F}}_2)\big) = \alpha(\overline{\mathbf{E}}_1,\overline{\mathbf{F}}_1) \vee \alpha(\overline{\mathbf{E}}_2,\overline{\mathbf{F}}_2),$$
$$\beta(\overline{H}_1 \cup \overline{H}_2) = \beta(\overline{H}_1) \cap \beta(\overline{H}_2),$$
$$\beta(\overline{H}_1 \cap \overline{H}_2) = \beta(\overline{H}_1) \cup \beta(\overline{H}_2).$$

The proof of each assertion of Theorem 2.3 can be considered as a simple exercise.

§3. Kernels and Hulls of Sets of Transportation Matrices of Zeros and Ones

In this section, we present a method for analyzing sets of transportation matrices of zeros and ones by means of extremal matrices. For the set of transportation matrices of zeros and ones from a certain transportation polyhedron, we define and construct the matrices called the kernel and the hull, which satisfy the following conditions:

(a) the kernel and hull are extremal matrices satisfying an order relation;

(b) the kernel and hull bound any matrix of zeros and ones from the transportation polyhedron;

(c) the metric function of the kernel and hull has the minimal value among all pairs of extremal matrices that satisfy condition (b);

(d) in terms of the order relation for matrices, the kernel is the "largest" extremal matrix contained in each matrix in question, whereas the hull is the "smallest" extremal matrix containing the matrix in question.

3.1. Definitions of the Kernel and Hull of a Matrix Set and a Criterion for their Nontriviality

Let $X = (x_{ij})$ be an arbitrary n-by-m matrix of zeros and ones. There always exist extremal matrices $\overline{H}'$, $\overline{H}'' \in \overline{E}_{n,m}$ such that $\overline{H}' \subseteq X \subseteq \overline{H}''$. For example, $H(0) \subseteq X \subseteq H(1)$ in the trivial case; here, the extreme left and right matrices contain, respectively, zeros and ones only. If the quantity

$$r(\overline{H}', \overline{H}'') = \sum_{i=1}^{n} \sum_{j=1}^{m} (h''_{ij} - h'_{ij}),$$

where $\overline{H}' = (h'_{ij})$ and $\overline{H}'' = (h''_{ij})$, is sufficiently small, then various properties of the matrix X are determined by those of the extremal matrices $\overline{H}'$ and $\overline{H}''$.

Consider arbitrary vectors $\mathbf{A} = (a_1, \ldots, a_n) \in \overline{Z}_+^n$ and $\mathbf{B} = (b_1, \ldots, b_m) \in \overline{Z}_+^m$ such that $M_Z(\mathbf{A}, \mathbf{B}; 1) \neq \varnothing$. For the matrix set $M_Z(\mathbf{A}, \mathbf{B}; 1)$, we define two extremal matrices, $\overline{H}_0$, $\overline{H}_1 \in \overline{E}_{n,m}$, such that $\overline{H}_0 \subseteq X \subseteq \overline{H}_1$ for any matrix $X \in M_Z(\mathbf{A}, \mathbf{B}; 1)$ and $r(\overline{H}_0, \overline{H}_1)$ has the minimal value; that is,

$$r(\overline{H}_0, \overline{H}_1) = \min_{\overline{H}', \overline{H}'' \in \overline{E}_{n,m}} \left\{ r(\overline{H}', \overline{H}'') : \overline{H}'' \subseteq X \subseteq \overline{H}'',\ X \in M_Z(\mathbf{A}, \mathbf{B}; 1) \right\}.$$

In the following definitions, we use the union and intersection of matrices of zeros and ones.

Definition 3.1. Suppose that $\mathbf{A} \in \overline{Z}^n_+$, $\mathbf{B} \in \overline{Z}^m_+$, and $M_Z(\mathbf{A}, \mathbf{B}; 1) \neq \varnothing$. The n-by-m matrix

$$R(\mathbf{A}, \mathbf{B}) = \bigcap_{X(\mathbf{A},\mathbf{B}) \in M_Z(\mathbf{A},\mathbf{B};1)} X(\mathbf{A}, \mathbf{B})$$

is called the *kernel of the matrix set* $M_Z(\mathbf{A}, \mathbf{B}; 1)$.

Definition 3.2. Suppose that $\mathbf{A} \in \overline{Z}^n_+$, $\mathbf{B} \in \overline{Z}^m_+$, and $M_Z(\mathbf{A}, \mathbf{B}; 1) \neq \varnothing$. The n-by-m matrix

$$O(\mathbf{A}, \mathbf{B}) = \bigcup_{X(\mathbf{A},\mathbf{B}) \in M_Z(\mathbf{A},\mathbf{B};1)} X(\mathbf{A}, \mathbf{B})$$

is called the *hull of the matrix set* $M_Z(\mathbf{A}, \mathbf{B}; 1)$.

Let us show that the kernel $R(\mathbf{A}, \mathbf{B})$ and hull $O(\mathbf{A}, \mathbf{B})$ are extremal matrices identical to the matrices $\overline{H}_0$ and $\overline{H}_1$ defined above, respectively; that is,

$$r(R(\mathbf{A}, \mathbf{B}), O(\mathbf{A}, \mathbf{B})) = \min_{\overline{H}', \overline{H}'' \in \overline{E}_{n,m}} \left\{ r(\overline{H}', \overline{H}'') : \overline{H}'' \subseteq X \subseteq \overline{H}'',\ X \in M_Z(\mathbf{A}, \mathbf{B}; 1) \right\}.$$

First, we analyze the conditions under which the matrices $R(\mathbf{A}, \mathbf{B})$ and $O(\mathbf{A}, \mathbf{B})$ are not trivial, that is, $R(\mathbf{A}, \mathbf{B}) \neq H(0)$ and $O(\mathbf{A}, \mathbf{B}) \neq H(1)$. We demonstrate that these conditions are determined by the fact that the matrices from $M_Z(\mathbf{A}, \mathbf{B}; 1)$ are minimax matrices for the matrix set $M(\mathbf{A}, \mathbf{B})$.

Theorem 3.1. *Suppose that* $\mathbf{A} = (a_1, \ldots, a_n) \in \overline{Z}^n_+$, $\mathbf{B} = (b_1, \ldots, b_m) \in \overline{Z}^m_+$, *and* $M_Z(\mathbf{A}, \mathbf{B}; 1) \neq \varnothing$.

(a) *If* $b_1 \geq 1$ (or $a_1 \geq 1$), *then the kernel* $R(\mathbf{A}, \mathbf{B})$ *is not a trivial matrix (i.e.,* $R(\mathbf{A}, \mathbf{B}) \neq H(0)$*) if and only if there exists* $p \in Z$ $(1 \leq p \leq n)$ *such that* $p \leq b_1$ *and* $\delta_p(\mathbf{A}, \mathbf{B}; 1) = 0$.

(b) *If* $b_m < n$ *(or* $a_n < m$*), then for* $b_m \geq 1$ *and* $a_n \geq 1$ *the hull* $O(\mathbf{A}, \mathbf{B})$ *is not a trivial matrix (i.e.,* $O(\mathbf{A}, \mathbf{B}) \neq H(1)$*) if and only if there exists* $p \in Z$ $(1 \leq p \leq n-1)$ *such that* $p < b_1$, $l_p(\mathbf{B}; 1) < m$, *and* $\delta_p(\mathbf{A}, \mathbf{B}; 1) = 0$.

Proof. Let $R(\mathbf{A}, \mathbf{B}) = (r_{ij})$ and $O(\mathbf{A}, \mathbf{B}) = (o_{ij})$ $(1 \leq i \leq n,\ 1 \leq j \leq m)$, where each r_{ij} and o_{ij} is zero or one, because the matrices $R(\mathbf{A}, \mathbf{B})$ and $O(\mathbf{A}, \mathbf{B})$ are, respectively, the union and intersection of matrices of zeros and ones.

(a) **Necessity.** Assume that $R(\mathbf{A}, \mathbf{B}) \neq H(0)$ and $\delta_p(\mathbf{A}, \mathbf{B}; 1) > 0$ for any $p \in Z$ $(1 \leq p \leq n,\ p \leq b_1)$. Then, there exists a matrix $X(\mathbf{A}, \mathbf{B}) = (x_{ij}) \in$

$M_Z(\mathbf{A}, \mathbf{B}; 1)$ with $x_{kq} = 0$ for any k, $q \in Z$ $(1 \leq k \leq n,\ 1 \leq q \leq m)$, which contradicts the assumption that $R(\mathbf{A}, \mathbf{B}) = (r_{ij}) \neq H(0)$.

Indeed, constructing the matrix $X(\mathbf{A}, \mathbf{B}) = (x_{ij})$ with $x_{kq} = 0$ for arbitrary k, $q \in Z$ $(1 \leq k \leq n,\ 1 \leq q \leq m)$ and assuming that $q > a_k$, we can apply Theorem 1.1 from Chapter 3 with $c = 1$ to show that there exists a principal $(1; k)$-reductive pair of vectors $\mathbf{A}^{(1)}(k)$ and $\mathbf{B}_*^{(1)} = (b_1^{(1)}, \ldots, b_m^{(1)})$ such that $b_j^{(1)} = b_j - 1$ for $1 \leq j \leq a_k$, $b_j^{(1)} = b_j$ for $a_k < j \leq m$, and $M_Z(\mathbf{A}^{(1)}(k), \mathbf{B}_*^{(1)}; 1) \neq \varnothing$. Therefore, the reductive vector pair $\mathbf{A}^{(1)}(k)$, $\mathbf{B}_*^{(1)}$ defines the matrix $X(\mathbf{A}, \mathbf{B}) = (x_{ij})$ with $x_{kj} = 1$ for $1 \leq j \leq a_k$ and $x_{kj} = 0$ for $a_k < j \leq m$. Accordingly, for $q > a_k$ there exists a matrix $X(\mathbf{A}, \mathbf{B}) = (x_{ij})$ such that $x_{kq} = 0$.

Assume that $q \leq a_k$. Since rows characterized by equal sums of elements are equivalent, we can assume without loss of generality that k is the minimal index among those of the components of $\mathbf{A}$ equal to a_k (setting $a = a_k$, we have $k = \min\{i : a_i = a,\ 1 \leq i \leq n\}$). Consider the characteristic function of $\mathbf{A}$ and $\mathbf{B}$ as defined by (1.4):

$$\delta_p(\mathbf{A}, \mathbf{B}; 1) = \sum_{j=1}^{m} b_j - \sum_{b_j \geq p} (b_j - p) - \sum_{i=1}^{n} a_i, \quad 1 \leq p \leq n.$$

It is obvious that when any component of $\mathbf{A}$ and any component of $\mathbf{B}$ are increased by unity, the value of $\delta_p(\mathbf{A}, \mathbf{B}; 1)$ either decreases by unity or remains unchanged (if $p > b_1$, then the value of $\delta_p(\mathbf{A}, \mathbf{B}; 1)$ may increase by unity). Let $\mathbf{A}' = (a_1', \ldots, a_n')$ and $\mathbf{B}' = (b_1', \ldots, b_m')$ be vectors with components such that

$$a_i' = \begin{cases} a_i, & 1 \leq i \leq n, \quad i \neq k, \\ a_k + 1, & i = k, \end{cases} \qquad b_j' = \begin{cases} b_j, & 1 \leq j \leq m, \quad j \neq q, \\ b_q + 1, & j = q. \end{cases}$$

The condition for $k \in Z$ implies that $\mathbf{A}' \in \overline{Z}_+^n$. The components of $\mathbf{B}'$ may not be arranged in nonascending order, but this possibility is of no importance for the present analysis. Since it is assumed that $\delta_p(\mathbf{A}, \mathbf{B}; 1) \geq 1$ for any p such that $1 \leq p \leq n$ and $p \leq b_1$, the inequality $\delta_p(\mathbf{A}', \mathbf{B}'; 1) \geq 0$ holds for all p such that $1 \leq p \leq n$ and $p \leq b_1'$. If $b_1' < p \leq n$, then we obviously have

$$\delta_p(\mathbf{A}', \mathbf{B}'; 1) = \sum_{j=1}^{m} b_j' - \sum_{i=1}^{p} a_i' \geq 0.$$

Consequently, $M_Z(\mathbf{A}', \mathbf{B}'; 1) \neq \varnothing$. Let us apply Theorem 1.1 from Chapter 3, with $c = 1$. For vectors $\mathbf{A}'$ and $\mathbf{B}'$, there exists a principal $(1; k)$-reductive pair of vectors $\mathbf{A}'^{(1)}(k)$ and $\mathbf{B}_*'^{(1)} = (b_1'^{(1)}, \ldots, b_m'^{(1)})$ such that $b_j'^{(1)} = b_j' - 1$ for $1 \leq j \leq a_k' = a_k + 1$, $b_j'^{(1)} = b_j'$ for $a_k + 1 < j \leq m$, and $M_Z(\mathbf{A}'^{(1)}(k), \mathbf{B}_*'^{(1)}; 1) \neq \varnothing$.

Therefore, the pair $\mathbf{A}'^{(1)}(k)$, $\mathbf{B}_*'^{(1)}$ defines the matrix $X'(\mathbf{A}', \mathbf{B}') = (x'_{ij})$ with $x'_{kj} = 1$ for $1 \leq j \leq a_k + 1$ and $x'_{kj} = 0$ for $a_k + 1 < j \leq m$. Hence, $x'_{kq} = 1$ in the matrix $X'(\mathbf{A}', \mathbf{B}')$. Substituting zero for this element, we obtain the matrix $X(\mathbf{A}, \mathbf{B}) = (x_{ij}) \in M_Z(\mathbf{A}, \mathbf{B}; 1)$ with $x_{kq} = 0$.

As a result, for any k, $q \in Z$ ($1 \leq k \leq n$, $1 \leq q \leq m$) there exists a matrix in $M_Z(\mathbf{A}, \mathbf{B}; 1)$ whose element with indices k and q is zero. Definition 3.1 implies that $R(\mathbf{A}, \mathbf{B}) = (r_{ij}) = H(0)$, which contradicts the assumption. Therefore, the assumption that $\delta_p(\mathbf{A}, \mathbf{B}; 1) > 0$ for any p such that $1 \leq p \leq n$ and $p \leq b_1$ is not true, which proves the necessity of the assumption of proposition (a).

Sufficiency. Suppose that $\delta_p(\mathbf{A}, \mathbf{B}; 1) = 0$ for a certain p ($1 \leq p \leq n$, $p \leq b_1$). By Theorem 2.2 from Chapter 3, any matrix $X(\mathbf{A}, \mathbf{B}) = (x_{ij}) \in M_Z(\mathbf{A}, \mathbf{B}; 1)$ is such that $x_{ij} = 1$ for $1 \leq i \leq p$ and $1 \leq j \leq l_p(\mathbf{B}; 1)$, where $l_p(\mathbf{B}; 1) \geq 1$. By Definition 3.1, the kernel $R(\mathbf{A}, \mathbf{B})$ contains an element equal to unity, and $R(\mathbf{A}, \mathbf{B}) = (r_{ij}) \neq H(0)$.

(b) **Necessity.** Suppose that $O(\mathbf{A}, \mathbf{B}) \neq H(1)$; that is, the elements of the matrix $O(\mathbf{A}, \mathbf{B}) = (o_{ij})$ are not all equal to unity. Assume that $\delta_p(\mathbf{A}, \mathbf{B}; 1) > 0$ for any $p \in Z$ such that $1 \leq p \leq n - 1$, $p < b_1$ and $l_p(\mathbf{B}; 1) < m$. Let us show that, for any k, $q \in Z$ ($1 \leq k \leq n$, $1 \leq q \leq m$) there exists a matrix $X(\mathbf{A}, \mathbf{B}) = (x_{ij}) \in M_Z(\mathbf{A}, \mathbf{B}; 1)$ with $x_{kq} = 1$, which contradicts the assumption that $O(\mathbf{A}, \mathbf{B}) \neq H(1)$.

If $q \leq a_k$, then by Theorem 1.1 from Chapter 3, with $c = 1$, there exists a matrix $X(\mathbf{A}, \mathbf{B}) = (x_{ij}) \in M_Z(\mathbf{A}, \mathbf{B}; 1)$ with $x_{kq} = 1$ (see the proof of necessity for proposition (a)).

Suppose $q > a_k$. Without loss of generality, we assume that k is the largest index among those of the components of $\mathbf{A}$ equal to a_k (setting $a = a_k$, we have $k = \max\{i : a_i = a,\ 1 \leq i \leq n\}$). It is obvious that when any component of $\mathbf{A}$ and any component of $\mathbf{B}$ are increased by unity, the value of $\delta_p(\mathbf{A}, \mathbf{B}; 1)$ either decreases by unity or remains unchanged. Let the components of $\mathbf{A}' = (a'_1, \ldots, a'_n)$ and $\mathbf{B} = (b'_1, \ldots, b'_m)$ be such that

$$a'_i = \begin{cases} a_i, & 1 \leq i \leq n, \quad i \neq k, \\ a_k - 1, & i = k, \end{cases} \qquad b'_j = \begin{cases} b_j, & 1 \leq j \leq m, \quad j \neq q, \\ b_q - 1, & j = q. \end{cases}$$

Then, $\mathbf{A}' \in \overline{Z}^n_+$ and $\mathbf{B}' \in \overline{Z}^m_+$. Since $\delta_p(\mathbf{A}, \mathbf{B}; 1) \geq 1$ for any p such that $1 \leq p \leq n - 1$, $p < b_1$, and $l_p(\mathbf{B}; 1) < m$ by assumption, the inequality $\delta_p(\mathbf{A}', \mathbf{B}'; 1) \geq 0$ holds for any p ($1 \leq p \leq n$, $p \leq b'_1$, $l_p(\mathbf{B}'; 1) < m$). If $p > b'_1$ or $l_p(\mathbf{B}'; 1) = m$, then we have, respectively,

$$\delta_p(\mathbf{A}', \mathbf{B}'; 1) = \sum_{j=1}^{m} b'_j - \sum_{i=1}^{p} a'_i \geq 0,$$

$$\delta_p(\mathbf{A}', \mathbf{B}'; 1) = pm - \sum_{i=1}^{p} a_i' \geq 0,$$

and hence $M_Z(\mathbf{A}', \mathbf{B}'; 1) \neq \varnothing$. By Theorem 1.1 from Chapter 3, there exists a matrix $X'(\mathbf{A}', \mathbf{B}') = (x_{ij}')$ with $x_{kq}' = 0$ (see the proof of necessity for proposition (a)). Replacing this element with unity, we obtain a matrix $X(\mathbf{A}, \mathbf{B}) = (x_{ij}) \in M_Z(\mathbf{A}, \mathbf{B}; 1)$ with $x_{kq} = 1$.

Thus, we have shown that for any k, $q \in Z$ $(1 \leq k \leq n,\ 1 \leq q \leq m)$ there exists a matrix from $M_Z(\mathbf{A}, \mathbf{B}; 1)$ whose element with indices k and q is unity. Then, $O(\mathbf{A}, \mathbf{B}) = (o_{ij}) = H(1)$ by Definition 3.2, which contradicts the assumption. Consequently, the assumption that $\delta_p(\mathbf{A}, \mathbf{B}; 1) > 0$ for any p such that $1 \leq p \leq n-1$, $p < b_1$, and $l_p(\mathbf{B}; 1) < m$ is not valid, which proves the necessity of the assumptions of proposition (b).

Sufficiency. If $\delta_p(\mathbf{A}, \mathbf{B}; 1) = 0$ for a certain p such that $1 \leq p \leq n-1$, $p < b_1$, and $l_p(\mathbf{B}; 1) < m$, then any matrix $X(\mathbf{A}, \mathbf{B}) = (x_{ij}) \in M_Z(\mathbf{A}, \mathbf{B}; 1)$ is such that $x_{ij} = 0$ for $p < i \leq n$ and $l_p(\mathbf{B}; 1) < j \leq m$ by Theorem 2.2 from Chapter 3. Definition 3.2 implies that the hull $O(\mathbf{A}, \mathbf{B})$ contains a zero element, and $O(\mathbf{A}, \mathbf{B}) = (o_{ij}) \neq H(1)$. This completes the proof.

3.2. Mutually Complementary Transportation Matrices

Now, we establish a binary relation between vector pairs such that the respective values of the characteristic function defined by (1.4) are equal. Thus, we show that the problems of construction of a kernel and a hull are equivalent.

With the pair of vectors $\mathbf{A} = (a_1, \ldots, a_n) \in \overline{Z}_+^n$ and $\mathbf{B} = (b_1, \ldots, b_m) \in \overline{Z}_+^m$, where $a_1 \leq m$ and $b_1 \leq n$, we associate the pair of vectors whose components are $\overline{\mathbf{A}}^1 = (\overline{a}_1^1, \ldots, \overline{a}_n^1) \in \overline{Z}_+^n$ $(\overline{a}_i^1 = m - a_{n-i+1},\ 1 \leq i \leq n)$ and $\overline{\mathbf{B}}^1 = (\overline{b}_1^1, \ldots, \overline{b}_m^1) \in \overline{Z}_+^m$ $(\overline{b}_j^1 = n - b_{m-j+1},\ 1 \leq j \leq m)$.

It is obvious that the matrix sets $M_Z(\mathbf{A}, \mathbf{B}; 1)$ and $M_Z(\overline{\mathbf{A}}^1, \overline{\mathbf{B}}^1; 1)$ are simultaneously either empty or nonempty. Furthermore, if $M_Z(\mathbf{A}, \mathbf{B}; 1) \neq \varnothing$, then the matrices from $M_Z(\mathbf{A}, \mathbf{B}; 1)$ and $M_Z(\overline{\mathbf{A}}^1, \overline{\mathbf{B}}^1; 1)$ are in one-to-one correspondence; that is, there exists a mapping $f : M_Z(\mathbf{A}, \mathbf{B}; 1) \to M_Z(\overline{\mathbf{A}}^1, \overline{\mathbf{B}}^1; 1)$ such that $f\left(X(\mathbf{A}, \mathbf{B})\right) = \overline{X}^1(\overline{\mathbf{A}}^1, \overline{\mathbf{B}}^1)$ and if $X(\mathbf{A}, \mathbf{B}) = (x_{ij})$, then the elements of $\overline{X}^1(\overline{\mathbf{A}}^1, \overline{\mathbf{B}}^1) = (\overline{x}_{ij}^1)$ are related as follows:

$$\overline{x}_{ij}^1 = 1 - x_{n-i+1\, m-j+1}, \quad 1 \leq i \leq n, \quad 1 \leq j \leq m.$$

Therefore, analysis of the matrix sets $M_Z(\mathbf{A}, \mathbf{B}; 1)$ and $M_Z(\overline{\mathbf{A}}^1, \overline{\mathbf{B}}^1; 1)$ is equivalent to analysis of matrices in these sets, and the concepts of kernel and hull may be interpreted as dual (see proposition (g) of Theorem 2.3). The kernel and hull of $M_Z(\mathbf{A}, \mathbf{B}; 1) \neq \varnothing$ are equivalent to the hull and kernel of $M_Z(\overline{\mathbf{A}}^1, \overline{\mathbf{B}}^1; 1)$, respectively. This is established by the following theorem about the equivalence of characteristic functions (1.4) of the vector pairs **A**, **B** and $\overline{\mathbf{A}}^1$, $\overline{\mathbf{B}}^1$.

Theorem 3.2′. *Suppose that vectors* $\mathbf{A} = (a_1, \dots, a_n) \in \overline{Z}^n_+$ *and* $\mathbf{B} = (b_1, \dots, b_m) \in \overline{Z}^m_+$, *where* $a_1 \le m$ *and* $b_1 \le n$, *satisfy the balance condition. Let* $\overline{\mathbf{A}}^1 = (\overline{a}^1_1, \dots, \overline{a}^1_n)$ *and* $\overline{\mathbf{B}}^1 = (\overline{b}^1_1, \dots, \overline{b}^1_m)$, *where*

$$\overline{a}^1_i = m - a_{n-i+1}, \quad 1 \le i \le n, \quad \overline{b}^1_j = n - b_{m-j+1}, \quad 1 \le j \le m.$$

Then, the vectors $\overline{\mathbf{A}}^1 \in \overline{Z}^n_+$ *and* $\overline{\mathbf{B}}^1 \in \overline{Z}^m_+$ *are characterized by equal sums of components,* $\overline{a}^1_1 \le m$, $\overline{b}^1_1 \le n$, *and function* (1.4) *of the vector pairs* **A**, **B** *and* $\overline{\mathbf{A}}^1$, $\overline{\mathbf{B}}^1$ *is such that*

$$\delta_k(\overline{\mathbf{A}}^1, \overline{\mathbf{B}}^1; 1) = \delta_{n-k}(\mathbf{A}, \mathbf{B}; 1), \quad 0 \le k \le n.$$

This theorem is a special case of Theorem 3.2, which is proved below.

Remark 3.1. By Theorem 3.2′, proposition (a) of Theorem 3.1 is a corollary of proposition (b) of Theorem 3.1, and vice versa.

Remark 3.2. Comparing the matrix $\overline{X}^1$ and vectors $\overline{\mathbf{A}}^1$ and $\overline{\mathbf{B}}^1$ with the matrix $\overline{X}'$ and vectors $\overline{\mathbf{A}}'$ and $\overline{\mathbf{B}}'$ (recall the proof of Theorem 1.13), we see that the former can be obtained by permuting the rows, columns, and components of the latter.

Suppose $c \in R$ and $c \ge 0$. In addition to vectors $\mathbf{A} = (a_1, \dots, a_n) \in \overline{R}^n_+$ and $\mathbf{B} = (b_1, \dots, b_m) \in \overline{R}^m_+$ characterized by equal sums of components ($a_1 \le cm$, $b_1 \le cn$), define vectors $\overline{\mathbf{A}}^c = (\overline{a}^c_1, \dots, \overline{a}^c_n)$ and $\overline{\mathbf{B}}^c = (\overline{b}^c_1, \dots, \overline{b}^c_m)$, where

$$\overline{a}^c_i = cm - a_{n-i+1}, \quad 1 \le i \le n, \quad \overline{b}^c_j = cn - b_{m-j+1}, \quad 1 \le j \le m.$$

It is clear that

$$\overline{\mathbf{A}}^c \in \overline{R}^n_+, \quad \overline{\mathbf{B}}^c \in \overline{R}^m_+, \quad \overline{a}^c_1 \le cm, \quad \overline{b}^c_1 \le cn, \quad \sum_{i=1}^n \overline{a}^c_i = \sum_{j=1}^m \overline{b}^c_j.$$

Obviously, $M(\mathbf{A}, \mathbf{B}; c) \ne \varnothing$ if and only if $M(\overline{\mathbf{A}}^c, \overline{\mathbf{B}}^c; c) \ne \varnothing$. Let $X(\mathbf{A}, \mathbf{B}) = (x_{ij}) \in M(\mathbf{A}, \mathbf{B}; c)$. With a matrix $X(\mathbf{A}, \mathbf{B})$ from $M(\mathbf{A}, \mathbf{B}; c)$, we associate the matrix $\overline{X}^c(\overline{\mathbf{A}}^c, \overline{\mathbf{B}}^c) = (\overline{x}^c_{ij})$ from $M(\overline{\mathbf{A}}^c, \overline{\mathbf{B}}^c; c)$ defined as follows:

$$\overline{x}^c_{ij} = c - x_{n-i+1\, m-j+1}, \quad 1 \le i \le n, \quad 1 \le j \le m.$$

The mapping $f : M(\mathbf{A},\mathbf{B};c) \to M(\overline{\mathbf{A}}^c,\overline{\mathbf{B}}^c;c)$, where $f\left(X(\mathbf{A},\mathbf{B})\right) = \overline{X}^c(\overline{\mathbf{A}}^c,\overline{\mathbf{B}}^c)$, is a one-to-one correspondence; therefore, analysis of a matrix from $M(\mathbf{A},\mathbf{B};c)$ is equivalent to analysis of a matrix from $M(\overline{\mathbf{A}}^c,\overline{\mathbf{B}}^c;c)$. In particular, it is obvious that $X(\mathbf{A},\mathbf{B})$ and $\overline{X}^c(\overline{\mathbf{A}}^c,\overline{\mathbf{B}}^c)$ are simultaneously uniform and extremal matrices (when $\mathbf{A} \in \overline{Z}^n_+$, $\mathbf{B} \in \overline{Z}^m_+$, and $c = 1$). It is also clear that both are minimax matrices with common minimax value c if there exists $p \in Z$ such that $1 \le p \le n-1$, $b_1 > cp$, $l_p(\mathbf{B};c) < m$, and $\delta_p(\mathbf{A},\mathbf{B};c) = 0$. The equivalence of analyses of the matrix sets $M(\mathbf{A},\mathbf{B};c)$ and $M(\overline{\mathbf{A}}^c,\overline{\mathbf{B}}^c;c)$ is established by the following theorem about the equivalence of the characteristic functions of the vector pairs $\mathbf{A}$, $\mathbf{B}$ and $\overline{\mathbf{A}}^c$, $\overline{\mathbf{B}}^c$ defined by (1.22) in Chapter 1.

Theorem 3.2. *Suppose that* $c \in R$, $c \ge 0$; *and vectors* $\mathbf{A} = (a_1,\dots,a_n) \in \overline{R}^n_+$ *and* $\mathbf{B} = (b_1,\dots,b_m) \in \overline{R}^m_+$, *where* $a_1 \le cm$ and $b_1 \le cn$, *are characterized by equal sums of components. Then, the function of the vector pairs* $\mathbf{A}$, $\mathbf{B}$ *and* $\overline{\mathbf{A}}^c$, $\overline{\mathbf{B}}^c$ *defined by* (1.22) *in Chapter* 1 *satisfies the condition*

$$\delta_k(\overline{\mathbf{A}}^c,\overline{\mathbf{B}}^c;c) = \delta_{n-k}(\mathbf{A},\mathbf{B};c), \quad 0 \le k \le n. \tag{3.1}$$

Proof. Consider the right-hand and left-hand sides of (3.1) separately, representing the right-hand side as

$$\delta_{n-k}(\mathbf{A},\mathbf{B};c) = \sum_{j=1}^{m} b_j - \sum_{b_j \ge c(n-k)} \left(b_j - c(n-k)\right) - \sum_{i=1}^{n-k} a_i =$$
$$\sum_{j=1}^{m} b_j - \sum_{b_j > c(n-k)} \left(b_j - c(n-k)\right) - \sum_{i=1}^{n-k} a_i =$$
$$\sum_{j > l'_{n-k}(\mathbf{B};c)} b_j + c(n-k)l'_{n-k}(\mathbf{B};c) - \sum_{i=1}^{n-k} a_i,$$

where

$$l'_{n-k}(\mathbf{B};c) = \begin{cases} \max\left\{j : b_j > c(n-k)\right\}, & b_1 > c(n-k), \\ 0, \quad b_1 \le c(n-k); \end{cases} \tag{3.2}$$

and the left-hand side as

$$\delta_k(\overline{\mathbf{A}}^c,\overline{\mathbf{B}}^c;c) = \sum_{j=1}^{m} \overline{b}^c_j - \sum_{\overline{b}^c_j \ge ck} (\overline{b}^c_j - ck) - \sum_{i=1}^{k} \overline{a}^c_i =$$
$$\sum_{j > l_k(\overline{\mathbf{B}}^c;c)} \overline{b}^c_j + ckl_k(\overline{\mathbf{B}}^c;c) - \sum_{i=1}^{k} \overline{a}^c_i,$$

where, by convention,

$$l_k(\overline{\mathbf{B}}^c; c) = \begin{cases} \max\{j : \overline{b}_j^c \geq ck\}, & \overline{b}_1^c \geq ck, \\ 0, & \overline{b}_1^c < ck. \end{cases} \tag{3.3}$$

In $\delta_k(\overline{\mathbf{A}}^c, \overline{\mathbf{B}}^c; c)$, we express the sums of components of $\overline{\mathbf{A}}^c$ and $\overline{\mathbf{B}}^c$ in terms of components of $\mathbf{A}$ and $\mathbf{B}$:

$$\sum_{j>l_k(\overline{\mathbf{B}}^c;c)} \overline{b}_j^c = \sum_{j>l_k(\overline{\mathbf{B}}^c;c)} (cn - b_{m-j+1}) = cn\left(m - l_k(\overline{\mathbf{B}}^c;c)\right) - \sum_{j>l_k(\overline{\mathbf{B}}^c;c)} b_{m-j+1} =$$
$$cn\left(m - l_k(\overline{\mathbf{B}}^c;c)\right) - \sum_{j\leq m-l_k(\overline{\mathbf{B}}^c;c)} b_j,$$

$$\sum_{i=1}^{k} \overline{a}_i^c = \sum_{i=1}^{k}(cm - a_{n-i+1}) = cmk - \sum_{i=1}^{k} a_{n-i+1} = cmk - \sum_{i=n-k+1}^{n} a_i.$$

As a result, we have

$$\delta_k(\overline{\mathbf{A}}^c, \overline{\mathbf{B}}^c; c) = cn\left(m - l_k(\overline{\mathbf{B}}^c;c)\right) - \sum_{j\leq m-l_k(\overline{\mathbf{B}}^c;c)} b_j + ckl_k(\overline{\mathbf{B}}^c;c) - cmk+$$
$$\sum_{i=n-k+1}^{n} a_i = c(n-k)\left(m - l_k(\overline{\mathbf{B}}^c;c)\right) - \sum_{j\leq m-l_k(\overline{\mathbf{B}}^c;c)} b_j + \sum_{i=n-k+1}^{n} a_i =$$
$$c(n-k)\left(m - l_k(\overline{\mathbf{B}}^c;c)\right) - \sum_{j=1}^{m} b_j + \sum_{j>m-l_k(\overline{\mathbf{B}}^c;c)} b_j + \sum_{i=1}^{n} a_i - \sum_{i=1}^{n-k} a_i.$$

The balance condition implies that

$$\delta_k(\overline{\mathbf{A}}^c, \overline{\mathbf{B}}^c; c) = c(n-k)\left(m - l_k(\overline{\mathbf{B}}^c;c)\right) + \sum_{j>m-l_k(\overline{\mathbf{B}}^c;c)} b_j - \sum_{i=1}^{n-k} a_i. \tag{3.4}$$

Comparing (3.2) with (3.3), we now show that

$$l'_{n-k}(\mathbf{B}; c) = m - l_k(\overline{\mathbf{B}}^c; c).$$

Suppose $l_k(\overline{\mathbf{B}}^c; c) = 0$. Then, $\overline{b}_1^c < ck$ and $cn - b_m < ck$, $b_m > c(n-k)$. By virtue of (3.2), $l'_{n-k}(\mathbf{B}; c) = m$.

Consider the case $l_k(\overline{\mathbf{B}}^c; c) > 0$. Let $l_k(\overline{\mathbf{B}}^c; c) < m$. Then, we have the following inequalities:

$$\overline{b}^c_{l_k(\overline{\mathbf{B}}^c;c)} \geq ck, \quad \overline{b}^c_{l_k(\overline{\mathbf{B}}^c;c)+1} < ck,$$

$$cn - b_{m-l_k(\overline{\mathbf{B}}^c;c)+1} \geq ck, \quad cn - b_{m-l_k(\overline{\mathbf{B}}^c;c)} < ck,$$

$$b_{m-l_k(\overline{\mathbf{B}}^c;c)+1} \leq c(n-k), \quad b_{m-l_k(\overline{\mathbf{B}}^c;c)} > c(n-k).$$

Hence, when $0 < l_k(\overline{\mathbf{B}}^c; c) < m$, it holds that $l'_{n-k}(\mathbf{B}; c) = m - l_k(\overline{\mathbf{B}}^c; c)$ by virtue of (3.2).

If $l_k(\overline{\mathbf{B}}^c; c) = m$, then

$$\overline{b}^c_m \geq ck, \quad cn - b_1 \geq ck, \quad b_1 \leq c(n-k).$$

Hence, (3.2) entails the equality $l'_{n-k}(\mathbf{B}; c) = 0$ when $l_k(\overline{\mathbf{B}}^c; c) = m$. Therefore,

$$l'_{n-k}(\mathbf{B}; c) = \begin{cases} m - l_k(\overline{\mathbf{B}}^c; c), & 0 < l_k(\overline{\mathbf{B}}^c; c) < m, \\ 0, & l_k(\overline{\mathbf{B}}^c; c) = m, \\ m, & l_k(\overline{\mathbf{B}}^c; c) = 0. \end{cases}$$

Thus, $l'_{n-k}(\mathbf{B}; c) = m - l_k(\overline{\mathbf{B}}^c; c)$ for any admissible value of $l_k(\overline{\mathbf{B}}^c; c)$. Combining (3.4) with the value of $\delta_{n-k}(\mathbf{A}, \mathbf{B}; c)$ expressed in terms of $l'_{n-k}(\mathbf{B}; c)$, we then obtain

$$\delta_k(\overline{\mathbf{A}}^c, \overline{\mathbf{B}}^c; c) = c(n-k) l'_{n-k}(\mathbf{B}; c) + \sum_{j > l'_{n-k}(\mathbf{B};c)} b_j - \sum_{i=1}^{n-k} a_i = \delta_{n-k}(\mathbf{A}, \mathbf{B}; c),$$

$0 \leq k \leq n$.

This completes the proof.

3.3. Extremality of the Kernels and Hulls of Matrix Sets

The kernels and hulls have remarkable properties. They are extremal matrices. In what follows, we demonstrate this and develop procedures for their construction.

Consider the kernel $R(\mathbf{A}, \mathbf{B})$ and hull $O(\mathbf{A}, \mathbf{B})$ of a matrix set $M_Z(\mathbf{A}, \mathbf{B}; 1) \neq \varnothing$, where $\mathbf{A} = (a_1, \ldots, a_n) \in \overline{Z}^n_+$ and $\mathbf{B} = (b_1, \ldots, b_m) \in \overline{Z}^m_+$. The matrices

$$R(\mathbf{A}, \mathbf{B}) = \bigcap_{X(\mathbf{A},\mathbf{B}) \in M_Z(\mathbf{A},\mathbf{B};1)} X(\mathbf{A}, \mathbf{B}),$$

$$O(\mathbf{A}, \mathbf{B}) = \bigcup_{X(\mathbf{A},\mathbf{B}) \in M_Z(\mathbf{A},\mathbf{B};1)} X(\mathbf{A}, \mathbf{B})$$

can be constructed without using matrices from $M_Z(\mathbf{A}, \mathbf{B}; 1)$. By Theorem 3.1, the kernel and hull are not trivial matrices when the matrices from $M_Z(\mathbf{A}, \mathbf{B}; 1)$ are minimax matrices. Using this fact in combination with

Theorem 2.2 from Chapter 3, we now develop a procedure for constructing the kernels and hulls of matrix sets.

Consider vectors $\mathbf{A} = (a_1, \ldots, a_n) \in \overline{Z}_+^n$ and $\mathbf{B} = (b_1, \ldots, b_m) \in \overline{Z}_+^m$ with equal sums of components, such that $M_Z(\mathbf{A}, \mathbf{B}; 1) \neq \varnothing$. If $b_1 = 0$ (which implies that $a_1 = 0$), then $R(\mathbf{A}, \mathbf{B}) = H(0)$ is a matrix of zeros. Let $b_1 \geq 1$ and, therefore, $a_1 \geq 1$). If $\delta_k(\mathbf{A}, \mathbf{B}; 1) > 0$ for any k such that $1 \leq k \leq n$ and $k \leq b_1$, then $R(\mathbf{A}, \mathbf{B}) = H(0)$ by Theorem 3.1. Accordingly, assume that there exists $k \in Z$ $(1 \leq k \leq n)$ such that $k \leq b_1$ and $\delta_k(\mathbf{A}, \mathbf{B}; 1) = 0$.

Denote by $p_k \in Z$ $(1 \leq p_k \leq n,\ 1 \leq k \leq t)$ every number such that $p_k \leq b_1$ and $\delta_{p_k}(\mathbf{A}, \mathbf{B}; 1) = 0$. For any p_k, there exists an integer $l_{p_k}(\mathbf{B}; 1) = \max\{j : b_j \geq p_k\}$. The numbers p_k and $l_{p_k}(\mathbf{B}; 1)$ define certain elements of every matrix in $M_Z(\mathbf{A}, \mathbf{B}; 1)$ by Theorem 2.2 from Chapter 3; that is, for any matrix $X(\mathbf{A}, \mathbf{B}) = (x_{ij}) \in M_Z(\mathbf{A}, \mathbf{B}; 1)$, we have

$$x_{ij} = \begin{cases} 1, & 1 \leq i \leq p_k, \quad 1 \leq j \leq l_{p_k}(\mathbf{B}; 1), \\ 0, & p_k < i \leq n, \quad l_{p_k}(\mathbf{B}; 1) < j \leq m. \end{cases}$$

Using the values of p_k and $l_{p_k}(\mathbf{B}; 1)$, we define an n-by-m matrix $R_{p_k}(\mathbf{A}, \mathbf{B}) = (r_{ij}^{(p_k)})$ such that

$$r_{ij}^{(p_k)} = \begin{cases} 1, & 1 \leq i \leq p_k, \quad 1 \leq j \leq l_{p_k}(\mathbf{B}; 1), \\ 0, & 1 \leq i \leq p_k, \quad l_{p_k}(\mathbf{B}; 1) < j \leq m, \\ 0, & p_k < i \leq n, \quad 1 \leq j \leq m. \end{cases}$$

It is obvious that every $X(\mathbf{A}, \mathbf{B}) \in M_Z(\mathbf{A}, \mathbf{B}; 1)$ satisfies the relation

$$R_{p_k}(\mathbf{A}, \mathbf{B}) \subseteq X(\mathbf{A}, \mathbf{B}), \quad 1 \leq k \leq t.$$

Therefore,

$$\bigcup_{k=1}^{t} R_{p_k}(\mathbf{A}, \mathbf{B}) \subseteq X(\mathbf{A}, \mathbf{B}), \quad X(\mathbf{A}, \mathbf{B}) \in M_Z(\mathbf{A}, \mathbf{B}; 1).$$

The definition of the kernel $R(\mathbf{A}, \mathbf{B})$ implies that

$$R'(\mathbf{A}, \mathbf{B}) = \bigcup_{k=1}^{t} R_{p_k}(\mathbf{A}, \mathbf{B}) \subseteq R(\mathbf{A}, \mathbf{B}).$$

Since every $R_{p_k}(\mathbf{A}, \mathbf{B})$ is an extremal matrix, $R'(\mathbf{A}, \mathbf{B})$ is an extremal matrix as well by Lemma 2.5.

Define the n-by-m extremal vector pair $(\overline{\mathbf{E}}^R, \overline{\mathbf{F}}^R)$ for which $R'(\mathbf{A}, \mathbf{B}) = \overline{H}(\overline{\mathbf{E}}^R, \overline{\mathbf{F}}^R)$. Every extremal matrix $R_{p_k}(\mathbf{A}, \mathbf{B})$ is defined by the extremal pair of vectors

$$\overline{\mathbf{E}}^R_{p_k} = \big(\underbrace{l_{p_k}(\mathbf{B};1),\ldots,l_{p_k}(\mathbf{B};1)}_{p_k},\underbrace{0,\ldots,0}_{n-p_k}\big),$$

$$\overline{\mathbf{F}}^R_{p_k} = (\underbrace{p_k,\ldots,p_k}_{l_{p_k}(\mathbf{B};1)},\underbrace{0,\ldots,0}_{m-l_{p_k}(\mathbf{B};1)});$$

$$R_{p_k}(\mathbf{A},\mathbf{B}) = \overline{H}(\overline{\mathbf{E}}^R_{p_k},\overline{\mathbf{F}}^R_{p_k}), \quad 1 \le k \le t.$$

By Lemma 2.4, we then have

$$(\overline{\mathbf{E}}^R,\overline{\mathbf{F}}^R) = \bigvee_{k=1}^{t}(\overline{\mathbf{E}}^R_{p_k},\overline{\mathbf{F}}^R_{p_k}) = \Big(\bigvee_{k=1}^{t}\overline{\mathbf{E}}^R_{p_k},\bigvee_{k=1}^{t}\overline{\mathbf{F}}^R_{p_k}\Big),$$

$$R'(\mathbf{A},\mathbf{B}) = \overline{H}\Big(\bigvee_{k=1}^{t}\overline{\mathbf{E}}^R_{p_k},\bigvee_{k=1}^{t}\overline{\mathbf{F}}^R_{p_k}\Big) = \overline{H}(\overline{\mathbf{E}}^R,\overline{\mathbf{F}}^R).$$

Let us show that $R(\mathbf{A},\mathbf{B}) = R'(\mathbf{A},\mathbf{B})$.

Theorem 3.3. *Suppose that $M_Z(\mathbf{A},\mathbf{B};1) \neq \varnothing$, where*

$$\mathbf{A} = (a_1,\ldots,a_n) \in \overline{Z}^n_+ \text{ and } \mathbf{B} = (b_1,\ldots,b_m) \in \overline{Z}^m_+.$$

(a) *If $b_1 = 0$ or $\delta_k(\mathbf{A},\mathbf{B};1) > 0$ for any k $(1 \le k \le b_1)$, then the kernel $R(\mathbf{A},\mathbf{B}) = H(0)$ is a matrix of zeros.*

(b) *If $b_1 \ge 1$ and there exists $p \in Z$ $(1 \le p \le b_1)$ such that $\delta_p(\mathbf{A},\mathbf{B};1) = 0$, then the kernel $R(\mathbf{A},\mathbf{B})$ is not the trivial matrix $H(0)$, and*

$$R(\mathbf{A},\mathbf{B}) = \bigcap_{X(\mathbf{A},\mathbf{B})\in M_Z(\mathbf{A},\mathbf{B};1)} X(\mathbf{A},\mathbf{B}) = \bigcup_{k=1}^{t} R_{p_k}(\mathbf{A},\mathbf{B}) = R'(\mathbf{A},\mathbf{B}),$$

where p_k $(1 \le k \le t)$ denote every number such that $p_k \le b_1$ and $\delta_{p_k}(\mathbf{A},\mathbf{B};1) = 0$.

Proof.

(a) If $b_1 = 0$, then the triviality of $R(\mathbf{A},\mathbf{B})$ is obvious. When $b_1 \ge 1$ and $\delta_k(\mathbf{A},\mathbf{B};1) > 0$ for any k $(1 \le k \le b_1)$, the triviality of $R(\mathbf{A},\mathbf{B})$ follows from Theorem 3.1.

(b) It is sufficient to show that $R(\mathbf{A},\mathbf{B}) \subseteq R'(\mathbf{A},\mathbf{B})$. Assume that this relation does not hold. Then, there exist indices p and q for which the corresponding elements of the matrices $R(\mathbf{A},\mathbf{B}) = (r_{ij})$ and $R'(\mathbf{A},\mathbf{B}) = (r'_{ij})$

are $r_{pq} = 1$ and $r'_{pq} = 0$, respectively. Therefore, $r_{ij} = 1$ for i and j such that $a_i = a_p$ and $b_j = b_q$. Furthermore, we have $r'_{ij} = 0$ for i and j such that $e_i = e_p$ and $f_j = f_q$, where $R'(\mathbf{A},\mathbf{B}) = \overline{H}(\overline{\mathbf{E}},\overline{\mathbf{F}})$ is defined by the extremal vector pair $(\overline{\mathbf{E}},\overline{\mathbf{F}}) = (\overline{\mathbf{E}}^R,\overline{\mathbf{F}}^R)$ constructed above, and $\overline{\mathbf{E}} = (e_1,\ldots,e_n)$, and $\overline{\mathbf{F}} = (f_1,\ldots,f_m)$ ($R'(\mathbf{A},\mathbf{B})$ is an extremal, and therefore uniform, matrix). Without loss of generality, we can now assume that p and q are the minimal indices of the components of $\mathbf{A}$ and $\mathbf{B}$ equal to a_p and b_q, respectively; that is introducing $a = a_p$ and $b = b_q$, we have $p = \min\{i : a_i = a,\ 1 \le i \le n\}$ and $q = \min\{j : b_j = b,\ 1 \le j \le m\}$. Note also that, if $a_i = a_p$ and $b_j = b_q$, then the components of $\overline{\mathbf{E}}$ and $\overline{\mathbf{F}}$ are such that $e_i = e_p$ and $f_j = f_q$.

Since $r_{pq} = 1$, it holds that $a_p \ge q$; otherwise, the principal $(1;p)$-reductive pair of vectors $\mathbf{A}^{(1)}(p)$ and $\mathbf{B}_*^{(1)}$ would define a matrix $X(\mathbf{A},\mathbf{B}) = (x_{ij}) \in M_Z(\mathbf{A},\mathbf{B};1)$ with $x_{pq} = 0$. Since $r'_{pq} = 0$, it holds that $l_{p_k}(\mathbf{B};1) < q$ for every $p_k \ge p$; otherwise, the equality $r_{pq}^{(p_k)} = 1$ for the matrix $R_{p_k}(\mathbf{A},\mathbf{B})$ would entail $r'_{pq} = 1$.

Consider vectors $\mathbf{A}' = (a'_1,\ldots,a'_n)$ and $\mathbf{B}' = (b'_1,\ldots,b'_m)$, where

$$a'_i = \begin{cases} a_i, & 1 \le i \le n, \quad i \ne p, \\ a_p + 1, & i = p, \end{cases} \qquad b'_j = \begin{cases} b_j, & 1 \le j \le m, \quad j \ne q, \\ b_q + 1, & j = q. \end{cases}$$

The conditions for indices p and q imply that $\mathbf{A}' \in \overline{Z}^n_+$ and $\mathbf{B}' \in \overline{Z}^m_+$. When $\mathbf{A}'$ and $\mathbf{B}'$ are derived from $\mathbf{A}$ and $\mathbf{B}$, the value of function (1.4) (i.e., function (1.22) from Chapter 1 with $c = 1$) may decrease by not more than unity. Therefore, if $\delta_k(\mathbf{A},\mathbf{B};1) > 0$, then $\delta_k(\mathbf{A}',\mathbf{B}';1) \ge 0$ for $1 \le k \le n$. Let us show that the value of function (1.4) for the vectors $\mathbf{A}'$ and $\mathbf{B}'$ is not less than that for $\mathbf{A}$ and $\mathbf{B}$ if $\delta_k(\mathbf{A},\mathbf{B};1) = 0$ for $k \le b_1$. Let p_k be such that

$$\delta_{p_k}(\mathbf{A},\mathbf{B};1) = \sum_{j=1}^{m} b_j - \sum_{b_j \ge p_k} (b_j - p_k) - \sum_{i=1}^{p_k} a_i = 0.$$

If $p_k < p$, then it can be readily shown that the value of $\delta_{p_k}(\mathbf{A},\mathbf{B};1)$ either remains invariant or increases by unity; that is, $\delta_{p_k}(\mathbf{A}',\mathbf{B}';1) \ge 0$.

Let $p_k \ge p$. Then, $l_{p_k}(\mathbf{B};1) < q$ as indicated above. Therefore, $b_q < p_k$ and

$$\sum_{b_j \ge p_k} (b_j - p_k) = \sum_{b'_j \ge p_k} (b'_j - p_k).$$

Since this entails

$$\sum_{j=1}^{m} b_j - \sum_{i=1}^{p_k} a_i = \sum_{j=1}^{m} b'_j - \sum_{i=1}^{p_k} a'_i,$$

we have $\delta_{p_k}(\mathbf{A}', \mathbf{B}'; 1) = \delta_{p_k}(\mathbf{A}, \mathbf{B}; 1)$ for $p_k \geq p$. Therefore, the vectors $\mathbf{A}'$ and $\mathbf{B}'$ are such that $\delta_k(\mathbf{A}', \mathbf{B}'; 1) \geq 0$ for any k $(1 \leq k \leq n)$, and $M_Z(\mathbf{A}', \mathbf{B}'; 1) \neq \varnothing$.

We noted above that $a_p \geq q$. Accordingly, $a'_p > q$, and there exists a matrix $X'(\mathbf{A}', \mathbf{B}') = (x'_{ij})$ from $M_Z(\mathbf{A}', \mathbf{B}'; 1)$, defined by the principal $(1;p)$-reductive pair of vectors $\mathbf{A}'^{(1)}(p)$ and $\mathbf{B}'^{(1)}_*$, such that $x'_{pq} = 1$. Replacing the element $x'_{pq} = 1$ of matrix $X'(\mathbf{A}', \mathbf{B}')$ by zero, we obviously obtain a matrix $X(\mathbf{A}, \mathbf{B}) = (x_{ij})$ from $M_Z(\mathbf{A}, \mathbf{B}; 1)$ with $x_{pq} = 0$. By Definition 3.1, the element r_{pq} of matrix $R(\mathbf{A}, \mathbf{B}) = (r_{ij})$ is zero, which contradicts the assumption. This proves that $r'_{pq} = 1$ in the matrix $R'(\mathbf{A}, \mathbf{B}) = (r'_{ij})$. Therefore, $R(\mathbf{A}, \mathbf{B}) \subseteq R'(\mathbf{A}, \mathbf{B})$. Since $R'(\mathbf{A}, \mathbf{B}) \subseteq R(\mathbf{A}, \mathbf{B})$, then $R(\mathbf{A}, \mathbf{B}) = R'(\mathbf{A}, \mathbf{B})$, which completes the proof.

Remark 3.3. The function defined by (1.4) in Theorem 2.2 from Chapter 3 determines conditions ensuring that $x_{pq} = 1$ for any matrix $X(\mathbf{A}, \mathbf{B}) = (x_{ij}) \in M_Z(\mathbf{A}, \mathbf{B}; 1)$.

In the proof of Theorem 3.3, it was shown that these conditions are necessary for the equality $x_{pq} = 1$ to hold for any matrix in $M_Z(\mathbf{A}, \mathbf{B}; 1)$. Thus, we have constructed a criterion that guarantees $x_{pq} = 1$ (or $x_{pq} = 0$) for any matrix from $M_Z(\mathbf{A}, \mathbf{B}; 1)$.

Theorem 3.4. *Suppose that* $M_Z(\mathbf{A}, \mathbf{B}; 1) \neq \varnothing$, *where* $\mathbf{A} = (a_1, \ldots, a_n) \in \overline{Z}^n_+$, $\mathbf{B} = (b_1, \ldots, b_m) \in \overline{Z}^m_+$, *and* p, $q \in Z$ $(1 \leq p \leq n,\ 1 \leq q \leq m)$.

(a) $x_{pq} = 1$ *for any* $X(\mathbf{A}, \mathbf{B}) = (x_{ij}) \in M_Z(\mathbf{A}, \mathbf{B}; 1)$ *if and only if there exists* $k \in Z$ $(1 \leq k \leq b_1)$ *such that* $k \geq p$, $l_k(\mathbf{B}; 1) \geq q$, *and* $\delta_k(\mathbf{A}, \mathbf{B}; 1) = 0$.

(b) *If* a_p, $b_q > 0$, *then the equality* $x_{pq} = 0$ *is valid for any* $X(\mathbf{A}, \mathbf{B}) = (x_{ij}) \in M_Z(\mathbf{A}, \mathbf{B}; 1)$ *if and only if there exists* $k \in Z$ $(1 \leq k < p)$ *such that* $l_k(\mathbf{B}; 1) < q$ *and* $\delta_k(\mathbf{A}, \mathbf{B}; 1) = 0$.

We now construct the hull $O(\mathbf{A}, \mathbf{B}) = (o_{ij})$ of the matrix set $M_Z(\mathbf{A}, \mathbf{B}; 1) \neq \varnothing$, where $\mathbf{A} = (a_1, \ldots, a_n) \in \overline{Z}^n_+$ and $\mathbf{B} = (b_1, \ldots, b_m) \in \overline{Z}^m_+$:

$$O(\mathbf{A}, \mathbf{B}) = \bigcup_{X(\mathbf{A}, \mathbf{B}) \in M_Z(\mathbf{A}, \mathbf{B}; 1)} X(\mathbf{A}, \mathbf{B}).$$

Theorem 3.4 states the conditions for $O(\mathbf{A}, \mathbf{B}) \neq H(1)$. Denote by $p_k \in Z$ $(1 \leq p_k < b_1,\ 1 \leq k \leq t)$ every number such that $\delta_{p_k}(\mathbf{A}, \mathbf{B}; 1) = 0$ and $l_{p_k}(\mathbf{B}; c) < m$. If the set $\{p_k : 1 \leq k \leq t\}$ is empty, then $t = 0$. We set $p_{t+1} = \max\{i : a_i > 0,\ 1 \leq i \leq n\}$ if $a_1 \geq 1$ and $p_1 = 0$ if $a_1 = 0$.

With each value of p_k $(1 \leq k \leq t + 1)$, we associate the n-by-m matrix $O_{p_k}(\mathbf{A}, \mathbf{B}) = (o^{(p_k)}_{ij})$. If $t > 0$, then Theorem 2.2 from Chapter 3 defines the elements of any matrix $X(\mathbf{A},\mathbf{B}) = (x_{ij}) \in M_Z(\mathbf{A},\mathbf{B};1)$ for every p_k $(1 \le k \le t)$:

$$x_{ij} = \begin{cases} 1, & 1 \le i \le p_k, \quad 1 \le j \le l_{p_k}(\mathbf{B};1), \\ 0, & p_k + 1 \le i \le n, \quad l_{p_k}(\mathbf{B};1) + 1 \le j \le m. \end{cases}$$

We define the elements of $O_{p_k}(\mathbf{A},\mathbf{B}) = (o_{ij}^{(p_k)})$, where $1 \le k \le t$, as follows:

$$o_{ij}^{(p_k)} = \begin{cases} 1, & 1 \le i \le p_k, \quad 1 \le j \le m, \\ 1, & p_k + 1 \le i \le n, \quad 1 \le j \le l_{p_k}(\mathbf{B};1), \\ 0, & p_k + 1 \le i \le n, \quad l_{p_k}(\mathbf{B};1) + 1 \le j \le m. \end{cases}$$

The elements of $O_{p_{t+1}}(\mathbf{A},\mathbf{B}) = (o_{ij}^{(p_{t+1})})$ are defined as

$$o_{ij}^{(p_{t+1})} = \begin{cases} 1, & 1 \le i \le p_{t+1}, \quad 1 \le j \le \max\{j : b_j > 0,\ 1 \le j \le m\}, \\ 0, & 1 \le i \le p_{t+1}, \quad \max\{j : b_j > 0,\ 1 \le j \le m\} < j \le m, \\ 0, & p_{t+1} < i \le n, \quad 1 \le j \le m. \end{cases}$$

For each $X(\mathbf{A},\mathbf{B}) \in M_Z(\mathbf{A},\mathbf{B};1)$, it is obvious that

$$X(\mathbf{A},\mathbf{B}) \subseteq O_{p_k}(\mathbf{A},\mathbf{B}), \quad 1 \le k \le t+1.$$

Therefore,

$$X(\mathbf{A},\mathbf{B}) \subseteq \bigcap_{k=1}^{t+1} O_{p_k}(\mathbf{A},\mathbf{B}), \quad X(\mathbf{A},\mathbf{B}) \in M_Z(\mathbf{A},\mathbf{B};1).$$

Note the following trivial case: if $a_1 = 0$, then $t = 0$, $p_1 = 0$, and $O_1(\mathbf{A},\mathbf{B}) = H(0) = O(\mathbf{A},\mathbf{B})$.

The definition of hull $O(\mathbf{A},\mathbf{B})$ implies that

$$O(\mathbf{A},\mathbf{B}) \subseteq \bigcap_{k=1}^{t+1} O_{p_k}(\mathbf{A},\mathbf{B}) = O'(\mathbf{A},\mathbf{B}).$$

Clearly, each $O_{p_k}(\mathbf{A},\mathbf{B})$ is an extremal matrix, and so is $O'(\mathbf{A},\mathbf{B})$ by Lemma 2.5.

We define the n-by-m extremal vector pair $(\overline{\mathbf{E}}^O, \overline{\mathbf{F}}^O)$ for which $O'(\mathbf{A},\mathbf{B}) = \overline{H}(\overline{\mathbf{E}}^O, \overline{\mathbf{F}}^O)$. Each extremal matrix $O_{p_k}(\mathbf{A},\mathbf{B})$ is determined by an extremal vector pair:

$$\overline{\mathbf{E}}^O_{p_k} = \Big(\underbrace{m, \ldots, m}_{p_k}, \underbrace{l_{p_k}(\mathbf{B};1), \ldots, l_{p_k}(\mathbf{B};1)}_{n-p_k}\Big),$$

$$\overline{\mathbf{F}}^O_{p_k} = (\underbrace{n, \ldots, n}_{l_{p_k}(\mathbf{B};1)}, \underbrace{p_k, \ldots, p_k}_{m-l_{p_k}(\mathbf{B};1)}),$$

where $t > 0$ and $1 \le k \le t$; and

$$\overline{\mathbf{E}}^O_{p_{t+1}} = \Big(\underbrace{\max\{j : b_j > 0\}, \ldots, \max\{j : b_j > 0\}}_{p_{t+1}}, \underbrace{0, \ldots, 0}_{n-p_{t+1}}\Big),$$

$$\overline{\mathbf{F}}^O_{p_{t+1}} = (\underbrace{p_{t+1}, \ldots, p_{t+1}}_{\max\{j:b_j>0\}}, \underbrace{0, \ldots, 0}_{m-\max\{j:b_j>0\}});$$

$$O_{p_k}(\mathbf{A}, \mathbf{B}) = \overline{H}(\overline{\mathbf{E}}^O_{p_k}, \overline{\mathbf{F}}^O_{p_k}), \quad 1 \le k \le t+1.$$

Hence, by Lemma 2.4,

$$(\overline{\mathbf{E}}^O, \overline{\mathbf{F}}^O) = \bigwedge_{k=1}^{t+1} (\overline{\mathbf{E}}^O_{p_k}, \overline{\mathbf{F}}^O_{p_k}) = \Big(\bigwedge_{k=1}^{t+1} \overline{\mathbf{E}}^O_{p_k}, \bigwedge_{k=1}^{t+1} \overline{\mathbf{F}}^O_{p_k}, \Big),$$

$$O'(\mathbf{A}, \mathbf{B}) = \overline{H}\Big(\bigwedge_{k=1}^{t+1} \overline{\mathbf{E}}^O_{p_k}, \bigwedge_{k=1}^{t+1} \overline{\mathbf{F}}^O_{p_k}, \Big) = \overline{H}(\overline{\mathbf{E}}^O, \overline{\mathbf{F}}^O).$$

Let us prove that $O(\mathbf{A}, \mathbf{B}) = O'(\mathbf{A}, \mathbf{B})$.

Theorem 3.5. *Suppose that $M_Z(\mathbf{A}, \mathbf{B}; 1) \neq \varnothing$, where $\mathbf{A} = (a_1, \ldots, a_n) \in \overline{Z}^n_+$ and $\mathbf{B} = (b_1, \ldots, b_m) \in \overline{Z}^m_+$. Then, the hull $O(\mathbf{A}, \mathbf{B})$ of $M_Z(\mathbf{A}, \mathbf{B}; 1)$ can be represented as*

$$O(\mathbf{A}, \mathbf{B}) = \bigcup_{X(\mathbf{A},\mathbf{B}) \in M_Z(\mathbf{A},\mathbf{B};1)} X(\mathbf{A}, \mathbf{B}) = \bigcap_{k=1}^{t+1} O_{p_k}(\mathbf{A}, \mathbf{B}) = O'(\mathbf{A}, \mathbf{B}).$$

Proof. It will suffice to show that $O'(\mathbf{A}, \mathbf{B}) \subseteq O(\mathbf{A}, \mathbf{B})$. Assume that p, $q \in Z$, where $1 \le p \le n$ and $1 \le q \le m$. Let us show that the matrices $O'(\mathbf{A}, \mathbf{B}) = (o'_{ij})$ and $O(\mathbf{A}, \mathbf{B}) = (o_{ij})$ are such that if $o_{pq} = 0$, then $o'_{pq} = 0$. Indeed, this is obvious when $a_p = 0$ or $b_q = 0$, because $(o^{(p_{t+1})}_{pq}) = 0$. When $a_p > 0$ and $b_q > 0$, we apply Theorem 3.4 to show that there exists $r \in Z$ $(1 \le r < p)$ for which $l_r(\mathbf{B}; 1) < q$ and $\delta_r(\mathbf{A}, \mathbf{B}; 1) = 0$. Therefore, $p_k = r$ for a certain k $(1 \le k \le t)$. For every p_k, there exists a matrix $O_{p_k}(\mathbf{A}, \mathbf{B}) = (o^{(p_k)}_{ij})$ with zero element $o^{(p_k)}_{pq}$. The definition of $O'(\mathbf{A}, \mathbf{B}) = (o'_{ij})$ implies that $o'_{pq} = 0$. Hence, $O'(\mathbf{A}, \mathbf{B}) \subseteq O(\mathbf{A}, \mathbf{B})$ and $O(\mathbf{A}, \mathbf{B}) \subseteq O'(\mathbf{A}, \mathbf{B})$ then entails $O(\mathbf{A}, \mathbf{B}) = O'(\mathbf{A}, \mathbf{B})$, which completes the proof.

Remark 3.4. In Theorem 3.3, the trivial case $R(\mathbf{A},\mathbf{B}) = H(0)$ specified by proposition (a) could be left out. Then, setting $p_0 = 0$ and $R_0(\mathbf{A},\mathbf{B}) = (r_{ij}^{(0)}) = H(0)$ we would have

$$R(\mathbf{A},\mathbf{B}) = \bigcup_{k=0}^{t} R_{p_k}(\mathbf{A},\mathbf{B}),$$

where $R_0(\mathbf{A},\mathbf{B}) = \overline{H}(\overline{\mathbf{E}}_0^R, \overline{\mathbf{F}}_0^R)$ with $\overline{\mathbf{E}}_0^R = (0,\dots,0) \in \overline{Z}_+^n$ and $\overline{\mathbf{F}}_0^R = (0,\dots,0) \in \overline{Z}_+^m$.

Let us write out expressions for the kernel $R(\mathbf{A},\mathbf{B})$ and hull $O(\mathbf{A},\mathbf{B})$. If $M_Z(\mathbf{A},\mathbf{B};1) \neq \varnothing$, where $\mathbf{A} \in \overline{Z}_+^n$ and $\mathbf{B} \in \overline{Z}_+^m$, then

$$R(\mathbf{A},\mathbf{B}) = \bigcap_{X(\mathbf{A},\mathbf{B}) \in M_Z(\mathbf{A},\mathbf{B};1)} X(\mathbf{A},\mathbf{B}) = \bigcup_{k=0}^{t} R_{p_k}(\mathbf{A},\mathbf{B}) = \overline{H}(\overline{\mathbf{E}}^R, \overline{\mathbf{F}}^R) = \overline{H}\Big(\bigvee_{k=0}^{t} \overline{\mathbf{E}}_{p_k}^R, \bigvee_{k=0}^{t} \overline{\mathbf{F}}_{p_k}^R\Big),$$

$$O(\mathbf{A},\mathbf{B}) = \bigcup_{X(\mathbf{A},\mathbf{B}) \in M_Z(\mathbf{A},\mathbf{B};1)} X(\mathbf{A},\mathbf{B}) = \bigcap_{k=1}^{t+1} O_{p_k}(\mathbf{A},\mathbf{B}) = \overline{H}(\overline{\mathbf{E}}^O, \overline{\mathbf{F}}^O) = \overline{H}\Big(\bigwedge_{k=1}^{t+1} \overline{\mathbf{E}}_{p_k}^O, \bigwedge_{k=1}^{t+1} \overline{\mathbf{F}}_{p_k}^O\Big).$$

The following theorem is valid by virtue of Theorems 3.3 and 3.5, Lemma 2.5, and the extremality of $R'(\mathbf{A},\mathbf{B})$ and $O'(\mathbf{A},\mathbf{B})$.

Theorem 3.6. *If $M_Z(\mathbf{A},\mathbf{B};1) \neq \varnothing$, where $\mathbf{A} \in \overline{Z}_+^n$ and $\mathbf{B} \in \overline{Z}_+^m$, then $R(\mathbf{A},\mathbf{B})$, $O(\mathbf{A},\mathbf{B}) \in \overline{E}_{n,m}$; that is, the kernel and hull of the matrix set $M_Z(\mathbf{A},\mathbf{B};1)$ are extremal.*

The following proposition is obviously true.

Theorem 3.7. *Suppose $X(\mathbf{A},\mathbf{B}) \in M_Z(\mathbf{A},\mathbf{B};1)$.*

(a) *If $\mathbf{A} \in \overline{Z}_+^n$ and $\mathbf{B} \in \overline{Z}_+^m$, then $X(\mathbf{A},\mathbf{B})$ is an extremal matrix if and only if $X(\mathbf{A},\mathbf{B}) = R(\mathbf{A},\mathbf{B}) = O(\mathbf{A},\mathbf{B})$.*

(b) *If $\mathbf{A} \in Z_+^n$ and $\mathbf{B} \in Z_+^m$, then $X(\mathbf{A},\mathbf{B})$ is a perfect matrix if and only if $\widetilde{X}(\widetilde{\mathbf{A}},\widetilde{\mathbf{B}}) = R(\widetilde{\mathbf{A}},\widetilde{\mathbf{B}}) = O(\widetilde{\mathbf{A}},\widetilde{\mathbf{B}})$, where $\widetilde{X}(\widetilde{\mathbf{A}},\widetilde{\mathbf{B}})$ is the matrix obtained by permuting the rows and columns of $X(\mathbf{A},\mathbf{B})$, with $\widetilde{\mathbf{A}} \in \overline{Z}_+^n$ and $\widetilde{\mathbf{B}} \in \overline{Z}_+^m$.*

Consider extremal matrices $\overline{H}_1$, $\overline{H}_2 \in \overline{E}_{n,m}$ such that $\overline{H}_1 \subseteq X(\mathbf{A},\mathbf{B}) \subseteq \overline{H}_2$ for any $X(\mathbf{A},\mathbf{B}) \in M_Z(\mathbf{A},\mathbf{B};1) \neq \varnothing$, where $\mathbf{A} \in \overline{Z}^n_+$ and $\mathbf{B} \in \overline{Z}^m_+$. Then,

$$\overline{H}_1 \subseteq \bigcap_{X(\mathbf{A},\mathbf{B})\in M_Z(\mathbf{A},\mathbf{B};1)}, \qquad \bigcup_{X(\mathbf{A},\mathbf{B})\in M_Z(\mathbf{A},\mathbf{B};1)} \subseteq \overline{H}_2;$$

hence, $\overline{H}_1 \subseteq R(\mathbf{A},\mathbf{B})$, $O(\mathbf{A},\mathbf{B}) \subseteq \overline{H}_2$. Therefore, the metric on the set of extremal matrices defined by (2.1) is such that $r\left(R(\mathbf{A},\mathbf{B}),\ O(\mathbf{A},\mathbf{B})\right) \leq r(\overline{H}_1, \overline{H}_2)$, and the following theorem is true.

Theorem 3.8. *For any matrix set $M_Z(\mathbf{A},\mathbf{B};1) \neq \varnothing$, where $\mathbf{A} \in \overline{Z}^n_+$ and $\mathbf{B} \in \overline{Z}^m_+$, it holds that*

$$r\left(R(\mathbf{A},\mathbf{B}),\ O(\mathbf{A},\mathbf{B})\right) = \min_{\overline{H}',\overline{H}''\in\overline{E}_{n,m}} \left\{ r(\overline{H}', \overline{H}'') : \overline{H}' \subseteq X(\mathbf{A},\mathbf{B}) \subseteq \overline{H}'',\ X(\mathbf{A},\mathbf{B}) \in M_Z(\mathbf{A},\mathbf{B};1) \right\}.$$

3.4. Algorithms for Constructing Kernels and Hulls

Algorithm 3.1. Construction of the kernel $R(\mathbf{A},\mathbf{B})$ of a matrix set $M_Z(\mathbf{A},\mathbf{B};1)$ and an extremal vector pair $(\overline{\mathbf{E}}^R, \overline{\mathbf{F}}^R)$ such that $R(\mathbf{A},\mathbf{B}) = \overline{H}(\overline{\mathbf{E}}^R, \overline{\mathbf{F}}^R)$.

Suppose that $\mathbf{A} = (a_1, \ldots, a_n) \in \overline{Z}^n_+$, $\mathbf{B} = (b_1, \ldots, b_m) \in \overline{Z}^m_+$, and $M_Z(\mathbf{A},\mathbf{B};1) \neq \varnothing$.

Step 1. Set $k = 0$ and $t = 0$.

Step 2. Introduce the n-by-m zero matrix $R'_t(\mathbf{A},\mathbf{B}) = (r'^{(t)}_{ij})$ and the n-by-m zero vector pair $(\overline{\mathbf{E}}'^R_t, \overline{\mathbf{F}}'^R_t) = \left((e'^{(t)}_1, \ldots, e'^{(t)}_n), (f'^{(t)}_1, \ldots, f'^{(t)}_m)\right)$.

Step 3. Increase k by unity.

Step 4. If $k \leq b_1$, then go to the next step. If $k > b_1$, go to Step 11.

Step 5. Calculate

$$l_k(\mathbf{B};1) = \max\{j : b_j \geq k,\ 1 \leq j \leq m\},$$

$$\delta_k(\mathbf{A},\mathbf{B};1) = \sum_{j>l_k(\mathbf{B};1)} b_j + k l_k(\mathbf{B};1) - \sum_{i=1}^{k} a_i.$$

Step 6. If $\delta_k(\mathbf{A},\mathbf{B};1) = 0$, then go to the next step. If $\delta_k(\mathbf{A},\mathbf{B};1) > 0$, then go to Step 3.

Step 7. Construct the n-by-m matrix $R_k(\mathbf{A},\mathbf{B}) = (r_{ij}^{(k)})$ and vector pair $(\overline{\mathbf{E}}_k^R, \overline{\mathbf{F}}_k^R) = \left((e_1^{(k)},\dots,e_n^{(k)}),(f_1^{(k)},\dots,f_m^{(k)})\right)$ defined as follows:

$$r_{ij}^{(k)} = \begin{cases} 1, & 1 \le i \le k, \quad 1 \le j \le l_k(\mathbf{B};1), \\ 0, & 1 \le i \le k, \quad l_k(\mathbf{B};1) < j \le m, \\ 0, & k < i \le n, \quad 1 \le j \le m, \end{cases}$$

$$e_i^{(k)} = \begin{cases} l_k(\mathbf{B};1), & 1 \le i \le k, \\ 0, & k < i \le n, \end{cases} \qquad f_j^{(k)} = \begin{cases} k, & 1 \le j \le l_k(\mathbf{B};1), \\ 0, & l_k(\mathbf{B};1) < j \le m. \end{cases}$$

Step 8. Increase t by unity.

Step 9. For the known matrices

$$R'_{t-1}(\mathbf{A},\mathbf{B}) = (r_{ij}'^{(t-1)}), \quad R_k(\mathbf{A},\mathbf{B}) = (r_{ij}^{(k)})$$

and vector pairs

$$(\overline{\mathbf{E}}'^R_{t-1}, \overline{\mathbf{F}}'^R_{t-1}) = \left((e_1'^{(t-1)},\dots,e_n'^{(t-1)}),(f_1'^{(t-1)},\dots,f_m'^{(t-1)})\right),$$

$$(\overline{\mathbf{E}}_k^R, \overline{\mathbf{F}}_k^R) = \left((e_1^{(k)},\dots,e_n^{(k)}),(f_1^{(k)},\dots,f_m^{(k)})\right),$$

construct the matrix

$$R'_t(\mathbf{A},\mathbf{B}) = R'_{t-1}(\mathbf{A},\mathbf{B}) \cup R_k(\mathbf{A},\mathbf{B}) = (r_{ij}'^{(t)})$$

and vector pair

$$(\overline{\mathbf{E}}'^R_t, \overline{\mathbf{F}}'^R_t) = (\overline{\mathbf{E}}'^R_{t-1}, \overline{\mathbf{F}}'^R_{t-1}) \vee (\overline{\mathbf{E}}_k^R, \overline{\mathbf{F}}_k^R) = (\overline{\mathbf{E}}'^R_{t-1} \vee \overline{\mathbf{E}}_k^R, \overline{\mathbf{F}}'^R_{t-1} \vee \overline{\mathbf{F}}_k^R) =$$
$$\left((e_1'^{(t)},\dots,e_n'^{(t)}),(f_1'^{(t)},\dots,f_m'^{(t)})\right).$$

The elements of $R'_t(\mathbf{A},\mathbf{B})$ and components of $(\overline{\mathbf{E}}'^R_t, \overline{\mathbf{F}}'^R_t)$ are calculated as follows:

$$r_{ij}'^{(t)} = \max(r_{ij}'^{(t-1)}, r_{ij}^{(k)}), \quad 1 \le i \le n, \quad 1 \le j \le m,$$
$$e_i'^{(t)} = \max(e_i'^{(t-1)}, e_i^{(k)}), \quad 1 \le i \le n,$$
$$f_j'^{(t)} = \max(f_j'^{(t-1)}, f_j^{(k)}), \quad 1 \le j \le m.$$

Step 10. Go to Step 3.

Step 11. Define the kernel $R(\mathbf{A},\mathbf{B})$ of the matrix set $M_Z(\mathbf{A},\mathbf{B};1)$ and the corresponding vector pair as

$$R(\mathbf{A},\mathbf{B}) = R'_t(\mathbf{A},\mathbf{B}), \quad (\overline{\mathbf{E}}^R, \overline{\mathbf{F}}^R) = (\overline{\mathbf{E}}'^R_t, \overline{\mathbf{F}}'^R_t), \quad R(\mathbf{A},\mathbf{B}) = \overline{H}(\overline{\mathbf{E}}^R, \overline{\mathbf{F}}^R).$$

Algorithm 3.1 relies on Theorem 3.3.

Algorithm 3.2. Construction of the hull $O(\mathbf{A},\mathbf{B})$ of a matrix set $M_Z(\mathbf{A},\mathbf{B};1)$ and the extremal vector pair $(\overline{\mathbf{E}}^O,\overline{\mathbf{F}}^O)$ such that $O(\mathbf{A},\mathbf{B}) = \overline{H}(\overline{\mathbf{E}}^O,\overline{\mathbf{F}}^O)$.

Suppose that $\mathbf{A} = (a_1,\dots,a_n) \in \overline{Z}_+^n$, $\mathbf{B} = (b_1,\dots,b_m) \in \overline{Z}_+^m$, and $M_Z(\mathbf{A},\mathbf{B};1) \neq \varnothing$.

Step 1. If $b_1 = 0$, go to Step 17. If $b_1 \geq 1$, go to the next step.

Step 2. Calculate $n_1 = \max\{i : a_i > 0,\ 1 \leq i \leq n\}$ and $m_1 = \max\{j : b_j > 0,\ 1 \leq j \leq m\}$.

Step 3. Set $t = 0$.

Step 4. Construct the n-by-m matrix $O'_t(\mathbf{A},\mathbf{B}) = (o'^{(t)}_{ij})$ and vector pair $(\overline{\mathbf{E}}'^O_t,\overline{\mathbf{F}}'^O_t) = \left((e'^{(t)}_1,\dots,e'^{(t)}_n),(f'^{(t)}_1,\dots,f'^{(t)}_m)\right)$ defined as follows:

$$o'^{(t)}_{ij} = \begin{cases} 1, & 1 \leq i \leq n_1, \quad 1 \leq j \leq m_1, \\ 0, & 1 \leq i \leq n_1, \quad m_1 < j \leq m, \\ 0, & n_1 < i \leq n, \quad 1 \leq j \leq m, \end{cases}$$

$$e'^{(t)}_i = \begin{cases} m_1, & 1 \leq i \leq n_1, \\ 0, & n_1 < i \leq n, \end{cases} \qquad f'^{(t)}_j = \begin{cases} n_1, & 1 \leq j \leq m_1, \\ 0, & m_1 < j \leq m. \end{cases}$$

Step 5. Set $k = 1$.

Step 6. If $k \leq b_1$, go to the next step. If $k > b_1$, then go to Step 16.

Step 7. Calculate

$$l_k(\mathbf{B};1) = \max\{j : b_j \geq k,\ 1 \leq j \leq m\}.$$

Step 8. If $l_k(\mathbf{B};1) < m$, go to the next step. If $l_k(\mathbf{B};1) = m$, then go to Step 14.

Step 9. Calculate

$$\delta_k(\mathbf{A},\mathbf{B};1) = \sum_{j > l_k(\mathbf{B};1)} b_j + k l_k(\mathbf{B};1) - \sum_{i=1}^{k} a_i.$$

Step 10. If $\delta_k(\mathbf{A},\mathbf{B};1) = 0$, go to the next step. If $\delta_k(\mathbf{A},\mathbf{B};1) > 0$, then go to Step 14.

Step 11. Construct the n-by-m matrix $O_k(\mathbf{A},\mathbf{B}) = (o^{(k)}_{ij})$ and vector pair $(\overline{\mathbf{E}}^O_k,\overline{\mathbf{F}}^O_k) = \left((e^{(k)}_1,\dots,e^{(k)}_n),(f^{(k)}_1,\dots,f^{(k)}_m)\right)$ defined as follows:

$$o^{(k)}_{ij} = \begin{cases} 1, & 1 \leq i \leq k, \quad 1 \leq j \leq m, \\ 1, & k+1 \leq i \leq n, \quad 1 \leq j \leq l_k(\mathbf{B};1), \\ 0, & k+1 \leq i \leq n, \quad l_k(\mathbf{B};1)+1 \leq j \leq m, \end{cases}$$

$$e_i^{(k)} = \begin{cases} m, & 1 \le i \le k, \\ l_k(\mathbf{B};1), & k+1 \le i \le n, \end{cases} \qquad f_j^{(k)} = \begin{cases} n, & 1 \le j \le l_k(\mathbf{B};1), \\ k, & l_k(\mathbf{B};1)+1 \le j \le m. \end{cases}$$

Step 12. Increase t by unity.

Step 13. For the constructed matrices

$$O'_{t-1}(\mathbf{A},\mathbf{B}) = (o_{ij}^{\prime(t-1)}), \quad O_k(\mathbf{A},\mathbf{B}) = (o_{ij}^{(k)})$$

and vector pairs

$$(\overline{\mathbf{E}}'^{O}_{t-1}, \overline{\mathbf{F}}'^{O}_{t-1}) = \left((e_1^{\prime(t-1)},\dots,e_n^{\prime(t-1)}),(f_1^{\prime(t-1)},\dots,f_m^{\prime(t-1)})\right),$$

$$(\overline{\mathbf{E}}^{O}_{k}, \overline{\mathbf{F}}^{O}_{k}) = \left((e_1^{(k)},\dots,e_n^{(k)}),(f_1^{(k)},\dots,f_m^{(k)})\right),$$

construct the matrix

$$O'_t(\mathbf{A},\mathbf{B}) = O'_{t-1}(\mathbf{A},\mathbf{B}) \cap O_k(\mathbf{A},\mathbf{B}) = (o_{ij}^{\prime(t)})$$

and vector pair

$$(\overline{\mathbf{E}}'^{O}_{t}, \overline{\mathbf{F}}'^{O}_{t}) = (\overline{\mathbf{E}}'^{O}_{t-1}, \overline{\mathbf{F}}'^{O}_{t-1}) \wedge (\overline{\mathbf{E}}^{O}_{k}, \overline{\mathbf{F}}^{O}_{k}) = (\overline{\mathbf{E}}'^{O}_{t-1} \wedge \overline{\mathbf{E}}^{O}_{k}, \overline{\mathbf{F}}'^{O}_{t-1} \wedge \overline{\mathbf{F}}^{O}_{k}) =$$
$$\left((e_1^{\prime(t)},\dots,e_n^{\prime(t)}),(f_1^{\prime(t)},\dots,f_m^{\prime(t)})\right).$$

The elements of $O'_t(\mathbf{A},\mathbf{B})$ and components of $(\overline{\mathbf{E}}'^{O}_{t}, \overline{\mathbf{F}}'^{O}_{t})$ are calculated as follows:

$$o_{ij}^{\prime(t)} = \min(o_{ij}^{\prime(t-1)}, o_{ij}^{(k)}), \quad 1 \le i \le n, \quad 1 \le j \le m,$$
$$e_i^{\prime(t)} = \min(e_i^{\prime(t-1)}, e_i^{(k)}), \quad 1 \le i \le n,$$
$$f_j^{\prime(t)} = \min(f_j^{\prime(t-1)}, f_j^{(k)}), \quad 1 \le j \le k.$$

Note here that the hull $O(\mathbf{A},\mathbf{B})$ is constructed as the intersection of special matrices with the source matrix $O'_0(\mathbf{A},\mathbf{B}) = (o_{ij}^{\prime(0)})$. Therefore, Step 11 does not involve construction of the matrix corresponding to $l_k(\mathbf{B};1) = 0$ and $\delta_k(\mathbf{A},\mathbf{B};1) = 0$. Note also that n and m can be replaced by n_1 and m_1 in determining the unit elements of the matrices constructed at this step.

Step 14. Increase k by unity.

Step 15. Go to Step 6.

Step 16. Define the hull $O(\mathbf{A},\mathbf{B})$ of the matrix set $M_Z(\mathbf{A},\mathbf{B};1)$ and the corresponding vector pair as

$$O(\mathbf{A},\mathbf{B}) = O'_t(\mathbf{A},\mathbf{B}), \quad (\overline{\mathbf{E}}^{O}, \overline{\mathbf{F}}^{O}) = (\overline{\mathbf{E}}'^{O}_{t}, \overline{\mathbf{F}}'^{O}_{t}), \quad O(\mathbf{A},\mathbf{B}) = \overline{H}(\overline{\mathbf{E}}^{O}, \overline{\mathbf{F}}^{O}).$$

Step 17. The hull $O(\mathbf{A},\mathbf{B})$ is the matrix $H(0)$, whose elements are zeros.

Algorithm 3.2 relies of Theorem 3.5.

Example 3.1. Consider the matrix

$$X = \begin{pmatrix} 1&1&1&1&1&1&1&1&1&0 \\ 1&1&1&1&1&1&1&1&0&1 \\ 1&1&1&1&1&1&1&0&0&0 \\ 1&1&1&1&1&1&0&1&0&0 \\ 1&1&1&1&1&1&1&0&0&0 \\ 1&1&1&1&0&1&0&0&0&0 \\ 1&1&1&1&1&0&0&0&0&0 \\ 1&0&1&0&0&0&0&0&0&0 \\ 0&1&0&1&0&0&0&0&0&0 \end{pmatrix}.$$

Obviously, $X = X(\mathbf{A},\mathbf{B}) \in M_Z(\mathbf{A},\mathbf{B};1)$, where

$$\mathbf{A} = (9,9,7,7,7,5,5,2,2) \in \overline{Z}^9_+, \mathbf{B} = (8,8,8,8,6,6,4,3,1,1) \in \overline{Z}^{10}_+.$$

Since the dimension is low, there is no need to invoke Algorithm 3.1 or 3.2. To construct the kernel $R(\mathbf{A},\mathbf{B})$ and hull $O(\mathbf{A},\mathbf{B})$, it suffices to apply Theorem 3.4, which states necessary and sufficient conditions for any element x_{pq} of every $X(\mathbf{A},\mathbf{B}) = (x_{ij})$ to be either zero or one. These conditions depend on the value of $k \in Z$ for which $\delta_k(\mathbf{A},\mathbf{B};1) = 0$ and on the value of $l_k(\mathbf{B};1)$. Let us determine all $k \in Z$ $(1 \le k \le b_1)$ for which function (1.4) is zero and the corresponding values of $l_k(\mathbf{B};1)$. Evaluating

$$\delta_k(\mathbf{A},\mathbf{B};1) = \sum_{j=1}^{10} b_j - \sum_{b_j \ge k} (b_j - k) - \sum_{i=1}^{k} a_i, \quad 1 \le k \le b_1 = 8,$$

we obtain $\delta_k(\mathbf{A},\mathbf{B};1) = 0$ for $k = 2,\ 5,\ 7$. Then, for $l_k(\mathbf{B};1) = \max\{j : b_j \ge k,\ 1 \le j \le 10\}$ with $k = 2,\ 5,\ 7$ we have $l_2(\mathbf{B};1) = 8$, $l_5(\mathbf{B};1) = 6$, and $l_7(\mathbf{B};1) = 4$. The pairs 2, 8; 5, 6; and 7, 4 define the zero or unit elements of every $X(\mathbf{A},\mathbf{B}) \in M_Z(\mathbf{A},\mathbf{B};1)$. Applying Theorem 3.4, we consider the elements of every $X(\mathbf{A},\mathbf{B}) = (x_{ij}) \in M_Z(\mathbf{A},\mathbf{B};1)$ that are always equal to either zero or one:

$$X(\mathbf{A},\mathbf{B}) = \begin{pmatrix} 1&1&1&1&1&1&1&1&x_{19}&x_{1\,10} \\ 1&1&1&1&1&1&1&1&x_{29}&x_{2\,10} \\ 1&1&1&1&1&1&x_{37}&x_{38}&0&0 \\ 1&1&1&1&1&1&x_{47}&x_{48}&0&0 \\ 1&1&1&1&1&1&x_{57}&x_{58}&0&0 \\ 1&1&1&1&x_{65}&x_{66}&0&0&0&0 \\ 1&1&1&1&x_{75}&x_{76}&0&0&0&0 \\ x_{81}&x_{82}&x_{83}&x_{84}&0&0&0&0&0&0 \\ x_{91}&x_{92}&x_{93}&x_{94}&0&0&0&0&0&0 \end{pmatrix}. \tag{3.5}$$

For each unknown element x_{ij} of the constructed matrix, there exist at least two matrices in $M_Z(\mathbf{A},\mathbf{B};1)$ such that the corresponding element is unity in one of them and zero in the other. Therefore, if all unknown elements of $X(\mathbf{A},\mathbf{B})$ are replaced by zeros, the resulting matrix is the kernel $R(\mathbf{A},\mathbf{B})$; and if they are replaced by ones, the resulting matrix is the hull $O(\mathbf{A},\mathbf{B})$:

$$R(\mathbf{A},\mathbf{B}) = \begin{pmatrix} 1&1&1&1&1&1&1&1&0&0\\ 1&1&1&1&1&1&1&1&0&0\\ 1&1&1&1&1&1&0&0&0&0\\ 1&1&1&1&1&1&0&0&0&0\\ 1&1&1&1&1&1&0&0&0&0\\ 1&1&1&1&0&0&0&0&0&0\\ 1&1&1&1&0&0&0&0&0&0\\ 0&0&0&0&0&0&0&0&0&0\\ 0&0&0&0&0&0&0&0&0&0 \end{pmatrix},$$

$$O(\mathbf{A},\mathbf{B}) = \begin{pmatrix} 1&1&1&1&1&1&1&1&1&1\\ 1&1&1&1&1&1&1&1&1&1\\ 1&1&1&1&1&1&1&1&0&0\\ 1&1&1&1&1&1&1&1&0&0\\ 1&1&1&1&1&1&1&1&0&0\\ 1&1&1&1&1&1&0&0&0&0\\ 1&1&1&1&1&1&0&0&0&0\\ 1&1&1&1&0&0&0&0&0&0\\ 1&1&1&1&0&0&0&0&0&0 \end{pmatrix}.$$

Counting ones and zeros in the rows and columns of the extremal matrices $R(\mathbf{A},\mathbf{B})$ and $O(\mathbf{A},\mathbf{B})$, we readily calculate the components of the corresponding vectors in the extremal pairs:

$$(\overline{\mathbf{E}}^R,\overline{\mathbf{F}}^R) = \left((8,8,6,6,6,4,4,0,0),(7,7,7,7,5,5,2,2,0,0)\right),$$

$$(\overline{\mathbf{E}}^O,\overline{\mathbf{F}}^O) = \left((10,10,8,8,8,6,6,4,4),(9,9,9,9,7,7,5,5,2,2)\right),$$

$$R(\mathbf{A},\mathbf{B}) = \overline{H}(\overline{\mathbf{E}}^R,\overline{\mathbf{F}}^R), \quad O(\mathbf{A},\mathbf{B}) = \overline{H}(\overline{\mathbf{E}}^O,\overline{\mathbf{F}}^O).$$

Since every $X(\mathbf{A},\mathbf{B}) = (x_{ij}) \in M_Z(\mathbf{A},\mathbf{B};1)$ satisfies the condition $R(\mathbf{A},\mathbf{B}) \subseteq X(\mathbf{A},\mathbf{B}) \subseteq O(\mathbf{A},\mathbf{B})$, it is defined by the following submatrices:

$$\begin{matrix} & 1 & 1 \\ 1 & x_{19} & x_{1\,10} \\ 1 & x_{29} & x_{2\,10} \end{matrix}, \quad \begin{matrix} & 2 & 1 \\ 1 & x_{37} & x_{38} \\ 1 & x_{47} & x_{48} \\ 1 & x_{57} & x_{58} \end{matrix}, \quad \begin{matrix} & 1 & 1 \\ 1 & x_{65} & x_{66} \\ 1 & x_{75} & x_{76} \end{matrix}, \quad \begin{matrix} & 1 & 1 & 1 & 1 \\ 2 & x_{81} & x_{82} & x_{83} & x_{84} \\ 2 & x_{91} & x_{92} & x_{93} & x_{94} \end{matrix}. \tag{3.6}$$

Here, the sums of row and column entries are shown.

Treating the numbers shown on the left and on top of these matrices as the components of vectors $\mathbf{A}_1$ and $\mathbf{B}_1$ and arranging them in appropriate order, we write out $\mathbf{A}_1 = \mathbf{A} - \overline{\mathbf{E}}^R = \overline{\mathbf{E}}^O - \mathbf{A}$ and $\mathbf{B}_1 = \mathbf{B} - \overline{\mathbf{F}}^R = \overline{\mathbf{F}}^O - \mathbf{B}$:

$$(1,1,1,1,1,1,1,2,2) = (9,9,7,7,7,5,5,2,2) - (8,8,6,6,6,4,4,0,0) =$$
$$(10,10,8,8,8,6,6,4,4) - (9,9,7,7,7,5,5,2,2),$$
$$(1,1,1,1,1,1,2,1,1,1) = (8,8,8,8,6,6,4,3,1,1) - (7,7,7,7,5,5,2,2,0,0) =$$
$$(9,9,9,9,7,7,5,5,2,2) - (8,8,8,8,6,6,4,3,1,1).$$

The number of unknown elements in (3.5) determines the values of the metric function for both kernel and hull. It can be readily shown that

$$r\left(R(\mathbf{A},\mathbf{B}), O(\mathbf{A},\mathbf{B})\right) = 22.$$

Note that the diameter of the set (metric space) $\overline{E}_{9\,10}$ of extremal matrices is $9 \times 10 = 90$. We can easily count their number for $M_Z(\mathbf{A},\mathbf{B};1)$. Taking into account the vector components shown on the left and top of the matrices in (3.6), we see that the corresponding numbers of submatrices of the matrix in (3.5) are 2, 3, 2, and 6. Thus, the set $M_Z(\mathbf{A},\mathbf{B};1)$ considered here contains $2 \times 3 \times 2 \times 6 = 72$ matrices.

Note that (3.5) is a general expression for any matrix in $M(\mathbf{A},\mathbf{B};1)$.

Using the pair of vectors $\mathbf{A}$ and $\mathbf{B}$ from this example and recalling Theorem 3.2′, we consider the vectors $\overline{\mathbf{A}}^1 = (\overline{a}_1^1, \ldots, \overline{a}_n^1)$ and $\overline{\mathbf{B}}^1 = (\overline{b}_1^1, \ldots, \overline{b}_m^1)$, where $\overline{a}_i^1 = m - a_{n-i+1}$ $(1 \le i \le n = 9)$ and $\overline{b}_j^1 = n - b_{m-j+1}$ $(1 \le j \le m = 10)$:

$$\overline{\mathbf{A}}^1 = (8,8,5,5,3,3,3,1,1), \quad \overline{\mathbf{B}}^1 = (8,8,6,5,3,3,1,1,1,1).$$

With each $X(\mathbf{A},\mathbf{B}) = (x_{ij}) \in M_Z(\mathbf{A},\mathbf{B};1)$, we associate the matrix $\overline{X}^1(\overline{\mathbf{A}}^1,\overline{\mathbf{B}}^1) = (\overline{x}_{ij}^1) \in M_Z(\overline{\mathbf{A}}^1,\overline{\mathbf{B}}^1;1)$, where

$$\overline{x}_{ij}^1 = 1 - x_{n-i+1\, m-j+1}, \quad 1 \le i \le n = 9, \quad 1 \le j \le m = 10. \tag{3.7}$$

It is obvious that the matrices thus constructed are in one-to-one correspondence with the source matrices, and the vector pairs $\mathbf{A}, \mathbf{B}$ and $\overline{\mathbf{A}}^1, \overline{\mathbf{B}}^1$ are equivalent with respect to the characteristic function defined by (1.4) (recall Theorem 3.2′). It was pointed out above that the concepts of kernel and hull can be considered as dual: the kernel and hull of $M_Z(\mathbf{A},\mathbf{B};1)$ correspond to the hull and kernel of $M_Z(\overline{\mathbf{A}}^1,\overline{\mathbf{B}}^1;1)$, respectively. Therefore, every matrix $\overline{X}^1(\overline{\mathbf{A}}^1,\overline{\mathbf{B}}^1)$ in $M_Z(\overline{\mathbf{A}}^1,\overline{\mathbf{B}}^1;1)$, as well as the kernel and hull of the matrix set, can be derived from (3.5) by applying relations (3.7):

$$\overline{X}^1(\overline{\mathbf{A}}^1,\overline{\mathbf{B}}^1) = \begin{pmatrix} 1 & 1 & 1 & 1 & 1 & 1 & \overline{x}^1_{17} & \overline{x}^1_{18} & \overline{x}^1_{19} & \overline{x}^1_{1\,10} \\ 1 & 1 & 1 & 1 & 1 & 1 & \overline{x}^1_{27} & \overline{x}^1_{28} & \overline{x}^1_{29} & \overline{x}^1_{2\,10} \\ 1 & 1 & 1 & 1 & \overline{x}^1_{35} & \overline{x}^1_{36} & 0 & 0 & 0 & 0 \\ 1 & 1 & 1 & 1 & \overline{x}^1_{45} & \overline{x}^1_{46} & 0 & 0 & 0 & 0 \\ 1 & 1 & \overline{x}^1_{53} & \overline{x}^1_{54} & 0 & 0 & 0 & 0 & 0 & 0 \\ 1 & 1 & \overline{x}^1_{63} & \overline{x}^1_{64} & 0 & 0 & 0 & 0 & 0 & 0 \\ 1 & 1 & \overline{x}^1_{73} & \overline{x}^1_{74} & 0 & 0 & 0 & 0 & 0 & 0 \\ \overline{x}^1_{81} & \overline{x}^1_{82} & 0 & 0 & 0 & 0 & 0 & 0 & 0 & 0 \\ \overline{x}^1_{91} & \overline{x}^1_{92} & 0 & 0 & 0 & 0 & 0 & 0 & 0 & 0 \end{pmatrix}. \tag{3.8}$$

If all unknown entries in (3.8) are set equal to zeros or ones, the resulting matrix will be the kernel $R(\overline{\mathbf{A}}^1,\overline{\mathbf{B}}^1)$ or the hull $O(\overline{\mathbf{A}}^1,\overline{\mathbf{B}}^1)$ of the matrix set $M_Z(\overline{\mathbf{A}}^1,\overline{\mathbf{B}}^1;1)$, respectively:

$$R(\overline{\mathbf{A}}^1,\overline{\mathbf{B}}^1) = \begin{pmatrix} 1&1&1&1&1&1&0&0&0&0 \\ 1&1&1&1&1&1&0&0&0&0 \\ 1&1&1&1&0&0&0&0&0&0 \\ 1&1&1&1&0&0&0&0&0&0 \\ 1&1&0&0&0&0&0&0&0&0 \\ 1&1&0&0&0&0&0&0&0&0 \\ 1&1&0&0&0&0&0&0&0&0 \\ 0&0&0&0&0&0&0&0&0&0 \\ 0&0&0&0&0&0&0&0&0&0 \end{pmatrix},$$

$$O(\overline{\mathbf{A}}^1,\overline{\mathbf{B}}^1) = \begin{pmatrix} 1&1&1&1&1&1&1&1&1&1 \\ 1&1&1&1&1&1&1&1&1&1 \\ 1&1&1&1&1&1&0&0&0&0 \\ 1&1&1&1&1&1&0&0&0&0 \\ 1&1&1&1&0&0&0&0&0&0 \\ 1&1&1&1&0&0&0&0&0&0 \\ 1&1&1&1&0&0&0&0&0&0 \\ 1&1&0&0&0&0&0&0&0&0 \\ 1&1&0&0&0&0&0&0&0&0 \end{pmatrix}.$$

Counting ones in the rows and columns of these matrices, we obtain the components of vectors in the extremal pairs that correspond to the kernel $R(\overline{\mathbf{A}}^1,\overline{\mathbf{B}}^1)$ and hull $O(\overline{\mathbf{A}}^1,\overline{\mathbf{B}}^1)$, respectively:

$$(\overline{\mathbf{E}}^{1}_{R},\overline{\mathbf{F}}^{1}_{R}) = \big((6,6,4,4,2,2,2,0,0),(7,7,4,4,2,2,0,0,0,0)\big),$$

$$(\overline{\mathbf{E}}^{O}_{}{}^{\!\!-1}, \overline{\mathbf{F}}^{O}{}^{\!\!-1}) = ((10,10,6,6,4,4,4,2,2),(9,9,7,7,4,4,2,2,2,2)),$$

$$R(\overline{\mathbf{A}}^1, \overline{\mathbf{B}}^1) = \overline{H}(\overline{\mathbf{E}}^{R}{}^{\!\!-1}, \overline{\mathbf{F}}^{R}{}^{\!\!-1}), \quad O(\overline{\mathbf{A}}^1, \overline{\mathbf{B}}^1) = \overline{H}(\overline{\mathbf{E}}^{O}{}^{\!\!-1}, \overline{\mathbf{F}}^{O}{}^{\!\!-1}).$$

By analogy with the kernels and hulls of matrix sets, the extremal vector pairs

$$(\overline{\mathbf{E}}^R, \overline{\mathbf{F}}^R), \quad (\overline{\mathbf{E}}^O, \overline{\mathbf{F}}^O), \quad (\overline{\mathbf{E}}^{R}{}^{\!\!-1}, \overline{\mathbf{F}}^{R}{}^{\!\!-1}), \quad (\overline{\mathbf{E}}^{O}{}^{\!\!-1}, \overline{\mathbf{F}}^{O}{}^{\!\!-1})$$

may be treated as dual to one another (if the order relation for the lattice is reversed, we obtain a lattice for which the operations "∩" and "∪" are changed to "∪" and "∩", respectively [15]).

Denoting by e_i^R, e_i^O, $\overline{e}_i^{1R}$, and $\overline{e}_i^{1O}$ ($1 \le i \le 9$) the components of the vectors $\overline{\mathbf{E}}^R$, $\overline{\mathbf{E}}^O$, $\overline{\mathbf{E}}^{R}{}^{-1}$, and $\overline{\mathbf{E}}^{O}{}^{-1}$, respectively, and by f_j^R, f_j^O, $\overline{f}_j^{1R}$, and $\overline{f}_j^{1O}$ ($1 \le j \le 10$) those of $\overline{\mathbf{F}}^R$, $\overline{\mathbf{F}}^O$, $\overline{\mathbf{F}}^{R}{}^{-1}$, and $\overline{\mathbf{F}}^{O}{}^{-1}$, respectively, we readily obtain (for $n = 9$ and $m = 10$)

$$\overline{e}_i^{1R} = 10 - e_{10-i}^O, \quad \overline{e}_i^{1O} = 10 - e_{10-i}^R, \quad 1 \le i \le 9,$$

$$\overline{f}_j^{1R} = 9 - e_{11-j}^O, \quad \overline{f}_j^{1O} = 9 - e_{11-j}^R, \quad 1 \le j \le 10.$$

Finally, note that the matrix expressions in (3.5) and (3.8) define not only the matrices in $M_Z(\mathbf{A}, \mathbf{B}; 1)$ and $M_Z(\overline{\mathbf{A}}^1, \overline{\mathbf{B}}^1; 1)$, but also every matrix in $M(\mathbf{A}, \mathbf{B}; 1)$ and $M(\overline{\mathbf{A}}^1, \overline{\mathbf{B}}^1; 1)$ respectively.

The results presented in this section can be extended to the case of arbitrary minimax matrices. Consider the following example.

Example 3.2. Extension of Example 3.1. For

$$\mathbf{C} = (27,27,21,21,21,15,15,6,6) \in \overline{Z}_+^9,$$

$$\mathbf{D} = (24,24,24,24,18,18,12,9,3,3) \in \overline{Z}_+^{10},$$

it can be readily shown that $\mathbf{C} = 3\mathbf{A}$ and $\mathbf{D} = 3\mathbf{B}$, where $\mathbf{A}$ and $\mathbf{B}$ are the vectors considered in Example 3.1. Therefore, the minimax values of these vector pairs are such that $c(\mathbf{C}, \mathbf{D}) = 3c(\mathbf{A}, \mathbf{B})$ and $c(\mathbf{C}, \mathbf{D}) = 3$. Since any $Y(\mathbf{C}, \mathbf{D}) \in M_Z(\mathbf{C}, \mathbf{D}; 3)$ (or $Y(\mathbf{C}, \mathbf{D}) \in M(\mathbf{C}, \mathbf{D}; 3)$) can be represented as

$Y(\mathbf{C},\mathbf{D}) = 3X(\mathbf{A},\mathbf{B})$, where $X(\mathbf{A},\mathbf{B})$ is expressed by (3.5), it holds that

$$Y(\mathbf{C},\mathbf{D}) =$$
$$\begin{pmatrix} 3 & 3 & 3 & 3 & 3 & 3 & 3 & 3 & 3x_{19} & 3x_{1\,10} \\ 3 & 3 & 3 & 3 & 3 & 3 & 3 & 3 & 3x_{29} & 3x_{2\,10} \\ 3 & 3 & 3 & 3 & 3 & 3 & 3x_{37} & 3x_{38} & 0 & 0 \\ 3 & 3 & 3 & 3 & 3 & 3 & 3x_{47} & 3x_{48} & 0 & 0 \\ 3 & 3 & 3 & 3 & 3 & 3 & 3x_{57} & 3x_{58} & 0 & 0 \\ 3 & 3 & 3 & 3 & 3x_{65} & 3x_{66} & 0 & 0 & 0 & 0 \\ 3 & 3 & 3 & 3 & 3x_{75} & 3x_{76} & 0 & 0 & 0 & 0 \\ 3x_{81} & 3x_{82} & 3x_{83} & 3x_{84} & 0 & 0 & 0 & 0 & 0 & 0 \\ 3x_{91} & 3x_{92} & 3x_{93} & 3x_{94} & 0 & 0 & 0 & 0 & 0 & 0 \end{pmatrix} \in M(\mathbf{C},\mathbf{D};3).$$

Here the unknown elements are taken from (3.6). Consequently, any matrix $Y(\mathbf{C},\mathbf{D})$ in $M(\mathbf{C},\mathbf{D};3)$ satisfies the condition $3R(\mathbf{A},\mathbf{B}) \subset Y(\mathbf{C},\mathbf{D}) \subset 3O(\mathbf{A},\mathbf{B})$, where $R(\mathbf{A},\mathbf{B})$ and $O(\mathbf{A},\mathbf{B})$ are those constructed in Example 3.1.

With $\mathbf{C}$ and $\mathbf{D}$, we associate the vectors

$$\overline{\mathbf{C}}^3 = (24,24,15,15,9,9,9,3,3) \in \overline{Z}^9_+,$$

$$\overline{\mathbf{D}}^3 = (24,24,18,15,9,9,3,3,3,3) \in \overline{Z}^{10}_+$$

(recall Theorem 3.2). For any $\overline{Y}^3(\overline{\mathbf{C}}^3,\overline{\mathbf{D}}^3) \in M(\overline{\mathbf{C}}^3,\overline{\mathbf{D}}^3;3)$, it holds that

$$\overline{Y}^3(\overline{\mathbf{C}}^3,\overline{\mathbf{D}}^3) = 3\overline{X}^1(\overline{\mathbf{A}}^1,\overline{\mathbf{B}}^1),$$

where $\overline{X}^1(\overline{\mathbf{A}}^1,\overline{\mathbf{B}}^1)$ is defined by (3.8), and

$$3R(\overline{\mathbf{A}}^1,\overline{\mathbf{B}}^1) \subset \overline{Y}^3(\overline{\mathbf{C}}^3,\overline{\mathbf{D}}^3) \subset 3O(\overline{\mathbf{A}}^1,\overline{\mathbf{B}}^1),$$

where $R(\overline{\mathbf{A}}^1,\overline{\mathbf{B}}^1)$ and $O(\overline{\mathbf{A}}^1,\overline{\mathbf{B}}^1)$ are the matrices constructed in Example 3.1.

3.5. Nontriviality Criterion for a Matrix Set with Prescribed Kernel and Hull

A matrix set is said to be *nontrivial* if it contains more than one matrix. Otherwise, the set is called *trivial.*

It is obvious that every extremal matrix $\overline{H} = \overline{H}(\overline{\mathbf{E}},\overline{\mathbf{F}})$ is simultaneously the kernel and hull of the matrix set $M_Z(\overline{\mathbf{E}},\overline{\mathbf{F}};1)$: $\overline{H}(\overline{\mathbf{E}},\overline{\mathbf{F}}) = R(\overline{\mathbf{E}},\overline{\mathbf{F}}) = O(\overline{\mathbf{E}},\overline{\mathbf{F}})$; thus, $M_Z(\overline{\mathbf{E}},\overline{\mathbf{F}};1) = M(\overline{\mathbf{E}},\overline{\mathbf{F}};1)$ contains only one matrix, $\overline{H}(\overline{\mathbf{E}},\overline{\mathbf{F}})$. However, an extremal matrix $\overline{H} = \overline{H}(\overline{\mathbf{E}},\overline{\mathbf{F}})$ can be neither kernel nor hull of the matrix set $M(\mathbf{A},\mathbf{B};1)$, where the pair of vectors $\mathbf{A}$ and $\mathbf{B}$ differs from the pair of $\overline{\mathbf{E}}$ and $\overline{\mathbf{F}}$. This is illustrated by the following example.

Example 3.3. Consider the following 3-by-3 matrices:

$$\begin{pmatrix} 1&1&1\\ 1&1&0\\ 1&0&0 \end{pmatrix}, \quad \begin{pmatrix} 1&1&1\\ 1&1&1\\ 1&0&0 \end{pmatrix}, \quad \begin{pmatrix} 1&1&1\\ 1&1&0\\ 1&1&0 \end{pmatrix}, \quad \begin{pmatrix} 1&1&1\\ 1&1&0\\ 1&0&1 \end{pmatrix},$$

$$\overline{H} = \overline{H}(\overline{\mathbf{E}}, \overline{\mathbf{F}}), \quad X_1 = X_1(\mathbf{A}_1, \mathbf{B}_1), \quad X_2 = X_2(\mathbf{A}_2, \mathbf{B}_2), \quad X_3 = X_3(\mathbf{A}_3, \mathbf{B}_3),$$

$$\begin{pmatrix} 1&1&1\\ 1&1&1\\ 1&0&1 \end{pmatrix}, \quad \begin{pmatrix} 1&1&1\\ 1&1&0\\ 1&1&1 \end{pmatrix}, \quad \begin{pmatrix} 1&1&1\\ 1&1&1\\ 1&1&0 \end{pmatrix}, \quad \begin{pmatrix} 1&1&1\\ 1&1&1\\ 1&1&1 \end{pmatrix},$$

$$X_4 = X_4(\mathbf{A}_4, \mathbf{B}_4), \; X_5 = X_5(\mathbf{A}_5, \mathbf{B}_5), \; X_6 = X_6(\mathbf{A}_6, \mathbf{B}_6), \; X_7 = X_7(\mathbf{A}_7, \mathbf{B}_7).$$

Here, the first matrix is singled out by using special notation.

It is obvious that $\overline{H} = \overline{H}\left((3,2,1),(3,2,1)\right)$ is an extremal matrix, $\overline{H} \subseteq X_i$, and $\overline{H} \neq X_i$ $(1 \leq i \leq 7)$. It can be readily demonstrated that $X_1 = X_1\left((3,3,1),(3,2,2)\right)$, $X_2 = X_2\left((3,2,2),(3,3,1)\right)$, $X_6 = X_6\left((3,3,2),(3,3,2)\right)$, and $X_7 = X_7\left((3,3,3),(3,3,3)\right)$ are extremal matrices. Therefore, $\overline{H}$ cannot be the kernel of the matrix set defined by the pair of vectors $\mathbf{A}_i$ and $\mathbf{B}_i$, where i = 1, 2, 6, 7: $\overline{H}(\overline{\mathbf{E}}, \overline{\mathbf{F}}) = R(\overline{\mathbf{E}}, \overline{\mathbf{F}}) \neq X_i(\mathbf{A}_i, \mathbf{B}_i) = R(\mathbf{A}_i, \mathbf{B}_i)$ for i = 1, 2, 6, 7. It is easy to check that $X_3 = X_3\left((3,2,2),(3,2,2)\right)$ is not an extremal matrix; however, it contains the nontrivial kernel $R(\mathbf{A}_3, \mathbf{B}_3)$ of the matrix set $M_Z(\mathbf{A}_3, \mathbf{B}_3; 1)$:

$$R\left((3,2,2),(3,2,2)\right) \subseteq X_3\left((3,2,2),(3,2,2)\right),$$

$$R\left((3,2,2),(3,2,2)\right) \neq X_3\left((3,2,2),(3,2,2)\right)$$

and

$$R\left((3,2,2),(3,2,2)\right) = \overline{H}\left((3,1,1),(3,1,1)\right) = \begin{pmatrix} 1&1&1\\ 1&0&0\\ 1&0&0 \end{pmatrix} \neq \overline{H}\left((3,2,1),(3,2,1)\right).$$

The matrices

$$X_4 = X_4\left((3,3,2),(3,2,3)\right), \quad X_5 = X_5\left((3,2,3),(3,3,2)\right)$$

can be obtained by permuting rows or columns of the extremal matrix $X_6 = X_6\left((3,3,2),(3,3,2)\right)$. For both X_4 and X_5, the components of one of the basis vectors are not arranged in nonascending order. Therefore, these matrices are not extremal, but are perfect. Since the kernels (and hulls) are defined not only for matrix sets with vectors whose components are arranged in nonascending order, the matrices X_4 and X_5 may be ignored in this example.

Because $X_i = X_i(\mathbf{A}_i, \mathbf{B}_i)$ ($i = 1, 2, 3, 6, 7$) are all the matrices for which $\overline{H} \subseteq X_i$, $\overline{H} \neq X_i$, and $\mathbf{A}_i$, $\mathbf{B}_i \in \overline{Z}_+^3$, the extremal matrix $\overline{H}$ is not the kernel of the matrix sets $M_Z(\mathbf{A}_i, \mathbf{B}_i; 1)$ ($i = 1, 2, 3, 6, 7$).

Therefore, $\overline{H} = \overline{H}(\overline{\mathbf{E}}, \overline{\mathbf{F}})$ cannot be the kernel of any matrix set defined by a vector pair different from the extremal pair $(\overline{\mathbf{E}}, \overline{\mathbf{F}})$. Similarly, it can be shown that the extremal matrix $\overline{H}(\overline{\mathbf{E}}, \overline{\mathbf{F}})$ considered here is the kernel $O(\mathbf{A}, \mathbf{B})$ of the matrix set $M_Z(\mathbf{A}, \mathbf{B}; 1)$ only if the vector pair $\mathbf{A}$, $\mathbf{B}$ is identical with $\overline{\mathbf{E}}$, $\overline{\mathbf{F}}$.

Consider now vectors $\mathbf{A} \in \overline{Z}_+^n$ and $\mathbf{B} \in \overline{Z}_+^m$ for which $M_Z(\mathbf{A}, \mathbf{B}; 1) \neq \varnothing$. Propositions (a) and (b) of Theorem 3.4 define the zero and unit elements for all matrices in $M_Z(\mathbf{A}, \mathbf{B}; 1)$, that is, the kernel $R(\mathbf{A}, \mathbf{B})$ and hull $O(\mathbf{A}, \mathbf{B})$ of the matrix set. Moreover, Theorem 3.4 determines the elements that are zeros and ones for different matrices. An element x_{pq} ($1 \leq p \leq n$, $1 \leq q \leq m$) is called *indeterminate* for the set $M_Z(\mathbf{A}, \mathbf{B}; 1)$ if the set contains matrices $X'(\mathbf{A}, \mathbf{B}) = (x'_{ij})$ and $X''(\mathbf{A}, \mathbf{B}) = (x''_{ij})$ such that $x'_{pq} = 1$ and $x''_{pq} = 0$. It is clear that if some element x_{pq} ($1 \leq p \leq n$, $1 \leq q \leq m$) is indeterminate in the matrix set $M_Z(\mathbf{A}, \mathbf{B}; 1)$, then the kernel $R(\mathbf{A}, \mathbf{B}) = (r_{ij})$ and hull $O(\mathbf{A}, \mathbf{B}) = (O_{ij})$ are such that $r_{pq} = 0$ and $o_{pq} = 1$. This is demonstrated by expressions (3.5) and (3.8) in Example 3.1.

We have repeatedly pointed out that if either $\mathbf{A}$ or $\mathbf{B}$ is a one-component vector, then the set $M(\mathbf{A}, \mathbf{B})$ contains only one matrix, which is completely defined. Note that in (3.5) and (3.8) all indeterminate elements belong to n-by-m submatrices of $X(\mathbf{A}, \mathbf{B})$ and $\overline{X}^1(\overline{\mathbf{A}}^1, \overline{\mathbf{B}}^1)$, where n, $m \geq 2$. This provides necessary and sufficient conditions for the existence of such a vector pair $(\mathbf{A}, \mathbf{B})$ different from the extremal pair $(\overline{\mathbf{E}}, \overline{\mathbf{F}})$ that $\overline{H}(\overline{\mathbf{E}}, \overline{\mathbf{F}}) = R(\mathbf{A}, \mathbf{B})$ or $\overline{H}(\overline{\mathbf{E}}, \overline{\mathbf{F}}) = O(\mathbf{A}, \mathbf{B})$. The following proposition shows that this condition is satisfied for a broad class of matrices.

Lemma 3.1. *Suppose that $\overline{H}(\overline{\mathbf{E}}, \overline{\mathbf{F}})$ is an extremal matrix from $\overline{E}_{n,m}$ and e is the value of a component of $\overline{\mathbf{E}} = (e_1, \ldots, e_n)$ such that $k(e) = \min\{i : e_i = e,\ 1 \leq i \leq n\}$ and $p(e) = \max\{i : e_i = e,\ 1 \leq i \leq n\}$. Then, there exist vectors $\mathbf{A} \in \overline{R}_+^n$ and $\mathbf{B} \in \overline{R}_+^m$ such that $M_Z(\mathbf{A}, \mathbf{B}; 1) \neq \varnothing$ and $(\mathbf{A}, \mathbf{B}) \neq (\overline{\mathbf{E}}, \overline{\mathbf{F}})$, for which*

(a) $\overline{H}(\overline{\mathbf{E}}, \overline{\mathbf{F}}) = R(\mathbf{A}, \mathbf{B})$,

(b) $\overline{H}(\overline{\mathbf{E}}, \overline{\mathbf{F}}) = O(\mathbf{A}, \mathbf{B})$

if and only if the vector $\overline{\mathbf{E}}$ has a component equal to e such that $p(e) > k(e)$ and, respectively,

(c) *either $l_{k(e)-1}(\overline{\mathbf{F}}; 1) - l_{k(e)}(\overline{\mathbf{F}}; 1) \geq 2$, when $k(e) \geq 2$, or $m - l_1(\overline{\mathbf{F}}; 1) \geq 2$, when $k(e) = 1$;* and

(d) *either* $l_{p(e)}(\overline{\mathbf{F}};1) - l_{p(e)+1}(\overline{\mathbf{F}};1) \geq 2$, *when* $p(e) < n$, *or* $l_n(\overline{\mathbf{F}};1) \geq 2$, *when* $p(e) = n$.

Proof.
Necessity.

(a) Suppose that $\overline{H}(\overline{\mathbf{E}}, \overline{\mathbf{F}}) = R(\mathbf{A}, \mathbf{B})$, where $\mathbf{A} = (a_1, \ldots, a_n) \in \overline{Z}_+^n$ and $\mathbf{B} = (b_1, \ldots, b_m) \in \overline{Z}_+^m$. It is obvious that $(\overline{\mathbf{E}}, \overline{\mathbf{F}}) \leq (\mathbf{A}, \mathbf{B})$, which implies that $e_i \leq a_i$ $(1 \leq i \leq n)$ and $f_j \leq b_j$ $(1 \leq j \leq m)$, where $\overline{\mathbf{F}} = (f_1, \ldots, f_m)$. Assume that for every component e_i of $\overline{\mathbf{E}}$ either $p(e_i) = k(e_i)$ or condition (c) for $p(e_i) > k(e_i)$ is not satisfied. Let us show that this entails the equality $(\mathbf{A}, \mathbf{B}) = (\overline{\mathbf{E}}, \overline{\mathbf{F}})$.

Recall the structure of a kernel matrix. By Theorem 3.3 and Remark 3.4,

$$R(\mathbf{A}, \mathbf{B}) = \bigcap_{X(\mathbf{A},\mathbf{B}) \in M_Z(\mathbf{A},\mathbf{B};1)} X(\mathbf{A}, \mathbf{B}) = \bigcup_{k=0}^{t} R_{p_k}(\mathbf{A}, \mathbf{B}), \tag{3.9}$$

where p_k $(1 \leq k \leq t)$ denotes every number such that $p_k \leq b_1$ and $\delta_{p_k}(\mathbf{A}, \mathbf{B}; 1) = 0$, $p_0 = 0$, $R_0(\mathbf{A}, \mathbf{B})$ is the matrix of zeros, and $R_{p_k} = (r_{ij}^{(p_k)})$ $(1 \leq k \leq t)$ is expressed as

$$r_{ij}^{(p_k)} = \begin{cases} 1, & 1 \leq i \leq p_k, \quad 1 \leq j \leq l_{p_k}(\mathbf{B}; 1), \\ 0, & 1 \leq i \leq p_k, \quad l_{p_k}(\mathbf{B}; 1) < j \leq m, \\ 0, & p_k < i \leq n, \quad 1 \leq j \leq m. \end{cases} \tag{3.10}$$

Suppose $p(e_i) = k(e_i)$. Then, $e_{i-1} > e_i > e_{i+1}$ for $i \geq 2$ and $e_1 > e_2$ for $i = 1$.

Consider the case of $i = 1$. From relation (1.6) it follows that $l_1(\overline{\mathbf{F}}; 1) > l_2(\overline{\mathbf{F}}; 1)$. Then, (3.9) with $p_k = 1$ implies that $\delta_1(\mathbf{A}, \mathbf{B}; 1) = 0$. Otherwise, if $\delta_1(\mathbf{A}, \mathbf{B}; 1) > 0$, then $\delta_k(\mathbf{A}, \mathbf{B}; 1) = 0$ for some $k > 1$, and we would have $l_1(\overline{\mathbf{F}}; 1) = l_2(\overline{\mathbf{F}}; 1) = e_1 = e_2$. Equality $\delta_1(\mathbf{A}, \mathbf{B}; 1) = 0$ implies that $a_1 = l_1(\mathbf{B}; 1)$. Combining (3.9) and (3.10), where $p_k = 1$, we obtain $a_1 = l_1(\mathbf{B}; 1) = l_1(\overline{\mathbf{F}}; 1) = e_1$. Thus, if $i = 1$, then $a_1 = e_1$.

Let $i > 1$ and $e_{i-1} > e_i > e_{i+1}$; then $l_{i-1}(\overline{\mathbf{F}}; 1) > l_i(\overline{\mathbf{F}}; 1) > l_{i+1}(\overline{\mathbf{F}}; 1)$. Combined with (3.9), these inequalities imply that $\delta_{i-1}(\mathbf{A}, \mathbf{B}; 1) = \delta_i(\mathbf{A}, \mathbf{B}; 1) = 0$. Otherwise, if $\delta_{i-1}(\mathbf{A}, \mathbf{B}; 1) > 0$, then $l_{i-1}(\overline{\mathbf{F}}; 1) = l_i(\overline{\mathbf{F}}; 1)$; and if $\delta_i(\mathbf{A}, \mathbf{B}; 1) > 0$, then $l_i(\overline{\mathbf{F}}; 1) = l_{i+1}(\overline{\mathbf{F}}; 1)$. For the function defined by (1.4) (i.e., (1.22) from Chapter 1 with $c = 1$), it holds that

$$\delta_{i-1}(\mathbf{A}, \mathbf{B}; 1) = \sum_{j=1}^{m} b_j - \sum_{b_j \geq i-1} (b_j - i + 1) - \sum_{p=1}^{i-1} a_p = 0,$$

$$\delta_i(\mathbf{A},\mathbf{B};1) = \sum_{j=1}^{m} b_j - \sum_{b_j \geq i}(b_j - i) - \sum_{p=1}^{i} a_p = 0.$$

Consider the remainder $\delta_{i-1}(\mathbf{A},\mathbf{B};1) - \delta_i(\mathbf{A},\mathbf{B};1)$. We have

$$\delta_{i-1}(\mathbf{A},\mathbf{B};1) - \delta_i(\mathbf{A},\mathbf{B};1) = \sum_{b_j \geq i}(b_j - i) - \sum_{b_j \geq i-1}(b_j - i + 1) + a_i = 0.$$

This yields

$$a_i = \sum_{b_j \geq i-1}(b_j - i + 1) - \sum_{b_j \geq i}(b_j - i) =$$
$$\sum_{b_j \geq i}(b_j - i + 1) + \sum_{b_j = i-1}(b_j - i + 1) - \sum_{b_j \geq i}(b_j - i) =$$
$$\sum_{j=1}^{l_i(\mathbf{B};1)} b_j - (i-1)l_i(\mathbf{B};1) - \sum_{j=1}^{l_i(\mathbf{B};1)} b_j + il_i(\mathbf{B};1) = l_i(\mathbf{B};1).$$

Combining (3.9) with (3.10), we obtain $a_i = l_i(\mathbf{B};1) = l_i(\overline{\mathbf{F}};1) = e_i$. Thus, we have shown that if $p(e_i) = k(e_i)$ (i.e., e_i is the only component of $\overline{\mathbf{E}}$ equal to e_i), then $a_i = e_i$.

Let $p(e_i) > k(e_i)$; then e_r, where $k(e_i) \leq r \leq p(e_i)$, are the components of $\overline{\mathbf{E}}$ equal to e_i, and their number is at least two. From (1.6), it follows that $l_{p(e_i)}(\overline{\mathbf{F}};1) = l_{k(e_i)}(\overline{\mathbf{F}};1)$. Assume that $l_{k(e_i)-1}(\overline{\mathbf{F}};1) - l_{k(e_i)}(\overline{\mathbf{F}};1) \leq 1$ for $k(e_i) \geq 2$ (the left-hand side of this inequality is obviously nonnegative) and $m - l_1(\overline{\mathbf{F}};1) \leq 1$ for $k(e_i) = 1$.

Consider the case of $k(e_i) = 1$. If $l_1(\overline{\mathbf{F}};1) = m$, then (1.6) implies that $e_1 = l_1(\overline{\mathbf{F}};1)$ and $a_1 = m = e_1$, because $\overline{\mathbf{E}} \leq \mathbf{A}$ (the order relation "$\leq$" for vectors was introduced in Section 2). Accordingly, $e_i = l_1(\overline{\mathbf{F}};1) = m = a_i$ for $1 \leq i \leq p(e_i)$. Suppose $l_1(\overline{\mathbf{F}};1) = m-1 = e_1$. Assuming that $a_1 = m$, we have $\delta_1(\mathbf{A},\mathbf{B};1) = 0$ and $l_1(\mathbf{B};1) = m$. Then $e_1 = m$ by virtue of (3.9) combined with (3.10). Therefore, $a_1 = m-1$, and $a_i = m-1 = e_i$ $(1 \leq i \leq p(e_i))$, because $\mathbf{A} \in \overline{Z}_+^n$ and $\overline{\mathbf{E}} \leq \mathbf{A}$.

Let $l_{k(e_i)-1}(\overline{\mathbf{F}};1) - l_{k(e_i)}(\overline{\mathbf{F}};1) \leq 1$ and $k(e_i) \geq 2$. Assuming that $l_{k(e_i)-1}(\overline{\mathbf{F}};1) = l_{k(e_i)}(\overline{\mathbf{F}};1)$, we see that $e_{k(e_i)-1} = e_{k(e_i)}$ by virtue of (1.6). This contradicts the definition of $k(e_i) = \min\{r : e_r = e_i,\ 1 \leq r \leq n\}$. Therefore, $l_{k(e_i)-1}(\overline{\mathbf{F}};1) - l_{k(e_i)}(\overline{\mathbf{F}};1) = 1$ and $e_{k(e_i)-1} = e_{k(e_i)} + 1$. Note that $l_k(\mathbf{B};1) \geq l_k(\overline{\mathbf{F}};1)$ since $\overline{\mathbf{F}} \leq \mathbf{B}$. By virtue of (3.9) and (3.10), $\delta_{k(e_i)-1}(\mathbf{A},\mathbf{B};1) = 0$ and $l_{k(e_i)-1}(\mathbf{B};1) = l_{k(e_i)-1}(\overline{\mathbf{F}};1)$. Otherwise, if $\delta_{k(e_i)-1}(\mathbf{A},\mathbf{B};1) > 0$, then there would exist $k \geq k(e_i)$ such that $\delta_k(\mathbf{A},\mathbf{B};1) = 0$ and $l_k(\mathbf{B};1) = l_{k(e_i)-1}(\overline{\mathbf{F}};1)$, in which case the equality $l_{k(e_i)-1}(\overline{\mathbf{F}};1) = l_{k(e_i)}(\overline{\mathbf{F}};1)$ leads to contradiction.

Similarly, we obtain $\delta_{p(e_i)}(\mathbf{A},\mathbf{B};1) = 0$ and $l_{p(e_i)}(\mathbf{B};1) = l_{p(e_i)}(\overline{\mathbf{F}};1)$. Suppose now that $a_{k(e_i)} \geq e_{k(e_i)} + 1$ or, equivalently, $a_{k(e_i)} \geq l_{k(e_i)}(\overline{\mathbf{F}};1) + 1 = l_{k(e_i)-1}(\overline{\mathbf{F}};1)$. Consider the values of function (1.4) for $k(e_i) - 1$ and $k(e_i)$:

$$\delta_{k(e_i)-1}(\mathbf{A},\mathbf{B};1) = \sum_{j=1}^{m} b_j - \sum_{j=1}^{l_{k(e_i)-1}(\mathbf{B};1)} (b_j - k(e_i) + 1) - \sum_{p=1}^{k(e_i)-1} a_p,$$

$$\delta_{k(e_i)}(\mathbf{A},\mathbf{B};1) = \sum_{j=1}^{m} b_j - \sum_{j=1}^{l_{k(e_i)}(\mathbf{B};1)} (b_j - k(e_i)) - \sum_{p=1}^{k(e_i)} a_p.$$

We have

$$\delta_{k(e_i)-1}(\mathbf{A},\mathbf{B};1) - \delta_{k(e_i)}(\mathbf{A},\mathbf{B};1) = a_{k(e_i)} + \sum_{j=1}^{l_{k(e_i)}(\mathbf{B};1)} (b_j - k(e_i)) - \sum_{j=1}^{l_{k(e_i)-1}(\mathbf{B};1)} (b_j - k(e_i) + 1).$$

Since $l_{k(e_i)-1}(\mathbf{B};1) = l_{k(e_i)-1}(\overline{\mathbf{F}};1)$, and inequality $a_{k(e_i)} \geq e_{k(e_i)} + 1$ implies that $l_{k(e_i)}(\mathbf{B};1) = l_{k(e_i)-1}(\overline{\mathbf{F}};1)$, this remainder can be expressed as

$$\delta_{k(e_i)-1}(\mathbf{A},\mathbf{B};1) - \delta_{k(e_i)}(\mathbf{A},\mathbf{B};1) = a_{k(e_i)} + \sum_{j=1}^{l_{k(e_i)-1}(\overline{\mathbf{F}};1)} (b_j - k(e_i)) - \sum_{j=1}^{l_{k(e_i)-1}(\overline{\mathbf{F}};1)} (b_j - k(e_i) + 1) = a_{k(e_i)} - l_{k(e_i)-1}(\overline{\mathbf{F}};1).$$

Therefore, since $\delta_{k(e_i)-1}(\mathbf{A},\mathbf{B};1) = 0$ and $\delta_{k(e_i)}(\mathbf{A},\mathbf{B};1) \geq 0$, it holds that

$$a_{k(e_i)} - l_{k(e_i)-1}(\overline{\mathbf{F}};1) \leq 0, \quad a_{k(e_i)} \leq l_{k(e_i)-1}(\overline{\mathbf{F}};1).$$

By assumption, $a_{k(e_i)} \geq l_{k(e_i)-1}(\overline{\mathbf{F}};1)$; hence $a_{k(e_i)} = l_{k(e_i)-1}(\overline{\mathbf{F}};1)$ and, therefore, $\delta_{k(e_i)}(\mathbf{A},\mathbf{B};1) = 0$ and $l_{k(e_i)}(\mathbf{B};1) = l_{k(e_i)-1}(\overline{\mathbf{F}};1) = l_{k(e_i)}(\overline{\mathbf{F}};1) + 1$. But then we have $l_{k(e_i)}(\overline{\mathbf{F}};1) = l_{k(e_i)}(\mathbf{B};1) = l_{k(e_i)}(\overline{\mathbf{F}};1) + 1$, which leads to contradiction. The contradiction means that $a_{k(e_i)} = e_{k(e_i)}$, and since $\mathbf{A} \in \overline{Z}^n_+$ and $\overline{\mathbf{E}} \leq \mathbf{A}$, we have $a_r = e_r$ for $k(e_i) \leq r \leq p(e_i)$. Consequently, the components of $\mathbf{A}$ are identical with those of $\overline{\mathbf{E}}$ in this case as well.

Assuming that every component e_i of $\overline{\mathbf{E}}$ is such that either $p(e_i) = k(e_i)$ or assumption (c) for the vectors $\overline{\mathbf{E}}$ and $\overline{\mathbf{F}}$ is violated when $p(e_i) > k(e_i)$, we showed that $\mathbf{A} = \overline{\mathbf{E}}$. Simultaneously, we demonstrated that $l_i(\mathbf{B};1) = l_i(\overline{\mathbf{F}};1)$ for any i $(1 \leq i \leq n)$. Since $\overline{\mathbf{F}} \leq \mathbf{B}$, for $f_j < b_j$ we have the inequality

$l_{b_j}(\mathbf{B};1) > l_{b_j}(\overline{\mathbf{F}};1)$ for some j $(1 \le j \le m)$, which is inconsistent with the equality $l_{b_j}(\mathbf{B};1) = l_{b_j}(\overline{\mathbf{F}};1)$. Therefore, $\mathbf{B} = \overline{\mathbf{F}}$. Consequently, $(\mathbf{A},\mathbf{B}) = (\overline{\mathbf{E}},\overline{\mathbf{F}})$, which proves the necessity for proposition (a) of the lemma.

(b) Suppose that $\overline{H}(\overline{\mathbf{E}},\overline{\mathbf{F}}) = O(\mathbf{A},\mathbf{B})$. Then, $(\mathbf{A},\mathbf{B}) \le (\overline{\mathbf{E}},\overline{\mathbf{F}})$; that is, $a_i \le e_i$ $(1 \le i \le n)$, $b_j \le f_j$ $(1 \le j \le m)$. Assume that every component e_i of the vector $\overline{\mathbf{E}}$ is such that either $p(e_i) = k(e_i)$, or assumption (d) for $\overline{\mathbf{E}}$ and $\overline{\mathbf{F}}$ is violated when $p(e_i) > k(e_i)$. Now, we show that this entails $(\mathbf{A},\mathbf{B}) = (\overline{\mathbf{E}},\overline{\mathbf{F}})$.

To do this, we invoke Theorem 3.2′ about the equivalence of the characteristic functions (1.22) of the pairs $(\mathbf{C},\mathbf{D})$ and $(\overline{\mathbf{C}}^1,\overline{\mathbf{D}}^1)$ in the case when $c = 1$; here, the components of $\mathbf{C} = (c_1,\ldots,c_n)$, $\mathbf{D} = (d_1,\ldots,d_m)$ and $\overline{\mathbf{C}}^1 = (\overline{c}_1^1,\ldots,\overline{c}_n^1)$, $\overline{\mathbf{D}}^1 = (\overline{d}_1^1,\ldots,\overline{d}_m^1)$ are related as follows:

$$\begin{aligned} \overline{c}_i^1 &= m - c_{n-i+1}, \quad 1 \le i \le n, \\ \overline{d}_j^1 &= n - d_{m-j+1}, \quad 1 \le j \le m. \end{aligned} \tag{3.11}$$

By Theorem 3.2′, extremality is retained under this relation. We have already pointed out that the matrix sets $M_Z(\mathbf{C},\mathbf{D};1)$ and $M_Z(\overline{\mathbf{C}}^1,\overline{\mathbf{D}}^1;1)$ are in one-to-one correspondence, so that (if $M_Z(\mathbf{C},\mathbf{D};1) \ne \varnothing$) the elements of $X(\mathbf{C},\mathbf{D}) = (x_{ij}) \in M_Z(\mathbf{C},\mathbf{D};1)$ and $\overline{X}^1(\overline{\mathbf{C}}^1,\overline{\mathbf{D}}^1) = (\overline{x}_{ij}^1) \in M_Z(\overline{\mathbf{C}}^1,\overline{\mathbf{D}}^1;1)$ are related as

$$\overline{x}_{ij}^1 = 1 - x_{n-i+1\, m-j+1}, \quad 1 \le i \le n, \quad 1 \le j \le m. \tag{3.12}$$

To prove assertion (b) of the lemma, we state the following simple proposition, which is then used in proofs concerning the hulls of matrix sets.

Lemma 3.2. *Suppose* $M_Z(\mathbf{C},\mathbf{D};1) \ne \varnothing$.

(a) *Matrices* X, $Y \in M_Z(\mathbf{C},\mathbf{D};1)$ *and* $\overline{X}^1$, $\overline{Y}^1 \in M_Z(\overline{\mathbf{C}}^1,\overline{\mathbf{D}}^1;1)$ *satisfy the relations*

$$\overline{X \cup Y}^1 = \overline{X}^1 \cap \overline{Y}^1, \quad \overline{X \cap Y}^1 = \overline{X}^1 \cup \overline{Y}^1,$$
$$\overline{\overline{X}^1 \cup \overline{Y}^1}^1 = X \cap Y, \quad \overline{\overline{X}^1 \cap \overline{Y}^1}^1 = X \cup Y.$$

(b) *Kernels and hulls satisfy the relations*

$$\overline{R}^1(\mathbf{C},\mathbf{D}) = O(\overline{\mathbf{C}}^1,\overline{\mathbf{D}}^1), \quad \overline{O}^1(\mathbf{C},\mathbf{D}) = R(\overline{\mathbf{C}}^1,\overline{\mathbf{D}}^1),$$
$$\overline{R}^1(\overline{\mathbf{C}}^1,\overline{\mathbf{D}}^1) = O(\mathbf{C},\mathbf{D}), \quad \overline{O}^1(\overline{\mathbf{C}}^1,\overline{\mathbf{D}}^1) = R(\mathbf{C},\mathbf{D}).$$

The proof follows from (3.11), (3.12), and Definitions 3.1 and 3.2.

We now return to the proof of Lemma 3.1. Since $\overline{H}(\overline{\mathbf{E}}, \overline{\mathbf{F}}) = O(\mathbf{A}, \mathbf{B})$, by Lemma 3.2 it holds that

$$\overset{-1}{\overline{H}}(\overline{\mathbf{E}}, \overline{\mathbf{F}}) = \overline{H}(\overset{-1}{\overline{\mathbf{E}}}, \overset{-1}{\overline{\mathbf{F}}}) = \overline{O}^1(\mathbf{A}, \mathbf{B}) = R(\overline{\mathbf{A}}^1, \overline{\mathbf{B}}^1).$$

Proposition (a) of Lemma 3.2 implies that $(\overline{\mathbf{A}}^1, \overline{\mathbf{B}}^1) = (\overset{-1}{\overline{\mathbf{E}}}, \overset{-1}{\overline{\mathbf{F}}})$; hence, $(\mathbf{A}, \mathbf{B}) = (\overline{\mathbf{E}}, \overline{\mathbf{F}})$. The necessity is thus proved.

Sufficiency.

(a) Suppose that there exists a component e_i of $\overline{\mathbf{E}}$ such that $p(e_i) > k(e_i)$, and the vector $\overline{\mathbf{F}}$ is such that $l_{k(e_i)-1}(\overline{\mathbf{F}}; 1) - l_{k(e_i)}(\overline{\mathbf{F}}; 1) \geq 2$ for $k(e_i) \geq 2$ and $m - l_1(\overline{\mathbf{F}}; 1) \geq 2$ for $k(e_i) = 1$. Changing two components in both $\overline{\mathbf{E}}$ and $\overline{\mathbf{F}}$, we construct the vectors $\mathbf{A} = (a_1, \ldots, a_n)$ and $\mathbf{B} = (b_1, \ldots, b_m)$:

$$\begin{aligned} a_p &= \begin{cases} e_p, & 1 \leq p \leq n, \quad p \neq k(e_i),\ k(e_i)+1, \\ e_p + 1, & p = k(e_i),\ k(e_i)+1, \end{cases} \\ b_j &= \begin{cases} f_j, & 1 \leq j \leq m, \quad j \neq l_{k(e_i)}(\overline{\mathbf{F}}; 1)+1,\ l_{k(e_i)}(\overline{\mathbf{F}}; 1)+2, \\ f_j + 1, & j \neq l_{k(e_i)}(\overline{\mathbf{F}}; 1)+1,\ l_{k(e_i)}(\overline{\mathbf{F}}; 1)+2. \end{cases} \end{aligned} \tag{3.13}$$

It is clear that $\mathbf{A} \in \overline{Z}^n_+$ and $\mathbf{B} \in \overline{Z}^m_+$. From (3.13), it follows that

$$l_k(\mathbf{B}; 1) = l_k(\overline{\mathbf{F}}; 1), \quad 1 \leq k \leq n, \quad k \neq k(e_i). \tag{3.14}$$

Let us show that $\delta_k(\mathbf{A}, \mathbf{B}; 1) = 0$ for any k such that $1 \leq k \leq n$, $k \neq k(e_i)$, and $\delta_{k(e_i)}(\mathbf{A}, \mathbf{B}; 1) > 0$. Then, proposition (a) of the lemma will be valid by virtue of (3.14), (3.9), and (3.10). Consider the function (1.4) of the pair of vectors $\mathbf{A}$ and $\mathbf{B}$:

$$\delta_k(\mathbf{A}, \mathbf{B}; 1) = \sum_{j=1}^{m} b_j - \sum_{b_j \geq k} (b_j - k) - \sum_{p=1}^{k} a_p.$$

From (3.13), it follows that

$$\sum_{j=1}^{m} b_j = \sum_{j=1}^{m} f_j + 2, \quad \sum_{p=1}^{k} a_p = \begin{cases} \displaystyle\sum_{p=1}^{k} e_p, & p < k(e_i), \\ \displaystyle\sum_{p=1}^{k} e_p + 1, & p = k(e_i), \\ \displaystyle\sum_{p=1}^{k} e_p + 2, & p \geq k(e_i) + 1. \end{cases} \tag{3.15}$$

If $k < k(e_i)$ (for $k(e_i) \geq 2$), then

$$\sum_{b_j \geq k} (b_j - k) = \sum_{f_j \geq k} (f_j - k) + 2,$$

and (3.15) entails the equality $\delta_k(\mathbf{A}, \mathbf{B}; 1) = \delta_k(\overline{\mathbf{E}}, \overline{\mathbf{F}}; 1) = 0$. If $k = k(e_i)$, the inequality $e_{l_{k(e_i)}(\overline{\mathbf{F}};1)} > e_{l_{k(e_i)}(\overline{\mathbf{F}};1)+1}$ implies that

$$\sum_{b_j \geq k(e_i)} (b_j - k(e_i)) = \sum_{f_j \geq k(e_i)} (f_j - k(e_i)) .$$

Accordingly, (3.15) implies that $\delta_{k(e_i)}(\mathbf{A}, \mathbf{B}; 1) = \delta_{k(e_i)}(\overline{\mathbf{E}}, \overline{\mathbf{F}}; 1) = 1$. For $k \geq k(e_i) + 1$, it is clear that

$$\sum_{b_j \geq k} (b_j - k) = \sum_{f_j \geq k} (f_j - k)$$

and $\delta_k(\mathbf{A}, \mathbf{B}; 1) = \delta_k(\overline{\mathbf{E}}, \overline{\mathbf{F}}; 1) = 0$. Consequently, $\overline{H}(\overline{\mathbf{E}}, \overline{\mathbf{F}}) = R(\mathbf{A}, \mathbf{B})$, where $(\mathbf{A}, \mathbf{B}) \neq (\overline{\mathbf{E}}, \overline{\mathbf{F}})$.

(b) Suppose that there exists a component e_i of $\overline{\mathbf{E}}$ such that $p(e_i) > k(e_i)$, and $l_{p(e_i)}(\overline{\mathbf{F}}; 1) - l_{p(e_i)+1}(\overline{\mathbf{F}}; 1) \geq 2$ for $p(e_i) < n$ and $l_n(\overline{\mathbf{F}}; 1) \geq 2$ for $p(e_i) = n$. Then, the existence of a vector pair $(\mathbf{A}, \mathbf{B})$ such that $\overline{H}(\overline{\mathbf{E}}, \overline{\mathbf{F}}) = O(\mathbf{A}, \mathbf{B})$ and $(\mathbf{A}, \mathbf{B}) \neq (\overline{\mathbf{E}}, \overline{\mathbf{F}})$ follows from the proof of proposition (a) combined with Lemma 3.2.

The conditions for the components of vectors in an extremal pair stated by Lemma 3.1 define the diversity of the matrix sets with prescribed kernel and hull. We formulate here respective theorems for the kernel and hull.

Theorem 3.9. *Let $\overline{H}(\overline{\mathbf{E}}, \overline{\mathbf{F}})$ be an extremal matrix, where $\overline{\mathbf{E}} = (e_1, \ldots, e_n)$ and $\overline{\mathbf{F}} = (f_1, \ldots, f_m)$; and let e be the value of some component of $\overline{\mathbf{E}}$ for which*

$$p(e) = \max\{i : e_i = e,\ 1 \leq i \leq n\} > k(e) = \min\{i : e_i = e,\ 1 \leq i \leq n\}.$$

Suppose that the vector $\overline{\mathbf{F}}$ and the value of $k(e)$ are such that

$$l_{k(e)}(\overline{\mathbf{F}}; 1) \leq \begin{cases} l_{k(e)-1}(\overline{\mathbf{F}}; 1) - 2, & k(e) > 1, \\ m - 2, & k(e) = 1. \end{cases}$$

Let $\mathbf{A}'$ and $\mathbf{B}'$ be arbitrary vectors, with components numbered by indices ranging from $k(e)$ to $p(e)$ and from $l_{k(e)}(\overline{\mathbf{F}}; 1) + 1$ to $l_{k(e)-1}(\overline{\mathbf{F}}; 1)$, respectively, that satisfy the relations

$$\mathbf{A}' = (a'_{k(e)}, \ldots, a'_{p(e)}) \in \overline{Z}_+^{p(e)-k(e)+1},$$

$$\mathbf{B}' = (b'_{l_{k(e)}(\overline{\mathbf{F}};1)}, \dots, b'_{l_{k(e)-1}(\overline{\mathbf{F}};1)}) \in \overline{Z}_+^{l_{k(e)-1}(\overline{\mathbf{F}};1) - l_{k(e)}(\overline{\mathbf{F}};1)},$$

$$a'_{k(e)} < \begin{cases} m - e, & k(e) = 1, \\ e_{k(e)-1} - e, & k(e) > 1, \end{cases}$$

$$b'_{l_{k(e)}(\overline{\mathbf{F}};1)+1} < \begin{cases} n - b_{l_{k(e)}(\overline{\mathbf{F}};1)+1}, & l_{k(e)}(\overline{\mathbf{F}};1) = 0, \\ b_{l_{k(e)}(\overline{\mathbf{F}};1)} - b_{l_{k(e)}(\overline{\mathbf{F}};1)+1}, & l_{k(e)}(\overline{\mathbf{F}};1) > 0, \end{cases}$$

$$M_Z(\mathbf{A}', \mathbf{B}'; 1) \neq \varnothing;$$

and $\delta_k(\mathbf{A}', \mathbf{B}'; 1) > 0$ *for all* k *such that* $1 \leq k \leq p(e) - k(e) + 1$, *where* $k \leq b'_{l_{k(e)}(\overline{\mathbf{F}};1)+1}$. *Then, the vectors* $\mathbf{A} = (a_1, \dots, a_n)$ *and* $\mathbf{B} = (b_1, \dots, b_m)$ *whose components are defined as*

$$a_i = \begin{cases} e_i, & i < k(e), \\ e_i + a'_i, & k(e) \leq i \leq p(e), \\ e_i, & i > p(e), \end{cases}$$

$$b_j = \begin{cases} f_j, & j \leq l_{k(e)}(\overline{\mathbf{F}};1), \\ f_j + b'_j, & l_{k(e)}(\overline{\mathbf{F}};1) + 1 \leq j \leq l_{k(e)-1}(\overline{\mathbf{F}};1), \\ f_j, & j > l_{k(e)-1}(\overline{\mathbf{F}};1), \end{cases}$$

satisfy the condition $R(\mathbf{A}, \mathbf{B}) = \overline{H}(\overline{\mathbf{E}}, \overline{\mathbf{F}})$.

Proof. Obviously, $\mathbf{A} \in \overline{Z}_+^n$ and $\mathbf{B} \in \overline{Z}_+^m$. Consider the function (1.4) of $\mathbf{A}$ and $\mathbf{B}$:

$$\delta_k(\mathbf{A}, \mathbf{B}; 1) = \sum_{j=1}^{m} b_j - \sum_{b_j \geq k} (b_j - k) - \sum_{i=1}^{k} a_i, \quad 1 \leq k \leq n.$$

Let us show that for any $k \in Z$ $(1 \leq k \leq n)$ wew have $\delta_k(\mathbf{A}, \mathbf{B}; 1) = 0$ and $l_k(\mathbf{B}; 1) = l_k(\overline{\mathbf{F}}; 1)$ if $k < k(e)$ or $k \geq p(e)$. Then, by virtue of (3.9) and (3.10) we will have $\overline{H}(\overline{\mathbf{E}}, \overline{\mathbf{F}}) \subseteq R(\mathbf{A}, \mathbf{B})$; moreover, if $k(e) \leq k \leq p(e) - 1$, then $\delta_k(\mathbf{A}, \mathbf{B}; 1) > 0$ for $k \leq b_{l_{k(e)}(\overline{\mathbf{F}};1)+1}$ and $\delta_k(\mathbf{A}, \mathbf{B}; 1) \geq 0$, $l_k(\mathbf{B}; 1) = l_k(\overline{\mathbf{F}}; 1)$ for $k > b_{l_{k(e)}(\overline{\mathbf{F}};1)+1}$. Then the equality $\overline{H}(\overline{\mathbf{E}}, \overline{\mathbf{F}}) = R(\mathbf{A}, \mathbf{B})$ will follow from (3.9) and (3.10).

The components of $\mathbf{A}$, $\mathbf{B}$, $\overline{\mathbf{E}}$, and $\overline{\mathbf{F}}$ are obviously related as follows:

$$\sum_{j=1}^{m} b_j = \sum_{j=1}^{m} f_j + \sum_{j=l_{k(e)}(\overline{\mathbf{F}};1)+1}^{l_{k(e)-1}(\overline{\mathbf{F}};1)} b'_j = \sum_{i=1}^{n} a_i = \sum_{i=1}^{n} e_i + \sum_{i=k(e)}^{p(e)} a'_i,$$

$$\sum_{i=1}^{k} a_i = \begin{cases} \sum_{i=1}^{k} e_i, & k < k(e), \\ \sum_{i=1}^{k} e_i + \sum_{i=k(e)}^{k} a'_i, & k(e) \le k \le p(e)-1, \\ \sum_{i=1}^{k} e_i + \sum_{i=k(e)}^{p(e)} a'_i, & k \ge p(e). \end{cases} \tag{3.16}$$

If $k < k(e)$ (for $k(e) \ge 2$), then

$$\sum_{b_j \ge k} (b_j - k) = \sum_{f_j \ge k} (f_j - k) + \sum_{j=l_{k(e)}(\overline{\mathbf{F}};1)+1}^{l_{k(e)-1}(\overline{\mathbf{F}};1)} b'_j,$$

(3.16) implies that $\delta_k(\mathbf{A}, \mathbf{B}; 1) = \delta_k(\overline{\mathbf{E}}, \overline{\mathbf{F}}; 1) = 0$, and the equality $l_k(\mathbf{B}; 1) = l_k(\overline{\mathbf{F}}; 1)$ is obvious. Let $k > p(e)$; then $l_k(\mathbf{B}; 1) = l_k(\overline{\mathbf{F}}; 1)$ and

$$\sum_{b_j \ge k} (b_j - k) = \sum_{f_j \ge k} (f_j - k). \tag{3.17}$$

Therefore, $\delta_k(\mathbf{A}, \mathbf{B}; 1) = \delta_k(\overline{\mathbf{E}}, \overline{\mathbf{F}}; 1) = 0$ by virtue of (3.17). If $k(e) \le k \le p(e) - 1$, then $l_{k(e)}(\overline{\mathbf{F}}; 1) < l_k(\mathbf{B}; 1) \le l_{k(e)-1}(\overline{\mathbf{F}}; 1)$ for $b'_{l_{k(e)}(\overline{\mathbf{F}};1)+1} \ge 0$. Accordingly, we have

$$\sum_{b_j \ge k} (b_j - k) = \sum_{b'_j \ge k-k(e)+1} \left(b'_j - k + k(e) - 1\right),$$

and the value of the function (1.4) of $\mathbf{A}$ and $\mathbf{B}$ can be expressed in terms of its value for $\mathbf{A}'$ and $\mathbf{B}'$:

$$\delta_k(\mathbf{A}, \mathbf{B}; 1) =$$

$$\delta_k(\overline{\mathbf{E}}, \overline{\mathbf{F}}; 1) + \sum_{j=l_{k(e)}(\overline{\mathbf{F}};1)+1}^{l_{k(e)-1}(\overline{\mathbf{F}};1)} b'_j - \sum_{b'_j \ge k-k(e)+1} \left(b'_j - k + k(e) - 1\right) - \sum_{i=k(e)}^{k} a'_i.$$

Since $\delta_k(\overline{\mathbf{E}}, \overline{\mathbf{F}}; 1) = 0$ and

$$\sum_{j=l_{k(e)}(\overline{\mathbf{F}};1)+1}^{l_{k(e)-1}(\overline{\mathbf{F}};1)} b'_j - \sum_{b'_j \geq k-k(e)+1} (b'_j - k + k(e) - 1) - \sum_{i=k(e)}^{k} a'_i = \delta_{k-k(e)+1}(\mathbf{A}', \mathbf{B}'; 1),$$

we have $\delta_k(\mathbf{A}, \mathbf{B}; 1) = \delta_{k-k(e)+1}(\mathbf{A}', \mathbf{B}'; 1)$. If $k - k(e) + 1 \leq b'_{l_{k(e)}(\overline{\mathbf{F}};1)+1}$, then the assumptions of the theorem imply that $\delta_k(\mathbf{A}, \mathbf{B}; 1) > 0$. If $k - k(e) + 1 > b'_{l_{k(e)}(\overline{\mathbf{F}};1)+1}$, then $\delta_k(\mathbf{A}, \mathbf{B}; 1) \geq 0$ and $l_k(\mathbf{B}; 1) = l_k(\overline{\mathbf{F}}; 1)$, which completes the proof.

Remark 3.5. Under the assumptions of Theorem 3.9, the set $\{(\mathbf{A}', \mathbf{B}')\}$ is not empty. In the proof of Lemma 3.1, it was shown that the vectors $\mathbf{A}'$ and $\mathbf{B}'$ whose first two components are ones and the remaining components are zeros satisfy the assumptions of Theorem 3.9.

Remark 3.6. Assuming that the function (1.4) of $\mathbf{A}'$ and $\mathbf{B}'$ in the statement of Theorem 3.9 is positive semidefinite rather than positive definite, we have $\overline{H}(\overline{\mathbf{E}}, \overline{\mathbf{F}}) \subseteq R(\mathbf{A}, \mathbf{B})$.

Lemma 3.2 and Theorem 3.9 entail the following proposition about an extremal matrix that is the hull of a matrix set.

Theorem 3.10. *Let the matrix $\overline{H}(\overline{\mathbf{E}}, \overline{\mathbf{F}})$, vectors $\overline{\mathbf{E}}$ and $\overline{\mathbf{F}}$, and the values of $p(e)$ and $k(e)$, where e is a component of $\overline{\mathbf{E}}$, be as defined in Theorem 3.9. Let $p(e) \geq k(e)$, and $\overline{\mathbf{F}}$ and $p(e)$ be such that*

$$l_{p(e)}(\overline{\mathbf{F}}; 1) \geq \begin{cases} l_{p(e)+1}(\overline{\mathbf{F}}; 1) + 2, & p(e) < n, \\ 2, & p(e) = n. \end{cases}$$

Suppose that vectors $\mathbf{A}'$ and $\mathbf{B}'$, with components numbered by indices ranging from $k(e)$ to $p(e)$ and from $l_{p(e)+1}(\overline{\mathbf{F}}; 1) + 1$ to $l_{p(e)}(\overline{\mathbf{F}}; 1)$, respectively, satisfy the conditions

$$\mathbf{A}' = (a'_{k(e)}, \ldots, a'_{p(e)}), \quad a'_i \leq a'_{i+1}, \quad k(e) \leq i \leq p(e) - 1,$$

$$\mathbf{B}' = (b'_{l_{p(e)+1}(\overline{\mathbf{F}};1)+1}, \ldots, b'_{l_{p(e)}(\overline{\mathbf{F}};1)}), \quad b'_j \leq b'_{j+1},$$

$$l_{p(e)+1}(\overline{\mathbf{F}}; 1) + 1 \leq j \leq l_{p(e)}(\overline{\mathbf{F}}; 1) - 1,$$

$$a'_{p(e)} < \begin{cases} a_{p(e)} - a_{p(e)+1}, & p(e) < n, \\ a_n, & p(e) = n, \end{cases}$$

$$b'_{l_{p(e)}(\overline{\mathbf{F}};1)} < \begin{cases} b_{l_{p(e)}(\overline{\mathbf{F}};1)} - b_{l_{p(e)+1}(\overline{\mathbf{F}};1)}, & l_{p(e)}(\overline{\mathbf{F}}; 1) < m, \\ b_m, & l_{p(e)}(\overline{\mathbf{F}}; 1) = m, \end{cases}$$

$$M_Z(\mathbf{A}', \mathbf{B}'; 1) \neq \varnothing,$$

and the vectors $\widetilde{\mathbf{A}}'$ *and* $\widetilde{\mathbf{B}}'$ *constructed by arranging the components of* $\mathbf{A}'$ *and* $\mathbf{B}'$ *in nonascending order are such that* $\delta_k(\widetilde{\mathbf{A}}', \widetilde{\mathbf{B}}'; 1) > 0$ *for any* k *in the range* $1 \leq k \leq p(e) - k(e) + 1$, *where* $k \leq b'_{l_{p(e)}(\overline{\mathbf{F}};1)}$. *Then, the vectors* $\mathbf{A} = (a_1, \ldots, a_n)$ *and* $\mathbf{B} = (b_1, \ldots, b_m)$, *whose components are defined as*

$$a_i = \begin{cases} e_i, & i < k(e), \\ e_i - a'_i, & k(e) \leq i \leq p(e), \\ e_i, & i > p(e), \end{cases}$$

$$b_j = \begin{cases} f_j, & j \leq l_{p(e)+1}(\overline{\mathbf{F}}; 1), \\ f_j - b'_j, & l_{p(e)+1}(\overline{\mathbf{F}}; 1) + 1 \leq j \leq l_{p(e)}(\overline{\mathbf{F}}; 1), \\ f_j, & j > l_{p(e)}(\overline{\mathbf{F}}; 1), \end{cases}$$

satisfy the condition $O(\mathbf{A}, \mathbf{B}) = \overline{H}(\overline{\mathbf{E}}, \overline{\mathbf{F}})$.

Remark 3.7. By Lemma 3.1, the set of vector pairs $\{(\mathbf{A}', \mathbf{B}')\}$ is not empty under the assumptions of Theorem 3.10.

Remark 3.8. Replacing the assumption that the function (1.4) of $\mathbf{A}'$ and $\mathbf{B}'$ is positive in the statement of Theorem 3.10 with the assumption that the function is nonnegative, we obtain $O(\mathbf{A}, \mathbf{B}) \subseteq \overline{H}(\overline{\mathbf{E}}, \overline{\mathbf{F}})$.

In what follows, we introduce a new extremality criterion to simplify the statement of Lemma 3.1.

3.6. An Alternative Extremality Criterion for a Pair of Vectors

To construct a matrix set with prescribed kernel or hull, it is convenient to use an extremality criterion different from those formulated in Section 1 of Chapter 4.

To this end, we introduce the following notation for the components of an arbitrary vector $\mathbf{A} = (a_1, \ldots, a_n)$ in $\overline{Z}^n_+$: if there exists at least one component of $\mathbf{A}$ whose value is a, we say that $l(\mathbf{A}; a) = \left|\{i : a_i = a,\ 1 \leq i \leq n\}\right|$ is the number of components of $\mathbf{A}$ whose values are equal to a. We denote by $\alpha_1 > \cdots > \alpha_t$ and $\beta_1 > \cdots > \beta_r$ the distinct components of $\mathbf{A} = (a_1, \ldots, a_n) \in \overline{Z}^n_+$ and $\mathbf{B} = (b_1, \ldots, b_m) \in \overline{Z}^m_+$, respectively. Then, the pairs α_p, $l(\mathbf{A}; \alpha_p)$ $(1 \leq p \leq t)$ and β_q, $l(\mathbf{B}; \beta_q)$ $(1 \leq q \leq r)$ define the vectors $\mathbf{A}$ and $\mathbf{B}$, respectively. Obviously, it suffices to consider only their positive components. Now, we formulate an extremality criterion that determines nontrivial matrix sets with prescribed kernel or hull.

Theorem 3.11. *Suppose that* $\mathbf{A} = (a_1, \ldots, a_n) \in \overline{Z}^n_+$; $\mathbf{B} = (b_1, \ldots, b_m) \in \overline{Z}^m_+$; *and* $\alpha_1 > \cdots > \alpha_t$ *and* $\beta_1 > \cdots > \beta_r$, *where* t, $r \geq 1$ *are the different* *positive components of* **A** *and* **B**, *respectively. The vectors* **A** *and* **B** *make up an extremal pair if and only if* $t = r$ *and the following relations are valid:*

$$l(\mathbf{A};\alpha_p) = \beta_{t-p+1} - \beta_{t-p+2}, \quad 1 \le p \le t, \tag{3.18}$$

$$l(\mathbf{B};\beta_q) = \alpha_{t-q+1} - \alpha_{t-q+2}, \quad 1 \le q \le t, \tag{3.19}$$

where $\alpha_{t+1} = \beta_{t+1} = 0$.

Proof.

Necessity. We use induction on the number of components of the first vector in the extremal pair **A**, **B**. If $n = 1$, then it is obvious that $\alpha_1 = a_1$, $t = 1$, $l(\mathbf{A};\alpha_1) = 1$, $b_j = 1$ for $1 \le j \le a_1 = l_1(\mathbf{B};1)$, $b_j = 0$ for $j > l_1(\mathbf{B};1)$, $\beta_1 = 1$, $r = 1$, and $l(\mathbf{B};\beta_1) = \alpha_1$. Therefore, $t = r = 1$ and since it is assumed that $\alpha_2 = \beta_2 = 0$, we have $l(\mathbf{A};\alpha_1) = \beta_1 - \beta_2$ and $l(\mathbf{B};\beta_1) = \alpha_1 - \alpha_2$.

Suppose that $t = r$ and relations (3.18) and (3.19) apply to an arbitrary extremal pair of vectors **A** and **B** when the number of components of the first vector is less than n, and let $\mathbf{A} = (a_1, \dots, a_n)$.

Suppose $a_n = 0$. Taking out the last component of **A**, one obtains an extremal vector pair such that $t = r$ and relations (3.18) and (3.19) are valid by the induction hypothesis. As a result, all parameters in (3.18) and (3.19) remain the same as for the starting pair of vectors **A** and **B**, which completes the proof in this case.

Suppose now that $a_n > 1$. Construct the (extremal) principai $(1;n)$-reductive pair $\mathbf{A}^{(1)}(n)$, $\mathbf{B}_*^{(1)}$ corresponding to the vector pair **A**, **B**. Take out the last zero component of $\mathbf{A}^{(1)}(n)$. The resulting vector pair $\mathbf{A}' = (a_1, \dots, a_{n-1})$, $\mathbf{B}' = (b'_1, \dots, b'_m) = \mathbf{B}_*^{(1)}$ is also extremal, $\mathbf{A}' \in \overline{Z}_+^{n-1}$, and

$$b'_j = \begin{cases} b_j - 1, & 1 \le j \le a_n = l_n(\mathbf{B};1), \\ b_j, & j > a_n. \end{cases}$$

Consider the values of parameters in (3.18) and (3.19) that correspond to the vector pairs **A**, **B** and **A**$'$, **B**$'$: α_p and $l(\mathbf{A};\alpha_p)$ $(1 \le p \le t)$; β_q and $l(\mathbf{B};\beta_q)$ $(1 \le q \le r)$; α'_p and $l(\mathbf{A}';\alpha'_p)$ $(1 \le p \le t')$; and β'_q and $l(\mathbf{B}';\beta'_q)$ $(1 \le q \le r')$, where t' and r' are the numbers of distinct positive values of the components of **A**$'$ and **B**$'$, and $\alpha'_p > \alpha'_{p+1}$ and $\beta'_q > \beta'_{q+1}$ for $1 \le p \le t' - 1$ and $1 \le q \le r' - 1$ when $t' > 1$ and $r' > 1$, respectively. By the induction hypothesis, (3.18) and (3.19) apply to **A**$'$ and **B**$'$. The components of **A** may be such that $a_{n-1} > a_n$ or $a_{n-1} = a_n$. These cases are analyzed separately in what follows.

Let $a_{n-1} > a_n$. Then, the components of **A** and **A**$'$ are such that

$$\begin{aligned} &l(\mathbf{A};\alpha_t) = 1, \quad t' = t - 1, \\ &\alpha'_p = \alpha_p, \quad l(\mathbf{A}';\alpha'_p) = l(\mathbf{A};\alpha_p), \quad 1 \le p \le t' = t - 1. \end{aligned} \tag{3.20}$$

Since $a_{n-1} > a_n$, the extremality of the vector pair **A**, **B** implies that the components of the second vector are such that

$$b_j = b_1 = \beta_1, \quad 1 \le j \le a_n, \quad b_{a_n+1} = \beta_2 = \beta_1 - 1. \tag{3.21}$$

Therefore, the components of **B** and **B**$'$ are related as follows: $\beta_1 = \beta_2 + 1$, $r' = r - 1$,

$$\begin{aligned} \beta'_q &= \begin{cases} \beta_1 - 1 = \beta_2, & q = 1, \\ \beta_{q+1}, & 2 \le q \le r - 1 = r', \end{cases} \\ l(\mathbf{B}'; \beta'_q) &= \begin{cases} l(\mathbf{B}; \beta_1) + l(\mathbf{B}; \beta_2), & q = 1, \\ l(\mathbf{B}; \beta_{q+1}), & 2 \le q \le r - 1 = r'. \end{cases} \end{aligned} \tag{3.22}$$

By the induction hypothesis, the vector pair **A**$'$, **B**$'$ obtained is such that $t' = r'$,

$$\begin{aligned} l(\mathbf{A}'; \alpha'_p) &= \beta'_{t'-p+1} - \beta'_{t'-p+2}, \quad 1 \le p \le t', \\ l(\mathbf{B}'; \beta'_q) &= \alpha'_{t'-q+1} - \alpha'_{t'-q+2}, \quad 1 \le q \le t' = r', \end{aligned} \tag{3.23}$$

where $\alpha_{t'+1} = \beta_{t'+1} = 0$. Therefore, the numbers of distinct components of **A** and **B** are $t = t' + 1$ and $r = r' + 1$, respectively, and $t = r$. Applying (3.20)–(3.23), we obtain

$$l(\mathbf{A}; \alpha_p) = \begin{cases} l(\mathbf{A}'; \alpha'_p), & 1 \le p \le t - 1, \\ 1, & p = t - 1, \end{cases}$$

$$l(\mathbf{A}; \alpha_p) = \begin{cases} \beta'_{t'-p+1} - \beta'_{t'-p+2}, & 1 \le p \le t - 1 = t', \\ 1, & p = t, \end{cases}$$

$$l(\mathbf{A}; \alpha_p) = \begin{cases} \beta_{t'-p+2} - \beta_{t'-p+3}, & 1 \le p \le t - 2 = t - 1', \\ \beta_2 - \beta_3, & p = t - 1 = t', \\ 1, & p = t, \end{cases}$$

$$l(\mathbf{A}; \alpha_p) = \begin{cases} \beta_{t-p+1} - \beta_{t-p+2}, & 1 \le p \le t - 2, \\ \beta_2 - \beta_3, & p = t - 1, \\ \beta_1 - \beta_2, & p = t, \end{cases}$$

$$l(\mathbf{A}; \alpha_p) = \beta_{t-p+1} - \beta_{t-p+2}, \quad 1 \le p \le t;$$

$$l(\mathbf{B}; \beta_q) = \begin{cases} a_n - 0, & q = 1, \\ l(\mathbf{B}'; \beta'_1) - l(\mathbf{B}; \beta_1), & q = 2, \\ l(\mathbf{B}'; \beta'_{q-1}), & 2 \le q \le t, \end{cases}$$

$$l(\mathbf{B};\beta_q) = \begin{cases} \alpha_t - \alpha_{t+1}, & q = 1, \\ a_{n-1} - a_n, & q = 2, \\ \alpha'_{t'-q} - \alpha'_{t'-q+1}, & 2 < q \le t, \end{cases}$$

$$l(\mathbf{B};\beta_q) = \begin{cases} \alpha_t - \alpha_{t+1}, & q = 1, \\ \alpha_{t-1} - \alpha_t, & q = 2, \\ \alpha_{t-q+1} - \alpha_{t-q+2}, & 2 < q \le t, \end{cases}$$

$$l(\mathbf{B};\beta_q) = \alpha_{t-q+1} - \alpha_{t-q+2}, \quad 1 \le q \le t.$$

This completes the proof of the necessity of relations (3.18) and (3.19) for an extremal pair of vectors **A** and **B** in the case of $a_{n-1} > a_n$.

Let $a_{n-1} = a_n$. Then, the components of **A** and **A′** are such that

$$\begin{aligned} &l(\mathbf{A};\alpha_t) > 1, \quad t' = t, \quad l(\mathbf{A}';\alpha'_t) = l(\mathbf{A};\alpha_t) - 1, \quad \alpha'_t = \alpha_t, \\ &\alpha'_p = \alpha_p, \quad l(\mathbf{A}';\alpha'_p) = l(\mathbf{A};\alpha_p), \quad 1 \le p \le t-1 \end{aligned} \tag{3.24}$$

for $t > 1$. Since $a_{n-1} = a_n$, the components of **B** and **B′** are such that

$$\begin{aligned} &r' = r, \quad \beta'_1 = \beta_1 - 1, \quad l(\mathbf{B}';\beta'_q) = l(\mathbf{B};\beta_q), \quad 1 \le q \le r, \\ &\beta'_q = \beta_q, \quad 2 \le q \le r \end{aligned} \tag{3.25}$$

for $q \ge 2$. By the induction hypothesis, for **A′** and **B′** it holds that $t' = r'$ and

$$\begin{aligned} l(\mathbf{A}';\alpha'_p) &= \beta'_{t-p+1} - \beta'_{t-p+2}, \quad 1 \le p \le t = t', \\ l(\mathbf{B}';\beta'_q) &= \alpha'_{t-q+1} - \alpha'_{t-q+2}, \quad 1 \le q \le t = r = t'. \end{aligned} \tag{3.26}$$

Applying (3.24)–(3.26) to the components of **A** and **B**, we obtain $t = r$,

$$l(\mathbf{A};\alpha_p) = \begin{cases} l(\mathbf{A}';\alpha'_p), & p < t, \\ l(\mathbf{A}';\alpha'_p) + 1, & p = t, \end{cases}$$

$$l(\mathbf{A};\alpha_p) = \begin{cases} \beta'_{t-p+1} - \beta'_{t-p+2}, & p < t, \\ \beta'_1 - \beta'_2 + 1, & p = t, \end{cases}$$

$$l(\mathbf{A};\alpha_p) = \begin{cases} \beta_{t-p+1} - \beta_{t-p+2}, & p < t, \\ \beta_1 - \beta_2, & p = t, \end{cases}$$

$$l(\mathbf{A};\alpha_p) = \beta_{t-p+1} - \beta_{t-p+2}, \quad 1 \le p \le t;$$

$$l(\mathbf{B};\beta_q) = l(\mathbf{B}';\beta'_q) = \alpha'_{t-q+1} - \alpha'_{t-q+2}, \quad 1 \le q \le t,$$

$$l(\mathbf{B};\beta_q) = \alpha_{t-q+1} - \alpha_{t-q+2}, \quad 1 \le q \le t.$$

This completes the proof of the necessity.

Sufficiency. Suppose that the vectors **A** and **B** are such that $t = r \geq 1$ and relations (3.18) and (3.19) are valid. Let us show that the vectors **A** and **B** make up an extremal pair by Theorem 1.4. Consider an arbitrary component $b_j > 0$ of **B**. Assume that $b_j = \beta_q$ for some q $(1 \leq q \leq t)$. By (3.18), it holds that

$$\beta_q - \beta_{q+1} = l(\mathbf{A}; \alpha_{t-q+1}), \quad 1 \leq q \leq t. \tag{3.27}$$

Calculating the sum of expressions in (3.27) over the index from q to t and using the fact that $\beta_{t+1} = 0$, we obtain

$$\beta_q = \sum_{k=1}^{t-q+1} l(\mathbf{A}; \alpha_k) = b_j. \tag{3.28}$$

By the definition of β_q $(1 \leq q \leq t)$, the positive components of **B** are such that

$$b_j = \begin{cases} \beta_1, & 1 \leq j \leq l(\mathbf{B}; \beta_1), \\ \beta_q, & \sum_{k=1}^{q-1} l(\mathbf{B}; \beta_k) + 1 \leq j \leq \sum_{k=1}^{q} l(\mathbf{B}; \beta_k), \quad 2 \leq q \leq t. \end{cases} \tag{3.29}$$

Calculating the sum of expressions in (3.19), with q replaced by k, over k in the range $1 \leq k \leq q$ and using the fact that $\alpha_{t+1} = 0$, we obtain

$$\alpha_{t-q+1} = \sum_{k=1}^{q} l(\mathbf{B}; \beta_k). \tag{3.30}$$

Relations (3.29) and (3.30) show that if $b_j = \beta_q$, then $j \leq \alpha_{t-q+1}$. Moreover, the latter relation implies that $j \leq a_i$ and $a_i \geq \alpha_{t-q+1}$. Since $l_j(\mathbf{A}; 1) = \max\{i : a_i \geq j\}$, the definitions of α_p and $l(\mathbf{A}; \alpha_p)$ entails

$$l_j(\mathbf{A}; 1) \geq \sum_{k=1}^{t-q+1} l(\mathbf{A}; \alpha_k).$$

Let us show that this is a strict equality. Assuming that $j \leq \alpha_{t-q+2}$, we have

$$j \leq \alpha_{t-(q-1)+1} = \sum_{k=1}^{q-1} l(\mathbf{B}; \beta_k).$$

This contradicts the relation in (3.29), because $b_j = \beta_q$. Therefore,

$$l_j(\mathbf{A}; 1) = \sum_{k=1}^{t-q+1} l(\mathbf{A}; \alpha_k).$$

Combined with (3.28), this implies that $b_j = l_j(\mathbf{A};1)$ for any $b_j > 0$. If $b_j = 0$, then (3.30) entails

$$j > \sum_{k=1}^{t} l(\mathbf{B};\beta_k) = \alpha_1,$$

which implies that $l_j(\mathbf{A};1) = 0$. Thus, $b_j = l_j(\mathbf{A};1)$ for $1 \le j \le m$, and we obtain relation (1.7) of Theorem 1.4, which establishes the extremality conditions for the pair of vectors **A** and **B**. This completes the proof.

Theorem 3.11 can be used to simplify the statement of Lemma 3.1 about the nontriviality of a matrix set with prescribed kernel or hull. Note also that the numbers of distinct positive values of vectors in any extremal pair are equal by Theorem 3.11.

Theorem 3.12. *Let* $(\overline{\mathbf{E}},\overline{\mathbf{F}})$ *be an extremal vector pair in* $\overline{E}^{n,m}$*, where both* $\overline{\mathbf{E}}$ *and* $\overline{\mathbf{F}}$ *have* $t \ge 1$ *distinct positive values of components. Let* α_p *and* β_q $(1 \le p,\ q \le t)$ *be the distinct positive values of components of* $\overline{\mathbf{E}}$ *and* $\overline{\mathbf{F}}$*,* $\alpha_p > \alpha_{p+1}$*, and* $\beta_q > \beta_{q+1}$ ($p,q \le t-1$ if $t > 1$)*. Suppose that* $\alpha_{t+1} = \beta_{t+1} = 0$*; and* $l(\overline{\mathbf{E}};\alpha_{t+1})$ *and* $l(\overline{\mathbf{F}};\beta_{t+1})$ *are the numbers of zero components of* $\overline{\mathbf{E}}$ *and* $\overline{\mathbf{F}}$*, respectively. Then, the following propositions apply to the extremal matrix* $\overline{H}(\overline{\mathbf{E}},\overline{\mathbf{F}})$.

(a) *There exists a vector pair* $(\mathbf{A},\mathbf{B}) \ne (\overline{\mathbf{E}},\overline{\mathbf{F}})$ $(M_Z(\mathbf{A},\mathbf{B};1) \ne \varnothing$, $\mathbf{A} \in \overline{Z}^n_+$, $\mathbf{B} \in \overline{Z}^m_+)$ *for which* $R(\mathbf{A},\mathbf{B}) = \overline{H}(\overline{\mathbf{E}},\overline{\mathbf{F}})$ *if and only if there exists* $k \in Z$ $(1 \le k \le t+1)$ *such that* $l(\overline{\mathbf{E}};\alpha_k) \ge 2$ *and* $l(\overline{\mathbf{F}};\beta_{t-k+2}) \ge 2$ *for the vector pair* $(\overline{\mathbf{E}},\overline{\mathbf{F}})$.

(b) *There exists a vector pair* $(\mathbf{A},\mathbf{B}) \ne (\overline{\mathbf{E}},\overline{\mathbf{F}})$ $(M_Z(\mathbf{A},\mathbf{B};1) \ne \varnothing$, $\mathbf{A} \in \overline{Z}^n_+$, $\mathbf{B} \in \overline{Z}^m_+)$ *for which* $O(\mathbf{A},\mathbf{B}) = \overline{H}(\overline{\mathbf{E}},\overline{\mathbf{F}})$ *if and only if there exists* $k \in Z$ $(1 \le k \le t)$ *such that* $l(\overline{\mathbf{E}};\alpha_k) \ge 2$ *and* $l(\overline{\mathbf{F}};\beta_{t-k+1}) \ge 2$ *for the vector pair* $(\overline{\mathbf{E}},\overline{\mathbf{F}})$.

Theorem 3.12 is an alternative statement of Lemma 3.1.

The results presented in this chapter are based on [18–23].

References to Chapter 4

[1] Aleksandrov P. S., *Vvedenie v obshchuyu teoriyu mnozhestv i funktsii* (Introduction to the General Theory of Sets and Functions), Gostekhteorizdat, 1948.

[2] Busacker R. G. and Saaty T. L., *Finite Graphs and Networks: An Introduction with Applications,* New York: McGraw-Hill, 1965. Translated under the title *Konechnye grafy i seti,* Moscow: Nauka, 1974.

[3] Berge C., *Theorie des graphes et ses applications*, Paris: Dunod, 1958. Translated under the title *Teoriya grafov i ee primeneniya,* Moscow: Inostr. Literatura, 1962.

[4] Gavrilov G. P. and Sapozhenko A. A., *Zadachi i uprazhneniya po kursu diskretnoi matematiki* (Problems and Exercises to a Course in Discrete Mathematics), Moscow: Nauka, 1992.

[5] Gale D., A Theorem on Flow in Networks, *Pacific J. Math.*, 1957, vol. 7, pp. 1073–1082.

[6] Glazman I.M. and Lyubich Yu.I., *Konechnomernyi lineinyi analiz* (Finite-Dimensional Linear Analysis), Moscow: Nauka, 1969.

[7] Gol'shtein E. G. and Yudin D. B., *Zadachi lineinogo programmirovaniya transportnogo tipa* (Transportation Problems in Linear Programming), Moscow: Nauka, 1969.

[8] Dunford N. and Schwartz J. T., *Linear Operators, Part 1: General Theory,* New York: Interscience, 1958. Translated under the title *Lineinye operatory (obshchaya teoriya),* Moscow: Mir, 1962.

[9] Emelichev V. A., Kovalev M. M., and Kravtsov M. K., *Mnogogranniki, grafy, optimizatsiya* (Polyhedrons, Graphs, and Optimization), Moscow: Nauka, 1981.

[10] Zykov A. A., *Teoriya konechnykh grafov* (Theory of Finite Graphs), Novosibirsk: Nauka, 1969.

[11] Zhuravlev Yu. I., On a Class of Algorithms over Finite Sets, *Dokl. Akad. Nauk SSSR*, 1963, vol. 151, no. 5, pp. 1025–1028.

[12] Kaluzhnin L., *Vvedenie v obshchuyu algebru* (An Introduction to General Algebra), Moscow: Nauka, 1973.

[13] Kelley J. L., *General Topology*, New York: Van Nostrand, 1957. Translated under the title *Obshchaya topologiya,* Moscow: Nauka, 1968.

[14] Kleinrock R., *Communication Nets: Stochastic Message Flow and Delay,* New York: McGraw-Hill, 1964. Translated under the title *Kommunikatsionnye seti,* Moscow: Nauka, 1970.

[15] Cohn P. M., *Free Rings and Their Relations*, London: Academic, 1971. Translated under the title *Svobodnye kol'tsa i ikh svyazi*, Moscow: Mir, 1975.

[16] Melikhov A. N., *Orientirovannye grafy i konechnye avtomaty* (Directed Graphs and Finite Automata), Moscow: Nauka, 1971.

[17] Mizin I. A., Urinson L. S., and Khrameshin G. K., *Peredacha informatsii v setyakh s kommutatsiei soobshchenii* (Data Transmission in Switched Networks), Moscow: Svyaz, 1972.

[18] Mironov A. A., Geometry of the Points in the Space R^n that Can Be Realized in a Graph, *Usp. Mat. Nauk*, 1977, vol. 32, no. 6, pp. 231–232.

[19] Mironov A. A., Realizability of a Set of Numbers in a Graph and Properties of Graphs with Prescribed Sets of Degrees, *Tr. Gor'kovskogo Gos. Univ.*, 1981, Markov, A.A., Ed., pp. 76–97.

[20] Mironov A. A., Probabilistic Properties of a Random Graph with Prescribed Set of Degrees of Its Vertices, *Izv. Akad. Nauk SSSR, Ser. Tekh. Kibern.*, 1990, no. 4, pp. 180–188.

[21] Mironov A. A. and Tsurkov V. I., Graphic Representation of Multilevel Hierarchical Structures, *Izv. Akad. Nauk SSSR, Ser. Tekh. Kibern.*, 1991, no. 3, pp. 148–155.

[22] Mironov A. A. and Tsurkov V. I., Approximation and Decomposition by Means of Extremal Graphs, *Zh. Vychisl. Mat. Mat. Fiz.*, 1993, vol. 33, no. 2, pp. 283–298.

[23] Mironov A. A., Uniform Generalized Graphs, *Dokl. Akad. Nauk*, 1996, vol. 351, no. 4, pp. 465–468.

[24] Moisil G. K., *Algebraicheskaya teoriya diskretnykh avtomaticheskikh ustroistv* (Algebraic Theory of Discrete Automatic Devices), Moscow: Nauka, 1963.

[25] Moder J. and Elmaghraby S. S., *Handbook of Operations Research*, New York: Van Nostrand, 1978. Translated under the title *Issledovanie operatsii. Metodologicheskie osnovy i matematicheskie metody,* vol. 1, Moscow: Mir, 1981.

[26] *Algoritmy i programmy resheniya zadach na grafakh i setyakh* (Algorithms and Programs for Solving Problems on Graphs and Networks), Nechepurenko, M.I., Ed., Novosibirsk: Nauka, 1990.

[27] Propoi A. I., *Elementy teorii optimal'nykh diskretnykh protsessov* (Elements of the Theory of Optimal Discrete Processes), Moscow: Nauka, 1973.

[28] Ryser H. J., Combinatorial Properties of Matrices of Zeros and Ones, *Can. J. Math.*, 1957, vol. 9, pp. 371–377.

[29] Ryser H. J., The Term Rank of a Matrix, *Can. J. Math.*, 1958, vol. 10, pp. 57–65.

[30] Ryser H. J., Traces of Matrices of Zeros and Ones, *Can. J. Math.*, 1960, vol. 12, pp. 463–476.

[31] *Teoriya setei svyazi* (Theory of Communication Networks), Roginskii V. N., Ed., Moscow: Radio i Svyaz, 1981.

[32] *Kombinatornyi analiz. Zadachi i uprazhneniya* (Combinatorial Analysis: Problems and Exercises), Rybnikov K. A., Ed., Moscow: Nauka, 1982.

[33] Saaty T., *Optimization in Integers and Related Extremal Problems*, New York: McGraw-Hill, 1970. Translated under the title *Tselochislennye metody optimizatsii i svyazannye s nimi ekstremal'nye problemy,* Moscow: Mir, 1973.

[34] Sachkov V. N., *Vvedenie v kombinatornye metody diskretnoi matematiki* (An Introduction to Combinatorial Methods of Discrete Mathematics), Moscow: Nauka, 1982.

[35] Sikorski R., *Boolean Algebras*, Berlin: Springer-Verlag, 1964, 2nd ed. Translated under the title *Bulevy algebry,* Moscow: Mir, 1969.

[36] Trius E. B., *Zadachi matematicheskogo programmirovaniya transportnogo tipa* (Transportation Problems in Mathematical Programming), Moscow: Sovetskoe Radio, 1967.

[37] Ford L. R. and Fulkerson D., *Flows in Networks*, Princeton (New Jersay): Princeton Univ. Press, 1962. Translated under the title *Potoki v setyakh,* Moscow: Mir, 1966.

[38] Harary F., *Graph Theory*, Reading (Mass.): Addison-Wesley, 1969. Translated under the title *Teoriya grafov,* Moscow: Inostr. Literatura, 1963.

[39] Hall M., Jr., *Combinatorial Theory*, Toronto: Blaisdell, 1967. Translated under the title *Kombinatorika,* Moscow: Mir, 1970.

[40] Hu T., *Integer Programming and Network Flows*, Reading (Mass): Addison-Wesley, 1969. Translated under the title *Tselochislennoe programmirovanie i potoki v setyakh,* Moscow: Mir, 1974.

Chapter 5

Fundamental Polyhedrons

In this chapter, we construct convex hulls of the set of extremal vector pairs and on the set of the corresponding extremal matrices. We analyze optimization problems posed on sets of uniform matrices.

In Section 1, we show that every extremal element is a vertex of a hull. We apply the well-known Krein–Mil'man theorem about extreme points [1, 2, 4] to vector pairs that define nonempty transportation polyhedrons [5]. We show that every vector pair belongs, up to a factor, to the hull of extremal vector pairs. We also show that any uniform matrix belongs, up to a factor, to the hull of extremal matrices.

Some minimax problems have nonunique solutions (e.g., recall Theorem 2.5 from Chapter 2). Therefore, optimization problems may be posed on the corresponding sets of solutions. In Section 2, we use decompositions in the form of convex linear combinations of extremal matrices to solve such problems. The optimization method is thereby reduced to choosing an optimal combination.

§1. Theorems about Vertices

The extremal vector pairs of equal dimensions obtained for different minimax problems have a remarkable property: Their hull is complete in a certain sense, and every extremal pair is a vertex of the hull. In other words, any pair of vectors can be written in the form of a convex linear combination of extremal vector pairs. This proposition applies to sets of uniform matrices of uqual dimensions as well.

1.1. Convex Hulls of Fundamental Systems of Vector Pairs and Matrices

Recall that one definition of the extremal vector pair is as follows: A vector pair $(\mathbf{A}, \mathbf{B})$, where $\mathbf{A} = (a_1, \ldots, a_n) \in \overline{Z}^n_+$ and $\mathbf{B} = (b_1, \ldots, b_m) \in \overline{Z}^m_+$, is said to be extremal if

$$\delta_k(\mathbf{A}, \mathbf{B}; 1) = \sum_{j=1}^{m} b_j - \sum_{b_j \geq k} (b_j - k) - \sum_{i=1}^{k} a_i = 0$$

for any $k \in Z$ $(1 \leq k \leq n)$, where $\delta_k(\mathbf{A}, \mathbf{B}; c)$ is the characteristic function (1.22) from Chapter 1 with $c = 1$, or function (1.4) from Chapter 4. When

$(\mathbf{A}, \mathbf{B})$ is an extremal vector pair, the set $M_Z(\mathbf{A}, \mathbf{B}; 1)$ contains a single matrix (by Theorem 1.1 from Chapter 4), which is also called extremal. To single out an extremal pair of vectors, we denote it by $(\overline{\mathbf{E}}, \overline{\mathbf{F}})$, where $\overline{\mathbf{E}} = (e_1, \ldots, e_n)$ and $\overline{\mathbf{F}} = (f_1, \ldots, f_m)$; the only extremal matrix in $M_Z(\overline{\mathbf{E}}, \overline{\mathbf{F}}; 1)$ is denoted by $\overline{H} = \overline{H}(\overline{\mathbf{E}}, \overline{\mathbf{F}}) = (h_{ij})$ $(1 \leq i \leq n,\ 1 \leq j \leq m)$. When n and m are fixed natural numbers, the corresponding extremal vector pairs and matrices are denoted, respectively, by

$$\overline{E}^{n,m} = \left\{(\overline{\mathbf{E}}, \overline{\mathbf{F}}) : \overline{\mathbf{E}} \in \overline{Z}_+^n,\ \overline{\mathbf{F}} \in \overline{Z}_+^m,\ \delta_k(\overline{\mathbf{E}}, \overline{\mathbf{F}}; 1) = 0,\ 1 \leq k \leq n\right\},$$

$$\overline{E}_{n,m} = \left\{\overline{H}(\overline{\mathbf{E}}, \overline{\mathbf{F}}) : (\overline{\mathbf{E}}, \overline{\mathbf{F}}) \in \overline{E}^{n,m},\ \overline{H}(\overline{\mathbf{E}}, \overline{\mathbf{F}}) \in M_Z(\overline{\mathbf{E}}, \overline{\mathbf{F}}; 1)\right\}.$$

By Definition 1.6 from Chapter 4, the sets $\overline{E}^{n,m}$ and $\overline{E}_{n,m}$ are referred to as fundamental systems. By Theorem 1.7 from Chapter 4, the numbers of elements in both $\overline{E}^{n,m}$ and $\overline{E}_{n,m}$ are equal to $C_{n+m}^n = C_{n+m}^m$ $(|\overline{E}^{n,m}| = |\overline{E}_{n,m}| = C_{n+m}^n)$. We introduced a special notation for fundamental subsystems:

$$\begin{aligned}
\overline{E}^{n^+,m} &= \left\{(\overline{\mathbf{E}}, \overline{\mathbf{F}}) : (\overline{\mathbf{E}}, \overline{\mathbf{F}}) \in \overline{E}^{n,m},\ e_n \geq 1\right\}, \\
\overline{E}^{n,m^+} &= \left\{(\overline{\mathbf{E}}, \overline{\mathbf{F}}) : (\overline{\mathbf{E}}, \overline{\mathbf{F}}) \in \overline{E}^{n,m},\ e_1 = m\right\}, \\
\overline{E}^{n^+,m^+} &= \left\{(\overline{\mathbf{E}}, \overline{\mathbf{F}}) : (\overline{\mathbf{E}}, \overline{\mathbf{F}}) \in \overline{E}^{n^+,m} \cap \overline{E}^{n,m^+}\right\}, \\
\overline{E}_{n^+,m} &= \left\{\overline{H}(\overline{\mathbf{E}}, \overline{\mathbf{F}}) : (\overline{\mathbf{E}}, \overline{\mathbf{F}}) \in \overline{E}^{n^+,m},\ \overline{H}(\overline{\mathbf{E}}, \overline{\mathbf{F}}) \in M_Z(\overline{\mathbf{E}}, \overline{\mathbf{F}}; 1)\right\}, \\
\overline{E}_{n,m^+} &= \left\{\overline{H}(\overline{\mathbf{E}}, \overline{\mathbf{F}}) : (\overline{\mathbf{E}}, \overline{\mathbf{F}}) \in \overline{E}^{n,m^+},\ \overline{H}(\overline{\mathbf{E}}, \overline{\mathbf{F}}) \in M_Z(\overline{\mathbf{E}}, \overline{\mathbf{F}}; 1)\right\}, \\
\overline{E}_{n^+,m^+} &= \left\{\overline{H}(\overline{\mathbf{E}}, \overline{\mathbf{F}}) : (\overline{\mathbf{E}}, \overline{\mathbf{F}}) \in \overline{E}^{n^+,m^+},\ \overline{H}(\overline{\mathbf{E}}, \overline{\mathbf{F}}) \in M_Z(\overline{\mathbf{E}}, \overline{\mathbf{F}}; 1)\right\}.
\end{aligned}$$

By Theorem 1.8 from Chapter 4, it holds that

$$\left|\overline{E}^{n^+,m}\right| = \left|\overline{E}_{n^+,m}\right| = C_{n+m-1}^n,$$

$$\left|\overline{E}^{n,m^+}\right| = \left|\overline{E}_{n,m^+}\right| = C_{n+m-1}^{n-1},$$

$$\left|\overline{E}^{n^+,m^+}\right| = \left|\overline{E}_{n^+,m^+}\right| = C_{n+m-2}^{n-1}.$$

Suppose that the elements of the fundamental systems $\overline{E}^{n,m}$ and $\overline{E}_{n,m}$ are denoted by numbers ranging from 0 to $C_{n+m}^n - 1$:

$$\overline{E}^{n,m} = \left\{(\overline{\mathbf{E}}_k, \overline{\mathbf{F}}_k) : 0 \leq k \leq C_{n+m}^n - 1\right\},$$

$$\overline{E}_{n,m} = \left\{\overline{H}_k(\overline{\mathbf{E}}_k, \overline{\mathbf{F}}_k) : 0 \leq k \leq C_{n+m}^n - 1\right\},$$

where

$$(\overline{\mathbf{E}}_0, \overline{\mathbf{F}}_0) = \Big((\underbrace{0,\dots,0}_{n}), (\underbrace{0,\dots,0}_{m}) \Big),$$

$$\Big(\overline{\mathbf{E}}_{C^n_{n+m}-1}, \overline{\mathbf{F}}_{C^n_{n+m}-1}\Big) = \Big((\underbrace{m,\dots,m}_{n}), (\underbrace{n,\dots,n}_{m}) \Big).$$

Then, $\overline{H}_0 = \overline{H}_0(\overline{\mathbf{E}}_0, \overline{\mathbf{F}}_0)$ is a zero matrix, and each element of $\overline{H}_{C^n_{n+m}-1} = \overline{H}_{C^n_{n+m}-1}(\overline{\mathbf{E}}_{C^n_{n+m}-1}, \overline{\mathbf{F}}_{C^n_{n+m}-1})$ is one.

Denote the polyhedrons that are the closed convex hulls of fundamental systems of extremal vector pairs $\overline{E}^{n,m}$ and extremal matrices $\overline{E}_{n,m}$, respectively, by $L(n,m)$ and $M(n,m)$:

$$L(n,m) = \Big\{ (\mathbf{A},\mathbf{B}) : (\mathbf{A},\mathbf{B}) = \sum_{k=0}^{C^n_{n+m}-1} \alpha_k (\overline{\mathbf{E}}_k, \overline{\mathbf{F}}_k) \Big\}, \tag{1.1}$$

$$M(n,m) = \Big\{ X : X = \sum_{k=0}^{C^n_{n+m}-1} \alpha_k \overline{H}_k (\overline{\mathbf{E}}_k, \overline{\mathbf{F}}_k) \Big\}, \tag{1.2}$$

where α_k $(0 \le k \le C^n_{n+m} - 1)$ denotes the coefficients that satisfy the conditions

$$\alpha_k \ge 0, \quad 0 \le k \le C^n_{n+m} - 1, \quad \sum_{k=0}^{C^n_{n+m}-1} \alpha_k = 1. \tag{1.3}$$

Let us show that every element (every extremal vector pair) in $\overline{E}^{n,m}$ is a vertex of $L(n,m)$. We use mathematical induction on the sum of components of vectors in the pair $(\overline{\mathbf{E}}, \overline{\mathbf{F}}) \in \overline{E}^{n,m}$. If $n+m=2$, then $|\overline{E}^{1,1}| = C^1_2 = 2$ and $\overline{E}^{1,1} = \{((0),(0)), ((1),(1))\}$ are pairs of 1-component vectors. Consider

$$L(1,1) = \Big\{ (\mathbf{A},\mathbf{B}) : (\mathbf{A},\mathbf{B}) = \alpha_0 \big((0),(0)\big) + \alpha_1 \big((1),(1)\big) \Big\},$$

where $\alpha_0 \ge 0$, $\alpha_1 \ge 0$, and $\alpha_0 + \alpha_1 = 1$. It is obvious that $L(1,1)$ is a one-dimensional polyhedron (i.e., segment) every point of which represents a pair of 1-component vectors; the points (extremal vector pairs) $(\overline{\mathbf{E}}_0, \overline{\mathbf{F}}_0) = \big((0),(0)\big)$ and $(\overline{\mathbf{E}}_1, \overline{\mathbf{F}}_1) = \big((1),(1)\big)$ are the vertices of $L(1,1)$.

Suppose that every element in $\overline{E}^{n,m}$, where $n+m \le r$, is a vertex of a polyhedron $L(n,m)$. Consider the fundamental system $\overline{E}^{n,m}$ and the polyhedron $L(n,m)$, where $n+m = r+1$. In what follows, we show that every element in $\overline{E}^{n,m}$ is a vertex of $L(n,m)$.

Assume that the converse is true: not every element of the fundamental system is a vertex of the corresponding polyhedron. Then, since the element belongs to the polyhedron, it can be represented as a linear combination of the remaining elements of the fundamental system, with nonnegative coefficients whose sum is unity.

Let us introduce a special notation for the elements of $\overline{E^{n,m}}$. To do this, we represent the fundamental system as the union of two subsystems:

$$\overline{E^{n,m}} = \overline{E^{n^0,m}} \cup \overline{E^{n^+,m}}, \tag{1.4}$$

where the subsystem $\overline{E^{n^+,m}}$ is as defined above and the first vector of every vector pair in the subsystem $\overline{E^{n^0,m}}$ has a zero component:

$$\overline{E^{n^0,m}} = \Big\{(\overline{\mathbf{E}},\overline{\mathbf{F}}) = ((e_1,\dots,e_n),(f_1,\dots,f_m)) \in \overline{E^{n,m}} : e_n = 0\Big\}.$$

It has been shown that the numbers of elements in $\overline{E^{n,m}}$ and $\overline{E^{n^+,m}}$ are C^n_{n+m} and C^n_{n+m-1}. Taking out the last component of the first vector of each pair in $\overline{E^{n^0,m}}$, we see that $|\overline{E^{n^0,m}}| = |\overline{E^{n-1,m}}| = C^{n-1}_{n+m-1}$. Using the decomposition in (1.4), we arrange the elements of $\overline{E^{n,m}}$ so that the first and last indices denote the elements of $\overline{E^{n^0,m}}$ and $\overline{E^{n^+,m}}$, respectively. Then, we substitute the calculated numbers of the elements of the fundamental system and its subsystems to rewrite the decomposition in (1.4) as

$$\begin{aligned}\overline{E^{n,m}} = \Big\{(\overline{\mathbf{E}}_k,\overline{\mathbf{F}}_k) : 0 \le k \le C^n_{n+m} - 1\Big\} = \\ \Big\{(\overline{\mathbf{E}}_k,\overline{\mathbf{F}}_k) : 0 \le k \le C^{n-1}_{n+m-1} - 1\Big\} \cup \\ \Big\{(\overline{\mathbf{E}}_k,\overline{\mathbf{F}}_k) : C^{n-1}_{n+m-1} \le k \le C^n_{n+m} - 1\Big\},\end{aligned} \tag{1.5}$$

$$(\overline{\mathbf{E}}_k,\overline{\mathbf{F}}_k) \in \begin{cases} \overline{E^{n^0,m}}, & 0 \le k \le C^{n-1}_{n+m-1} - 1, \\ \overline{E^{n^+,m}}, & C^{n-1}_{n+m-1} \le k \le C^n_{n+m} - 1. \end{cases}$$

We assume that there exists an element (extremal vector pair) $(\overline{\mathbf{E}}_p,\overline{\mathbf{F}}_p) \in \overline{E^{n,m}}$, where $n+m = r+1$, that is not a vertex of $L(n,m)$. Then, using the fact that $(\overline{\mathbf{E}}_p,\overline{\mathbf{F}}_p) \in L(n,m)$, we represent the element $(\overline{\mathbf{E}}_p,\overline{\mathbf{F}}_p)$ as

$$(\overline{\mathbf{E}}_p,\overline{\mathbf{F}}_p) = \sum_{\substack{k=0\\k\neq p}}^{C^n_{n+m}-1} \alpha_k(\overline{\mathbf{E}}_k,\overline{\mathbf{F}}_k), \quad (\overline{\mathbf{E}}_k,\overline{\mathbf{F}}_k) \in \overline{E^{n,m}},$$

$$\alpha_k \ge 0, \quad 0 \le k \le C^n_{n+m} - 1, \quad k \ne p, \quad \sum_{\substack{k=0\\k\neq p}}^{C^n_{n+m}-1} \alpha_k = 1.$$

By virtue of (1.5), we have

$$(\overline{\mathbf{E}}_p, \overline{\mathbf{F}}_p) = \sum_{\substack{k=0\\k\neq p}}^{C_{n+m-1}^{n-1}-1} \alpha_k(\overline{\mathbf{E}}_k, \overline{\mathbf{F}}_k) + \sum_{\substack{k=C_{n+m-1}^{n-1}\\k\neq p}}^{C_{n+m}^{n}-1} \alpha_k(\overline{\mathbf{E}}_k, \overline{\mathbf{F}}_k), \tag{1.6}$$

Now, we consider the cases $(\overline{\mathbf{E}}_p, \overline{\mathbf{F}}_p) \in \overline{E}^{n^0,m}$ and $(\overline{\mathbf{E}}_p, \overline{\mathbf{F}}_p) \in \overline{E}^{n^+,m}$, which are equivalent to the conditions $0 \leq p \leq C_{n+m-1}^{n-1} - 1$ and $C_{n+m-1}^{n-1} \leq p \leq C_{n+m}^{n} - 1$, respectively.

Suppose $(\overline{\mathbf{E}}_p, \overline{\mathbf{F}}_p) \in \overline{E}^{n^0,m}$. Then the last component of $\overline{\mathbf{E}}_p = (e_1^{(p)}, \ldots, e_n^{(p)})$ is $e_n^{(p)} = 0$. Since the first vector $\overline{\mathbf{E}}_k = (e_1^{(k)}, \ldots, e_n^{(k)})$ of every pair $(\overline{\mathbf{E}}_k, \overline{\mathbf{F}}_k)$ in $\overline{E}^{n^+,m}$ has a positive last component and the last component of the first vector of every pair $(\overline{\mathbf{E}}_k, \overline{\mathbf{F}}_k)$ in $\overline{E}^{n^0,m}$ is zero, equation (1.6) implies that $\alpha_k = 0$ for any k, $C_{n+m-1}^{n-1} \leq k \leq C_{n+m}^{n} - 1$ and

$$(\overline{\mathbf{E}}_p, \overline{\mathbf{F}}_p) = \sum_{\substack{k=0\\k\neq p}}^{C_{n+m-1}^{n-1}-1} \alpha_k(\overline{\mathbf{E}}_k, \overline{\mathbf{F}}_k), \quad (\overline{\mathbf{E}}_k, \overline{\mathbf{F}}_k) \in \overline{E}^{n^0,m}. \tag{1.7}$$

The last component of the first vector of every vector pair $(\overline{\mathbf{E}}_k, \overline{\mathbf{F}}_k)$ contained in (1.7) $(0 \leq k \leq C_{n+m}^{n-1} - 1)$ is zero. Therefore, equation (1.7) remains valid when the last components of the first vectors of all pairs $(\overline{\mathbf{E}}_k, \overline{\mathbf{F}}_k)$ are taken out. We denote by $\overline{\mathbf{E}}'_k = (e_1^{(k)}, \ldots, e_{n-1}^{(k)})$ the vector obtained by removing the last (zero) component from the vector $\overline{\mathbf{E}}_k = (e_1^{(k)}, \ldots, e_{n-1}^{(k)}, e_n^{(k)})$. Then, by virtue of (1.7), the element $(\overline{\mathbf{E}}'_p, \overline{\mathbf{F}}_p)$ in $\overline{E}^{n-1,m}$ is not a vertex of the polyhedron $L(n-1, m)$, where $n-1+m = r$, which contradicts the induction hypothesis. Thus, every element $(\overline{\mathbf{E}}_k, \overline{\mathbf{F}}_k)$ in $\overline{E}^{n^0,m}$ is a vertex of $L(n, m)$.

Suppose now that $(\overline{\mathbf{E}}_p, \overline{\mathbf{F}}_p) \in \overline{E}^{n^+,m}$. Then, $e_n^{(p)} \geq 1$ in $\overline{\mathbf{E}}_p = (e_1^{(p)}, \ldots, e_n^{(p)})$. By theorem 1.4 from Chapter 4, this implies that $f_1^{(p)} = n$ in $\overline{\mathbf{F}}_p = (f_1^{(p)}, \ldots, f_m^{(p)})$. Similarly, $e_n^{(k)} \geq 1$ in every vector pair $(\overline{\mathbf{E}}_k, \overline{\mathbf{F}}_k) = ((e_1^{(k)}, \ldots, e_n^{(k)}), (f_1^{(k)}, \ldots, f_m^{(k)}))$ in $\overline{E}^{n^+,m}$, and $f_1^{(k)} = n$. If $(\overline{\mathbf{E}}_k, \overline{\mathbf{F}}_k) = ((e_1^{(k)}, \ldots, e_n^{(k)}), (f_1^{(k)}, \ldots, f_m^{(k)})) \in \overline{E}^{n^0,m}$, then $e_n^{(k)} = 0$, and $f_1^{(k)} \leq n - 1$ by Theorem 1.4 from Chapter 4. Therefore, equation (1.6) implies that $\alpha_k = 0$ for any k $(0 \leq k \leq C_{n+m-1}^{n-1} - 1)$, because we would have $f_1^{(p)} < n$ otherwise. Consequently,

$$(\overline{\mathbf{E}}_p, \overline{\mathbf{F}}_p) = \sum_{k=C_{n+m-1}^{n-1}}^{C_{n+m}^{n}-1} \alpha_k(\overline{\mathbf{E}}_k, \overline{\mathbf{F}}_k), \quad (\overline{\mathbf{E}}_k, \overline{\mathbf{F}}_k) \in \overline{E}^{n^+,m}. \tag{1.8}$$

In (1.8), the number of summands is $C^n_{n+m} - C^{n-1}_{n+m-1} = C^n_{n+m-1} = |\overline{E}^{n,m-1}|$. We use this fact in equation (1.9) below. Let us establish a one-to-one correspondence between the sets $\overline{E}^{n^+,m}$ and $\overline{E}^{n,m-1}$, that is, determine an isomorphism $\varphi : \overline{E}^{n^+,m} \to \overline{E}^{n,m-1}$. Suppose $(\overline{\mathbf{E}}, \overline{\mathbf{F}}) = ((e_1, \ldots, e_n), (f_1, \ldots, f_m)) \in \overline{E}^{n^+,m}$. Then, $e_n \geq 1$; therefore, $e_i \geq 1$ $(1 \leq i \leq n)$ and $f_1 = n$. Decreasing each component of $\overline{\mathbf{E}}$ by unity and taking out the first component of $\overline{\mathbf{F}}$, we obtain the vector pair $(\overline{\mathbf{E}}', \overline{\mathbf{F}}') = ((e'_1, \ldots, e'_n), (f'_1, \ldots, f'_{m-1}))$, where $e'_i = e_i - 1$ $(1 \leq i \leq n)$ and $f'_j = f_{j+1}$ $(1 \leq j \leq m-1)$. By Definition 1.2 from Chapter 4, $(\overline{\mathbf{E}}', \overline{\mathbf{F}}')$ is an extremal pair of vectors, and $(\overline{\mathbf{E}}', \overline{\mathbf{F}}') \in \overline{E}^{n,m-1}$. It is obvious that the mapping $\varphi(\overline{\mathbf{E}}, \overline{\mathbf{F}}) = (\overline{\mathbf{E}}', \overline{\mathbf{F}}')$ defines an isomorphism between $\overline{E}^{n^+,m}$ and $\overline{E}^{n,m-1}$. We now renumber the summands in (1.8), setting $q = p - C^{n-1}_{n+m-1}$ and $t = k - C^{n-1}_{n+m-1}$, where $C^{n-1}_{n+m-1} \leq k \leq C^n_{n+m} - 1$ and $k \neq p$. Making use of the isomorphism $\varphi : \overline{E}^{n^+,m} \to \overline{E}^{n,m-1}$, $\varphi(\overline{\mathbf{E}}, \overline{\mathbf{F}}) = (\overline{\mathbf{E}}', \overline{\mathbf{F}}')$, we rewrite equation (1.8) as

$$(\overline{\mathbf{E}}'_q, \overline{\mathbf{F}}'_q) = \sum_{\substack{t=0 \\ t \neq q}}^{C^n_{n+m-1}-1} \alpha_t (\overline{\mathbf{E}}'_t, \overline{\mathbf{F}}'_t), \quad (\overline{\mathbf{E}}'_t, \overline{\mathbf{F}}'_t) \in \overline{E}^{n,m-1}, \tag{1.9}$$

$$\alpha_t \geq 0, \quad 0 \leq t \leq C^n_{n+m-1} - 1, \quad t \neq q, \quad \sum_{\substack{t=0 \\ t \neq q}}^{C^n_{n+m-1}-1} \alpha_t = 1.$$

Equation (1.9) shows that the element $(\overline{\mathbf{E}}'_q, \overline{\mathbf{F}}'_q) \in \overline{E}^{n,m-1}$ is not a vertex of $L(n, m-1)$, where $n+m-1 = r$, which contradicts the induction hypothesis. Therefore, if (an extremal vector pair) $(\overline{\mathbf{E}}_p, \overline{\mathbf{F}}_p) \in \overline{E}^{n^+,m}$, then $(\overline{\mathbf{E}}_p, \overline{\mathbf{F}}_p)$ is a vertex of the polyhedron $L(n, m)$. Thus, we have proved the following proposition.

Theorem 1.1. *A vector pair belongs to the fundamental system $\overline{E}^{n,m}$ if and only if it is a vertex of the polyhedron $L(n, m)$ defined by* (1.1) *and* (1.3).

Consider the fundamental system $\overline{E}_{n,m} = \{\overline{H}_k(\overline{\mathbf{E}}_k, \overline{\mathbf{F}}_k) : 0 \leq k \leq C^n_{n+m} - 1\}$ of (extremal) matrices and the polyhedron $M(n, m)$ defined by (1.2) and (1.3). If an element $\overline{H}_p(\overline{\mathbf{E}}_p, \overline{\mathbf{F}}_p)$ is not a vertex of $M(n, m)$, then we can use the fact that $\overline{H}_p(\overline{\mathbf{E}}_p, \overline{\mathbf{F}}_p) \in M(n, m)$ to represent the extremal matrix $\overline{H}_p(\overline{\mathbf{E}}_p, \overline{\mathbf{F}}_p)$ as

$$\overline{H}_p(\overline{\mathbf{E}}_p, \overline{\mathbf{F}}_p) = \sum_{\substack{k=0 \\ k \neq p}}^{C^n_{n+m}-1} \alpha_k \overline{H}_k(\overline{\mathbf{E}}_k, \overline{\mathbf{F}}_k), \quad (\overline{\mathbf{E}}_k, \overline{\mathbf{F}}_k) \in \overline{E}^{n,m}, \tag{1.10}$$

$$\alpha_k \geq 0, \quad 0 \leq k \leq C_{n+m}^n - 1, \quad k \neq p, \quad \sum_{\substack{k=1 \\ k\neq p}}^{C_{n+m}^n - 1} \alpha_k = 1.$$

Because the sum of the elements of every row and column of an extremal matrix $\overline{H}_k(\overline{\mathbf{E}}_k, \overline{\mathbf{F}}_k)$ is a component of a vector in the extremal pair $(\overline{\mathbf{E}}_k, \overline{\mathbf{F}}_k)$, equation (1.10) implies that a similar equation holds for the elements of the fundamental system $\overline{E}^{n,m} = \{(\overline{\mathbf{E}}_k, \overline{\mathbf{F}}_k) : 0 \leq k \leq C_{n+m}^n - 1\}$, which contradicts Theorem 1.1. Thus, we have proved the following proposition.

Theorem 1.2. *A matrix belongs to the fundamental system $\overline{E}_{n,m}$ if and only if it is a vertex of the polyhedron $M(n,m)$ defined by* (1.2) *and* (1.3).

In convex analysis, the vertices of convex polyhedrons are called extremal points. Theorems 1.1 and 1.2 elucidate the concepts of extremal vector pair and extremal matrix.

Lemma 1.1. *Every matrix in the polyhedron $M(n,m)$ is uniform, and its elements are not greater than unity.*

The proof follows from conditions (1.2) and (1.3) combined with Theorem 1.6 about the uniformity of extremal matrices from Chapter 4.

The zero components of vectors in a pair that define the plan matrices of a closed transportation problem may be unimportant. Accordingly, we define the polyhedrons that correspond, respectively, to the subsystems $\overline{E}^{n^+,m}$, $\overline{E}^{n,m^+}$, $\overline{E}^{n^+,m^+}$ and $\overline{E}_{n^+,m}$, $\overline{E}_{n,m^+}$, $\overline{E}_{n^+,m^+}$ of the fundamental systems of vector pairs and matrices, $\overline{E}^{n,m}$ and $\overline{E}_{n,m}$ (their elements are counted in Theorem 1.8 from Chapter 4):

$$L(n^+, m) = \Big\{(\mathbf{A}, \mathbf{B}) : (\mathbf{A}, \mathbf{B}) = \sum_{k=0}^{C_{n+m-1}^n - 1} \alpha_k (\overline{\mathbf{E}}_k, \overline{\mathbf{F}}_k), \quad (\overline{\mathbf{E}}_k, \overline{\mathbf{F}}_k) \in \overline{E}^{n^+,m}\Big\},$$

$$M(n^+, m) = \Big\{X : X = \sum_{k=0}^{C_{n+m-1}^n - 1} \alpha_k \overline{H}_k(\overline{\mathbf{E}}_k, \overline{\mathbf{F}}_k), \quad \overline{H}_k(\overline{\mathbf{E}}_k, \overline{\mathbf{F}}_k) \in \overline{E}_{n^+,m}\Big\},$$

$$\sum_{k=0}^{C_{n+m-1}^n - 1} \alpha_k = 1;$$

$$L(n, m^+) = \Big\{(\mathbf{A}, \mathbf{B}) : (\mathbf{A}, \mathbf{B}) = \sum_{k=0}^{C_{n+m-1}^{n-1} - 1} \alpha_k (\overline{\mathbf{E}}_k, \overline{\mathbf{F}}_k), \quad (\overline{\mathbf{E}}_k, \overline{\mathbf{F}}_k) \in \overline{E}^{n,m^+}\Big\},$$

$$M(n,m^+) = \left\{X : X = \sum_{k=0}^{C_{n+m-1}^{n-1}-1} \alpha_k \overline{H}_k(\overline{\mathbf{E}}_k, \overline{\mathbf{F}}_k), \quad \overline{H}_k(\overline{\mathbf{E}}_k, \overline{\mathbf{F}}_k) \in \overline{E}_{n,m^+}\right\},$$

$$\sum_{k=0}^{C_{n+m-1}^{n-1}-1} \alpha_k = 1;$$

$$L(n^+,m^+) = \left\{(\mathbf{A},\mathbf{B}) : (\mathbf{A},\mathbf{B}) = \sum_{k=0}^{C_{n+m-2}^{n-1}-1} \alpha_k (\overline{\mathbf{E}}_k, \overline{\mathbf{F}}_k), \quad (\overline{\mathbf{E}}_k, \overline{\mathbf{F}}_k) \in \overline{E}^{n^+,m^+}\right\},$$

$$M(n^+,m^+) = \left\{X : X = \sum_{k=0}^{C_{n+m-2}^{n-1}-1} \alpha_k \overline{H}_k(\overline{\mathbf{E}}_k, \overline{\mathbf{F}}_k), \quad \overline{H}_k(\overline{\mathbf{E}}_k, \overline{\mathbf{F}}_k) \in \overline{E}_{n^+,m^+}\right\},$$

$$\sum_{k=0}^{C_{n+m-2}^{n-1}-1} \alpha_k = 1;$$

where the coefficients α_k are nonnegative. It is obvious that

$$L(n^+,m^+) = L(n^+,m) \cap L(n,m^+),$$

$$L(n^+,m^+),\ L(n^+,m),\ L(n,m^+) \subset L(n,m);$$

$$M(n^+,m^+) = M(n^+,m) \cap M(n,m^+),$$

$$M(n^+,m^+),\ M(n^+,m),\ M(n,m^+) \subset M(n,m).$$

The following theorem is a corollary to Theorems 1.1 and 1.2.

Theorem 1.3. *Every element of the subsystems* $\overline{E}^{n^+,m}$, $\overline{E}^{n,m^+}$, $\overline{E}^{n^+,m^+}$ *and* $\overline{E}_{n^+,m}$, $\overline{E}_{n,m^+}$, $\overline{E}_{n^+,m^+}$ *is a vertex of one of the polyhedrons* $L(n^+,m)$, $L(n,m^+)$, $L(n^+,m^+)$ *and* $M(n^+,m)$, $M(n,m^+)$, $M(n^+,m^+)$, *respectively.*

1.2. Representation of Vector Pairs by Convex Linear Combinations of Extremal Pairs

Now, we show that any vector pair $(\mathbf{A},\mathbf{B})$ such that $\mathbf{A} \in \overline{R}_+^n$, $\mathbf{B} \in \overline{R}_+^m$, and $M(\mathbf{A},\mathbf{B};1) \neq \varnothing$ belongs to a polyhedron whose vertices are extremal pairs.

Consider the following set of vector pairs $(\mathbf{A},\mathbf{B})$ with components arranged in nonascending order:

$$L'(n,m) = \left\{(\mathbf{A},\mathbf{B}) : \mathbf{A} \in \overline{R}_+^n,\ \mathbf{B} \in \overline{R}_+^m,\ M(\mathbf{A},\mathbf{B};1) \neq \varnothing\right\}.$$

It is obvious that the set $L'(n,m)$ can be alternatively expressed by using the characteristic function (1.2) from Chapter 4 (i.e., function (1.22) from Chapter 1 with $c = 1$):

$$L'(n,m) = \Big\{(\mathbf{A},\mathbf{B}) = \big((a_1,\dots,a_n),(b_1,\dots,b_m)\big) : \mathbf{A} \in \overline{R}^n_+,\ \mathbf{B} \in \overline{R}^m_+,$$
$$\sum_{i=1}^{n} a_i = \sum_{j=1}^{m} b_j,\ \delta_k(\mathbf{A},\mathbf{B};1) \geq 0,\ 1 \leq k \leq n\Big\}.$$

The following proposition does not call for a proof.

Lemma 1.2. *The set $L'(n,m)$ of vector pairs is a closed, convex set such that $\overline{E}^{n,m} \subset L(n,m) \subseteq L'(n,m)$.*

Let us show that actually $L(n,m) = L'(n,m)$. To do this, we invoke the concept of extreme point and recall the Krein–Mil'man theorem about extreme points [1, 2, 4].

Let U be a convex set. A point $u \in U$ is said to be an extreme point of a set U if the representation $u = \alpha u_1 + \beta u_2$ is valid only for $u_1 = u_2 = u$; here $0 < \alpha, \beta < 1$, $\alpha + \beta = 1$, and $u_1, u_2 \in U$. Since a convex polyhedron is identical with the hull of its extreme points, we see that the set of elements of the fundamental system $\overline{E}^{n,m}$ is the set of extreme points of the polyhedron $L(n,m)$. The Krein–Mil'man theorem states that any compact convex set is identical with the closed convex hull of its extreme points.

Theorem 1.4. $L(n,m) = L'(n,m)$.

The proof is based on the following lemmas.

Lemma 1.3. *Every point (element) in (the fundamental system) $\overline{E}^{n,m}$ is an extreme point of the set $L'(n,m)$.*

Lemma 1.4. *Every point in the difference $L'(n,m)\backslash\overline{E}^{n,m}$ is not an extreme point of $L'(n,m)$.*

Let $(\mathbf{A}_1,\mathbf{B}_1)$ and $(\mathbf{A}_2,\mathbf{B}_2)$ be arbitrary pairs of vectors having equal dimensions. We introduce

$$[(\mathbf{A}_1,\mathbf{B}_1),(\mathbf{A}_2,\mathbf{B}_2)] =$$
$$\Big\{(\mathbf{A},\mathbf{B}) : (\mathbf{A},\mathbf{B}) = \alpha(\mathbf{A}_1,\mathbf{B}_1) + \beta(\mathbf{A}_2,\mathbf{B}_2),\ \alpha,\ \beta \geq 0,\ \alpha+\beta = 1\Big\},$$

that is, a segment whose points are vector pairs and endpoints are the vector pairs $(\mathbf{A}_1,\mathbf{B}_1)$ and $(\mathbf{A}_2,\mathbf{B}_2)$.

Both lemmas are proved by induction on n, with initial step determined by the following lemma.

Lemma 1.5. *A point in the set $L'(1,m)$ belongs to $\overline{E}^{1,m}$ if and only if it is an extreme point of $L'(1,m)$ and $L(1,m) = L'(1,m)$.*

Proof. Consider the set

$$L'(1,m) = \Big\{ \big((a_1), (b_1, \dots, b_m)\big) : a_1 \geq 0,\ (b_1, \dots, b_m) \in \overline{R}^m_+,\ a_1 = \sum_{j=1}^{m} b_j,\ M\big((a_1), (b_1, \dots, b_m); 1\big) \neq \varnothing \Big\}.$$

Every pair of vectors with nonnegative components in which one is a 1-component vector whose single component is equal to the sum of components of the other uniquely defines a 1-row matrix. Therefore,

$$L'(1,m) = \Big\{ \big((a_1), (b_1, \dots, b_m)\big) : a_1 \geq 0,\ (b_1, \dots, b_m) \in \overline{R}^m_+,\ b_1 \leq 1,\ a_1 = \sum_{j=1}^{m} b_j \Big\}.$$

It is obvious that the number of points (extremal vector pairs) in $\overline{E}^{1,m}$ is $C^1_{m+1} = m + 1$. It is also clear that every element of $\overline{E}^{1,m}$ can be expressed as

$$\overline{E}^{1,m} = \Big\{ \Big((k), (\underbrace{1, \dots, 1}_{k}, \underbrace{0, \dots, 0}_{m-k})\Big) : 0 \leq k \leq m \Big\}.$$

Combining the expressions above, we see that the sets $L'(1,m)$, $\overline{E}^{1,m}$, and $L(1,m)$ can be treated as identical with the sets $L'(m)$, E^m, and $L(m)$ in $\overline{R}^m_+ \subset R^m$, respectively:

$$L'(m) = \Big\{ \mathbf{B} = (b_1, \dots, b_m) : \mathbf{B} \in \overline{R}^m_+,\ b_1 \leq 1 \Big\},$$

$$E^m = \Big\{ \mathbf{F}_k = (\underbrace{1, \dots, 1}_{k}, \underbrace{0, \dots, 0}_{m-k}) : 0 \leq k \leq m \Big\},$$

$$L(m) = \Big\{ \mathbf{B} : \mathbf{B} \in \overline{R}^m_+,\ \mathbf{B} = \sum_{k=0}^{m} \alpha_k \mathbf{F}_k,\ \alpha_k \geq 0,\ 0 \leq k \leq m,\ \sum_{k=0}^{m} \alpha_k = 1 \Big\}.$$

It is obvious that $L(m) \subseteq L'(m)$; moreover, $L(m)$ is an m-dimensional simplex in R^m, and E^m is the set of vertices (extreme points) of the simplex.

Let $\mathbf{B} = (b_1, \dots, b_m)$ be an arbitrary vector from $L'(m)$. We set

$$\alpha_m = b_m, \quad \alpha_k = b_k - b_{k+1}, \quad 1 \leq k \leq m-1, \quad \alpha_0 = 1 - b_1.$$

Since $\mathbf{B} \in \overline{R}^m_+$ and $b_1 \leq 1$, every α_k $(0 \leq k \leq m)$ is a nonnegative number, and

$$\sum_{k=0}^{m} \alpha_k = \sum_{k=1}^{m-1} (b_k - b_{k+1}) + b_m + 1 - b_1 = 1.$$

Consider now the vector $\mathbf{C} = (c_1, \ldots, c_m)$ defined as

$$\mathbf{C} = \sum_{k=0}^{m} \alpha_k \mathbf{F}_k, \quad \mathbf{F}_k = (\underbrace{1, \ldots, 1}_{k}, \underbrace{0, \ldots, 0}_{m-k}).$$

Since

$$\mathbf{C} = b_m \mathbf{F}_m + (b_{m-1} - b_m)\mathbf{F}_{m-1} + \cdots + (b_1 - b_2)\mathbf{F}_1,$$

we can use the properties of vectors from E^m to show that

$$c_j = \sum_{k=j}^{m} \alpha_k = b_m + (b_{m-1} - b_m) + \cdots + (b_j - b_{j+1}) = b_j$$

for any j $(1 \le j \le m)$. Then, we have

$$\mathbf{B} = \sum_{k=0}^{m} \alpha_k \mathbf{F}_k,$$

which is an identity when the vectors belong to $L'(m)$. Consequently, $L'(m) \subseteq L(m)$ and $L(m) = L'(m)$. Since $L(m)$ is an m-dimensional simplex and $L'(m) = L(m)$, it holds that $L'(1,m) = L(1,m)$ and only all of the points in $\overline{E}^{1,m}$ are the extreme points of $L'(1,m)$, which completes the proof.

Proof of Lemma 1.3. For $n = 1$, the proposition of the lemma is true by the preceding lemma. Suppose that Lemma 1.3 applies to the set $\overline{E}^{n,m}$: every point in $\overline{E}^{n,m}$ (every element of a fundamental system) is an extreme point of $L'(n,m)$. Consider the set $\overline{E}^{n+1,m}$ and assume that there exists a point (extremal vector pair) $(\overline{\mathbf{E}}, \overline{\mathbf{F}})$ that is not an extreme point of $L'(n+1,m)$. Then, there exists such a segment $\big[(\mathbf{A},\mathbf{B}),(\mathbf{C},\mathbf{D})\big]$ that $(\mathbf{A},\mathbf{B}),(\mathbf{C},\mathbf{D}) \in L'(n+1,m)$ and $(\overline{\mathbf{E}}, \overline{\mathbf{F}}) \in \big[(\mathbf{A},\mathbf{B}),(\mathbf{C},\mathbf{D})\big]$, where $(\overline{\mathbf{E}}, \overline{\mathbf{F}}) \ne (\mathbf{A},\mathbf{B}),(\mathbf{C},\mathbf{D})$. This implies that $(\overline{\mathbf{E}}, \overline{\mathbf{F}}) = \alpha(\mathbf{A},\mathbf{B}) + \beta(\mathbf{C},\mathbf{D})$, where α, $\beta > 0$ and $\alpha + \beta = 1$. In the first vector of the pair $(\overline{\mathbf{E}}, \overline{\mathbf{F}}) = \big((e_1, \ldots, e_n, e_{n+1}), (f_1, \ldots, f_m)\big)$, either $e_{n+1} = 0$ or $e_{n+1} \ge 1$. Let us consider these cases separately.

Suppose $e_{n+1} = 0$. Then, vectors $\mathbf{A} = (a_1, \ldots, a_n, a_{n+1})$ and $\mathbf{C} = (c_1, \ldots, c_n, c_{n+1})$ are such that $a_{n+1} = c_{n+1} = 0$, because we would have $e_{n+1} = \alpha a_{n+1} + \beta c_{n+1} > 0$ otherwise. We define $\mathbf{A}' = (a_1, \ldots, a_n)$, $\mathbf{C}' = (c_1, \ldots, c_n)$, and $\overline{\mathbf{E}}' = (e_1, \ldots, e_n)$ as the vectors obtained by taking out their last components. It is clear that $(\mathbf{A}',\mathbf{B}),(\mathbf{C}',\mathbf{D}) \in L'(n,m)$, $(\overline{\mathbf{E}}', \overline{\mathbf{F}}) \in \overline{E}^{n,m}$, and $(\overline{\mathbf{E}}', \overline{\mathbf{F}}) = \alpha(\mathbf{A}',\mathbf{B}) + \beta(\mathbf{C}',\mathbf{D})$, where α, $\beta > 0$ and $\alpha + \beta = 1$. And this means that the point $(\overline{\mathbf{E}}', \overline{\mathbf{F}})$ in $\overline{E}^{n,m}$ is not an extreme point of $L'(n,m)$, which contradicts the induction hypothesis. Thus, if $(\overline{\mathbf{E}}, \overline{\mathbf{F}}) \in \overline{E}^{n+1,m}$ and $e_{n+1} = 0$, then $(\overline{\mathbf{E}}, \overline{\mathbf{F}})$ is an extreme point of the set $L'(n+1,m)$.

When $e_{n+1} \geq 1$, the first component of the vector $\overline{\mathbf{F}} = (f_1, \ldots, f_m)$ is such that $f_1 = n+1$, and $f_j = n+1$ for $1 \leq j \leq e_{n+1}$ by Theorem 1.4. Then, it also holds that the first components of $\mathbf{B} = (b_1, \ldots, b_m)$ and $\mathbf{D} = (d_1, \ldots, d_m)$ are $b_j = d_j = n+1$ $(1 \leq j \leq e_{n+1})$, because we would have $f_j = \alpha b_j + \beta d_j < n+1$ $(1 \leq j \leq e_{n+1})$ otherwise. Now, consider the vectors $\mathbf{A}$ and $\mathbf{C}$. The conditions for the first components of $\mathbf{B}$ and $\mathbf{D}$ imply that $a_{n+1}, c_{n+1} \geq e_{n+1}$. Otherwise, we would have $M(\mathbf{A}, \mathbf{B}; 1), M(\mathbf{C}, \mathbf{D}; 1) = \varnothing$. Then, the equation $e_{n+1} = \alpha a_{n+1} + \beta c_{n+1}$, where α, $\beta > 0$ and $\alpha + \beta = 1$, yields $a_{n+1} = c_{n+1} = e_{n+1}$. Thus, given vector pairs $(\mathbf{A}, \mathbf{B})$, $(\mathbf{C}, \mathbf{D})$, and $(\overline{\mathbf{E}}, \overline{\mathbf{F}})$, we can construct the pairs $(\mathbf{A}', \mathbf{B}') = ((a'_1, \ldots, a'_{n+1}), (b'_1, \ldots, b'_m))$, $(\mathbf{C}', \mathbf{D}') = ((c'_1, \ldots, c'_{n+1}), (d'_1, \ldots, d'_m))$, and $(\overline{\mathbf{E}}', \overline{\mathbf{F}}') = ((e'_1, \ldots, e'_{n+1}), (f'_1, \ldots, f'_m))$ such that

$$a'_i = a_i - e_{n+1}, \quad c'_i = c_i - e_{n+1}, \quad e'_i = e_i - e_{n+1}, \quad 1 \leq i \leq n+1,$$

$$b'_j = \begin{cases} b_{j+e_{n+1}}, & 1 \leq j \leq m - e_{n+1}, \\ 0, & m - e_{n+1} < j \leq m, \end{cases}$$

$$d'_j = \begin{cases} d_{j+e_{n+1}}, & 1 \leq j \leq m - e_{n+1}, \\ 0, & m - e_{n+1} < j \leq m, \end{cases}$$

$$f'_j = \begin{cases} f_{j+e_{n+1}}, & 1 \leq j \leq m - e_{n+1}, \\ 0, & m - e_{n+1} < j \leq m. \end{cases}$$

The vectors $\mathbf{A}'$, $\mathbf{C}'$, and $\overline{\mathbf{E}}'$ are obtained by decreasing each component of $\mathbf{A}$, $\mathbf{C}$, and $\overline{\mathbf{E}}$ by e_{n+1}, and the vectors $\mathbf{B}'$, $\mathbf{D}'$, and $\overline{\mathbf{F}}'$ are obtained by taking out the first e_{n+1} components and adding e_{n+1} zero components. It is obvious that $(\mathbf{A}', \mathbf{B}'), (\mathbf{C}', \mathbf{D}') \in L'(n+1, m)$ and $(\overline{\mathbf{E}}', \overline{\mathbf{F}}') \in \overline{E}^{n+1,m}$ while $e'_{n+1} = 0$.

The vector equations $\alpha\mathbf{A} + \beta\mathbf{C} = \overline{\mathbf{E}}$ and $\alpha\mathbf{B} + \beta\mathbf{D} = \overline{\mathbf{F}}$ entail the vector equations $\alpha\mathbf{A}' + \beta\mathbf{C}' = \overline{\mathbf{E}}'$ and $\alpha\mathbf{B}' + \beta\mathbf{D}' = \overline{\mathbf{F}}'$, because $\alpha + \beta = 1$. Therefore, the corresponding vector pairs satisfy the equation $\alpha(\mathbf{A}', \mathbf{B}') + \beta(\mathbf{C}', \mathbf{D}') = (\overline{\mathbf{E}}', \overline{\mathbf{F}}')$, where α, $\beta > 0$ and $\alpha + \beta = 1$. Recall that the last component of $\overline{\mathbf{E}}'$ is $e'_{n+1} = 0$ in this equation. According to the above analysis based on the induction hypothesis, $(\overline{\mathbf{E}}', \overline{\mathbf{F}}')$ is an extreme point of $L'(n+1, m)$ in this case. Therefore, the equality $\alpha(\mathbf{A}', \mathbf{B}') + \beta(\mathbf{C}', \mathbf{D}') = (\overline{\mathbf{E}}', \overline{\mathbf{F}}')$ with α, $\beta > 0$ and $\alpha + \beta = 1$ is violated, and so is the equality $\alpha(\mathbf{A}, \mathbf{B}) + \beta(\mathbf{C}, \mathbf{D}) = (\overline{\mathbf{E}}, \overline{\mathbf{F}})$ $(\alpha, \beta > 0, \alpha + \beta = 1)$. Therefore, $(\overline{\mathbf{E}}, \overline{\mathbf{F}}) \in \overline{E}^{n+1,m}$, where $e_{n+1} \geq 1$, is an extreme point of $L'(n+1, m)$, which completes the proof.

Proof of Lemma 1.4. The case $n = 1$ is proved by applying Lemma 1.5. Suppose that no vector pair $(\mathbf{A}, \mathbf{B})$ in $L'(n, m) \backslash \overline{E}^{n,m}$ is an extreme point and consider the sets $L'(n+1,m)$ and $\overline{E}^{n+1,m}$. Let

$$(\mathbf{A},\mathbf{B}) = \big((a_1,\ldots,a_n,a_{n+1}),(b_1,\ldots,b_m)\big)$$

be an arbitrary point in $L'(n+1,m)\backslash\overline{E}^{n+1,m}$. Let us show that there exists such a segment

$$\big[(\mathbf{A}_1,\mathbf{B}_1),(\mathbf{A}_2,\mathbf{B}_2)\big] \subseteq L'(n+1,m)$$

that $(\mathbf{A},\mathbf{B})$ can be expressed in terms of its endpoints as $(\mathbf{A},\mathbf{B})=\alpha(\mathbf{A}_1,\mathbf{B}_1)+\beta(\mathbf{A}_2,\mathbf{B}_2)$, where α, $\beta>0$ and $\alpha+\beta=1$. Thereby, it will be proved that $(\mathbf{A},\mathbf{B})$ is not an extreme point. The segment $\big[(\mathbf{A}_1,\mathbf{B}_1),(\mathbf{A}_2,\mathbf{B}_2)\big]$ depends on the values of the first and last components of $\mathbf{A}=(a_1,\ldots,a_{n+1})\in\overline{R}_+^{n+1}$.

We analyze the following three cases separately:

$$a_{n+1}=0;$$

$$a_{n+1}>0 \ \text{ and } \ a_1 = l_1(\mathbf{B};1)+\sum_{b_j<1} b_j;$$

$$a_{n+1}>0 \ \text{ and } \ a_1 < l_1(\mathbf{B};1)+\sum_{b_j<1} b_j.$$

Recall that

$$l_1(\mathbf{B};1) = \Big|\{j : b_j \geq 1,\ 1\leq j\leq m\}\Big| = \max\Big\{j : b_j\geq 1,\ 1\leq j\leq m\Big\},$$

when $b_1\geq 1$.

Consider the simplest case, $a_{n+1}=0$. Let $\mathbf{A}'=(a_1,\ldots,a_n)$ be the vector derived from $\mathbf{A}$ by taking out its last component. Since $(\mathbf{A},\mathbf{B})\in L'(n+1,m)\backslash\overline{E}^{n+1,m}$, we have $(\mathbf{A}',\mathbf{B})\in L'(n,m)\backslash\overline{E}^{n,m}$. By the induction hypothesis, $(\mathbf{A}',\mathbf{B})$ is not an extreme point of $L'(n,m)$. Consequently, there exist points $(\mathbf{A}_1',\mathbf{B}_1),(\mathbf{A}_2',\mathbf{B}_2)\in L'(n,m)$ such that $(\mathbf{A}',\mathbf{B})=\alpha(\mathbf{A}_1',\mathbf{B}_1)+\beta(\mathbf{A}_2',\mathbf{B}_2)$, where α, $\beta>0$ and $\alpha+\beta=1$. This implies that if vectors $\mathbf{A}_1$ and $\mathbf{A}_2$ are derived from $\mathbf{A}_1'$ and $\mathbf{A}_2'$, respectively, by adding zeros as the last components, then $(\mathbf{A},\mathbf{B})=\alpha(\mathbf{A}_1,\mathbf{B}_1)+\beta(\mathbf{A}_2,\mathbf{B}_2)$ and $(\mathbf{A},\mathbf{B})$ is not an extreme point of $L'(n+1,m)$.

Next, consider the case of $a_{n+1}>0$ and

$$a_1 = l_1(\mathbf{B};1)+\sum_{b_j<1} b_j.$$

Note that $\delta_1(\mathbf{A},\mathbf{B};1)=0$ in this case. Let us introduce the point $(\mathbf{A}',\mathbf{B}')$, where the components of $\mathbf{A}'=(a_1',\ldots,a_n')$ and $\mathbf{B}'=(b_1',\ldots,b_m')$ are defined as follows:

$$a_i' = a_{i+1},\quad 1\leq i\leq n,$$

$$b_j' = \begin{cases} b_j-1, & 1\leq j\leq l_1(\mathbf{B};1),\\ 0, & l_1(\mathbf{B};1)<j\leq m.\end{cases} \tag{1.11}$$

Two possibilities must be examined: $(\mathbf{A}', \mathbf{B}') \in \overline{E}^{n,m}$ and $(\mathbf{A}', \mathbf{B}') \notin \overline{E}^{n,m}$. (It is obvious that $(\mathbf{A}', \mathbf{B}') \in L'(n, m)$.)

If $(\mathbf{A}', \mathbf{B}') \in \overline{E}^{n,m}$, then $(\mathbf{A}, \mathbf{B}) \notin \overline{E}^{n+1,m}$; therefore, $l_1(\mathbf{B}; 1) < m$ and $b_{l_1(\mathbf{B};1)+1} > 0$; note also that $b_{l_1(\mathbf{B};1)+1} < 1$. To construct a segment $[(\mathbf{A}_1, \mathbf{B}_1), (\mathbf{A}_2, \mathbf{B}_2)] \subseteq L'(n+1, m)$ containing $(\mathbf{A}, \mathbf{B})$ as an interior point, we use the fact that $\delta_k(\mathbf{A}, \mathbf{B}; 1) = 0$, $1 \le k \le n+1$, $a_2 \in Z$, and $a_1 - a_2 > 0$ since $(\mathbf{A}', \mathbf{B}') \in \overline{E}^{n,m}$. Introducing $m_1 = \max\{j : b_j > 0,\ 1 \le j \le m\}$, we have $b_{m_1} < 1$. Suppose that

$$\varepsilon = \min\left\{m_1 - a_1,\ a_1 - a_2,\ b_{l_1(\mathbf{B};1)} - b_{l_1(\mathbf{B};1)+1},\ b_{m_1}\right\}.$$

Define the components of the vectors

$$\mathbf{A}_1 = (a_1^{(1)}, \ldots, a_{n+1}^{(1)}), \quad \mathbf{A}_2 = (a_1^{(2)}, \ldots, a_{n+1}^{(2)}),$$

$$\mathbf{B}_1 = (b_1^{(1)}, \ldots, b_{m+1}^{(1)}), \quad \mathbf{B}_2 = (b_1^{(2)}, \ldots, b_{m+1}^{(2)})$$

as

$$a_i^{(1)} = \begin{cases} a_1 - \varepsilon, & i = 1, \\ a_i, & 2 \le i \le n+1, \end{cases} \quad a_i^{(2)} = \begin{cases} a_1 + \varepsilon, & i = 1, \\ a_i, & 2 \le i \le n+1, \end{cases}$$

$$b_j^{(1)} = \begin{cases} b_j, & 1 \le j \le l_1(\mathbf{B}; 1),\ m_1 < j \le m, \\ b_j - \varepsilon/(m_1 - l_1(\mathbf{B}; 1)), & l_1(\mathbf{B}; 1) < j \le m_1, \end{cases}$$

$$b_j^{(2)} = \begin{cases} b_j, & 1 \le j \le l_1(\mathbf{B}; 1),\ m_1 < j \le m, \\ b_j + \varepsilon/(m_1 - l_1(\mathbf{B}; 1)), & l_1(\mathbf{B}; 1) < j \le m_1, \end{cases}$$

It is clear that $\mathbf{A}_1$, $\mathbf{A}_2 \in \overline{R}_+^{n+1}$, $\mathbf{B}_1$, $\mathbf{B}_2 \in \overline{R}_+^m$, and

$$\sum_{i=1}^{n+1} a_i^{(1)} = \sum_{j=1}^{m} b_j^{(1)}, \quad \sum_{i=1}^{n+1} a_i^{(2)} = \sum_{j=1}^{m} b_j^{(2)}.$$

Let us show that function (1.4) from Chapter 4 is zero for $(\mathbf{A}_1, \mathbf{B}_1)$ and $(\mathbf{A}_2, \mathbf{B}_2)$. Since

$$a_1^{(1)} = l_1(\mathbf{B}_1; 1) + \sum_{b_j^{(1)} < 1} b_j^{(1)}, \quad a_1^{(2)} = l_1(\mathbf{B}_2; 1) + \sum_{b_j^{(2)} < 1} b_j^{(2)},$$

it holds that $\delta_1(\mathbf{A}_1, \mathbf{B}_1; 1) = \delta_2(\mathbf{A}_2, \mathbf{B}_2; 1) = 0$. For $k \in Z$, $2 \le k \le n+1$, it follows from $\delta_k(\mathbf{A}, \mathbf{B}; 1) = 0$ that

$$\delta_k(\mathbf{A}_1, \mathbf{B}_1; 1) = \sum_{j=1}^{m} b_j^{(1)} - \sum_{b_j^{(1)} \ge k} (b_j^{(1)} - k) + \sum_{i=1}^{k} a_i^{(1)} =$$

$$\sum_{j=1}^{m} b_j - \sum_{b_j \ge k} (b_j - k) - \sum_{i=1}^{k} a_i = \sum_{j=1}^{m} b_j^{(2)} - \sum_{b_j^{(2)} \ge k} (b_j^{(2)} - k) - \sum_{i=1}^{k} a_i^{(2)} =$$

$$\delta_k(\mathbf{A}_2, \mathbf{B}_2; 1) = \delta_k(\mathbf{A}, \mathbf{B}; 1) = 0.$$

Because $\delta_k(\mathbf{A}_1, \mathbf{B}_1; 1)$ and $\delta_k(\mathbf{A}_2, \mathbf{B}_2; 1)$ are nonnegative functions, it is clear that $(\mathbf{A}_1, \mathbf{B}_1), (\mathbf{A}_2, \mathbf{B}_2) \in L'(n+1, m)$. The relation

$$(\mathbf{A}, \mathbf{B}) = \frac{1}{2}(\mathbf{A}_1, \mathbf{B}_1) + \frac{1}{2}(\mathbf{A}_2, \mathbf{B}_2)$$

implies that $(\mathbf{A}, \mathbf{B}) \in L'(n+1, m) \backslash \overline{E}^{n+1,m}$ is not an extreme point of the set $L'(n+1, m)$.

Now, consider the case $(\mathbf{A}', \mathbf{B}') \notin \overline{E}^{n,m}$. The induction hypothesis implies that there exists a segment $[(\mathbf{A}'_1, \mathbf{B}'_1), (\mathbf{A}'_2, \mathbf{B}'_2)] \subseteq L'(n, m)$ such that $(\mathbf{A}', \mathbf{B}') = \alpha(\mathbf{A}'_1, \mathbf{B}'_1) + \beta(\mathbf{A}'_2, \mathbf{B}'_2)$, α, $\beta > 0$, and $\alpha + \beta = 1$. Assume that $(\mathbf{A}', \mathbf{B}')$ is the midpoint of $[(\mathbf{A}'_1, \mathbf{B}'_1), (\mathbf{A}'_2, \mathbf{B}'_2)]$; that is

$$(\mathbf{A}', \mathbf{B}') = \frac{1}{2}(\mathbf{A}'_1, \mathbf{B}'_1) + \frac{1}{2}(\mathbf{A}'_2, \mathbf{B}'_2). \tag{1.12}$$

Since $\mathbf{A} \in \overline{R}_+^{n+1}$, it holds that $a_1 \ge a_2$. Let us map the segment $[(\mathbf{A}'_1, \mathbf{B}'_1), (\mathbf{A}'_2, \mathbf{B}'_2)]$ from $L'(n, m)$ onto the segment $[(\mathbf{A}_1, \mathbf{B}_1), (\mathbf{A}_2, \mathbf{B}_2)]$ in $L'(n+1, m)$ so that $a_1 = a_2$ or $a_1 > a_2$. The equality $a_1 = a_2$ is valid when either $l_1(\mathbf{B}; 1) = m$ or $b_{l_1(\mathbf{B};1)+1} = 0$ for $l_1(\mathbf{B}; 1) < m$; otherwise, we would have $a_1 = a_2$ only if

$$\sum_{b_j < 1} b_j = 0.$$

Suppose $a_1 = a_2$. Then, $a'_1 = a_2 = a_1$ in the vector $\mathbf{A}' = (a'_1, \ldots, a'_n)$ defined by (1.11). Consider relation (1.12), where $\mathbf{A}'_1 = (a_1'^{(1)}, \ldots, a_n'^{(1)})$, $\mathbf{B}'_1 = (b_1'^{(1)}, \ldots, b_m'^{(1)})$, $\mathbf{A}'_2 = (a_1'^{(2)}, \ldots, a_n'^{(2)})$, and $\mathbf{B}'_2 = (b_1'^{(2)}, \ldots, b_m'^{(2)})$. The validity conditions for $a_1 = a_2$ indicated above imply that either $a'_1 \in Z$ and $a'_1 = l_1(\mathbf{B}; 1)$ when $l_1(\mathbf{B}; 1) = m$, or $b_j = 0$ for $j > l_1(\mathbf{B}; 1)$ when $l_1(\mathbf{B}; 1) < m$. Therefore, the first components of $\mathbf{A}'$, $\mathbf{A}'_1$, and $\mathbf{A}'_2$ in (1.12) are equal: $a_1'^{(1)} = a_1'^{(2)} = a'_1$. Assuming that $a_1'^{(1)} > a'_1$ or $a_1'^{(2)} > a'_1$, we see that the values of function (1.4) with $k = 1$ from Chapter 4 that correspond to $(\mathbf{A}'_1, \mathbf{B}'_1)$ and $(\mathbf{A}'_2, \mathbf{B}'_2)$ are negative; accordingly, $M(\mathbf{A}'_1, \mathbf{B}'_1; 1) = \varnothing$ and $M(\mathbf{A}'_2, \mathbf{B}'_2; 1) = \varnothing$. Define the components of the vectors $\mathbf{A}_1 = (a_1^{(1)}, \ldots, a_{n+1}^{(1)})$, $\mathbf{B}_1 = (b_1^{(1)}, \ldots, b_m^{(1)})$, $\mathbf{A}_2 = (a_1^{(2)}, \ldots, a_{n+1}^{(2)})$, and $\mathbf{B}_2 = (b_1^{(2)}, \ldots, b_m^{(2)})$ as follows:

$$a_1^{(1)} = a'_1, \quad a_i^{(1)} = a_{i-1}^{\prime(1)}, \quad 2 \le i \le n+1,$$

$$a_1^{(2)} = a'_1, \quad a_i^{(2)} = a_{i-1}^{\prime(2)}, \quad 2 \le i \le n+1,$$

$$b_j^{(1)} = \begin{cases} b_j^{\prime(1)} + 1, & 1 \le j \le l_1(\mathbf{B}; 1), \\ 0, & l_1(\mathbf{B}; 1) < j \le m, \end{cases}$$

$$b_j^{(2)} = \begin{cases} b_j^{\prime(2)} + 1, & 1 \le j \le l_1(\mathbf{B}; 1), \\ 0, & l_1(\mathbf{B}; 1) < j \le m. \end{cases}$$

It is clear that $\mathbf{A}_1$, $\mathbf{A}_2 \in \overline{R}_+^{n+1}$ and $\mathbf{B}_1$, $\mathbf{B}_2 \in \overline{R}_+^m$. Adding a first row with ones as its first $l_1(\mathbf{B}; 1)$ elements and zeros as its remaining elements to any matrix in $M(\mathbf{A}'_1, \mathbf{B}'_1; 1)$ and $M(\mathbf{A}'_2, \mathbf{B}'_2; 1)$, we obtain a matrix belonging to $M(\mathbf{A}_1, \mathbf{B}_1; 1)$ or $M(\mathbf{A}_2, \mathbf{B}_2; 1)$. Thus, we have $M(\mathbf{A}_1, \mathbf{B}_1; 1) \ne \varnothing$, $M(\mathbf{A}_2, \mathbf{B}_2; 1) \ne \varnothing$, and $(\mathbf{A}_1, \mathbf{B}_1), (\mathbf{A}_2, \mathbf{B}_2) \in L'(n+1, m)$. Consequently,

$$\big[(\mathbf{A}_1, \mathbf{B}_1), (\mathbf{A}_2, \mathbf{B}_2)\big] \subseteq L'(n+1, m).$$

It can be readily shown that $b_j^{(1)}/2 + b_j^{(2)}/2 = 0 = b_j$ for $l_1(\mathbf{B}; 1) < j \le m$ and $b_j^{(1)}/2 + b_j^{(2)}/2 = b_j^{\prime(1)}/2 + b_j^{\prime(2)}/2 + 1 = b'_j + 1 = b_j$ for $1 \le j \le l_1(\mathbf{B}; 1)$. Therefore, $(\mathbf{A}, \mathbf{B})$ is the midpoint of the segment $\big[(\mathbf{A}_1, \mathbf{B}_1), (\mathbf{A}_2, \mathbf{B}_2)\big]$, and $(\mathbf{A}, \mathbf{B})$ is not an extreme point of the set $L'(n+1, m)$.

To complete the analysis of the case of $a_{n+1} > 0$ and

$$a_1 = l_1(\mathbf{B}; 1) + \sum_{b_j < 1} b_j,$$

it remains to examine the last possibility: $a_1 > a_2$ when $(\mathbf{A}', \mathbf{B}') \notin \overline{E}^{n,m}$. Again, assume that $(\mathbf{A}', \mathbf{B}')$ is the midpoint of a segment $\big[(\mathbf{A}'_1, \mathbf{B}'_1), (\mathbf{A}'_2, \mathbf{B}'_2)\big] \subseteq L'(n, m)$. Note that its endpoints can be defined so that the first components of $\mathbf{A}'_1$ and $\mathbf{A}'_2$ are close to the first component of $\mathbf{A}'$. Therefore, we suppose that relation (1.12) holds and $|a'_1 - a_1^{\prime(1)}| = |a'_1 - a_1^{\prime(2)}| < a_1 - a_2$, where $a'_1 = a_2$ by the definition (1.11) of the vector pair $(\mathbf{A}', \mathbf{B}')$. We introduce vectors $\mathbf{A}_1$, $\mathbf{A}_2$, $\mathbf{B}_1$, and $\mathbf{B}_2$ with components defined as follows:

$$a_1^{(1)} = a_1, \quad a_i^{(1)} = a_{i-1}^{\prime(1)}, \quad 2 \le i \le n+1,$$

$$a_1^{(2)} = a_1, \quad a_i^{(2)} = a_{i-1}'^{(2)}, \quad 2 \le i \le n+1,$$

$$b_j^{(1)} = \begin{cases} b_j'^{(1)} + 1, & 1 \le j \le l_1(\mathbf{B}; 1), \\ b_j, & l_1(\mathbf{B}; 1) < j \le m, \end{cases}$$

$$b_j^{(2)} = \begin{cases} b_j'^{(2)} + 1, & 1 \le j \le l_1(\mathbf{B}; 1), \\ b_j, & l_1(\mathbf{B}; 1) < j \le m. \end{cases}$$

It is clear that $\mathbf{B}_1, \mathbf{B}_2 \in \overline{R}_+^m$; since $|a_1' - a_1'^{(1)}|, |a_1' - a_1'^{(2)}| < a_1 - a_2 = a_1 - a_1'$, we have $\mathbf{A}_1, \mathbf{A}_2 \in \overline{R}_+^{n+1}$. Adding the vector

$$(\underbrace{1, \ldots, 1}_{l_1(\mathbf{B};1)}, \underbrace{0, \ldots, 0}_{m - l_1(\mathbf{B};1)}),$$

as the first row to any matrix from $M(\mathbf{A}_1', \mathbf{B}_1'; 1)$ or $M(\mathbf{A}_2', \mathbf{B}_2'; 1)$, we obtain a matrix from $M(\mathbf{A}_1, \mathbf{B}_1; 1)$ or $M(\mathbf{A}_2, \mathbf{B}_2; 1)$, respectively. Therefore, $M(\mathbf{A}_1, \mathbf{B}_1; 1), M(\mathbf{A}_2, \mathbf{B}_2; 1) \ne \varnothing$. Consequently, it holds that $[(\mathbf{A}_1, \mathbf{B}_1), (\mathbf{A}_2, \mathbf{B}_2)] \subseteq L'(n+1, m)$. Because

$$a_i^{(1)}/2 + a_i^{(2)}/2 = a_i, \quad 1 \le i \le n+1,$$

and

$$b_j^{(1)}/2 + b_j^{(2)}/2 = \begin{cases} b_j'^{(1)}/2 + b_j'^{(2)}/2 + 1, & 1 \le j \le l_1(\mathbf{B}; 1), \\ b_j, & l_1(\mathbf{B}; 1) < j \le m, \end{cases} =$$

$$\begin{cases} b_j' + 1, & 1 \le j \le l_1(\mathbf{B}; 1), \\ b_j, & l_1(\mathbf{B}; 1) < j \le m, \end{cases} = b_j, \quad 1 \le j \le m,$$

(by the definition in (1.11)), $(\mathbf{A}, \mathbf{B})$ is the midpoint of the segment $[(\mathbf{A}_1, \mathbf{B}_1), (\mathbf{A}_2, \mathbf{B}_2)]$. Therefore, $(\mathbf{A}, \mathbf{B})$ is not an extreme point of $L'(n+1, m)$.

Finally, to prove Lemma 1.4, it remains to analyze the case

$$0 < a_{n+1} \le a_1 < l_1(\mathbf{B}; 1) + \sum_{b_j < 1} b_j.$$

Define $p = \max\{i : a_i = a_1, \ 1 \le i \le n+1\}$ and $m_1 = \max\{j : b_j > 0, \ 1 \le j \le m\}$, and set $a_{n+2} = 0$ for $p = n+1$. Let us show that $(\mathbf{A}, \mathbf{B})$ is not an extreme point of $L'(n+1, m)$. When $l_p(\mathbf{B}; 1) = |\{j : b_j \ge p, \ 1 \le j \le m\}| = \max\{j : b_j \ge p, \ 1 \le j \le m\}$, two possibilities must be examined: $l_p(\mathbf{B}; 1) < m_1$ and $l_p(\mathbf{B}; 1) = m_1$.

Suppose $l_p(\mathbf{B}; 1) < m_1$. Let us show that $\delta_k(\mathbf{A}, \mathbf{B}; 1) > 0$ for $2 \le k \le p$, where $p \ge 2$. Note that the inequality $\delta_1(\mathbf{A}, \mathbf{B}; 1) > 0$ is valid by the condition for the first component of **A**. Assume that for some $k \in Z$ $(2 \leq k < p,\, p \geq 2)$ it holds that

$$\delta_k(\mathbf{A},\mathbf{B};1) = kl_k(\mathbf{B};1) + \sum_{b_j<k} b_j - a_1 k = 0. \tag{1.13}$$

Since $\delta_q(\mathbf{A},\mathbf{B};1) \geq 0$ for any q, $1 \leq q \leq n+1$, we have

$$\delta_p(\mathbf{A},\mathbf{B};1) = pl_p(\mathbf{B};1) + \sum_{b_j<p} b_j - a_1 p \geq 0.$$

These relations imply that

$$a_1 k = kl_k(\mathbf{B};1) + \sum_{b_j<k} b_j, \tag{1.14}$$

$$a_1 p \leq pl_p(\mathbf{B};1) + \sum_{b_j<p} b_j. \tag{1.15}$$

Consider equation (1.13). By Theorem 1.2′ from Chapter 4, every element in the first k rows and $l_k(\mathbf{B};1)$ columns of any matrix in $M(\mathbf{A},\mathbf{B};1)$ is unity. Therefore, $a_1 \geq l_k(\mathbf{B};1)$. Assuming that $a_1 = l_k(\mathbf{B};1)$, we see that $a_i = l_k(\mathbf{B};1)$ for any i $(1 \leq i \leq p)$. Then, equation (1.13) implies that

$$\sum_{b_j<k} b_j = 0.$$

Hence, $l_1(\mathbf{B};1) = l_k(\mathbf{B};1)$ and

$$a_1 = l_1(\mathbf{B};1) + \sum_{b_j<1} b_j,$$

which leads to contradiction. Thus, $a_1 > l_k(\mathbf{B};1)$. Combining (1.14) and (1.15), we obtain

$$a_1 p - a_1 k = pl_p(\mathbf{B};1) - kl_k(\mathbf{B};1) + \sum_{b_j<p} b_j - \sum_{b_j<k} b_j =$$

$$pl_p(\mathbf{B};1) - kl_k(\mathbf{B};1) + \sum_{k\leq b_j<p} b_j \leq$$

$$pl_p(\mathbf{B};1) - kl_k(\mathbf{B};1) + p\left(l_k(\mathbf{B};1) - l_p(\mathbf{B};1)\right) = (p-k)l_k(\mathbf{B};1).$$

The inequality $(p-k)a_1 \leq (p-k)l_k(\mathbf{B};1)$ implies that $a_1 \leq l_k(\mathbf{B};1)$, which contradicts the inequality $a_1 > l_k(\mathbf{B};1)$. This proves that $\delta_k(\mathbf{A},\mathbf{B};1) > 0$ for $2 \leq k < p$, where $p \geq 2$.

To construct a segment $\left[(\mathbf{A}_1, \mathbf{B}_1), (\mathbf{A}_2, \mathbf{B}_2)\right] \subseteq L'(n+1, m)$ whose midpoint is the vector pair $(\mathbf{A}, \mathbf{B})$, we set

$$\varepsilon = \min\Big\{ l_1(\mathbf{B}; 1) + \sum_{b_j < 1} b_j - a_1,\ a_1 - a_{p+1},\ b_{l_p(\mathbf{B};1)} - b_{l_p(\mathbf{B};1)+1},\ b_{m_1},$$
$$\delta_k(\mathbf{A}, \mathbf{B}; 1),\ 1 \le k < p,\ \text{ where } \ p \ge 2 \Big\}.$$

We define the components of $\mathbf{A}_1 = (a_1^{(1)}, \ldots, a_{n+1}^{(1)})$, $\mathbf{A}_2 = (a_1^{(2)}, \ldots, a_{n+1}^{(2)})$, $\mathbf{B}_1 = (b_1^{(1)}, \ldots, b_m^{(1)})$, and $\mathbf{B}_2 = (b_1^{(2)}, \ldots, b_m^{(2)})$ in terms of those of $\mathbf{A} = (a_1, \ldots, a_{n+1})$ and $\mathbf{B} = (b_1, \ldots, b_{n+1})$ and $\varepsilon > 0$:

$$a_i^{(1)} = \begin{cases} a_i - \varepsilon/p, & 1 \le i \le p, \\ a_i, & p < i \le n+1, \end{cases} \qquad a_i^{(2)} = \begin{cases} a_i + \varepsilon/p, & 1 \le i \le p, \\ a_i, & p < i \le n+1, \end{cases}$$

$$b_j^{(1)} = \begin{cases} b_j, & 1 \le j \le l_p(\mathbf{B}; 1), \\ b_j - \varepsilon/\left(m_1 - l_p(\mathbf{B}; 1)\right), & l_p(\mathbf{B}; 1) < j \le m_1, \\ 0, & m_1 < j \le m, \end{cases}$$

$$b_j^{(2)} = \begin{cases} b_j, & 1 \le j \le l_p(\mathbf{B}; 1), \\ b_j + \varepsilon/\left(m_1 - l_p(\mathbf{B}; 1)\right), & l_p(\mathbf{B}; 1) < j \le m_1, \\ 0, & m_1 < j \le m. \end{cases}$$

It is obvious that

$$\sum_{i=1}^{n+1} a_i^{(1)} = \sum_{j=1}^{m} b_j^{(1)}, \quad \sum_{i=1}^{n+1} a_i^{(2)} = \sum_{j=1}^{m} b_j^{(2)},$$

and the value of ε is chosen so that $\mathbf{A}_1$, $\mathbf{A}_2 \in \overline{R}_+^{n+1}$ and $\mathbf{B}_1$, $\mathbf{B}_2 \in \overline{R}_+^m$. It can be readily shown that

$$(\mathbf{A}, \mathbf{B}) = \frac{1}{2}(\mathbf{A}_1, \mathbf{B}_1) + \frac{1}{2}(\mathbf{A}_2, \mathbf{B}_2).$$

Let us now show that the characteristic function of $(\mathbf{A}_1, \mathbf{B}_1)$ and $(\mathbf{A}_2, \mathbf{B}_2)$, as defined by (1.4) in Chapter 4, is nonnegative for all admissible $k \in Z$. For the vector pair $(\mathbf{A}_1, \mathbf{B}_1)$, the characteristic function is

$$\delta_k(\mathbf{A}_1, \mathbf{B}_1; 1) = \sum_{j=1}^{m} b_j^{(1)} - \sum_{b_j^{(1)} \ge k} (b_j^{(1)} - k) - \sum_{i=1}^{k} a_i^{(1)}.$$

Let us express $\delta_k(\mathbf{A}_1, \mathbf{B}_1; 1)$ in terms of $\delta_k(\mathbf{A}, \mathbf{B}; 1)$, where $\delta_k(\mathbf{A}, \mathbf{B}; 1) \geq 0$ for any k, $1 \leq k \leq n+1$. When $k \geq p$, we have

$$\sum_{j=1}^{m} b_j^{(1)} = \sum_{j=1}^{m} b_j - \varepsilon, \quad \sum_{i=1}^{k} a_i^{(1)} = \sum_{i=1}^{k} a_i - \varepsilon, \quad \sum_{b_j^{(1)} \geq k} (b_j^{(1)} - k) = \sum_{b_j \geq k} (b_j - k).$$

Therefore, $\delta_k(\mathbf{A}_1, \mathbf{B}_1; 1) = \delta_k(\mathbf{A}, \mathbf{B}; 1) \geq 0$ in this case. Suppose that $p \geq 2$ and $1 \leq k < p$. Then,

$$\sum_{j=1}^{m} b_j^{(1)} = \sum_{j=1}^{m} b_j - \varepsilon,$$

$$\sum_{i=1}^{k} a_i^{(1)} = a_1^{(1)} k = a_1 k - \varepsilon k/p = \sum_{i=1}^{k} a_i - \varepsilon k/p,$$

$$\sum_{b_j^{(1)} \geq k} (b_j^{(1)} - k) \leq \sum_{b_j \geq k} (b_j - k),$$

$$\delta_k(\mathbf{A}_1, \mathbf{B}_1; 1) \geq \delta_k(\mathbf{A}, \mathbf{B}; 1) - \varepsilon(p-k)/p.$$

Since $\varepsilon \leq \min\{\delta_k(\mathbf{A}, \mathbf{B}; 1) : 1 \leq k < p\}$ and $\delta_k(\mathbf{A}, \mathbf{B}; 1) > 0$ for $1 \leq k < p$, it holds that $\delta_k(\mathbf{A}_1, \mathbf{B}_1; 1) > 0$. Thus, $\delta_k(\mathbf{A}_1, \mathbf{B}_1; 1) \geq 0$ for any k $(1 \leq k \leq n+1)$ and $(\mathbf{A}_1, \mathbf{B}_1) \in L'(n+1, m)$.

Similarly,

$$\delta_k(\mathbf{A}_2, \mathbf{B}_2; 1) = \sum_{j=1}^{m} b_j^{(2)} - \sum_{b_j^{(2)} \geq k} (b_j^{(2)} - k) - \sum_{i=1}^{k} a_i^{(2)}.$$

When $k \geq p$,

$$\sum_{j=1}^{m} b_j^{(2)} = \sum_{j=1}^{m} b_j + \varepsilon, \quad \sum_{i=1}^{k} a_i^{(2)} = \sum_{i=1}^{k} a_i + \varepsilon, \quad \sum_{b_j^{(2)} \geq k} (b_j^{(2)} - k) = \sum_{b_j \geq k} (b_j - k);$$

therefore, $\delta_k(\mathbf{A}_2, \mathbf{B}_2; 1) = \delta_k(\mathbf{A}, \mathbf{B}; 1) \geq 0$. For $p \geq 2$ and $1 \leq k < p$,

$$\sum_{j=1}^{m} b_j^{(2)} = \sum_{j=1}^{m} b_j + \varepsilon,$$

$$\sum_{i=1}^{k} a_i^{(2)} = a_1 k + \varepsilon k/p = \sum_{i=1}^{k} a_i + \varepsilon k/p,$$

$$\sum_{b_j^{(2)} \geq k} (b_j^{(2)} - k) \leq \sum_{b_j \geq k} (b_j - k) + \varepsilon.$$

Consequently, $\delta_k(\mathbf{A}_2, \mathbf{B}_2; 1) \geq \delta_k(\mathbf{A}, \mathbf{B}; 1) - \varepsilon k/p \geq \varepsilon - \varepsilon k/p > 0$. Thus, $(\mathbf{A}_2, \mathbf{B}_2) \in L'(n+1, m)$ and $[(\mathbf{A}_1, \mathbf{B}_1), (\mathbf{A}_2, \mathbf{B}_2)] \subseteq L'(n+1, m)$. Since $(\mathbf{A}, \mathbf{B})$ is the midpoint of this segment, $(\mathbf{A}, \mathbf{B})$ is not an extreme point of the set $L'(n+1, m)$.

As the last step in proving Lemma 1.4, we consider the case $l_p(\mathbf{B}; 1) = m_1$. Define $q = \min\{j : b_j = b_{m_1},\ 1 \leq j \leq m_1\}$. Again, we have $\delta_1(\mathbf{A}, \mathbf{B}; 1) > 0$. Let us show that $\delta_k(\mathbf{A}, \mathbf{B}; 1) > 0$ for every k such that $l_k(\mathbf{B}; 1) = m_1$. Assume that there exists $k \in Z$ such that $l_k(\mathbf{B}; 1) = m_1$ and $\delta_k(\mathbf{A}, \mathbf{B}; 1) = 0$. Since this implies that $l_k(\mathbf{B}; 1) = l_p(\mathbf{B}; 1) = m_1$, it holds that

$$\delta_k(\mathbf{A}, \mathbf{B}; 1) = km_1 - \sum_{i=1}^{k} a_i = 0$$

and $km_1 - a_1 k = 0$. Then, $a_1 = m_1$, which contradicts the inequality $\delta_1(\mathbf{A}, \mathbf{B}; 1) > 0$. Therefore, $\delta_k(\mathbf{A}, \mathbf{B}; 1) > 0$ for every k such that $l_k(\mathbf{B}; 1) = m_1$. As in the preceding case, let us construct a segment $[(\mathbf{A}_1, \mathbf{B}_1), (\mathbf{A}_2, \mathbf{B}_2)] \subseteq L'(n+1, m)$ whose midpoint is $(\mathbf{A}, \mathbf{B})$. Note that the inequality $\delta_p(\mathbf{A}, \mathbf{B}; 1) > 0$ entails $p < n+1$. We introduce $b_0 = n+1$ and set

$$\varepsilon = \min\Big\{ l_1(\mathbf{B}; 1) + \sum_{b_j < 1} b_j - a_1,\ a_1 - a_{p+1},\ b_{q-1} - b_q,\ b_{m_1},\ \delta_k(\mathbf{A}, \mathbf{B}; 1), \\ l_k(\mathbf{B}; 1) = m_1 \Big\}.$$

Using the components of $\mathbf{A} = (a_1, \ldots, a_{n+1})$ and $\mathbf{B} = (b_1, \ldots, b_m)$ and the value of $\varepsilon > 0$, we define $\mathbf{A}_1 = (a_1^{(1)}, \ldots, a_{n+1}^{(1)})$, $\mathbf{A}_2 = (a_1^{(2)}, \ldots, a_{n+1}^{(2)})$, $\mathbf{B}_1 = (b_1^{(1)}, \ldots, b_m^{(1)})$, $\mathbf{B}_2 = (b_1^{(2)}, \ldots, b_m^{(2)})$:

$$a_i^{(1)} = \begin{cases} a_i - \varepsilon/p, & 1 \leq i \leq p, \\ a_i, & p < i \leq n+1, \end{cases} \qquad a_i^{(2)} = \begin{cases} a_i + \varepsilon/p, & 1 \leq i \leq p, \\ a_i, & p < i \leq n+1, \end{cases}$$

$$b_j^{(1)} = \begin{cases} b_j, & 1 \leq j < q, \\ b_j - \varepsilon/(m_1 - q + 1), & q \leq j \leq m_1, \\ 0, & m_1 < j \leq m, \end{cases}$$

$$b_j^{(2)} = \begin{cases} b_j, & 1 \leq j < q, \\ b_j + \varepsilon/(m_1 - q + 1), & q \leq j \leq m_1, \\ 0, & m_1 < j \leq m. \end{cases}$$

It is clear that $\mathbf{A}_1, \mathbf{A}_2 \in \overline{R}_+^{n+1}$, $\mathbf{B}_1, \mathbf{B}_2 \in \overline{R}_+^m$,

$$\sum_{i=1}^{n+1} a_i^{(1)} = \sum_{j=1}^{m} b_j^{(1)}, \quad \sum_{i=1}^{n+1} a_i^{(2)} = \sum_{j=1}^{m} b_j^{(2)},$$

$$(\mathbf{A}, \mathbf{B}) = \frac{1}{2}(\mathbf{A}_1, \mathbf{B}_1) + \frac{1}{2}(\mathbf{A}_2, \mathbf{B}_2).$$

To show that $\delta_k(\mathbf{A}_1, \mathbf{B}_1; 1), \delta_k(\mathbf{A}_2, \mathbf{B}_2; 1) \geq 0$ for any k $(1 \leq k \leq n+1)$. Consider the vector pair $(\mathbf{A}_1, \mathbf{B}_1)$:

$$\delta_k(\mathbf{A}_1, \mathbf{B}_1; 1) = \sum_{j=1}^{m} b_j^{(1)} - \sum_{b_j^{(1)} \geq k} (b_j^{(1)} - k) - \sum_{i=1}^{k} a_i^{(1)}.$$

If $k \in Z$ and $l_k(\mathbf{B}; 1) = m_1$, then

$$\sum_{j=1}^{m} b_j^{(1)} = \sum_{j=1}^{m} b_j - \varepsilon, \quad \sum_{i=1}^{k} a_i^{(1)} = \sum_{i=1}^{k} a_i - \varepsilon\alpha, \quad \sum_{b_j^{(1)} \geq k} (b_j^{(1)} - k) \leq \sum_{b_j \geq k} (b_j - k),$$

where $\alpha = 1$ for $k \geq p$ and $\alpha = k/p$ for $k < p$; hence, $\delta_k(\mathbf{A}_1, \mathbf{B}_1; 1) \geq \delta_k(\mathbf{A}, \mathbf{B}; 1) - \varepsilon(p-k)/p \geq \varepsilon - \varepsilon(p-k)/p \geq 0$. If $k \in Z$ and $l_k(\mathbf{B}; 1) < m_1$, then

$$\sum_{j=1}^{m} b_j^{(1)} = \sum_{j=1}^{m} b_j - \varepsilon, \quad \sum_{i=1}^{k} a_i^{(1)} = \sum_{i=1}^{k} a_i - \varepsilon, \quad \sum_{b_j^{(1)} \geq k} (b_j^{(1)} - k) = \sum_{b_j \geq k} (b_j - k).$$

Hence, $\delta_k(\mathbf{A}_1, \mathbf{B}_1; 1) = \delta_k(\mathbf{A}, \mathbf{B}; 1) \geq 0$ and $(\mathbf{A}_1, \mathbf{B}_1) \in L'(n+1, m)$.

Consider the vector pair $(\mathbf{A}_2, \mathbf{B}_2)$:

$$\delta_k(\mathbf{A}_2, \mathbf{B}_2; 1) = \sum_{j=1}^{m} b_j^{(2)} - \sum_{b_j^{(2)} \geq k} (b_j^{(2)} - k) - \sum_{i=1}^{k} a_i^{(2)}.$$

If $k \in Z$ and $l_k(\mathbf{B}; 1) = m_1$, then

$$\sum_{j=1}^{m} b_j^{(2)} = \sum_{j=1}^{m} b_j + \varepsilon, \quad \sum_{i=1}^{k} a_i^{(2)} = \sum_{i=1}^{k} a_i + \varepsilon\alpha, \quad \sum_{b_j^{(2)} \geq k} (b_j^{(2)} - k) = \sum_{b_j \geq k} (b_j - k) + \varepsilon,$$

where $\alpha = 1$ for $k \geq p$ and $\alpha = k/p$ for $k < p$; hence, $\delta_k(\mathbf{A}_2, \mathbf{B}_2; 1) = \delta_k(\mathbf{A}, \mathbf{B}; 1) - \varepsilon\alpha \geq \varepsilon - \varepsilon\alpha \geq 0$. If $k \in Z$ and $l_k(\mathbf{B}; 1) < m_1$, then

$$\sum_{j=1}^{m} b_j^{(2)} = \sum_{j=1}^{m} b_j + \varepsilon, \quad \sum_{i=1}^{k} a_i^{(2)} = \sum_{i=1}^{k} a_i + \varepsilon, \quad \sum_{b_j^{(2)} \geq k} (b_j^{(2)} - k) = \sum_{b_j \geq k} (b_j - k).$$

Hence, $\delta_k(\mathbf{A}_2, \mathbf{B}_2; 1) = \delta_k(\mathbf{A}, \mathbf{B}; 1) \geq 0$ and $\delta_k(\mathbf{A}_2, \mathbf{B}_2; 1) \geq 0$ for any k ($1 \leq k \leq n+1$). Consequently, $(\mathbf{A}_2, \mathbf{B}_2) \in L'(n+1, m)$ and $[(\mathbf{A}_1, \mathbf{B}_1), (\mathbf{A}_2, \mathbf{B}_2)] \subseteq L'(n+1, m)$. Since $(\mathbf{A}, \mathbf{B})$ is the midpoint of this segment, $(\mathbf{A}, \mathbf{B})$ is not an extreme point of $L'(n+1, m)$. This completes the proof of Lemma 1.4.

Combining the Krein–Mil'man theorem about extreme points with Lemmas 1.3 and 1.4, we obtain the assertion of Theorem 1.4: $L(n, m) = L'(n, m)$. Therefore, it holds that

$$L(n, m) = \Big\{(\mathbf{A}, \mathbf{B}) : (\mathbf{A}, \mathbf{B}) = \sum_{k=0}^{C_{n+m}^{n}-1} \alpha_k(\overline{\mathbf{E}}_k, \overline{\mathbf{F}}_k), (\overline{\mathbf{E}}_k, \overline{\mathbf{F}}_k) \in \overline{E}^{n,m},\ \alpha_k \geq 0,$$

$$0 \leq k \leq C_{n+m}^{n} - 1, \sum_{k=0}^{C_{n+m}^{n}-1} \alpha_k = 1\Big\} =$$

$$\Big\{(\mathbf{A}, \mathbf{B}) : \mathbf{A} \in \overline{R}_{+}^{n},\ \mathbf{B} \in \overline{R}_{+}^{m},\ M(\mathbf{A}, \mathbf{B}; 1) \neq \varnothing\Big\} = L'(n, m).$$

Thus, the polyhedron $L(n, m)$ contains only the n-by-m pairs of vectors (with components arranged in nonascending order) that define the plan matrices of a transportation problem whose elements are not greater than unity.

Definition 1.1. The polyhedron $L(n, m)$ is called the *fundamental polyhedron* of vector pairs.

The constructions above can be used to formulate a geometric criterion for the existence of a plan matrix corresponding to a vector pair with elements not greater than unity.

Theorem 1.5. *An arbitrary vector pair* $(\mathbf{A}, \mathbf{B})$ *is such that* $M(\mathbf{A}, \mathbf{B}; 1) \neq \varnothing$ *or* $M_Z(\mathbf{A}, \mathbf{B}; 1) \neq \varnothing$ *if and only if the fundamental polyhedron* $L(n, m)$ *contains the point (vector pair)* $(\widetilde{\mathbf{A}}, \widetilde{\mathbf{B}})$, *where* $\widetilde{\mathbf{A}}$ *and* $\widetilde{\mathbf{B}}$ *are derived from* $\mathbf{A}$ *and* $\mathbf{B}$, *respectively, by arranging their components in nonascending order.*

We have already constructed the sets $L(n^+, m)$, $L(n, m^+)$, and $L(n^+, m^+)$, which are subpolyhedrons of the fundamental polyhedron $L(n, m)$. The following proposition is obviously true.

Theorem 1.6. *The polyhedrons* $L(n^+, m)$, $L(n, m^+)$, *and* $L(n^+, m^+)$ *defined by the vertex sets* $\overline{E}^{n^+,m}$, $\overline{E}^{n,m^+}$, *and* $\overline{E}^{n^+,m^+}$, *respectively, can be expressed as follows:*

$$L(n^+, m) = \Big\{(\mathbf{A}, \mathbf{B}) = ((a_1, \ldots, a_n), (b_1, \ldots, b_m)) :$$
$$\mathbf{A} \in \overline{R}_{+}^{n},\ \mathbf{B} \in \overline{R}_{+}^{m},\ b_1 = n,\ M(\mathbf{A}, \mathbf{B}; 1) \neq \varnothing\Big\},$$

$$L(n, m^+) = \Big\{(\mathbf{A}, \mathbf{B}) = ((a_1, \ldots, a_n), (b_1, \ldots, b_m)) : \\ \mathbf{A} \in \overline{R}^n_+,\ a_1 = m,\ \mathbf{B} \in \overline{R}^m_+,\ M(\mathbf{A}, \mathbf{B}; 1) \neq \varnothing\Big\},$$

$$L(n^+, m^+) = \Big\{(\mathbf{A}, \mathbf{B}) = ((a_1, \ldots, a_n), (b_1, \ldots, b_m)) : \\ \mathbf{A} \in \overline{R}^n_+,\ a_1 = m,\ \mathbf{B} \in \overline{R}^m_+,\ b_1 = n,\ M(\mathbf{A}, \mathbf{B}; 1) \neq \varnothing\Big\}.$$

It should be noted that if $(\mathbf{A}, \mathbf{B}) \in L(n, m)$, then the representation of the pair as

$$(\mathbf{A}, \mathbf{B}) = \sum_{k=0}^{C^n_{n+m}-1} \alpha_k (\overline{\mathbf{E}}_k, \overline{\mathbf{F}}_k), \quad (\overline{\mathbf{E}}_k, \overline{\mathbf{F}}_k) \in \overline{E}^{n,m}, \\ \alpha_k \geq 0, \quad 0 \leq k \leq C^n_{n+m} - 1, \quad \sum_{k=0}^{C^n_{n+m}-1} \alpha_k = 1, \tag{1.16}$$

is nonunique in the general case. Since every extremal pair of vectors defines a (unique) extremal matrix, relations (1.16) define a uniform matrix from $\widetilde{M}(\mathbf{A}, \mathbf{B}; 1)$. It is shown below that relations (1.16) define the entire matrix set $\widetilde{M}(\mathbf{A}, \mathbf{B}; 1)$ (see Theorem 1.9).

1.3. Representation of Uniform Matrices by Convex Linear Combinations of Extremal Matrices

Here, we show that every uniform matrix with elements not greater than unity belongs to a polyhedron whose vertices are extremal matrices.

We denote by $M'(n, m)$ the set of all n-by-m uniform matrices such that their elements are not greater than unity and the sums of row and column elements are the components of vectors from $\overline{R}^n_+$ and $\overline{R}^m_+$, respectively:

$$M'(n, m) = \Big\{X(\mathbf{A}, \mathbf{B}) : \mathbf{A} \in \overline{R}^n_+,\ \mathbf{B} \in \overline{R}^m_+,\ X(\mathbf{A}, \mathbf{B}) \in \widetilde{M}(\mathbf{A}, \mathbf{B}; 1), \\ X(\mathbf{A}, \mathbf{B}) \in \widetilde{M}(\mathbf{A}, \mathbf{B}; 1), M(\mathbf{A}, \mathbf{B}; 1) \neq \varnothing\Big\}.$$

Applying Theorem 1.4, we represent the matrix set $M'(n, m)$ as

$$M'(n, m) = \bigcup_{(\mathbf{A}, \mathbf{B}) \in L(n,m)} \widetilde{M}(\mathbf{A}, \mathbf{B}; 1).$$

The following proposition does not call for a proof.

Lemma 1.6. *The set $M'(n, m)$ of uniform matrices is a closed, convex set such that*

$$\overline{E}_{n,m} \subset M(n, m) \subseteq M'(n, m).$$

Let us show that $M(n,m) = M'(n,m)$ and then apply the Krein–Mil'man theorem about extreme points and Lemma 1.3.

Let X_1 and X_2 be arbitrary matrices with equal dimensions. We denote by $[X_1, X_2] = \{X : X = \alpha X_1 + \beta X_2,\ \alpha,\ \beta \geq 0,\ \alpha + \beta = 1\}$ the segment consisting of matrices with endpoints at X_1 and X_2.

We now prove that any point (extremal matrix) $\overline{H}$ in the set $\overline{E}_{n,m}$ (fundamental system of extremal matrices) is not an interior point of any segment $[X_1, X_2] \subseteq M'(n,m)$. We will also prove that any point (uniform matrix) X in $M'(n,m)\backslash\overline{E}_{n,m}$ is an interior point of a segment $[X_1, X_2] \subseteq M'(n,m)$. Thus, we will establish that $M(n,m) = M'(n,m)$, where the polyhedron $M(n,m)$ defined by (1.2) and (1.3) is a convex, closed hull of the set $\overline{E}_{n,m}$.

Let $\overline{H} = \overline{H}(\overline{\mathbf{E}}, \overline{\mathbf{F}})$ be an arbitrary point (extremal matrix) in $\overline{E}_{n,m}$. Suppose that $\overline{H}$ is an interior point of a segment $[X_1, X_2]$, where $X_1 = X_1(\mathbf{A}_1, \mathbf{B}_1) \in M'(n,m)$ and $X_2 = X_2(\mathbf{A}_2, \mathbf{B}_2) \in M'(n,m)$. Then,

$$\overline{H}(\overline{\mathbf{E}}, \overline{\mathbf{F}}) = \alpha X_1(\mathbf{A}_1, \mathbf{B}_1) + \beta X_2(\mathbf{A}_2, \mathbf{B}_2), \quad \alpha,\ \beta > 0, \quad \alpha + \beta = 1. \quad (1.17)$$

However, relation (1.17) also applies to the sums of both row and column elements of $\overline{H}$, X_1, and X_2. Therefore, it holds that $(\overline{\mathbf{E}}, \overline{\mathbf{F}}) = \alpha(\mathbf{A}_1, \mathbf{B}_1) + \beta(\mathbf{A}_2, \mathbf{B}_2)$, where $\alpha,\ \beta > 0$ and $\alpha + \beta = 1$, which contradicts Lemma 1.3. This proves the following proposition.

Lemma 1.7. *Every point (extremal matrix) in $\overline{E}_{n,m}$ is an extreme point of the set $M'(n,m)$ of uniform matrices.*

Now, let us show that every extreme point of the set $M'(n,m)$ is an extremal matrix.

Lemma 1.8. *Suppose that a point (extremal matrix) $X(\mathbf{A}, \mathbf{B}) \in M'(n,m)$ is an extreme point of the set $M'(n,m)$. Then, $X(\mathbf{A}, \mathbf{B}) \in \overline{E}_{n,m}$ (equivalently, $(\mathbf{A}, \mathbf{B}) \in \overline{E^{n,m}}$).*

Proof. The elements of the set of one-row uniform matrices $M'(1,m)$ can be treated as the elements of the set $L'(1,m) = L(1,m)$ of vector pairs. Therefore, the assertion of the lemma is obviously true for $n = 1$. We assume that it is also true for $\mathbf{A} \in \overline{R}^n_+$ and consider the case $\mathbf{A} \in \overline{R}^{n+1}_+$.

Let $X(\mathbf{A}, \mathbf{B})$ be an extreme point of $M'(n+1,m)$. Consider the matrix $X_i(\mathbf{A}^{(i)}, \mathbf{B}^{(i)})$ derived from $X(\mathbf{A}, \mathbf{B})$ by taking out its ith row. Since $X(\mathbf{A}, \mathbf{B}) \in M'(n+1,m)$, it holds that $X_i(\mathbf{A}^{(i)}, \mathbf{B}^{(i)}) \in M'(n,m)$. It can be readily shown that every point (matrix) $X_i(\mathbf{A}^{(i)}, \mathbf{B}^{(i)})$ $(1 \leq i \leq n+1)$ is an extreme point of the set $M'(n,m)$. Then $X_i(\mathbf{A}^{(i)}, \mathbf{B}^{(i)}) \in \overline{E}_{n,m}$ for $1 \leq i \leq n+1$ by the induction hypothesis. This means that every point $X_i(\mathbf{A}^{(i)}, \mathbf{B}^{(i)})$ $(1 \leq i \leq n+1)$ is an extremal matrix. Therefore, every element of any matrix $X_i(\mathbf{A}^{(i)}, \mathbf{B}^{(i)})$ is either zero or one. Consequently, every element of the source matrix $X(\mathbf{A},\mathbf{B}) \in M'(n+1,m)$ is also either zero or one. Since $X(\mathbf{A},\mathbf{B})$ is a uniform matrix, $X(\mathbf{A},\mathbf{B})$ is an extremal matrix by Theorem 1.6; that is, $X(\mathbf{A},\mathbf{B}) \in \overline{E}_{n+1,m}$ (equivalently, $(\mathbf{A},\mathbf{B}) \in \overline{E}^{n+1,m}$). Lemma is proved.

Using Lemmas 1.7 and 1.8 and invoking the Krein–Mil'man theorem about extreme points, we obtain the following proposition.

Theorem 1.7. $M(n,m) = M'(n,m)$.

By Theorem 1.7, every uniform matrix whose elements are nonnegative and are not greater than unity can be expressed in terms of extremal matrices:

$$M(n,m) = \Big\{X : X = \sum_{k=0}^{C_{n+m}^{n}-1} \alpha_k \overline{H}_k(\overline{\mathbf{E}}_k,\overline{\mathbf{F}}_k),\ \overline{H}_k(\overline{\mathbf{E}}_k,\overline{\mathbf{F}}_k) \in \overline{E}_{n,m},\ \alpha_k \geq 0,$$
$$0 \leq k \leq C_{n+m}^{n} - 1,\ \sum_{k=0}^{C_{n+m}^{n}-1} \alpha_k = 1\Big\} =$$
$$\Big\{X(\mathbf{A},\mathbf{B}) : \mathbf{A} \in \overline{R}_{+}^{n},\ \mathbf{B} \in \overline{R}_{+}^{m},\ X(\mathbf{A},\mathbf{B}) \in \widetilde{M}(\mathbf{A},\mathbf{B};1),$$
$$M(\mathbf{A},\mathbf{B};1) \neq \varnothing\Big\} = M'(n,m).$$

Consequently, the polyhedron $M(n,m)$ contains only all of the uniform plan matrices, which are defined by pairs of vectors from $\overline{R}_{+}^{n}$ and $\overline{R}_{+}^{m}$; their elements are not greater than unity.

Definition 1.2. The polyhedron $M(n,m)$ is called the *fundamental transportation polyhedron* of uniform matrices.

Note that if an arbitrary uniform matrix $X(\mathbf{A},\mathbf{B})$ belongs to $\widetilde{M}(\mathbf{A},\mathbf{B};1) \neq \varnothing$, where $\mathbf{A} \in R_{+}^{n}$ and $\mathbf{B} \in R_{+}^{m}$, then one can permute rows and columns to obtain the corresponding matrix $\widetilde{X}(\widetilde{\mathbf{A}},\widetilde{\mathbf{B}})$ that belongs to the fundamental transportation polyhedron. In the general case, representation of a uniform matrix in terms of a linear combination of extremal matrices,

$$X(\mathbf{A},\mathbf{B}) = \sum_{k=0}^{C_{n+m}^{n}-1} \alpha_k \overline{H}_k(\overline{\mathbf{E}}_k,\overline{\mathbf{F}}_k),\quad \overline{H}_k(\overline{\mathbf{E}}_k,\overline{\mathbf{F}}_k) \in \overline{E}_{n,m},$$
$$\alpha_k \geq 0,\quad 0 \leq k \leq C_{n+m}^{n} - 1,\quad \sum_{k=0}^{C_{n+m}^{n}-1} \alpha_k = 1, \tag{1.18}$$

is not unique. Note that each solution of equation (1.18) defines a solution of (1.16).

Example 1.1. Let us represent the pair of vectors $\mathbf{A} = (3/2, 1/2)$ and $\mathbf{B} = (5/4, 3/4)$ and the corresponding uniform matrix

$$X(\mathbf{A},\mathbf{B}) = \begin{pmatrix} 7/8 & 5/8 \\ 3/8 & 1/8 \end{pmatrix}$$

as linear combinations of extremal vector pairs and extremal matrices having the form of (1.16) and (1.18), respectively.

It is obvious that $(\mathbf{A},\mathbf{B}) \in L(2,2)$ and $X(\mathbf{A},\mathbf{B}) \in M(2,2)$. Let us write out all elements of the fundamental systems $\overline{E}^{2,2}$ and $\overline{E}_{2,2}$. They are the vertices of the polyhedrons $L(2,2)$ and $M(2,2)$. Here, $|\overline{E}^{2,2}| = |\overline{E}_{2,2}| = 6$ and

$$\begin{aligned}\overline{E}^{2,2} = \Big\{ &(\overline{\mathbf{E}}_0, \overline{\mathbf{F}}_0) = ((0,0),(0,0)), \ (\overline{\mathbf{E}}_1, \overline{\mathbf{F}}_1) = ((1,0),(1,0)), \\ &(\overline{\mathbf{E}}_2, \overline{\mathbf{F}}_2) = ((2,0),(1,1)), \ (\overline{\mathbf{E}}_3, \overline{\mathbf{F}}_3) = ((1,1),(2,0)), \\ &(\overline{\mathbf{E}}_4, \overline{\mathbf{F}}_4) = ((2,1),(2,1)), \ (\overline{\mathbf{E}}_5, \overline{\mathbf{F}}_5) = ((2,2),(2,2)) \Big\},\end{aligned}$$

$$\begin{aligned}\overline{E}_{2,2} = \Big\{ &\overline{H}_0 = \begin{pmatrix} 0 & 0 \\ 0 & 0 \end{pmatrix}, \ \overline{H}_1 = \begin{pmatrix} 1 & 0 \\ 0 & 0 \end{pmatrix}, \ \overline{H}_2 = \begin{pmatrix} 1 & 1 \\ 0 & 0 \end{pmatrix}, \\ &\overline{H}_3 = \begin{pmatrix} 1 & 0 \\ 1 & 0 \end{pmatrix}, \ \overline{H}_4 = \begin{pmatrix} 1 & 1 \\ 1 & 0 \end{pmatrix}, \ \overline{H}_5 = \begin{pmatrix} 1 & 1 \\ 1 & 1 \end{pmatrix} \Big\},\end{aligned}$$

$$\overline{H}_k = \overline{H}_k(\overline{\mathbf{E}}_k, \overline{\mathbf{F}}_k), \quad 0 \le k \le 5.$$

By Theorems 1.4 and 1.7,

$$(\mathbf{A},\mathbf{B}) = \sum_{k=0}^{5} \alpha_k (\overline{\mathbf{E}}_k, \overline{\mathbf{F}}_k), \tag{1.19}$$

$$X(\mathbf{A},\mathbf{B}) = \sum_{k=0}^{5} \alpha_k \overline{H}_k (\overline{\mathbf{E}}_k, \overline{\mathbf{F}}_k), \tag{1.20}$$

$$\alpha_k \ge 0, \quad 0 \le k \le 5, \quad \sum_{k=0}^{5} \alpha_k = 1.$$

Treating (1.19) and (1.20) as linear equations in α_k $(0 \le k \le 5)$, we construct general formulas for the vector pair $(\mathbf{A},\mathbf{B})$ and matrix $X(\mathbf{A},\mathbf{B})$:

$$\begin{aligned}((3/2,1/2),(5/4,3/4)) = {} & \alpha_1 ((1,0),(1,0)) + \alpha_2 ((2,0),(1,1)) + \\ & \alpha_3 ((1,1),(2,0)) + \alpha_4 ((2,1),(2,1)) + \alpha_5 ((2,2),(2,2)),\end{aligned}$$

$$\begin{pmatrix} 7/8 & 5/8 \\ 3/8 & 1/8 \end{pmatrix} =$$
$$\alpha_1 \begin{pmatrix} 1 & 0 \\ 0 & 0 \end{pmatrix} + \alpha_2 \begin{pmatrix} 1 & 1 \\ 0 & 0 \end{pmatrix} + \alpha_3 \begin{pmatrix} 1 & 0 \\ 1 & 0 \end{pmatrix} + \alpha_4 \begin{pmatrix} 1 & 1 \\ 1 & 0 \end{pmatrix} + \alpha_5 \begin{pmatrix} 1 & 1 \\ 1 & 1 \end{pmatrix}.$$

These expressions are equivalent to the linear equations

$$\begin{cases} \alpha_1 + 2\alpha_2 + \alpha_3 + 2\alpha_4 + 2\alpha_5 = 3/2, \\ \alpha_3 + \alpha_4 + 2\alpha_5 = 1/2, \\ \alpha_1 + \alpha_2 + 2\alpha_3 + 2\alpha_4 + 2\alpha_5 = 5/4, \\ \alpha_2 + \alpha_4 + 2\alpha_5 = 3/4, \end{cases} \tag{1.21}$$

$$\begin{cases} \alpha_1 + \alpha_2 + \alpha_3 + \alpha_4 + \alpha_5 = 7/8, \\ \alpha_2 + \alpha_4 + \alpha_5 = 5/8, \\ \alpha_3 + \alpha_4 + \alpha_5 = 3/8, \\ \alpha_5 = 1/8. \end{cases} \tag{1.22}$$

Each solution to (1.22) is a solution to (1.21), and the set of solutions to (1.21) defines the set $\widetilde{M}(\mathbf{A}, \mathbf{B}; 1)$, where $\mathbf{A} = (3/2, 1/2)$ and $\mathbf{B} = (5/4, 3/4)$. Solving the set of equations (1.21) for the independent variables α_1 and α_2 and (1.22) for α_1, we obtain

$$\alpha_3 = \alpha_2 - \frac{1}{4}, \quad \alpha_4 = 1 - \alpha_1 - 2\alpha_2, \quad \alpha_5 = \frac{1}{2}\alpha_1 + \frac{1}{2}\alpha_2 - \frac{1}{8};$$

$$\alpha_2 = \frac{1}{2} - \alpha_1, \quad \alpha_3 = \frac{1}{4} - \alpha_1, \quad \alpha_4 = \alpha_1, \quad \alpha_5 = \frac{1}{8}.$$

Taking into account the constraints for α_k ($0 \le k \le 5$), we write

$$\begin{gathered} ((3/2, 1/2), (5/4, 3/4)) = \\ \alpha_1 ((1,0),(1,0)) + \alpha_2 ((2,0),(1,1)) + (\alpha_2 - \tfrac{1}{4}) ((1,1),(2,0)) + \\ (1 - \alpha_1 - 2\alpha_2) ((2,1),(2,1)) + (\tfrac{1}{2}\alpha_1 + \tfrac{1}{2}\alpha_2 - \tfrac{1}{8}) ((2,2),(2,2)), \\ \alpha_1 \ge 0, \quad \alpha_2 \ge \frac{1}{4}, \quad \alpha_1 + 2\alpha_2 \le 1; \end{gathered} \tag{1.23}$$

$$\begin{gathered} \begin{pmatrix} 7/8 & 5/8 \\ 3/8 & 1/8 \end{pmatrix} = \alpha_1 \begin{pmatrix} 1 & 0 \\ 0 & 0 \end{pmatrix} + (\tfrac{1}{2} - \alpha_1) \begin{pmatrix} 1 & 1 \\ 0 & 0 \end{pmatrix} + \\ (\tfrac{1}{4} - \alpha_1) \begin{pmatrix} 1 & 0 \\ 1 & 0 \end{pmatrix} + \alpha_1 \begin{pmatrix} 1 & 1 \\ 1 & 0 \end{pmatrix} + \tfrac{1}{8} \begin{pmatrix} 1 & 1 \\ 1 & 1 \end{pmatrix}, \\ 0 \le \alpha_1 \le \frac{1}{4}. \end{gathered} \tag{1.24}$$

Equation (1.23) and constraints for α_1 and α_2 summarize all representations of the given vector pair in terms of extremal pairs and, therefore, determines all uniform matrices in the set $\widetilde{M}\left(((3/2,1/2),(5/4,3/4))\,;1\right)$. Equation (1.24) and constraints for α_1 summarize all representations of the given matrix in terms of extremal matrices. For the extreme values of α_1 ($\alpha_1 = 0$ and $\alpha_1 = 1/4$), we have

$$\begin{pmatrix} 7/8 & 5/8 \\ 3/8 & 1/8 \end{pmatrix} = \frac{1}{2}\begin{pmatrix} 1 & 1 \\ 0 & 0 \end{pmatrix} + \frac{1}{4}\begin{pmatrix} 1 & 0 \\ 1 & 0 \end{pmatrix} + \frac{1}{8}\begin{pmatrix} 1 & 1 \\ 1 & 1 \end{pmatrix},$$

$$\begin{pmatrix} 7/8 & 5/8 \\ 3/8 & 1/8 \end{pmatrix} = \frac{1}{4}\begin{pmatrix} 1 & 0 \\ 0 & 0 \end{pmatrix} + \frac{1}{4}\begin{pmatrix} 1 & 1 \\ 0 & 0 \end{pmatrix} + \frac{1}{4}\begin{pmatrix} 1 & 1 \\ 1 & 0 \end{pmatrix} + \frac{1}{8}\begin{pmatrix} 1 & 1 \\ 1 & 1 \end{pmatrix}.$$

The following proposition applies to the sets $M(n^+,m)$, $M(n,m^+)$, and $M(n^+,m^+)$, which are subpolyhedrons of the fundamental transportation polyhedron $M(n,m)$.

Theorem 1.8. *The polyhedrons $M(n^+,m)$, $M(n,m^+)$, and $M(n^+,m^+)$ defined by the vertex sets $\overline{E}_{n^+,m}$, $\overline{E}_{n,m^+}$, and $\overline{E}_{n^+,m^+}$, respectively, can be expressed as follows:*

$$M(n^+,m) = \Big\{X(\mathbf{A},\mathbf{B}) : \mathbf{A} \in \overline{R}^n_+,\ \mathbf{B} \in \overline{R}^m_+,\ b_1 = n,\\ X(\mathbf{A},\mathbf{B}) \in \widetilde{M}(\mathbf{A},\mathbf{B};1) \neq \varnothing\Big\},$$

$$M(n,m^+) = \Big\{X(\mathbf{A},\mathbf{B}) : \mathbf{A} \in \overline{R}^n_+,\ a_1 = m,\ \mathbf{B} \in \overline{R}^m_+,\\ X(\mathbf{A},\mathbf{B}) \in \widetilde{M}(\mathbf{A},\mathbf{B};1) \neq \varnothing\Big\},$$

$$M(n^+,m^+) = \Big\{X(\mathbf{A},\mathbf{B}) : \mathbf{A} \in \overline{R}^n_+,\ a_1 = m,\ \mathbf{B} \in \overline{R}^m_+,\ b_1 = n,\\ X(\mathbf{A},\mathbf{B}) \in \widetilde{M}(\mathbf{A},\mathbf{B};1) \neq \varnothing\Big\},$$

where $\mathbf{A} = (a_1,\dots,a_n)$ and $\mathbf{B} = (b_1,\dots,b_m)$.

For a vector pair $(\mathbf{A},\mathbf{B}) \in L(n,m)$, we denote by

$$\mathfrak{A}(\mathbf{A},\mathbf{B}) = \Big\{(\alpha_0,\dots,\alpha_{C^n_{n+m}-1}) : (\mathbf{A},\mathbf{B}) = \sum_{k=0}^{C^n_{n+m}-1} \alpha_k(\overline{\mathbf{E}}_k,\overline{\mathbf{F}}_k),\\ (\overline{\mathbf{E}}_k,\overline{\mathbf{F}}_k) \in \overline{E}^{n,m},\ \alpha_k \geq 0,\ 0 \leq k \leq C^n_{n+m}-1,\ \sum_{k=0}^{C^n_{n+m}-1} \alpha_k = 1\Big\}$$

the set of solutions $(\alpha_0,\dots,\alpha_{C^n_{n+m}-1})$ that define all representations of $(\mathbf{A},\mathbf{B})$ in terms of extremal pairs. The set $\mathfrak{A}(\mathbf{A},\mathbf{B})$ determines all uniform matrices in $\widetilde{M}(\mathbf{A}, \mathbf{B}; 1)$. Combining Theorems 1.4 and 1.7, we obtain the following proposition.

Theorem 1.9. *For an arbitrary vector pair* $(\mathbf{A}, \mathbf{B})$ *from* $L(n, m)$, *the set* $\mathfrak{A}(\mathbf{A}, \mathbf{B})$ *defines the matrix set* $\widetilde{M}(\mathbf{A}, \mathbf{B}; 1)$ *as*

$$\widetilde{M}(\mathbf{A}, \mathbf{B}; 1) = \Big\{ X(\mathbf{A}, \mathbf{B}) : X(\mathbf{A}, \mathbf{B}) = \sum_{k=0}^{C_{n+m}^{n}-1} \alpha_k \overline{H}_k(\overline{\mathbf{E}}_k, \overline{\mathbf{F}}_k),$$

$$\overline{H}_k(\overline{\mathbf{E}}_k, \overline{\mathbf{F}}_k) \in \overline{E}_{n,m},\ 0 \le k \le C_{n+m}^{n} - 1,\ (\alpha_0, \ldots, \alpha_{C_{n+m}^{n}-1}) \in \mathfrak{A}(\mathbf{A}, \mathbf{B}) \Big\}.$$

Note that different vectors in $\mathfrak{A}(\mathbf{A}, \mathbf{B})$ may define the same matrix $X(\mathbf{A}, \mathbf{B})$.

Applying Theorem 1.9 to the vector pair $\big((3/2, 1/2), (5/4, 3/4)\big)$ from Example 1.1, we see that (1.23) defines the set $\widetilde{M}\left(\left((3/2, 1/2), (5/4, 3/4)\right); 1\right)$ of uniform matrices:

$$\widetilde{M}\left(\left((3/2,1/2),(5/4,3/4)\right);1\right) = \Big\{ X\left((3/2,1/2),(5/4,3/4)\right) :$$

$$X\left((3/2,1/2),(5/4,3/4)\right) = \alpha_1 \begin{pmatrix} 1 & 0 \\ 0 & 0 \end{pmatrix} + \alpha_2 \begin{pmatrix} 1 & 1 \\ 0 & 0 \end{pmatrix} +$$

$$(\alpha_2 - \tfrac{1}{4}) \begin{pmatrix} 1 & 0 \\ 1 & 0 \end{pmatrix} + (1 - \alpha_1 - 2\alpha_2) \begin{pmatrix} 1 & 1 \\ 1 & 0 \end{pmatrix} + (\tfrac{1}{2}\alpha_1 + \tfrac{1}{2}\alpha_2 - \tfrac{1}{8}) \begin{pmatrix} 1 & 1 \\ 1 & 1 \end{pmatrix},$$

$$\alpha_1 \ge 0, \quad \alpha_2 \ge \tfrac{1}{4}, \quad \alpha_1 + 2\alpha_2 \le 1 \Big\}.$$

1.4. The Case of an Arbitrary Uniform Matrix

In the preceding subsections, we considered transportation matrices with elements not greater than unity. Now, we revoke this restriction.

Suppose that $c > 0$, the vectors

$$\mathbf{A} = (a_1, \ldots, a_n) \in \overline{R}_{+}^{n}, \quad \mathbf{B} = (b_1, \ldots, b_m) \in \overline{R}_{+}^{m}$$

are characterized by equal sums of components, and

$$\delta_k(\mathbf{A}, \mathbf{B}; c) = \sum_{j=1}^{m} b_j - \sum_{b_j \ge ck} (b_j - ck) - \sum_{i=1}^{k} a_i \ge 0.$$

The fact that the characteristic function $\delta_k(\mathbf{A}, \mathbf{B}; c)$ is nonnegative implies that $M(\mathbf{A}, \mathbf{B}; c) \ne \varnothing$ (recall Theorem 2.1 from Chapter 1). Then, the vectors

$\mathbf{A}/c = (a_1/c, \ldots, a_n/c) \in \overline{R}^n_+$ and $\mathbf{B}/c = (b_1/c, \ldots, b_m/c) \in \overline{R}^m_+$ are such that

$$\delta_k(\mathbf{A}/c, \mathbf{B}/c; 1) = \frac{1}{c}\delta_k(\mathbf{A}, \mathbf{B}; c) \geq 0, \quad M(\mathbf{A}/c, \mathbf{B}/c; 1) \neq \varnothing,$$

and the matrix sets $M(\mathbf{A}, \mathbf{B}; c)$ and $M(\mathbf{A}/c, \mathbf{B}/c; 1)$ are in one-to-one correspondence defined by the following condition: $(x_{ij}) \in M(\mathbf{A}, \mathbf{B}; c)$ if and only if $(x_{ij}/c) \in M(\mathbf{A}/c, \mathbf{B}/c; 1)$. Therefore, the results obtained in the preceding subsections of this section can be summarized to state three theorems, which establish that any pair of vectors and any uniform matrix belong to the respective fundamental polyhedrons.

Theorem 1.10.

(a) *For any $c > 0$ and any vector pair $(\mathbf{A}, \mathbf{B})$ such that $\mathbf{A} \in \overline{R}^n_+$, $\mathbf{B} \in \overline{R}^m_+$, and $M(\mathbf{A}, \mathbf{B}; c) \neq \varnothing$, it holds that $(\mathbf{A}/c, \mathbf{B}/c) \in L(n, m)$.*

(b) *Any vector pair $(\mathbf{A}, \mathbf{B})$ such that $\mathbf{A} \in \overline{R}^n_+$, $\mathbf{B} \in \overline{R}^m_+$, and $M(\mathbf{A}, \mathbf{B}; c) \neq \varnothing$ can be represented as a linear combination of extremal vector pairs:*

$$(\mathbf{A}, \mathbf{B}) = c\sum_{k=0}^{C^n_{n+m}-1} \alpha_k(\overline{\mathbf{E}}_k, \overline{\mathbf{F}}_k), \quad (\overline{\mathbf{E}}_k, \overline{\mathbf{F}}_k) \in \overline{E}^{n,m},$$

$$\alpha_k \geq 0, \quad 0 \leq k \leq C^n_{n+m} - 1, \quad \sum_{k=0}^{C^n_{n+m}-1} \alpha_k = 1.$$

Theorem 1.11.

(a) *Suppose that $M(\mathbf{A}, \mathbf{B}; c)$, where $\mathbf{A} \in \overline{R}^n_+$, $\mathbf{B} \in \overline{R}^m_+$, and $c > 0$, is a nonempty matrix set. Then, for any uniform matrix $X(\mathbf{A}/c, \mathbf{B}/c) \in \widetilde{M}(\mathbf{A}/c, \mathbf{B}/c; 1)$, it holds that $X(\mathbf{A}/c, \mathbf{B}/c) \in M(n, m)$.*

(b) *If $M(\mathbf{A}, \mathbf{B}; c) \neq \varnothing$, where $\mathbf{A} \in \overline{R}^n_+$ and $\mathbf{B} \in \overline{R}^m_+$, then any uniform matrix $X(\mathbf{A}, \mathbf{B}) \in \widetilde{M}(\mathbf{A}, \mathbf{B}; c)$ can be represented as a linear combination of extremal matrices:*

$$X(\mathbf{A}, \mathbf{B}) = c\sum_{k=0}^{C^n_{n+m}-1} \alpha_k \overline{H}_k(\overline{\mathbf{E}}_k, \overline{\mathbf{F}}_k), \quad \overline{H}_k(\overline{\mathbf{E}}_k, \overline{\mathbf{F}}_k) \in \overline{E}_{n,m},$$

$$\alpha_k \geq 0, \quad 0 \leq k \leq C^n_{n+m} - 1, \quad \sum_{k=0}^{C^n_{n+m}-1} \alpha_k = 1.$$

Theorem 1.12. *Let* $(\mathbf{A},\mathbf{B})$ *be an arbitrary vector pair such that* $\mathbf{A}\in\overline{R}^n_+$, $\mathbf{B}\in\overline{R}^m_+$, *and* $M(\mathbf{A},\mathbf{B};c)\neq\varnothing$, *where* $c>0$. *Then, the set* $\mathfrak{A}(\mathbf{A}/c,\mathbf{B}/c)$ *defines a set* $\widetilde{M}(\mathbf{A},\mathbf{B};c)$ *of uniform matrices as*

$$\widetilde{M}(\mathbf{A},\mathbf{B};c)=\Big\{X(\mathbf{A},\mathbf{B}):X(\mathbf{A},\mathbf{B})=c\sum_{k=0}^{C^n_{n+m}-1}\alpha_k\overline{H}_k(\overline{\mathbf{E}}_k,\overline{\mathbf{F}}_k),$$

$$\overline{H}_k(\overline{\mathbf{E}}_k,\overline{\mathbf{F}}_k)\in\overline{E}_{n,m},\ 0\le k\le C^n_{n+m}-1,$$

$$(\alpha_0,\dots,\alpha_{C^n_{n+m}-1})\in\mathfrak{A}(\mathbf{A}/c,\mathbf{B}/c)\Big\}.$$

We should make the following remark concerning Theorems 1.10–1.12. If $M(\mathbf{A},\mathbf{B};c)\neq\varnothing$, where $c<1$, then the vector pair $(\mathbf{A},\mathbf{B})$ and any matrix $X(\mathbf{A},\mathbf{B})$ in $\widetilde{M}(\mathbf{A},\mathbf{B};c)$ belong to the polyhedrons $L(n,m)$ and $M(n,m)$, respectively. Therefore, these theorems are also of interest in the case when $c>1$ and $M(\mathbf{A},\mathbf{B};1)=\varnothing$.

Example 1.2 (Extension of Example 1.1). Consider the following vector pair and matrix:

$$((12,4),(10,6))=8\,((3/2,1/2),(5/4,3/4))\,,$$

$$\begin{pmatrix}7&5\\3&1\end{pmatrix}=8\begin{pmatrix}7/8&5/8\\3/8&1/8\end{pmatrix}.$$

They are derived from the vector pair and matrix considered in Example 1.1. It follows from (1.23) and (1.24) that

$$\begin{aligned}((12,4),(10,6))=8\Big[&\alpha_1\,((1,0),(1,0))+\alpha_2\,((2,0),(1,1))+\\&(\alpha_2-\tfrac14)\,((1,1),(2,0))+(1-\alpha_1-2\alpha_2)\,((2,1),(2,1))+\\&(\tfrac12\alpha_1+\tfrac12\alpha_2-\tfrac18)\,((2,2),(2,2))\Big],\\&\alpha_1\ge0,\quad\alpha_2\ge\frac14,\quad\alpha_1+2\alpha_2\le1;\end{aligned}\tag{1.25}$$

$$\begin{aligned}\begin{pmatrix}7&5\\3&1\end{pmatrix}=8\Big[&\alpha_1\begin{pmatrix}1&0\\0&0\end{pmatrix}+(\tfrac12-\alpha_1)\begin{pmatrix}1&1\\0&0\end{pmatrix}+\\&(\tfrac14-\alpha_1)\begin{pmatrix}1&0\\1&0\end{pmatrix}+\alpha_1\begin{pmatrix}1&1\\1&0\end{pmatrix}+\tfrac18\begin{pmatrix}1&1\\1&1\end{pmatrix}\Big],\\&0\le\alpha_1\le\frac14.\end{aligned}\tag{1.26}$$

Equation (1.26) and constraints for the independent variable α_1 express all representations of the given uniform matrix in terms of extremal matrices with $c = 8$. Equation (1.25) and constraints for the independent variables α_1 and α_2 express the vector pair in terms of extremal pairs and defines the matrix set $\widetilde{M}\left(((12,4),(10,6))\,;8\right)$ as

$$\widetilde{M}\left(((12,4),(10,6))\,;8\right) = \Bigg\{X\left((12,4),(10,6)\right) : X\left((12,4),(10,6)\right) =$$

$$8\left[\alpha_1\begin{pmatrix}1&0\\0&0\end{pmatrix} + \alpha_2\begin{pmatrix}1&1\\0&0\end{pmatrix} + (\alpha_2 - \tfrac{1}{4})\begin{pmatrix}1&0\\1&0\end{pmatrix} + \right.$$

$$\left.(1-\alpha_1-2\alpha_2)\begin{pmatrix}1&1\\1&0\end{pmatrix} + (\tfrac{1}{2}\alpha_1 + \tfrac{1}{2}\alpha_1 - \tfrac{1}{8})\begin{pmatrix}1&1\\1&1\end{pmatrix}\right],$$

$$\alpha_1 \geq 0,\ \alpha_2 \geq \tfrac{1}{4},\ \alpha_1 + 2\alpha_2 \leq 1\Bigg\}.$$

1.5. The Case of a Minimax Uniform Matrix

We now demonstrate the decrease in the number of coefficients in the formula for the convex linear combination in the case of a minimax matrix.

Let $(\mathbf{A},\mathbf{B})$ be an arbitrary pair of vectors such that $\mathbf{A} \in \overline{R}^n_+$, $\mathbf{B} \in \overline{R}^m_+$, and $M(\mathbf{A},\mathbf{B};c) \neq \varnothing$. By Theorems 1.10 and 1.11, for the vector pair $(\mathbf{A},\mathbf{B})$ and any uniform matrix $X(\mathbf{A},\mathbf{B}) \in \widetilde{M}(\mathbf{A},\mathbf{B};c)$, it holds that

$$\begin{gathered}(\mathbf{A},\mathbf{B}) = c\sum_{k=0}^{C^n_{n+m}-1}\alpha_k(\overline{\mathbf{E}}_k,\overline{\mathbf{F}}_k), \quad (\overline{\mathbf{E}}_k,\overline{\mathbf{F}}_k) \in \overline{E}^{n,m}, \\ \alpha_k \geq 0, \quad 0 \leq k \leq C^n_{n+m}-1, \quad \sum_{k=0}^{C^n_{n+m}-1}\alpha_k = 1;\end{gathered} \tag{1.27}$$

$$\begin{gathered}X(\mathbf{A},\mathbf{B}) = c\sum_{k=0}^{C^n_{n+m}-1}\alpha_k\overline{H}_k(\overline{\mathbf{E}}_k,\overline{\mathbf{F}}_k), \quad \overline{H}_k(\overline{\mathbf{E}}_k,\overline{\mathbf{F}}_k) \in \overline{E}_{n,m}, \\ \alpha_k \geq 0, \quad 0 \leq k \leq C^n_{n+m}-1, \quad \sum_{k=0}^{C^n_{n+m}-1}\alpha_k = 1.\end{gathered} \tag{1.28}$$

Recall that each solution α_k $(0 \leq k \leq C^n_{n+m}-1)$ of (1.28) is a solution of (1.27). The set of solutions α_k to (1.27) $(0 \leq k \leq C^n_{n+m}-1)$ defines all uniform matrices in $\widetilde{M}(\mathbf{A},\mathbf{B};c)$ by Theorem 1.12.

Knowing the components of $\mathbf{A} = (a_1,\ldots,a_n)$ and $\mathbf{B} = (b_1,\ldots,b_m)$, one can determine the coefficients α_k $(0 \leq k \leq C^n_{n+m}-1)$ that vanish in any representations (1.27) and (1.28). The following proposition is obviously true.

Lemma 1.9. *Suppose that vectors* $\mathbf{A} = (a_1, \ldots, a_n) \in \overline{R}^n_+$ *and* $\mathbf{B} = (b_1, \ldots, b_m) \in \overline{R}^m_+$, *where* $M(\mathbf{A}, \mathbf{B}; c) \neq \varnothing$, *are such that* $a_p = a_{p+1}$ *or* $b_q = b_{q+1}$ *for some* p *or* q $(1 \leq p \leq n-1,\ 1 \leq q \leq m+1)$. *Then, a coefficient* α_k *in decompositions* (1.27) *and* (1.28) *is zero if the corresponding extremal vector pair* $(\overline{\mathbf{E}}_k, \overline{\mathbf{F}}_k) = ((e_1^{(k)}, \ldots, e_n^{(k)}), (f_1^{(k)}, \ldots, f_m^{(k)}))$ *is such that* $e_p^{(k)} > e_{p+1}^{(k)}$ *or* $f_q^{(k)} > f_{q+1}^{(k)}$.

Lemma 1.9 can be applied to reduce the number of coefficients in (1.27) and (1.28). The numbers of coefficients in these decompositions can be further reduced in the case of a minimax matrix (i.e., when $c = c(\mathbf{A}, \mathbf{B})$). Note that if $c = c(\mathbf{A}, \mathbf{B})$ in (1.27) and (1.28), then the vector pair $(\mathbf{A}/c, \mathbf{B}/c)$ and the uniform matrix $X(\mathbf{A}/c, \mathbf{B}/c) \in \widetilde{M}(\mathbf{A}/c, \mathbf{B}/c; 1)$ are boundary points of the fundamental polyhedrons $L(n, m)$ and $M(n, m)$, respectively (provided that $c(\mathbf{A}, \mathbf{B}) \neq 0$). Accordingly, decompositions (1.27) and (1.28) contain only the extremal vector pairs and extremal matrices that are the edges of the fundamental polyhedrons $L(n, m)$ and $M(n, m)$. These edges contain $(\mathbf{A}/c, \mathbf{B}/c)$ and $X(\mathbf{A}/c, \mathbf{B}/c)$, respectively.

Consider vectors $\mathbf{A} = (a_1, \ldots, a_n) \in \overline{R}^n_+$ and $\mathbf{B} = (b_1, \ldots, b_m) \in \overline{R}^m_+$ satisfying the balance condition, and suppose that $M(\mathbf{A}, \mathbf{B}; c) \neq \varnothing$, where $c = c(\mathbf{A}, \mathbf{B})$. Then, by Lemma 1.1 from Chapter 2, there exists $p \in Z$ $(1 \leq p \leq b_1)$ such that $\delta_p(\mathbf{A}, \mathbf{B}; c(\mathbf{A}, \mathbf{B})) = 0$ and

$$\delta_p(\mathbf{A}, \mathbf{B}; c(\mathbf{A}, \mathbf{B})) = \sum_{j > l_p(\mathbf{B}; c(\mathbf{A}, \mathbf{B}))} b_j + c(\mathbf{A}, \mathbf{B}) l_p(\mathbf{B}; c(\mathbf{A}, \mathbf{B})) - \sum_{i=1}^{p} a_i = 0,$$

where $l_p(\mathbf{B}; c(\mathbf{A}, \mathbf{B})) = \max\{j : b_j \geq c(\mathbf{A}, \mathbf{B})p,\ 1 \leq j \leq m\}$. The values of p and $l_p(\mathbf{B}; c(\mathbf{A}, \mathbf{B}))$ define the submatrices of every matrix in $M(\mathbf{A}, \mathbf{B}; c(\mathbf{A}, \mathbf{B}))$ whose elements are $c(\mathbf{A}, \mathbf{B})$ and 0, respectively. If $X(\mathbf{A}, \mathbf{B}) = (x_{ij})$ is an arbitrary matrix in $M(\mathbf{A}, \mathbf{B}; c(\mathbf{A}, \mathbf{B}))$, then

$$x_{ij} = \begin{cases} c(\mathbf{A}, \mathbf{B}), & 1 \leq i \leq p, \quad 1 \leq j \leq l_p(\mathbf{B}; c(\mathbf{A}, \mathbf{B})), \\ 0, & i > p, \quad j > l_p(\mathbf{B}; c(\mathbf{A}, \mathbf{B})) \end{cases}$$

by Theorem 1.2 from Chapter 1.

Let us construct the matrices $X' = (x'_{ij})$ and $X'' = (x''_{ij})$ $(1 \leq i \leq n,\ 1 \leq j \leq m)$ defined by $c(\mathbf{A}, \mathbf{B})$, p, and $l_p(\mathbf{B}; c(\mathbf{A}, \mathbf{B}))$ as follows:

$$x'_{ij} = \begin{cases} c(\mathbf{A}, \mathbf{B}), & 1 \leq i \leq p, \quad 1 \leq j \leq l_p(\mathbf{B}; c(\mathbf{A}, \mathbf{B})), \\ 0, & 1 \leq i \leq p, \quad j > l_p(\mathbf{B}; c(\mathbf{A}, \mathbf{B})), \\ 0, & i > p, \quad 1 \leq j \leq m, \end{cases}$$

$$x''_{ij} = \begin{cases} c(\mathbf{A},\mathbf{B}), & 1 \le i \le p, \quad 1 \le j \le m, \\ c(\mathbf{A},\mathbf{B}), & i > p, \quad 1 \le j \le l_p(\mathbf{B}; c(\mathbf{A},\mathbf{B})), \\ 0, & i > p, \quad j > l_p(\mathbf{B}; c(\mathbf{A},\mathbf{B})). \end{cases}$$

Replacing every element equal to $c(\mathbf{A},\mathbf{B})$ in both X' and X'' with unity, we obtain extremal matrices $R = (r_{ij})$ and $O = (o_{ij})$, which are used to construct kernels and hulls in Theorems 3.3 and 3.5 from Chapter 4:

$$r_{ij} = \begin{cases} 1, & 1 \le i \le p, \quad 1 \le j \le l_p(\mathbf{B}; c(\mathbf{A},\mathbf{B})), \\ 0, & 1 \le i \le p, \quad j > l_p(\mathbf{B}; c(\mathbf{A},\mathbf{B})), \\ 0, & i > p, \quad 1 \le j \le m, \end{cases}$$

$$o_{ij} = \begin{cases} 1, & 1 \le i \le p, \quad 1 \le j \le m, \\ 1, & qi > p, \quad 1 \le j \le l_p(\mathbf{B}; c(\mathbf{A},\mathbf{B})), \\ 0, & i > p, \quad j > l_p(\mathbf{B}; c(\mathbf{A},\mathbf{B})). \end{cases}$$

We introduce a new notation for the matrices R and O: $R = \overline{H}'$, $O = \overline{H}''$, $(r_{ij}) = (h'_{ij})$, $(o_{ij}) = (h''_{ij})$. It is clear that the four matrices thus constructed are such that $X' = X'(\mathbf{A}',\mathbf{B}')$, $X'' = X''(\mathbf{A}'',\mathbf{B}'')$, $\overline{H}' = \overline{H}'(\overline{\mathbf{E}}',\overline{\mathbf{F}}')$, and $\overline{H}'' = \overline{H}''(\overline{\mathbf{E}}'',\overline{\mathbf{F}}'')$, where

$$\mathbf{A}' = \Big(\underbrace{c(\mathbf{A},\mathbf{B})l_p(\mathbf{B}; c(\mathbf{A},\mathbf{B}))\,, \ldots,\, c(\mathbf{A},\mathbf{B})l_p(\mathbf{B}; c(\mathbf{A},\mathbf{B}))}_{p}\,,\, \underbrace{0, \ldots, 0}_{n-p}\Big),$$

$$\mathbf{B}' = \Big(\underbrace{c(\mathbf{A},\mathbf{B})p\,, \ldots,\, c(\mathbf{A},\mathbf{B})p}_{l_p(\mathbf{B};c(\mathbf{A},\mathbf{B}))}\,,\, \underbrace{0, \ldots\ldots, 0}_{m-l_p(\mathbf{B};c(\mathbf{A},\mathbf{B}))}\Big),$$

$$\mathbf{A}'' = \Big(\underbrace{c(\mathbf{A},\mathbf{B})m\,, \ldots,\, c(\mathbf{A},\mathbf{B})m}_{p}\,,\, \underbrace{c(\mathbf{A},\mathbf{B})l_p(\mathbf{B}; c(\mathbf{A},\mathbf{B}))\,, \ldots,\, c(\mathbf{A},\mathbf{B})l_p(\mathbf{B}; c(\mathbf{A},\mathbf{B}))}_{n-p}\Big),$$

$$\mathbf{B}'' = \Big(\underbrace{c(\mathbf{A},\mathbf{B})n\,, \ldots,\, c(\mathbf{A},\mathbf{B})n}_{l_p(\mathbf{B};c(\mathbf{A},\mathbf{B}))}\,,\, \underbrace{c(\mathbf{A},\mathbf{B})p\,, \ldots,\, c(\mathbf{A},\mathbf{B})p}_{m-l_p(\mathbf{B};c(\mathbf{A},\mathbf{B}))}\Big),$$

$$\overline{\mathbf{E}}' = \Big(\underbrace{l_p(\mathbf{B}; c(\mathbf{A},\mathbf{B}))\,, \ldots,\, l_p(\mathbf{B}; c(\mathbf{A},\mathbf{B}))}_{p}\,,\, \underbrace{0, \ldots, 0}_{n-p}\Big),$$

$$\overline{\mathbf{F}}' = \Big(\underbrace{p, \ldots\ldots, p}_{l_p(\mathbf{B};c(\mathbf{A},\mathbf{B}))}\,,\, \underbrace{0, \ldots\ldots, 0}_{m-l_p(\mathbf{B};c(\mathbf{A},\mathbf{B}))}\Big),$$

$$\overline{\mathbf{E}}'' = \big(\underbrace{m, \dots, m}_{p}, \underbrace{l_p(\mathbf{B}; c(\mathbf{A}, \mathbf{B})), \dots, l_p(\mathbf{B}; c(\mathbf{A}, \mathbf{B}))}_{n-p}\big),$$

$$\overline{\mathbf{F}}'' = \big(\underbrace{n, \dots\dots, n}_{l_p(\mathbf{B};c(\mathbf{A},\mathbf{B}))}, \underbrace{p, \dots\dots\dots, p}_{m-l_p(\mathbf{B};c(\mathbf{A},\mathbf{B}))}\big).$$

It is obvious that the four matrices and the vector pairs that define these matrices satisfy the relations

$$X' = c(\mathbf{A}, \mathbf{B})\overline{H}', \quad X'' = c(\mathbf{A}, \mathbf{B})\overline{H}'',$$

$$(\mathbf{A}', \mathbf{B}') = c(\mathbf{A}, \mathbf{B})(\overline{\mathbf{E}}', \overline{\mathbf{F}}'), \quad (\mathbf{A}'', \mathbf{B}'') = c(\mathbf{A}, \mathbf{B})(\overline{\mathbf{E}}'', \overline{\mathbf{F}}'')$$

and

$$(\overline{\mathbf{E}}', \overline{\mathbf{F}}') < (\overline{\mathbf{E}}'', \overline{\mathbf{F}}''), \quad \overline{H}' \subset \overline{H}''.$$

Consider representations (1.27) and (1.28) of the vector pair $(\mathbf{A}, \mathbf{B})$ and a uniform matrix $X(\mathbf{A}, \mathbf{B})$ from $\widetilde{M}(\mathbf{A}, \mathbf{B}; c(\mathbf{A}, \mathbf{B}))$ in terms of extremal pairs and matrices that correspond to the minimax value $c = c(\mathbf{A}, \mathbf{B})$:

$$\begin{gathered}(\mathbf{A}, \mathbf{B}) = c(\mathbf{A}, \mathbf{B}) \sum_{k=0}^{C_{n+m}^n - 1} \alpha_k (\overline{\mathbf{E}}_k, \overline{\mathbf{F}}_k), \quad (\overline{\mathbf{E}}_k, \overline{\mathbf{F}}_k) \in \overline{E}^{n,m}, \\ \alpha_k \geq 0, \quad 0 \leq k \leq C_{n+m}^n - 1, \quad \sum_{k=0}^{C_{n+m}^n - 1} \alpha_k = 1;\end{gathered} \tag{1.29}$$

$$\begin{gathered}X(\mathbf{A}, \mathbf{B}) = c(\mathbf{A}, \mathbf{B}) \sum_{k=0}^{C_{n+m}^n - 1} \alpha_k \overline{H}_k (\overline{\mathbf{E}}_k, \overline{\mathbf{F}}_k), \quad \overline{H}_k(\overline{\mathbf{E}}_k, \overline{\mathbf{F}}_k) \in \overline{E}_{n,m}, \\ \alpha_k \geq 0, \quad 0 \leq k \leq C_{n+m}^n - 1, \quad \sum_{k=0}^{C_{n+m}^n - 1} \alpha_k = 1.\end{gathered} \tag{1.30}$$

The following theorem is obviously true.

Theorem 1.13. *Suppose that* $c(\mathbf{A}, \mathbf{B})$ *is the minimax value for a vector pair* $\mathbf{A} = (a_1, \dots, a_n) \in \overline{R}_+^n$, $\mathbf{B} = (b_1, \dots, b_m) \in \overline{R}_+^m$ *and* $\delta_p(\mathbf{A}, \mathbf{B}; c(\mathbf{A}, \mathbf{B})) = 0$ *for some* $p \in Z$ $(1 \leq p \leq b_1)$. *Then, the following assertions are true.*

(a) *In representations* (1.29) *and* (1.30), $\alpha_k = 0$ *for* k *such that* $(\overline{\mathbf{E}}_k, \overline{\mathbf{F}}_k) < (\overline{\mathbf{E}}', \overline{\mathbf{F}}')$, $\overline{H}_k(\overline{\mathbf{E}}_k, \overline{\mathbf{F}}_k) \subset \overline{H}'(\overline{\mathbf{E}}', \overline{\mathbf{F}}')$ *or* $(\overline{\mathbf{E}}'', \overline{\mathbf{F}}'') < (\overline{\mathbf{E}}_k, \overline{\mathbf{F}}_k)$, $\overline{H}''(\overline{\mathbf{E}}'', \overline{\mathbf{F}}'') \subset \overline{H}_k(\overline{\mathbf{E}}_k, \overline{\mathbf{F}}_k)$.

(b) *In representations* (1.29) *and* (1.30), *some vertices (extreme points) of the fundamental polyhedrons can be taken out:*

$$(\mathbf{A},\mathbf{B}) = c(\mathbf{A},\mathbf{B}) \sum_{(\overline{\mathbf{E}}',\overline{\mathbf{F}}')\le(\overline{\mathbf{E}}_k,\overline{\mathbf{F}}_k)\le(\overline{\mathbf{E}}'',\overline{\mathbf{F}}'')} \alpha_k(\overline{\mathbf{E}}_k,\overline{\mathbf{F}}_k), \quad (\overline{\mathbf{E}}_k,\overline{\mathbf{F}}_k) \in \overline{E}^{n,m},$$

$$\alpha_k \ge 0, \quad (\overline{\mathbf{E}}',\overline{\mathbf{F}}') \le (\overline{\mathbf{E}}_k,\overline{\mathbf{F}}_k) \le (\overline{\mathbf{E}}'',\overline{\mathbf{F}}''), \qquad \sum_{(\overline{\mathbf{E}}',\overline{\mathbf{F}}')\le(\overline{\mathbf{E}}_k,\overline{\mathbf{F}}_k)\le(\overline{\mathbf{E}}'',\overline{\mathbf{F}}'')} \alpha_k = 1;$$

$$X(\mathbf{A},\mathbf{B}) = c(\mathbf{A},\mathbf{B}) \sum_{\overline{H}'\subseteq\overline{H}_k\subseteq\overline{H}''} \alpha_k \overline{H}_k(\overline{\mathbf{E}}_k,\overline{\mathbf{F}}_k), \quad \overline{H}_k = \overline{H}_k(\overline{\mathbf{E}}_k,\overline{\mathbf{F}}_k) \in \overline{E}_{n,m},$$

$$\alpha_k \ge 0, \quad \overline{H}' = \overline{H}'(\overline{\mathbf{E}}',\overline{\mathbf{F}}') \subseteq \overline{H}_k \subseteq \overline{H}'' = \overline{H}''(\overline{\mathbf{E}}'',\overline{\mathbf{F}}''), \qquad \sum_{\overline{H}'\subseteq\overline{H}_k\subseteq\overline{H}''} \alpha_k = 1.$$

Applying Lemma 1.9 and Theorem 1.13 (with various $p \in Z$, $1 \le p \le b_1$ for which $\delta_k(\mathbf{A},\mathbf{B};c(\mathbf{A},\mathbf{B})) = 0$), one can determine the coefficients α_k $(0 \le k \le C^n_{n+m} - 1)$ that vanish in any representation (1.29) or (1.30). Note that when the vectors that make up the pair have zero components, analysis can be restricted to polyhedrons of lower dimensions.

Example 1.3. Consider the vector pair $(\mathbf{A},\mathbf{B}) = ((12,4),(10,6))$ from Example 1.2. It can be readily checked that the minimax value is $c(\mathbf{A},\mathbf{B}) = 6$, $\delta_1(\mathbf{A},\mathbf{B};6) = 0$, and $l_1(\mathbf{B};6) = 2$. The extremal vector pairs and extremal matrices contained in $\overline{E}^{2,2}$ and $\overline{E}_{2,2}$, respectively, are written out in Example 1.1. Let us single out the extremal vector pairs $(\overline{\mathbf{E}}',\overline{\mathbf{F}}')$, $(\overline{\mathbf{E}}'',\overline{\mathbf{F}}'')$ and extremal matrices that determine zero coefficients in representations (1.29) and (1.30) by Theorem 1.13. In the present example (recall Example 1.1), we have

$$(\overline{\mathbf{E}}',\overline{\mathbf{F}}') = ((2,0),(1,1)) = (\overline{\mathbf{E}}_2,\overline{\mathbf{F}}_2),$$
$$(\overline{\mathbf{E}}'',\overline{\mathbf{F}}'') = ((2,1),(2,1)) = (\overline{\mathbf{E}}_4,\overline{\mathbf{F}}_4),$$
$$\overline{H}' = \begin{pmatrix} 1 & 1 \\ 0 & 0 \end{pmatrix} = \overline{H}_2, \quad \overline{H}'' = \begin{pmatrix} 1 & 1 \\ 1 & 0 \end{pmatrix} = \overline{H}_4.$$

It is readily checked that the following sets of extremal vector pairs and extremal matrices consist of two elements: $\{(\overline{\mathbf{E}}_k,\overline{\mathbf{F}}_k) \in \overline{E}^{2,2} : (\overline{\mathbf{E}}',\overline{\mathbf{F}}') \le (\overline{\mathbf{E}}_k,\overline{\mathbf{F}}_k) \le (\overline{\mathbf{E}}'',\overline{\mathbf{F}}'')\} = \{(\overline{\mathbf{E}}',\overline{\mathbf{F}}'),(\overline{\mathbf{E}}'',\overline{\mathbf{F}}'')\}$ and $\{\overline{H}_k \in \overline{E}_{2,2} : \overline{H}' \subseteq \overline{H}_k \subseteq \overline{H}''\} = \{\overline{H}',\overline{H}''\}$. Therefore, by Theorem 1.13,

$$(\mathbf{A},\mathbf{B}) = 6\left(\alpha_2(\overline{\mathbf{E}}_2,\overline{\mathbf{F}}_2) + \alpha_4(\overline{\mathbf{E}}_4,\overline{\mathbf{F}}_4)\right), \quad \alpha_2,\alpha_4 \ge 0, \quad \alpha_2+\alpha_4 = 1,$$

$$X(\mathbf{A},\mathbf{B}) = 6\left(\alpha_2\overline{H}_2 + \alpha_4\overline{H}_4\right), \quad \alpha_2,\alpha_4 \ge 0, \quad \alpha_2+\alpha_4 = 1,$$

where $X(\mathbf{A},\mathbf{B})$ is an arbitrary uniform matrix from $\widetilde{M}(\mathbf{A},\mathbf{B};6)$. The first relation for α_2 and α_4 is

$$((12,4),(10,6)) = 6\Big(\alpha_2\left((2,0),(1,1)\right) + \alpha_4\left((2,1),(2,1)\right)\Big).$$

The last relation can be rewritten as the system of linear equations

$$\begin{cases} 12\alpha_2 + 12\alpha_4 &= 12, \\ 6\alpha_4 &= 4, \\ 6\alpha_2 + 12\alpha_4 &= 10, \\ 6\alpha_2 + 6\alpha_4 &= 6, \end{cases}$$

which has a unique solution: $\alpha_2 = 1/3$ and $\alpha_4 = 2/3$. Hence,

$$((12,4),(10,6)) = 6\Big(\tfrac{1}{3}\left((2,0),(1,1)\right) + \tfrac{2}{3}\left((2,1),(2,1)\right)\Big)$$

is a unique representation of the given vector pair $(\mathbf{A},\mathbf{B})$ in terms of extremal pairs that corresponds to $c = c(\mathbf{A},\mathbf{B}) = 6$. Consequently, the uniform matrix $X(\mathbf{A},\mathbf{B})$ in $\widetilde{M}(\mathbf{A},\mathbf{B};6)$ is uniquely defined:

$$X(\mathbf{A},\mathbf{B}) = 6\left(\frac{1}{3}\begin{pmatrix} 1 & 1 \\ 0 & 0 \end{pmatrix} + \frac{2}{3}\begin{pmatrix} 1 & 1 \\ 1 & 0 \end{pmatrix}\right) = \begin{pmatrix} 6 & 6 \\ 4 & 0 \end{pmatrix}.$$

The representation

$$\begin{pmatrix} 6 & 6 \\ 4 & 0 \end{pmatrix} = 6\left(\frac{1}{3}\begin{pmatrix} 1 & 1 \\ 0 & 0 \end{pmatrix} + \frac{2}{3}\begin{pmatrix} 1 & 1 \\ 1 & 0 \end{pmatrix}\right)$$

of the uniform minimax matrix thus constructed in terms of extremal matrices is unique.

Note also another important property of the fundamental polyhedron $L(n,m)$ of vector pairs. Denote the set of transportation matrices defined by elements of $L(n,m)$ by

$$\mathfrak{N}(n,m) = \Big\{X(\mathbf{A},\mathbf{B}) \in M(\mathbf{A},\mathbf{B}) : (\mathbf{A},\mathbf{B}) \in L(n,m)\Big\}.$$

Recall that $\widetilde{X}(\widetilde{\mathbf{A}},\widetilde{\mathbf{B}})$ is a matrix derived from $X(\mathbf{A},\mathbf{B})$ by permuting its rows and columns and the vectors $\widetilde{\mathbf{A}}$ and $\widetilde{\mathbf{B}}$ are obtained by arranging components of $\mathbf{A}$ and $\mathbf{B}$ in nonascending order. Theorem 1.4 implies that every n-by-m transportation matrix belongs to $\mathfrak{N}(n,m)$ up to a permutation of its rows and columns and up to a positive factor.

Theorem 1.14. *Suppose that* $\mathbf{A} \in R^n_+$, $\mathbf{B} \in R^m_+$, $M(\mathbf{A},\mathbf{B}) \neq \varnothing$, $X(\mathbf{A},\mathbf{B}) = (x_{ij})$ *is an arbitrary matrix in* $M(\mathbf{A},\mathbf{B})$, *and* $\alpha = \max\{x_{ij} : 1 \le i \le n,\ 1 \le j \le m\} > 0$. *Then,* $\widetilde{X}(\widetilde{\mathbf{A}},\widetilde{\mathbf{B}}) \in \mathfrak{N}(n,m)$ *when* $\alpha \le 1$, *and* $\widetilde{X}(\widetilde{\mathbf{A}},\widetilde{\mathbf{B}})/\alpha \in \mathfrak{N}(n,m)$ *when* $\alpha > 1$.

§2. Applications of Fundamental Polyhedrons in Optimization

Here, we consider optimization problems on uniform transportation polyhedrons. Their solutions are given by the coefficients in representations of uniform matrices in terms of convex linear combinations of extremal matrices.

2.1. Optimization on Uniform Polyhedrons

Let $\Phi(X)$ be a functional defined on n-by-m matrices. Consider vectors $\mathbf{A} \in R_+^n$ and $\mathbf{B} \in R_+^m$ satisfying the balance condition. The vector pair $(\mathbf{A}, \mathbf{B})$ defines the domains of $\Phi(X)$: $M(\mathbf{A}, \mathbf{B})$ and $\widetilde{M}(\mathbf{A}, \mathbf{B})$ are the conventional and uniform transportation polyhedrons, respectively. Recall that the polyhedron $\widetilde{M}(\mathbf{A}, \mathbf{B})$, which consists of all uniform matrices, is referred to as the uniform one (see Definitions 1.1 and 2.1 in Chapter 1).

Consider the problem

$$\Phi(X(\mathbf{A}, \mathbf{B})) \to \min(\max), \quad X(\mathbf{A}, \mathbf{B}) \in \widetilde{M}(\mathbf{A}, \mathbf{B}).$$

Here, $\mathbf{A} \in R_+^n$ and $\mathbf{B} \in R_+^m$, where $n, m \geq 2$. When $\mathbf{A} = (a_1, \ldots, a_n) \in \overline{R}_+^n$ and $\mathbf{B} = (b_1, \ldots, b_m) \in \overline{R}_+^n$, the constraints for the elements of a matrix $X(\mathbf{A}, \mathbf{B}) = (x_{ij}) \in \widetilde{M}(\mathbf{A}, \mathbf{B})$ have the form

$$\sum_{j=1}^{m} x_{ij} = a_i, \quad 1 \leq i \leq n, \quad \sum_{i=1}^{n} x_{ij} = b_j, \quad 1 \leq j \leq m$$

(combined with the requirement that the elements of the matrices are nonnegative, these relations make up the set of constraints imposed in conventional transportation problems [3, 7, 11, 12]),

$$x_{p1} \geq \ldots \geq x_{pm} \geq 0, \quad 1 \leq p \leq n,$$

$$x_{1q} \geq \ldots \geq x_{nq} \geq 0, \quad 1 \leq q \leq m,$$

$$x_{pq} = x_{p\,q+1} \quad \text{for} \quad b_q = b_{q+1}, \quad 1 \leq q \leq m-1,$$

$$x_{pq} = x_{p+1\,q} \quad \text{for} \quad a_p = a_{p+1}, \quad 1 \leq p \leq n-1.$$

The solution of this problem is given by a convex linear combination of extremal matrices as demonstrated in the preceding section. When the components of $\mathbf{A}$ and $\mathbf{B}$ are not arranged in nonascending order, one should change to the vectors $\widetilde{\mathbf{A}} \in \overline{R}_+^n$ and $\widetilde{\mathbf{B}} \in \overline{R}_+^m$.

Definition 2.1. An optimization problem defined on the set $\widetilde{M}(\mathbf{A}, \mathbf{B})$ of uniform matrices is called a *uniform optimization problem.*

Suppose vectors $\mathbf{A} \in \overline{R}^n_+$ and $\mathbf{B} \in \overline{R}^m_+$ make up a closed system. We assume that $a_1 > 0$ (which implies that $b_1 > 0$ as well). Suppose $M(\mathbf{A}, \mathbf{B}; c) \neq \varnothing$; therefore, $c \geq c(\mathbf{A}, \mathbf{B})$. Recall that the minimax value can be readily calculated by Algorithm 1.1 described in Chapter 2. We can also assume that $c \leq \min(a_1, b_1)$, because $M(\mathbf{A}, \mathbf{B}; \min(a_1, b_1)) = M(\mathbf{A}, \mathbf{B})$. Applying Theorems 1.10–1.12 to the vector pair $(\mathbf{A}, \mathbf{B})$ and an arbitrary uniform matrix $X(\mathbf{A}, \mathbf{B}) \in \widetilde{M}(\mathbf{A}, \mathbf{B}; c)$, we have

$$(\mathbf{A}, \mathbf{B}) = c \sum_{k=0}^{C^n_{n+m}-1} \alpha_k (\overline{\mathbf{E}}_k, \overline{\mathbf{F}}_k), \quad X(\mathbf{A}, \mathbf{B}) = c \sum_{k=0}^{C^n_{n+m}-1} \alpha_k \overline{H}_k (\overline{\mathbf{E}}_k, \overline{\mathbf{F}}_k),$$

$$(\overline{\mathbf{E}}_k, \overline{\mathbf{F}}_k) \in \overline{E}^{n,m}, \quad \overline{H}_k(\overline{\mathbf{E}}_k, \overline{\mathbf{F}}_k) \in \overline{E}_{n,m},$$

$$\alpha_k \geq 0, \quad 0 \leq k \leq C^n_{n+m} - 1, \quad \sum_{k=0}^{C^n_{n+m}-1} \alpha_k = 1, \tag{2.1}$$

$$\widetilde{M}(\mathbf{A}, \mathbf{B}; c) = \left\{ X(\mathbf{A}, \mathbf{B}) = c \sum_{k=0}^{C^n_{n+m}-1} \alpha_k \overline{H}_k (\overline{\mathbf{E}}_k, \overline{\mathbf{F}}_k) : \right.$$
$$\left. (\alpha_0, \ldots, \alpha_{C^n_{n+m}-1}) \in \mathfrak{A}(\mathbf{A}/c, \mathbf{B}/c) \right\}.$$

Once again, we note that different vectors in $\mathfrak{A}(\mathbf{A}/c, \mathbf{B}/c)$ may correspond to the same matrix $X(\mathbf{A}, \mathbf{B})$. In (2.1) it can be assumed that $c = 1$ when $M(\mathbf{A}, \mathbf{B}; 1) \neq \varnothing$ and $c > 1$ when $M(\mathbf{A}, \mathbf{B}; 1) = \varnothing$.

Here, we consider linear, convex, and concave functionals $\Phi(X(\mathbf{A}, \mathbf{B}))$ on the set of uniform matrices. The following theorem gives expressions for the optimal solutions in terms of the values of the functional on extremal matrices. Upper and lower estimates are obtained for the optimums in the convex and concave cases, respectively.

Theorem 2.1. *Suppose that* $\mathbf{A} = (a_1, \ldots, a_n) \in \overline{R}^n_+$, $\mathbf{B} = (b_1, \ldots, b_m) \in \overline{R}^m_+$, *and* $c = \min(a_1, b_1) > 0$. *Then,* $\widetilde{M}(\mathbf{A}, \mathbf{B}) = \widetilde{M}(\mathbf{A}, \mathbf{B}; c)$ *and the following propositions are true:*

(a) *If* $\Phi(X(\mathbf{A}, \mathbf{B}))$ *is a linear functional, then*

$$\min_{X(\mathbf{A},\mathbf{B}) \in \widetilde{M}(\mathbf{A},\mathbf{B})} \Phi(X(\mathbf{A}, \mathbf{B})) = c \min \left\{ \sum_{k=0}^{C^n_{n+m}-1} \alpha_k \Phi(\overline{H}_k(\overline{\mathbf{E}}_k, \overline{\mathbf{F}}_k)) : \right.$$
$$\left. (\alpha_0, \ldots, \alpha_{C^n_{n+m}-1}) \in \mathfrak{A}(\mathbf{A}/c, \mathbf{B}/c) \right\}; \tag{2.2}$$

$$\max_{X(\mathbf{A},\mathbf{B})\in\widetilde{M}(\mathbf{A},\mathbf{B})} \Phi(X(\mathbf{A},\mathbf{B})) = c\max\left\{ \sum_{k=0}^{C_{n+m}^{n}-1} \alpha_k \Phi(\overline{H}_k(\overline{\mathbf{E}}_k,\overline{\mathbf{F}}_k)) : (\alpha_0,\ldots,\alpha_{C_{n+m}^{n}-1}) \in \mathfrak{A}(\mathbf{A}/c,\mathbf{B}/c)\right\}. \tag{2.3}$$

(b) *If* $\Phi(X(\mathbf{A},\mathbf{B}))$ *is a convex functional, then*

$$\min_{X(\mathbf{A},\mathbf{B})\in\widetilde{M}(\mathbf{A},\mathbf{B})} \Phi(X(\mathbf{A},\mathbf{B})) \le \min\left\{ \sum_{k=0}^{C_{n+m}^{n}-1} \alpha_k \Phi(c\overline{H}_k(\overline{\mathbf{E}}_k,\overline{\mathbf{F}}_k)) : (\alpha_0,\ldots,\alpha_{C_{n+m}^{n}-1}) \in \mathfrak{A}(\mathbf{A}/c,\mathbf{B}/c)\right\}. \tag{2.4}$$

(c) *If* $\Phi(X(\mathbf{A},\mathbf{B}))$ *is a concave functional, then*

$$\max_{X(\mathbf{A},\mathbf{B})\in\widetilde{M}(\mathbf{A},\mathbf{B})} \Phi(X(\mathbf{A},\mathbf{B})) \ge \max\left\{ \sum_{k=0}^{C_{n+m}^{n}-1} \alpha_k \Phi(c\overline{H}_k(\overline{\mathbf{E}}_k,\overline{\mathbf{F}}_k)) : (\alpha_0,\ldots,\alpha_{C_{n+m}^{n}-1}) \in \mathfrak{A}(\mathbf{A}/c,\mathbf{B}/c)\right\}. \tag{2.5}$$

(d) *Suppose that* α_k^* *and* α_k^{**}, *where* $0 \le k \le C_{n+m}^{n} - 1$, *are the coefficients that correspond to the minimum and maximum in* (2.2) *and* (2.3), *respectively. Then the matrices*

$$X^*(\mathbf{A},\mathbf{B}) = c\sum_{k=0}^{C_{n+m}^{n}-1} \alpha_k^* \overline{H}_k(\overline{\mathbf{E}}_k,\overline{\mathbf{F}}_k), \quad X^{**}(\mathbf{A},\mathbf{B}) = c\sum_{k=0}^{C_{n+m}^{n}-1} \alpha_k^{**} \overline{H}_k(\overline{\mathbf{E}}_k,\overline{\mathbf{F}}_k),$$

respectively, minimize and maximize the linear functional $\Phi(X(\mathbf{A},\mathbf{B}))$ *on the set* $\widetilde{M}(\mathbf{A},\mathbf{B})$:

$$\min_{X(\mathbf{A},\mathbf{B})\in\widetilde{M}(\mathbf{A},\mathbf{B})} \Phi(X(\mathbf{A},\mathbf{B})) = \Phi(X^*(\mathbf{A},\mathbf{B})),$$

$$\max_{X(\mathbf{A},\mathbf{B})\in\widetilde{M}(\mathbf{A},\mathbf{B})} \Phi(X(\mathbf{A},\mathbf{B})) = \Phi(X^{**}(\mathbf{A},\mathbf{B})).$$

Remark 2.1.

(a) If $c(\mathbf{A},\mathbf{B}) \le c \le \min(a_1, b_1)$ in Theorem 2.1, then (2.2)–(2.5) remain valid provided that $X(\mathbf{A},\mathbf{B}) \in \widetilde{M}(\mathbf{A},\mathbf{B};c)$.

(b) If n, $m \ge 2$ and a_2, $b_2 > 0$, then $\min\left\{c : \widetilde{M}(\mathbf{A},\mathbf{B};c) = \widetilde{M}(\mathbf{A},\mathbf{B})\right\} < \min(a_1, b_1)$.

(c) If $c = c(\mathbf{A}, \mathbf{B})$, or some components of the vectors $\mathbf{A}$ and $\mathbf{B}$ are equal, then some of the variables α_k $(0 \le k \le C_{n+m}^n - 1)$ are zeros by Theorem 1.13 and Lemma 1.9.

(d) If the positive components of $\mathbf{A}$ and $\mathbf{B}$ are equal, then $\left|\widetilde{M}(\mathbf{A}, \mathbf{B})\right| = 1$.

The following lemma compares the optimal values of functionals on matrix sets $M(\mathbf{A}, \mathbf{B})$ and $\widetilde{M}(\mathbf{A}, \mathbf{B})$.

Lemma 2.1. *Any functional $\Phi(X(\mathbf{A}, \mathbf{B}))$ on a transportation polyhedron $M(\mathbf{A}, \mathbf{B})$ is such that*

$$\min_{X(\mathbf{A},\mathbf{B}) \in M(\mathbf{A},\mathbf{B})} \Phi(X(\mathbf{A}, \mathbf{B})) \le \min_{X(\mathbf{A},\mathbf{B}) \in \widetilde{M}(\mathbf{A},\mathbf{B})} \Phi(X(\mathbf{A}, \mathbf{B})),$$

$$\max_{X(\mathbf{A},\mathbf{B}) \in M(\mathbf{A},\mathbf{B})} \Phi(X(\mathbf{A}, \mathbf{B})) \ge \max_{X(\mathbf{A},\mathbf{B}) \in \widetilde{M}(\mathbf{A},\mathbf{B})} \Phi(X(\mathbf{A}, \mathbf{B})).$$

It was shown in Chapter 2 that all functionals (1.7)–(1.16) defined on an arbitrary conventional transportation polyhedron in Chapter 1 are minimized by a uniform, hereditarily minimax matrix. Theorem 2.1 and Lemma 2.1 show that solutions to uniform transportation problems set bounds for the general transportation problem.

The following simple example illustrates Theorem 2.1 and Lemma 2.1.

Example 2.1. Consider transportation polyhedrons $M(\mathbf{A}, \mathbf{B})$ and $\widetilde{M}(\mathbf{A}, \mathbf{B})$, where $\mathbf{A} = (12, 4)$ and $\mathbf{B} = (10, 6)$. In Example 1.2, we constructed a set $\widetilde{M}(\mathbf{A}, \mathbf{B}; 8)$ of uniform matrices. Suppose that $X(\mathbf{A}, \mathbf{B}) = (x_{ij}) \in M(\mathbf{A}, \mathbf{B})$. If $x_{11} > 8$, then $x_{21} < x_{22}$; if $x_{11} = 8$, then $x_{21} = x_{22} = 2$. Therefore, $\widetilde{M}(\mathbf{A}, \mathbf{B}) = \widetilde{M}(\mathbf{A}, \mathbf{B}; 8)$, where $c = 8 < \min(a_1, b_1) = 10$. Hence (see Example 1.2),

$$\widetilde{M}(\mathbf{A}, \mathbf{B}) = \Big\{ X(\mathbf{A}, \mathbf{B}) : X(\mathbf{A}, \mathbf{B}) = X\left((12, 4), (10, 6)\right) =$$

$$8 \left[\alpha_1 \begin{pmatrix} 1 & 0 \\ 0 & 0 \end{pmatrix} + \alpha_2 \begin{pmatrix} 1 & 1 \\ 0 & 0 \end{pmatrix} + (\alpha_2 - \tfrac{1}{4}) \begin{pmatrix} 1 & 0 \\ 1 & 0 \end{pmatrix} + (1 - \alpha_1 - 2\alpha_2) \begin{pmatrix} 1 & 1 \\ 1 & 0 \end{pmatrix} + \right.$$

$$\left. (\tfrac{1}{2}\alpha_1 + \tfrac{1}{2}\alpha_1 - \tfrac{1}{8}) \begin{pmatrix} 1 & 1 \\ 1 & 1 \end{pmatrix} \right], \ \alpha_1 \ge 0, \ \alpha_2 \ge \tfrac{1}{4}, \ \alpha_1 + 2\alpha_2 \le 1 \Big\}.$$

Let us minimize and maximize the linear functional

$$\Phi(X(\mathbf{A}, \mathbf{B})) = 4x_{11} + x_{12} + 3x_{21} + 2x_{22}$$

on the matrix sets $M\left((12,4),(10,6)\right)$ and $\widetilde{M}\left((12,4),(10,6)\right)$.

Every matrix in $\widetilde{M}(\mathbf{A},\mathbf{B})$ has the form

$$X(\mathbf{A},\mathbf{B}) = 8\begin{pmatrix} \alpha_1/2+\alpha_2/2+5/8 & 7/8-\alpha_1/2-\alpha_2/2 \\ 5/8-\alpha_1/2-\alpha_2/2 & \alpha_1/2+\alpha_2/2-1/8 \end{pmatrix}.$$

This expression can be used to obtain the value of the functional on $X(\mathbf{A},\mathbf{B}) \in \widetilde{M}(\mathbf{A},\mathbf{B})$:

$$\Phi(X(\mathbf{A},\mathbf{B})) = 8(\alpha_1+\alpha_2)+40, \quad X(\mathbf{A},\mathbf{B}) \in \widetilde{M}(\mathbf{A},\mathbf{B}).$$

The constraints $\alpha_1 \geq 0$, $\alpha_2 \geq 1/4$, and $\alpha_1 + 2\alpha_2 \leq 1$ define a triangle on the plane (α_1,α_2), with vertices at $(0,1/4)$, $(0,1/2)$, and $(1/2,1/4)$. The first and third vertices, respectively, minimize and maximize the linear form $\alpha_1+\alpha_2$. Therefore,

$$\min_{X(\mathbf{A},\mathbf{B})\in\widetilde{M}(\mathbf{A},\mathbf{B})} \Phi(X(\mathbf{A},\mathbf{B})) = 42, \qquad \max_{X(\mathbf{A},\mathbf{B})\in\widetilde{M}(\mathbf{A},\mathbf{B})} \Phi(X(\mathbf{A},\mathbf{B})) = 46,$$

$$\Phi\left(\begin{pmatrix} 6 & 6 \\ 4 & 0 \end{pmatrix}\right) = 42, \quad \Phi\left(\begin{pmatrix} 8 & 4 \\ 2 & 2 \end{pmatrix}\right) = 46$$

are the solutions of the uniform linear transportation problems.

Every element of the matrix $X\left((12,4),(10,6)\right) = (x_{ij})$ from $M(\mathbf{A},\mathbf{B})$ defines the remaining elements of the matrix; for example, $x_{12} = 6 - x_{22}$, $x_{21} = 4 - x_{22}$, and $x_{11} = 6 + x_{22}$. Hence,

$$\Phi(X(\mathbf{A},\mathbf{B})) = 2x_{22}+42, \quad X(\mathbf{A},\mathbf{B}) = (x_{ij}) \in M(\mathbf{A},\mathbf{B}),$$

where $0 \leq x_{22} \leq 4$. Therefore,

$$\min_{X(\mathbf{A},\mathbf{B})\in M(\mathbf{A},\mathbf{B})} \Phi(X(\mathbf{A},\mathbf{B})) = 42, \qquad \max_{X(\mathbf{A},\mathbf{B})\in M(\mathbf{A},\mathbf{B})} \Phi(X(\mathbf{A},\mathbf{B})) = 50,$$

$$\Phi\left(\begin{pmatrix} 6 & 6 \\ 4 & 0 \end{pmatrix}\right) = 42, \quad \Phi\left(\begin{pmatrix} 10 & 2 \\ 0 & 4 \end{pmatrix}\right) = 50$$

are the solutions of the linear transportation problems.

The minimax matrix that minimizes the functional $\Phi(X(\mathbf{A},\mathbf{B}))$ on the matrix sets $\widetilde{M}(\mathbf{A},\mathbf{B})$ and $M(\mathbf{A},\mathbf{B})$, where $\mathbf{A} = (12,4)$ and $\mathbf{B} = (10,6)$, was constructed in Example 1.3. Since any 2-by-2 minimax matrix is a hereditarily minimax one, the functional in both cases is minimized by the same single hereditarily minimax matrix:

$$\overset{*}{X}\left((12,4),(10,6)\right) = \begin{pmatrix} 6 & 6 \\ 4 & 0 \end{pmatrix}.$$

The functional minimized in this example is linear, and the hereditarily minimax matrix $\overset{*}{X}(\mathbf{A},\mathbf{B})$ is a vertex of both transportation polyhedrons, $M(\mathbf{A},\mathbf{B})$ and $\widetilde{M}(\mathbf{A},\mathbf{B})$. This is also true for uniform transportation polyhedrons of general form.

One can now readily prove the following proposition.

Theorem 2.2. *Suppose that* $\mathbf{A} \in R^n_+$, $\mathbf{B} \in R^m_+$, *and* $M(\mathbf{A},\mathbf{B}) \neq \varnothing$. *Then,*

(a) *any minimax uniform matrix belongs to a face of the uniform transportation polyhedron* $\widetilde{M}(\mathbf{A},\mathbf{B})$;

(b) *the hereditarily minimax matrix is a vertex of the uniform transportation polyhedron* $\widetilde{M}(\mathbf{A},\mathbf{B})$.

Let us analyze an example, in which the uniform transportation problem is of higher dimension.

Example 2.2. Consider the following linear form on 20-by-20 matrices:

$$\Phi(X) = \sum_{i=1}^{20}\sum_{j=1}^{20} c_{ij}x_{ij},$$

where $X = (x_{ij})$ $(1 \le i,j \le 20)$, $c_{ij} \in R$ $(1 \le i,j \le 20)$, and

$$c_{ij} = \begin{cases} p, & 1 \le i \le 10, \quad 1 \le j \le 10, \\ q, & 11 \le i \le 20, \quad 1 \le j \le 10, \\ r, & 1 \le i \le 10, \quad 11 \le j \le 20, \\ t, & 11 \le i \le 20, \quad 11 \le j \le 20. \end{cases}$$

Then, the linear function $\Phi(X)$ can be represented as

$$\Phi(X) = p\sum_{i=1}^{10}\sum_{j=1}^{10} x_{ij} + q\sum_{i=11}^{20}\sum_{j=1}^{10} x_{ij} + r\sum_{i=1}^{10}\sum_{j=11}^{20} x_{ij} + t\sum_{i=11}^{20}\sum_{j=11}^{20} x_{ij}.$$

Let us minimize and maximize the functional $\Phi(X)$ on the set $\widetilde{M}(\mathbf{A},\mathbf{A})$ of uniform matrices with respect to the coefficients p, q, r, and t, where

$$\mathbf{A} = (\underbrace{40,\ldots,40}_{10},\ \underbrace{10,\ldots,10}_{10}) \in \overline{R}^{20}_+.$$

Both the numbers of extremal elements in the fundamental systems $\overline{E}^{20,20}$ and $\overline{E}_{20,20}$ and the numbers of vertices in the fundamental polyhedrons

$L(20,20)$ and $M(20,20)$ are C_{40}^{20} by Theorem 1.7 from Chapter 4 and by Theorems 1.2, 1.4, and 1.7. However, by applying Lemma 1.9, it can be readily shown that only six extremal vector pairs and six extremal matrices are required in (1.27) and (1.28). To number the extremal elements, we use the following ad hoc notation:

$$
\begin{aligned}
(\overline{\mathbf{E}}_0, \overline{\mathbf{F}}_0) &= \Big((\underbrace{0,\dots,0}_{20}),\ (\underbrace{0,\dots,0}_{20})\Big),\\
(\overline{\mathbf{E}}_1, \overline{\mathbf{F}}_1) &= \Big((\underbrace{10,\dots,10}_{10},\ \underbrace{0,\dots,0}_{10}),\ (\underbrace{10,\dots,10}_{10},\ \underbrace{0,\dots,0}_{10})\Big),\\
(\overline{\mathbf{E}}_2, \overline{\mathbf{F}}_2) &= \Big((\underbrace{10,\dots,10}_{20}),\ (\underbrace{20,\dots,20}_{10},\ \underbrace{0,\dots,0}_{10})\Big),\\
(\overline{\mathbf{E}}_3, \overline{\mathbf{F}}_3) &= \Big((\underbrace{20,\dots,20}_{10},\ \underbrace{0,\dots,0}_{10}),\ (\underbrace{10,\dots,10}_{20})\Big),\\
(\overline{\mathbf{E}}_4, \overline{\mathbf{F}}_4) &= \Big((\underbrace{20,\dots,20}_{10},\ \underbrace{10,\dots,10}_{10}),\ (\underbrace{20,\dots,20}_{10},\ \underbrace{10,\dots,10}_{10})\Big),\\
(\overline{\mathbf{E}}_5, \overline{\mathbf{F}}_5) &= \Big((\underbrace{20,\dots,20}_{20}),\ (\underbrace{20,\dots,20}_{20})\Big).
\end{aligned}
$$

Setting $\overline{H}_k(\overline{\mathbf{E}}_k, \overline{\mathbf{F}}_k) = (h_{ij}^{(k)})$, where $1 \le i, j \le 20$, and $0 \le k \le 5$, we express the extremal matrices as follows: $\overline{H}_0(\overline{\mathbf{E}}_0, \overline{\mathbf{F}}_0) = H(0)$ is the matrix of zeros,

$$
h_{ij}^{(1)} = \begin{cases} 1, & 1 \le i \le 10, \quad 1 \le j \le 10, \\ 0, & 1 \le i \le 10, \quad 11 \le j \le 20, \\ 0, & 11 \le i \le 20, \quad 1 \le j \le 20, \end{cases}
$$

$$
h_{ij}^{(2)} = \begin{cases} 1, & 1 \le i \le 20, \quad 1 \le j \le 10, \\ 0, & 1 \le i \le 20, \quad 11 \le j \le 20, \end{cases}
$$

$$
h_{ij}^{(3)} = \begin{cases} 1, & 1 \le i \le 10, \quad 1 \le j \le 20, \\ 0, & 11 \le i \le 20, \quad 1 \le j \le 20, \end{cases}
$$

$$
h_{ij}^{(4)} = \begin{cases} 1, & 1 \le i \le 10, \quad 1 \le j \le 20, \\ 1, & 11 \le i \le 20, \quad 1 \le j \le 10, \\ 0, & 11 \le i \le 20, \quad 11 \le j \le 20, \end{cases}
$$

and $\overline{H}_5(\overline{\mathbf{E}}_5, \overline{\mathbf{F}}_5) = H(1)$ is the matrix of ones.

We set $c = 40$, because we then have $\widetilde{M}(\mathbf{A}, \mathbf{A}) = \widetilde{M}(\mathbf{A}, \mathbf{A}; 40)$. Dropping the extreme elements $(\overline{\mathbf{E}}_0, \overline{\mathbf{F}}_0)$ and $\overline{H}_0(\overline{\mathbf{E}}_0, \overline{\mathbf{F}}_0)$, we apply (1.27) and (1.28) to the vector pair $(\mathbf{A},\mathbf{A})$ and any $X(\mathbf{A},\mathbf{A}) \in \widetilde{M}(\mathbf{A},\mathbf{A})$:

$$(\mathbf{A},\mathbf{A}) = 40\sum_{k=1}^{5} \alpha_k(\overline{\mathbf{E}}_k,\overline{\mathbf{F}}_k), \quad X(\mathbf{A},\mathbf{A}) = 40\sum_{k=1}^{5} \alpha_k \overline{H}_k(\overline{\mathbf{E}}_k,\overline{\mathbf{F}}_k),$$

$$\alpha_k \geq 0, \quad 1 \leq k \leq 5, \quad \sum_{k=1}^{5} \alpha_k \leq 1.$$

The linear functional $\Phi(X)$ is expressed as

$$\Phi(X(\mathbf{A},\mathbf{A})) = 40\sum_{k=1}^{5} \alpha_k \Phi(\overline{H}_k(\overline{\mathbf{E}}_k,\overline{\mathbf{F}}_k)),$$

where $X(\mathbf{A},\mathbf{A})$ is an arbitrary matrix from $\widetilde{M}(\mathbf{A},\mathbf{A})$. We readily calculate the values of the functional on the extremal matrices:

$$\Phi(\overline{H}_1(\overline{\mathbf{E}}_1,\overline{\mathbf{F}}_1)) = 100p, \quad \Phi(\overline{H}_2(\overline{\mathbf{E}}_2,\overline{\mathbf{F}}_2)) = 100(p+q),$$
$$\Phi(\overline{H}_3(\overline{\mathbf{E}}_3,\overline{\mathbf{F}}_3)) = 100(p+r), \quad \Phi(\overline{H}_4(\overline{\mathbf{E}}_4,\overline{\mathbf{F}}_4)) = 100(p+q+r),$$
$$\Phi(\overline{H}_5(\overline{\mathbf{E}}_5,\overline{\mathbf{F}}_5)) = 100(p+q+r+t).$$

The functional $\Phi(X)$ of any uniform matrix $X(\mathbf{A},\mathbf{A}) \in \widetilde{M}(\mathbf{A},\mathbf{A})$ is expressed as

$$\Phi(X(\mathbf{A},\mathbf{A})) =$$
$$4000\Big(p\alpha_1 + (p+q)\alpha_2 + (p+r)\alpha_3 + (p+q+r)\alpha_4 + (p+q+r+t)\alpha_5\Big).$$

Representing the vector pair $(\mathbf{A},\mathbf{A})$ as a convex linear combination, we obtain the system of linear equations

$$\begin{cases} 40 &= 40(10\alpha_1 + 10\alpha_2 + 20\alpha_3 + 20\alpha_4 + 20\alpha_5), \\ 10 &= 40(10\alpha_2 + 10\alpha_4 + 20\alpha_5), \\ 40 &= 40(10\alpha_1 + 20\alpha_2 + 10\alpha_3 + 20\alpha_4 + 20\alpha_5), \\ 10 &= 40(10\alpha_3 + 10\alpha_4 + 20\alpha_5). \end{cases}$$

Its solution is

$$\alpha_1 = \frac{1}{40} + \alpha_4 + 4\alpha_5, \quad \alpha_2 = \alpha_3 = \frac{1}{40} - \alpha_4 - 2\alpha_5,$$

where α_4 and α_5 are independent variables. These relations for α_k $(1 \leq k \leq 5)$ can be used to simplify the calculation of the functional of a uniform matrix. We write

$$\Phi(X(\mathbf{A},\mathbf{A})) = 100(3p+q+r) + 4000(p-q-r+t)\alpha_5.$$

Taking into account the constraints for α_k, $1 \leq k \leq 5$ (they are nonnegative, and their sum is not greater than unity), we obtain

$$\alpha_4, \alpha_5 \geq 0, \quad \alpha_4 + 2\alpha_5 \leq \frac{1}{40}.$$

These constraints define a triangle in the plane (α_4, α_5), with vertices at $(0,0)$, $(0,1/80)$, and $(1/40,0)$. The minimal and maximal values of the linear functional $\Phi(X(\mathbf{A},\mathbf{A}))$, where $X(\mathbf{A},\mathbf{A}) \in \widetilde{M}(\mathbf{A},\mathbf{A})$, are attained at vertices of the triangle. Theorem 1.1 determines the minimizing and maximizing matrices of a uniform transportation problem. The constraint for α_5 implies that

$$\min_{X(\mathbf{A},\mathbf{A}) \in \widetilde{M}(\mathbf{A},\mathbf{A})} \Phi(X(\mathbf{A},\mathbf{A})) = \min_{0 \leq \alpha_5 \leq 1/80} \Phi(X(\mathbf{A},\mathbf{A})),$$

$$\max_{X(\mathbf{A},\mathbf{A}) \in \widetilde{M}(\mathbf{A},\mathbf{A})} \Phi(X(\mathbf{A},\mathbf{A})) = \max_{0 \leq \alpha_5 \leq 1/80} \Phi(X(\mathbf{A},\mathbf{A})).$$

Therefore, the minimal and maximal values of $\Phi(X)$ on the set $\widetilde{M}(\mathbf{A},\mathbf{A})$ of uniform matrices are expressed as follows:

$$\min_{X(\mathbf{A},\mathbf{A}) \in \widetilde{M}(\mathbf{A},\mathbf{A})} \Phi(X(\mathbf{A},\mathbf{A})) = \begin{cases} 100(3p+q+r), & p+t \geq q+r, \\ 50(7p+q+r+t), & p+t < q+r, \end{cases}$$

$$\max_{X(\mathbf{A},\mathbf{A}) \in \widetilde{M}(\mathbf{A},\mathbf{A})} \Phi(X(\mathbf{A},\mathbf{A})) = \begin{cases} 50(7p+q+r+t), & p+t \geq q+r, \\ 100(3p+q+r), & p+t < q+r. \end{cases}$$

Let us use (1.28) to construct the matrices that optimize the functional. When $\alpha_5 = 0$, we have $\alpha_1 = 1/40 + \alpha_4$ and $\alpha_2 = \alpha_3 = 1/40 - \alpha_4$, where $0 \leq \alpha_4 \leq 1/40$. When $\alpha_5 = 1/80$, we have $\alpha_4 = \alpha_2 = \alpha_3 = 0$ and $\alpha_1 = 3/40$. The values of α_k $(1 \leq k \leq 5)$ thus obtained define two uniform matrices in $\widetilde{M}(\mathbf{A},\mathbf{A})$:

$$X_0(\mathbf{A},\mathbf{A}) = (x_{ij}^{(0)}), \quad X_{1/80}(\mathbf{A},\mathbf{A}) = (x_{ij}^{(1/80)}), \quad 1 \leq i,j \leq 20;$$

$$x_{ij}^{(0)} = \begin{cases} 3, & 1 \leq i \leq 10, \quad 1 \leq j \leq 10, \\ 1, & 11 \leq i \leq 20, \quad 1 \leq j \leq 10, \\ 1, & 1 \leq i \leq 10, \quad 11 \leq j \leq 20, \\ 0, & 11 \leq i \leq 20, \quad 11 \leq j \leq 20, \end{cases}$$

$$x_{ij}^{(1/80)} = \begin{cases} 7/2, & 1 \leq i \leq 10, \quad 1 \leq j \leq 10, \\ 1/2, & 11 \leq i \leq 20, \quad 1 \leq j \leq 10, \\ 1/2, & 1 \leq i \leq 10, \quad 11 \leq j \leq 20, \\ 1/2, & 11 \leq i \leq 20, \quad 11 \leq j \leq 20. \end{cases}$$

The solutions can be represented as follows:

$$\min_{X(\mathbf{A},\mathbf{A})\in\widetilde{M}(\mathbf{A},\mathbf{A})} \Phi(X(\mathbf{A},\mathbf{A})) = \begin{cases} \Phi(X_0(\mathbf{A},\mathbf{A})), & p+t \geq q+r, \\ \Phi(X_{1/80}(\mathbf{A},\mathbf{A})), & p+t < q+r, \end{cases}$$

$$\max_{X(\mathbf{A},\mathbf{A})\in\widetilde{M}(\mathbf{A},\mathbf{A})} \Phi(X(\mathbf{A},\mathbf{A})) = \begin{cases} \Phi(X_{1/80}(\mathbf{A},\mathbf{A})), & p+t \geq q+r, \\ \Phi(X_0(\mathbf{A},\mathbf{A})), & p+t < q+r. \end{cases}$$

2.2. Optimization on Sets of Uniform Polyhedrons

A linear optimization problem on uniform matrices can be either solved by using specific properties of uniform matrices or reduced to a simple mathematical programming problem in which the coefficients are sought in convex linear combinations of extremal matrices.

Consider a functional $\Phi(X)$ defined on n-by-m matrices and a vector pair $(\mathbf{A},\mathbf{B})$ such that $M(\mathbf{A},\mathbf{B}) \neq \varnothing$, $\mathbf{A} \in R_+^n$, and $\mathbf{B} \in R_+^m$. The minimal and maximal values of the functional $\Phi(X)$ on the matrix set $M'(\mathbf{A},\mathbf{B}) \subseteq M(\mathbf{A},\mathbf{B})$ are functions of $(\mathbf{A},\mathbf{B})$. We denote these functions as follows:

$$\widehat{\Phi}(\mathbf{A},\mathbf{B}) = \min\left\{\Phi(X(\mathbf{A},\mathbf{B})) : X(\mathbf{A},\mathbf{B}) \in M'(\mathbf{A},\mathbf{B})\right\},$$

$$\check{\Phi}(\mathbf{A},\mathbf{B}) = \max\left\{\Phi(X(\mathbf{A},\mathbf{B})) : X(\mathbf{A},\mathbf{B}) \in M'(\mathbf{A},\mathbf{B})\right\}.$$

Let $(\mathbf{A},\mathbf{B})$ be a set of vector pairs, where $\mathbf{A} \in R_+^n$, $\mathbf{B} \in R_+^m$, and $M(\mathbf{A},\mathbf{B}) \neq \varnothing$. Each vector pair is associated with a matrix set $M'(\mathbf{A},\mathbf{B})$. Consider the functionals

$$\widehat{\Phi}(\mathbf{A},\mathbf{B}),\ \check{\Phi}(\mathbf{A},\mathbf{B}), \quad (\mathbf{A},\mathbf{B}) \in \left\{(\mathbf{A},\mathbf{B})\right\}, \quad X(\mathbf{A},\mathbf{B}) \in M'(\mathbf{A},\mathbf{B}).$$

Here, a vector pair (i.e., a domain of a transportation model) that optimizes these functionals is required.

Define $K^{n,m} = \left\{k : k \in Z,\ 0 \leq k \leq C_{n+m}^n - 1\right\}$, where the number of indices in $K^{n,m}$ is equal to the numbers of vertices of the fundamental polyhedrons $L(n,m)$ and $M(n,m)$. Assume that the elements of the fundamental systems $\overline{E}^{n,m}$ and $\overline{E}_{n,m}$ are numbered as in (1.5). For $K \subseteq K^{n,m}$ and $c \in R$, where $c > 0$, consider the following sets of vector pairs $\left\{(\mathbf{A},\mathbf{B})\right\} = \left\{\left((a_1,\ldots,a_n),(b_1,\ldots,b_m)\right)\right\}$:

$$L^{n,m}(K;c) = \Big\{(\mathbf{A},\mathbf{B}) : \mathbf{A} \in \overline{R}_+^n,\ \mathbf{B} \in \overline{R}_+^m,\ (\mathbf{A},\mathbf{B}) = c\sum_{k\in K}\alpha_k(\overline{\mathbf{E}}_k,\overline{\mathbf{F}}_k),$$
$$\alpha_k \geq 0,\ k \in K,\ \sum_{k\in K}\alpha_k = 1\Big\};$$

$$L^{n,m}(c;K) = \Big\{(\mathbf{A},\mathbf{B}) : (\mathbf{A},\mathbf{B}) \in L^{n,m}(K;c),\ \min(a_1,b_1) \le c\Big\}.$$

Suppose that $\mathbf{A} = (a_1,\dots,a_n) \in \overline{R}^n_+$, $\mathbf{B} = (b_1,\dots,b_m) \in \overline{R}^m_+$, and $M(\mathbf{A},\mathbf{B}) \neq \varnothing$. Then, if $a_1, b_1 \le c$, where $c > 0$, then Theorem 1.5 implies that $(\mathbf{A}/c, \mathbf{B}/c) \in L(n,m)$. Consequently, if $(\mathbf{A},\mathbf{B}) \in L^{n,m}(c;K)$, then $(\mathbf{A}/c, \mathbf{B}/c) \in L(n,m)$.

Definition 2.2. Sets $L^{n,m}(K;c)$ and $L^{n,m}(c;K)$ are said to be *complete* if every vector pair $(\mathbf{A},\mathbf{B})$ in $L^{n,m}(K;c)$ or $L^{n,m}(c;K)$ is such that, in any representation in the form of (1.27),

$$(\mathbf{A},\mathbf{B}) = c\sum_{k=0}^{C^n_{n+m}-1} \alpha_k(\overline{\mathbf{E}}_k, \overline{\mathbf{F}}_k), \quad (\overline{\mathbf{E}}_k, \overline{\mathbf{F}}_k) \in \overline{E}^{n,m},$$

$$\alpha_k \ge 0, \quad 0 \le k \le C^n_{n+m} - 1, \quad \sum_{k=0}^{C^n_{n+m}-1} \alpha_k = 1,$$

α_k is zero for $k \in K^{n,m} \backslash K$.

The index set K defines the set of all extremal vector pairs that can be used to express elements from $L^{n,m}(K;c)$ and $L^{n,m}(c;K)$ as convex linear combinations.

It is obvious that the index set $\{0, C^n_{n+m} - 1\}$ and the index set corresponding to the extremal vector pairs considered in Example 2.2 define complete sets of vector pairs. Recalling Lemma 1.9, one can construct various complete sets $L^{n,m}(K;c)$ and $L^{n,m}(c;K)$.

The completeness of the vector set $\{(\mathbf{A},\mathbf{B})\}$ can be formalized by invoking the representation of a uniform matrix in terms of a convex linear combination of extremal matrices. A set of vector pairs is complete if and only if every pair $(\mathbf{A},\mathbf{B})$ belonging to the set is such that $\alpha_k = 0$ for $k \in K^{n,m}\backslash K$ in any representation (1.28) of the matrix $X(\mathbf{A},\mathbf{B})$ from $\widetilde{M}(\mathbf{A},\mathbf{B};c)$. Here are examples of complete sets:

$$L^{n,m}(K^{n,m};1) = L(n,m),$$

$$L^{n,m}(K^{n,m};c) = \Big\{(\mathbf{A},\mathbf{B}) : \mathbf{A} \in \overline{R}^n_+,\ \mathbf{B} \in \overline{R}^m_+,\ M(\mathbf{A},\mathbf{B};c) \neq \varnothing\Big\},$$

$$L^{n,m}(c;K^{n,m}) = \Big\{(\mathbf{A},\mathbf{B}) : \mathbf{A} \in \overline{R}^n_+,\ \mathbf{B} \in \overline{R}^m_+,\ \sum_{i=1}^{n} a_i = \sum_{j=1}^{m} b_j,\ \min(a_1,b_1) \le c\Big\}.$$

Here, we develop the methods of optimization on sets of uniform matrices. These methods are based on the representation of a uniform matrix in the form of a convex linear combination of extremal matrices. A complete set of vector pairs defines a set $\{X\}$ of uniform matrices with the following property (recall Theorem 1.12). If $(\mathbf{A},\mathbf{B}) \in L^{n,m}(K;c)$ or $(\mathbf{A},\mathbf{B}) \in L^{n,m}(c;K)$, then $\widetilde{M}(\mathbf{A},\mathbf{B};c) \subseteq \{X\}$ or $\widetilde{M}(\mathbf{A},\mathbf{B}) \subseteq \{X\}$, respectively. Thus, the completeness of a set of vector pairs ensures that the convex linear combinations of extremal matrices belong to $\{X\}$.

It is obvious that $L^{n,m}(K;c)$ and $L^{n,m}(c;K)$ are convex polyhedrons of vector pairs. We refer to any such set as $L^{n,m}$.

Let $\Phi(X)$ be a linear functional defined on n-by-m matrices. For a set of coefficients α_k ($k \in K$, $K \subseteq K^{n,m}$), set $\{\overline{H}_k(\overline{\mathbf{E}}_k,\overline{\mathbf{F}}_k) : k \in K\}$ of extremal matrices, and functional $\Phi(X)$, we define

$$\Phi(\alpha_k : k \in K) = \sum_{k\in K} \alpha_k \Phi(\overline{H}_k(\overline{\mathbf{E}}_k,\overline{\mathbf{F}}_k)). \tag{2.6}$$

Consider a functional $\Phi(X)$ on a matrix set

$$\{X(\mathbf{A},\mathbf{B})\} = \bigcup_{(\mathbf{A},\mathbf{B})\in L^{n,m}} \widetilde{M}(\mathbf{A},\mathbf{B};c),$$

where $\widetilde{M}(\mathbf{A},\mathbf{B};c) = \widetilde{M}(\mathbf{A},\mathbf{B})$ for $L^{n,m} = L^{n,m}(c;K)$.

The following proposition describes the construction of the optimal values of $\widehat{\Phi}(\mathbf{A},\mathbf{B})$ and $\breve{\Phi}(\mathbf{A},\mathbf{B})$ and the vector pairs $(\mathbf{A}^*,\mathbf{B}^*)$ and $(\mathbf{A}^{**},\mathbf{B}^{**})$ on which they are attained. The results are expressed as convex linear combinations of extremal elements.

Lemma 2.2. *Let $\Phi(X)$ be a linear functional of an n-by-m matrix X. Then, the functionals $\widehat{\Phi}(\mathbf{A},\mathbf{B})$ and $\breve{\Phi}(\mathbf{A},\mathbf{B})$, where $(\mathbf{A},\mathbf{B}) \in L^{n,m}$ and $L^{n,m}$ is a complete set of vector pairs, satisfy the following estimates:*

(a) it holds that

$$\begin{aligned}&\min\Big\{\widehat{\Phi}(\mathbf{A},\mathbf{B}) : (\mathbf{A},\mathbf{B}) \in L^{n,m}\Big\} = \\ &c\min\Big\{\Phi(\alpha_k : k \in K) : \alpha_k \ge 0,\ k \in K,\ \sum_{k\in K}\alpha_k = 1,\ (\mathbf{A},\mathbf{B}) \in L^{n,m}\Big\},\end{aligned} \tag{2.7}$$

and (2.6) *and* (2.7) *contain coefficients α_k^* ($k \in K$) such that*

$$c\Phi(\alpha_k^* : k \in K) = \min\Big\{\widehat{\Phi}(\mathbf{A},\mathbf{B}) : (\mathbf{A},\mathbf{B}) \in L^{n,m}\Big\},$$

and the vector pair

$$(\mathbf{A}^*,\mathbf{B}^*) = c\sum_{k\in K} \alpha_k^*(\overline{\mathbf{E}}_k,\overline{\mathbf{F}}_k)$$

minimizes the functional $\widehat{\Phi}(\mathbf{A}, \mathbf{B})$:

$$\widehat{\Phi}(\mathbf{A}^*, \mathbf{B}^*) = \min\left\{\widehat{\Phi}(\mathbf{A}, \mathbf{B}) : (\mathbf{A}, \mathbf{B}) \in L^{n,m}\right\};$$

(b) it holds that

$$\max\left\{\breve{\Phi}(\mathbf{A}, \mathbf{B}) : (\mathbf{A}, \mathbf{B}) \in L^{n,m}\right\} =$$

$$c \max\left\{\Phi(\alpha_k : k \in K) : \alpha_k \geq 0,\ k \in K,\ \sum_{k \in K} \alpha_k = 1,\ (\mathbf{A}, \mathbf{B}) \in L^{n,m}\right\}, \tag{2.8}$$

and (2.6) *and* (2.8) *contain coefficients* α_k^{**} $(k \in K)$ *such that*

$$c\Phi(\alpha_k^{**} : k \in K) = \min\left\{\breve{\Phi}(\mathbf{A}, \mathbf{B}) : (\mathbf{A}, \mathbf{B}) \in L^{n,m}\right\},$$

and the vector pair

$$(\mathbf{A}^{**}, \mathbf{B}^{**}) = c \sum_{k \in K} \alpha_k^{**}(\overline{\mathbf{E}}_k, \overline{\mathbf{F}}_k)$$

maximizes the functional $\breve{\Phi}(\mathbf{A}, \mathbf{B})$:

$$\breve{\Phi}(\mathbf{A}^{**}, \mathbf{B}^{**}) = \max\left\{\breve{\Phi}(\mathbf{A}, \mathbf{B}) : (\mathbf{A}, \mathbf{B}) \in L^{n,m}\right\}.$$

The proof follows from the relations in (2.1) and Definition 2.2.

Let the right-hand side of (2.6) be a convex linear combination. Unless other constraints for the coefficients of this combination are imposed, that is, $L^{n,m} = L^{n,m}(K; c)$, we have

$$\min \Phi(\alpha_k : k \in K) = \min\left\{\Phi(\overline{H}_k(\overline{\mathbf{E}}_k, \overline{\mathbf{F}}_k)) : k \in K\right\},$$

$$\max \Phi(\alpha_k : k \in K) = \max\left\{\Phi(\overline{H}_k(\overline{\mathbf{E}}_k, \overline{\mathbf{F}}_k)) : k \in K\right\}.$$

Consider the case $L^{n,m} = L^{n,m}(c; K)$. Then, the vectors

$$\mathbf{A}^* = (a_1^*, \ldots, a_n^*), \quad \mathbf{B}^* = (b_1^*, \ldots, b_m^*),$$

$$\mathbf{A}^{**} = (a_1^{**}, \ldots, a_n^{**}), \quad \mathbf{B}^{**} = (b_1^{**}, \ldots, b_m^{**})$$

defined in the statement of Lemma 2.2 are such that

$$a_1^*,\ b_1^*,\ a_1^{**},\ b_1^{**} \leq c.$$

Lemma 2.2 leads to the following theorem.

Theorem 2.3. *Suppose that a functional $\Phi(X)$ and a set $L^{n,m}$ of vector pairs satisfy the assumptions of Lemma 2.2. Then, the functionals $\widehat{\Phi}(\mathbf{A},\mathbf{B})$ and $\breve{\Phi}(\mathbf{A},\mathbf{B})$, where $(\mathbf{A},\mathbf{B}) \in L^{n,m}$, satisfy the following relations.*

(a) *If $L^{n,m} = L^{n,m}(K;c)$ and k^*, k^{**} are such indices from K that*

$$\min\left\{\Phi(\overline{H}_k(\overline{\mathbf{E}}_k,\overline{\mathbf{F}}_k)) : k \in K\right\} = \Phi(\overline{H}_{k^*}(\overline{\mathbf{E}}_{k^*},\overline{\mathbf{F}}_{k^*})),$$

$$\max\left\{\Phi(\overline{H}_k(\overline{\mathbf{E}}_k,\overline{\mathbf{F}}_k)) : k \in K\right\} = \Phi(\overline{H}_{k^{**}}(\overline{\mathbf{E}}_{k^{**}},\overline{\mathbf{F}}_{k^{**}})),$$

then the vector pairs $(\mathbf{A}^*,\mathbf{B}^*) = c(\overline{\mathbf{E}}_{k^*},\overline{\mathbf{F}}_{k^*})$ and $(\mathbf{A}^{**},\mathbf{B}^{**}) = c(\overline{\mathbf{E}}_{k^{**}},\overline{\mathbf{F}}_{k^{**}})$ optimize the functionals $\widehat{\Phi}(\mathbf{A},\mathbf{B})$ and $\breve{\Phi}(\mathbf{A},\mathbf{B})$:

$$\min\left\{\widehat{\Phi}(\mathbf{A},\mathbf{B}) : (\mathbf{A},\mathbf{B}) \in L^{n,m}\right\} = \widehat{\Phi}(c\overline{\mathbf{E}}_{k^*},c\overline{\mathbf{F}}_{k^*}) = c\Phi(\overline{H}_{k^*}(\overline{\mathbf{E}}_{k^*},\overline{\mathbf{F}}_{k^*})),$$

$$\max\left\{\breve{\Phi}(\mathbf{A},\mathbf{B}) : (\mathbf{A},\mathbf{B}) \in L^{n,m}\right\} = \breve{\Phi}(c\overline{\mathbf{E}}_{k^{**}},c\overline{\mathbf{F}}_{k^{**}}) = c\Phi(\overline{H}_{k^{**}}(\overline{\mathbf{E}}_{k^{**}},\overline{\mathbf{F}}_{k^{**}})).$$

(b) *If $L^{n,m} = L^{n,m}(c;K)$ and*

$$\min\left\{\widehat{\Phi}(\mathbf{A},\mathbf{B}) : (\mathbf{A},\mathbf{B}) \in L^{n,m}\right\} = \widehat{\Phi}(\mathbf{A}^*,\mathbf{B}^*),$$

$$\max\left\{\breve{\Phi}(\mathbf{A},\mathbf{B}) : (\mathbf{A},\mathbf{B}) \in L^{n,m}\right\} = \breve{\Phi}(\mathbf{A}^{**},\mathbf{B}^{**}),$$

where

$$(\mathbf{A}^*,\mathbf{B}^*) = c\sum_{k\in K}\alpha_k^*(\overline{\mathbf{E}}_k,\overline{\mathbf{F}}_k), \quad (\mathbf{A}^{**},\mathbf{B}^{**}) = c\sum_{k\in K}\alpha_k^{**}(\overline{\mathbf{E}}_k,\overline{\mathbf{F}}_k),$$

and $(\overline{\mathbf{E}}_k,\overline{\mathbf{F}}_k) = ((e_1^{(k)},\ldots,e_n^{(k)}),(f_1^{(k)},\ldots,f_m^{(k)}))$, then the coefficients α_k^*, α_k^{**} satisfy the relations

$$\alpha_k^*,\ \alpha_k^{**} \geq 0, \quad k \in K,$$

$$\sum_{k\in K}\alpha_k^* = 1, \quad \sum_{k\in K}\alpha_k^{**} = 1,$$

$$\sum_{k\in K}\alpha_k^* e_1^{(k)} \leq 1, \quad \sum_{k\in K}\alpha_k^* f_1^{(k)} \leq 1,$$

$$\sum_{k\in K}\alpha_k^{**} e_1^{(k)} \leq 1, \quad \sum_{k\in K}\alpha_k^{**} f_1^{(k)} \leq 1.$$

Remark 2.2.

(a) The key functionals in Lemma 2.2 and Theorem 2.3 are expressed as

$$\widehat{\Phi}(\mathbf{A},\mathbf{B}) = \min\Big\{\Phi(X(\mathbf{A},\mathbf{B})) : X(\mathbf{A},\mathbf{B}) \in M'(\mathbf{A},\mathbf{B}),\ (\mathbf{A},\mathbf{B}) \in L^{n,m}\Big\},$$

$$\breve{\Phi}(\mathbf{A},\mathbf{B}) = \max\Big\{\Phi(X(\mathbf{A},\mathbf{B})) : X(\mathbf{A},\mathbf{B}) \in M'(\mathbf{A},\mathbf{B}),\ (\mathbf{A},\mathbf{B}) \in L^{n,m}\Big\},$$

where $M'(\mathbf{A},\mathbf{B}) = \widetilde{M}(\mathbf{A},\mathbf{B};c)$ when $L^{n,m} = L^{n,m}(K;c)$ and $M'(\mathbf{A},\mathbf{B}) = \widetilde{M}(\mathbf{A},\mathbf{B})$ when $L^{n,m} = L^{n,m}(c;K)$.

(b) If $L^{n,m} = L^{n,m}(K;c)$, then we have the following coefficients in the convex linear combinations given in Lemma 2.2 and Theorem 2.3:

$$\alpha_k^* = \begin{cases} 1, & k = k^*, \\ 0, & k \in K, \quad k \neq k^*, \end{cases} \qquad \alpha_k^{**} = \begin{cases} 1, & k = k^{**}, \\ 0, & k \in K,\ k \neq k^{**}. \end{cases}$$

When $L^{n,m} = L^{n,m}(c;K)$, the coefficients α_k^* and α_k^{**} $(k \in K)$ that optimize $\widehat{\Phi}(\mathbf{A},\mathbf{B})$ and $\breve{\Phi}(\mathbf{A},\mathbf{B})$ in Theorem 2.3 are calculated by solving a simple mathematical programming problem [6, 7, 11].

(c) It is clear that $\left|M(\overline{\mathbf{E}},\overline{\mathbf{F}};1)\right| = \left|M(c\overline{\mathbf{E}},c\overline{\mathbf{F}};c)\right| = 1$ for any extremal vector pair $(\overline{\mathbf{E}},\overline{\mathbf{F}})$ and any $c > 0$. Note that the set $\widetilde{M}(\overline{\mathbf{E}},\overline{\mathbf{F}})$ may contain an infinite number of matrices. For example, any uniform matrix in $\widetilde{M}\left((2,1),(2,1)\right)$ can be expressed as

$$X\left((2,1),(2,1)\right) = \begin{pmatrix} 1+\varepsilon & 1-\varepsilon \\ 1-\varepsilon & \varepsilon \end{pmatrix},$$

where $0 \le \varepsilon \le 1/2$. When $\varepsilon = 0$, the matrix is extremal.

Example 2.3: Application of Theorem 2.3.
Consider the linear functional

$$\Phi(X) = 2\sum_{i=1}^{2}\sum_{j=1}^{2} x_{ij} - 3\sum_{i=3}^{4}\sum_{j=1}^{2} x_{ij} - 2\sum_{i=1}^{2}\sum_{j=3}^{4} x_{ij} + \sum_{i=3}^{4}\sum_{j=3}^{4} x_{ij},$$

where $X = (x_{ij})$ $(1 \le i,j \le 4)$. The extremal vector pairs

$$\begin{aligned}
(\overline{\mathbf{E}}_0,\overline{\mathbf{F}}_0) &= \left((0,0,0,0),(0,0,0,0)\right),\\
(\overline{\mathbf{E}}_1,\overline{\mathbf{F}}_1) &= \left((2,2,0,0),(2,2,0,0)\right),\\
(\overline{\mathbf{E}}_2,\overline{\mathbf{F}}_2) &= \left((2,2,2,2),(4,4,0,0)\right),\\
(\overline{\mathbf{E}}_3,\overline{\mathbf{F}}_3) &= \left((4,4,0,0),(2,2,2,2)\right),\\
(\overline{\mathbf{E}}_4,\overline{\mathbf{F}}_4) &= \left((4,4,2,2),(4,4,2,2)\right),\\
(\overline{\mathbf{E}}_5,\overline{\mathbf{F}}_5) &= \left((4,4,4,4),(4,4,4,4)\right),
\end{aligned}$$

constitute complete sets by Lemma 1.9. One can easily construct the extremal matrices $\overline{H}_k(\overline{\mathbf{E}}_k, \overline{\mathbf{F}}_k)$, where $k \in K = \{0, 1, 2, 3, 4, 5\}$. Denote by Φ_k the value of $\Phi(X)$ on $\overline{H}_k(\overline{\mathbf{E}}_k, \overline{\mathbf{F}}_k)$ $(k \in K)$. We have

$$\Phi_0 = 0, \quad \Phi_1 = 8, \quad \Phi_2 = -4, \quad \Phi_3 = 0, \quad \Phi_4 = -12, \quad \Phi_5 = -8.$$

Therefore, (2.6) takes the form

$$\Phi(\alpha_k : k \in K) = 8\alpha_1 - 4\alpha_2 - 12\alpha_4 - 8\alpha_5.$$

Consider the following complete sets of vector pairs: $L^{2,2} = L^{2,2}(K; 4)$ and $L^{2,2} = L^{2,2}(4; K)$.

Suppose that $L^{2,2} = L^{2,2}(K; 4)$, and $\widehat{\Phi}(\mathbf{A}, \mathbf{B})$ and $\breve{\Phi}(\mathbf{A}, \mathbf{B})$ are functionals of vector pairs, where $(\mathbf{A}, \mathbf{B}) \in L^{2,2}(K; 4)$. Then, the nonnegative coefficients α_k $(k \in K)$ whose sum is unity make up the set $L^{2,2}$. Therefore,

$$\min\Big\{\widehat{\Phi}(\mathbf{A}, \mathbf{B}) : (\mathbf{A}, \mathbf{B}) \in L^{2,2}\Big\} =$$
$$4 \min\Big\{8\alpha_1 - 4\alpha_2 - 12\alpha_4 - 8\alpha_5 : \alpha_k \geq 0,\ k \in K,\ \sum_{k \in K}^{5} \alpha_k = 1\Big\};$$

$$\max\Big\{\breve{\Phi}(\mathbf{A}, \mathbf{B}) : (\mathbf{A}, \mathbf{B}) \in L^{2,2}\Big\} =$$
$$4 \max\Big\{8\alpha_1 - 4\alpha_2 - 12\alpha_4 - 8\alpha_5 : \alpha_k \geq 0,\ k \in K,\ \sum_{k \in K}^{5} \alpha_k = 1\Big\}.$$

It is obvious that

$$\min\Big\{\widehat{\Phi}(\mathbf{A}, \mathbf{B}) : (\mathbf{A}, \mathbf{B}) \in L^{2,2}\Big\} = 4\Phi(\overline{H}_4(\overline{\mathbf{E}}_4, \overline{\mathbf{F}}_4)) = -48,$$

$$\max\Big\{\breve{\Phi}(\mathbf{A}, \mathbf{B}) : (\mathbf{A}, \mathbf{B}) \in L^{2,2}\Big\} = 4\Phi(\overline{H}_1(\overline{\mathbf{E}}_1, \overline{\mathbf{F}}_1)) = 32,$$

and

$$(\mathbf{A}^*, \mathbf{B}^*) = 4(\overline{\mathbf{E}}_4, \overline{\mathbf{F}}_4), \qquad (\mathbf{A}^{**}, \mathbf{B}^{**}) = 4(\overline{\mathbf{E}}_1, \overline{\mathbf{F}}_1).$$

Suppose $L^{2,2} = L^{2,2}(4; K)$. Then, the functionals $\widehat{\Phi}(\mathbf{A}, \mathbf{B})$ and $\breve{\Phi}(\mathbf{A}, \mathbf{B})$ must be optimized by optimizing the linear form $\Phi(\alpha_k : k \in K)$ under the following constraints:

$$\begin{cases} \alpha_0 + \alpha_1 + \alpha_2 + \alpha_3 + \alpha_4 + \alpha_5 = 1, \\ 2\alpha_1 + 2\alpha_2 + 4\alpha_3 + 4\alpha_4 + 4\alpha_5 \leq 1, \\ 2\alpha_1 + 4\alpha_2 + 2\alpha_3 + 4\alpha_4 + 4\alpha_5 \leq 1, \\ \alpha_k \geq 0, \quad 0 \leq k \leq 5. \end{cases}$$

This system yields $\alpha_1 \leq 1/2$ and $\alpha_2, \alpha_3, \alpha_4, \alpha_5 \leq 1/4$.

The functional $\widehat{\Phi}(\mathbf{A},\mathbf{B})$, where $(\mathbf{A},\mathbf{B}) \in L^{2,2}(4;K)$, is maximized when $\alpha_4 = 1/4$, $\alpha_0 = 3/4$, and $\alpha_k = 0$ ($k \in K$, $k \neq 0$, $k \neq 4$). The functional $\breve{\Phi}(\mathbf{A},\mathbf{B}) \in L^{2,2}(4;K)$ is maximized when $\alpha_1 = 1$ and $\alpha_k = 0$ ($k \in K$, $k \neq 1$). Thus,

$$\min\left\{\widehat{\Phi}(\mathbf{A},\mathbf{B}) : (\mathbf{A},\mathbf{B}) \in L^{2,2}\right\} = 4 \cdot \alpha_4 \Phi(\overline{H}_4(\overline{\mathbf{E}}_4, \overline{\mathbf{F}}_4)) = -12,$$

$$\max\left\{\breve{\Phi}(\mathbf{A},\mathbf{B}) : (\mathbf{A},\mathbf{B}) \in L^{2,2}\right\} = 4 \cdot \alpha_1 \Phi(\overline{H}_1(\overline{\mathbf{E}}_1, \overline{\mathbf{F}}_1)) = 32,$$

and

$$(\mathbf{A}^*, \mathbf{B}^*) = (\overline{\mathbf{E}}_4, \overline{\mathbf{F}}_4), \qquad (\mathbf{A}^{**}, \mathbf{B}^{**}) = 4(\overline{\mathbf{E}}_1, \overline{\mathbf{F}}_1).$$

Let us analyze the cases of convex and concave $\Phi(X)$. Let $\{(\mathbf{A},\mathbf{B})\}$ be a convex set of vector pairs, where $\mathbf{A} \in R^n_+$, $\mathbf{B} \in R^m_+$, and each pair defines a transportation polyhedron $M(\mathbf{A},\mathbf{B})$. For every $(\mathbf{A},\mathbf{B}) \in \{(\mathbf{A},\mathbf{B})\}$ define a convex matrix set $M'(\mathbf{A},\mathbf{B}) \subseteq M(\mathbf{A},\mathbf{B})$. Consider the convex matrix set

$$\{X(\mathbf{A},\mathbf{B})\} = \bigcup_{(\mathbf{A},\mathbf{B}) \in \{(\mathbf{A},\mathbf{B})\}} M'(\mathbf{A},\mathbf{B}) \tag{2.9}$$

and a functional $\Phi(X)$ ($X \in \{X(\mathbf{A},\mathbf{B})\}$), which may be a convex or concave function. Assume that $(\mathbf{A},\mathbf{B}) = \alpha(\mathbf{A}',\mathbf{B}') + \beta(\mathbf{A}'',\mathbf{B}'')$, where α, $\beta \geq 0$, $\alpha + \beta = 1$, and $(\mathbf{A}',\mathbf{B}')$, $(\mathbf{A}'',\mathbf{B}'') \in \{(\mathbf{A},\mathbf{B})\}$.

Let us examine the functional $\widehat{\Phi}(\mathbf{A},\mathbf{B})$ assuming that $\Phi(X(\mathbf{A},\mathbf{B}))$ is a convex function. We have the following sequence of relations:

$$\alpha\widehat{\Phi}(\mathbf{A}',\mathbf{B}') + \beta\widehat{\Phi}(\mathbf{A}'',\mathbf{B}'') =$$

$$\alpha \min\left\{\Phi(X(\mathbf{A}',\mathbf{B}')) : X(\mathbf{A}',\mathbf{B}') \in M'(\mathbf{A}',\mathbf{B}')\right\} +$$

$$\beta \min\left\{\Phi(X(\mathbf{A}'',\mathbf{B}'')) : X(\mathbf{A}'',\mathbf{B}'') \in M'(\mathbf{A}'',\mathbf{B}'')\right\} =$$

$$\alpha\Phi(X_1(\mathbf{A}',\mathbf{B}')) + \beta\Phi(X_2(\mathbf{A}'',\mathbf{B}'')) \geq \Phi\left(\alpha X_1(\mathbf{A}',\mathbf{B}') + \beta X_2(\mathbf{A}'',\mathbf{B}'')\right) =$$

$$\Phi(X(\mathbf{A},\mathbf{B})) \geq \min\left\{\Phi(X(\mathbf{A},\mathbf{B})) : X(\mathbf{A},\mathbf{B}) \in M'(\mathbf{A},\mathbf{B})\right\} = \widehat{\Phi}(\mathbf{A},\mathbf{B}).$$

Hence, the functional $\widehat{\Phi}(\mathbf{A},\mathbf{B})$ is a convex function.

Similarly, it is shown that $\breve{\Phi}(\mathbf{A},\mathbf{B})$ is a concave functional if $\Phi(X(\mathbf{A},\mathbf{B}))$ is a concave function.

Note that if $\Phi(X(\mathbf{A},\mathbf{B}))$ is a linear functional, then $\widehat{\Phi}(\mathbf{A},\mathbf{B})$ and $\breve{\Phi}(\mathbf{A},\mathbf{B})$ are convex and concave functionals, respectively.

For a functional $\Phi(X)$ and constant $c > 0$, we define

$$\Phi(\alpha_k : k \in K; c) = \sum_{k \in K} \alpha_k \Phi(c\overline{H}_k(\overline{\mathbf{E}}_k, \overline{\mathbf{F}}_k)),$$

where the coefficients α_k and extremal matrices are defined by (2.6). The following proposition is true by virtue of (2.1) combined with the convexity and concavity conditions for a functional.

Theorem 2.4. *Let $\Phi(X)$ be an arbitrary convex or concave functional defined on a convex set of n-by-m matrices. Then, the functionals $\widehat{\Phi}(\mathbf{A}, \mathbf{B})$ and $\breve{\Phi}(\mathbf{A}, \mathbf{B})$, where $(\mathbf{A}, \mathbf{B}) \in L^{n,m}$ $(L^{n,m} = L^{n,m}(K; c), L^{n,m}(c; K))$ satisfy the following relations.*

(a) *The functional $\widehat{\Phi}(\mathbf{A}, \mathbf{B})$ is convex, and*

$$\min\left\{\widehat{\Phi}(\mathbf{A}, \mathbf{B}) : (\mathbf{A}, \mathbf{B}) \in L^{n,m}\right\} \leq \min\left\{\Phi(\alpha_k : k \in K; c), (\mathbf{A}, \mathbf{B}) \in L^{n,m}\right\}.$$

(b) *The functional $\breve{\Phi}(\mathbf{A}, \mathbf{B})$ is concave, and*

$$\max\left\{\breve{\Phi}(\mathbf{A}, \mathbf{B}) : (\mathbf{A}, \mathbf{B}) \in L^{n,m}\right\} \geq \max\left\{\Phi(\alpha_k : k \in K; c), (\mathbf{A}, \mathbf{B}) \in L^{n,m}\right\}.$$

Note that the bounds given by the theorem are attainable.

Suppose that linear functionals $\Phi^{(i)}(X) = \Phi^{(i)}(X(\mathbf{A}, \mathbf{B}))$ $(1 \leq i \leq t)$ are defined on the convex matrix set in (2.9). Define the following convex set of linear functionals on the matrix set $\{X(\mathbf{A}, \mathbf{B}) : (\mathbf{A}, \mathbf{B}) \in \{(\mathbf{A}, \mathbf{B})\}, X(\mathbf{A}, \mathbf{B}) \in M'(\mathbf{A}, \mathbf{B})\}$:

$$\mathfrak{M}^{n,m}(\{X\}) = \mathfrak{M}^{n,m}(\{X(\mathbf{A}, \mathbf{B})\}) = \Big\{\Phi(X(\mathbf{A}, \mathbf{B})) : \Phi(X(\mathbf{A}, \mathbf{B})) = \sum_{i=1}^{t} \beta_i \Phi^{(i)}(X(\mathbf{A}, \mathbf{B})),\ \beta_i \geq 0,\ 1 \leq i \leq t,\ \sum_{i=1}^{t} \beta_i = 1\Big\}.$$

Here, an optimal transportation model is required. Any closed transportation model is formulated in terms of a vector pair, which defines a matrix set, and optimizable functional on that set. We seek the optimal transportation model $((\mathbf{A}, \mathbf{B}), \Phi(X))$, where $\Phi(X) = \Phi(X(\mathbf{A}, \mathbf{B}))$, in the set

$$Q = \Big\{((\mathbf{A}, \mathbf{B}), \Phi(X)) : (\mathbf{A}, \mathbf{B}) \in L^{n,m},\ \Phi(X) \in \mathfrak{M}^{n,m}(\{X(\mathbf{A}, \mathbf{B})\})\Big\}. \tag{2.10}$$

We assume that each matrix set $M'(\mathbf{A},\mathbf{B})$ contains uniform matrices only. In view of (2.9) and (2.10), this means that

$$\{X(\mathbf{A},\mathbf{B})\} = \bigcup_{(\mathbf{A},\mathbf{B})\in L^{n,m}} \widetilde{M}(\mathbf{A},\mathbf{B};c), \tag{2.11}$$

where $L^{n,m} = L^{n,m}(K;c)$ (or $L^{n,m} = L^{n,m}(c;K)$) is a complete set of vector pairs.

For any vector pair $(\mathbf{A},\mathbf{B})$, we can use (2.1) to express any linear functional $\Phi(X)$ from the set defined by (2.10) as

$$\Phi(X) = \sum_{i=1}^{t} \beta_i \Phi^{(i)}(X) = \sum_{i=1}^{t} \beta_i \Phi^{(i)}(X(\mathbf{A},\mathbf{B})) =$$
$$\sum_{i=1}^{t} \beta_i \Phi^{(i)}\Big(c\sum_{k\in K}\alpha_k \overline{H}_k(\overline{\mathbf{E}}_k,\overline{\mathbf{F}}_k)\Big) = c\sum_{i=1}^{t}\sum_{k\in K}\beta_i\alpha_k\Phi^{(i)}\left(\overline{H}_k(\overline{\mathbf{E}}_k,\overline{\mathbf{F}}_k)\right) =$$
$$c\sum_{k\in K}\alpha_k\sum_{i=1}^{t}\beta_i\Phi^{(i)}\left(\overline{H}_k(\overline{\mathbf{E}}_k,\overline{\mathbf{F}}_k)\right) = c\sum_{k\in K}\alpha_k\Phi\left(\overline{H}_k(\overline{\mathbf{E}}_k,\overline{\mathbf{F}}_k)\right).$$

Consider the particular expression

$$\Phi(X) = c\sum_{i=1}^{t}\sum_{k\in K}\beta_i\alpha_k\Phi^{(i)}\left(\overline{H}_k(\overline{\mathbf{E}}_k,\overline{\mathbf{F}}_k)\right). \tag{2.12}$$

The right-hand side of (2.12) is linear with respect to $\gamma_k^{(i)} = \alpha_k\beta_i$, where $k \in K$ and $1 \le i \le t$. Every functional $\Phi^{(i)}(X)$ of an extremal matrix $\overline{H}_k(\overline{\mathbf{E}}_k,\overline{\mathbf{F}}_k)$ can be easily calculated. Expression (2.12) contains $t|K|$ functionals of this type. Suppose that β_i^*, α_k^* and β_i^{**}, α_k^{**} ($1 \le i \le t$, $k \in K$) minimize and maximize the right-hand side of (2.12), respectively. Denote the corresponding minimal and maximal values of (2.12) by Φ^* and Φ^{**}, respectively. Express the vector pairs and functionals as

$$(\mathbf{A}^*,\mathbf{B}^*) = c\sum_{k\in K}\alpha_k^*(\overline{\mathbf{E}}_k,\overline{\mathbf{F}}_k), \quad (\mathbf{A}^{**},\mathbf{B}^{**}) = c\sum_{k\in K}\alpha_k^{**}(\overline{\mathbf{E}}_k,\overline{\mathbf{F}}_k),$$
$$\Phi^*(X) = \sum_{i=1}^{t}\beta_i^*\Phi^{(i)}(X), \quad \Phi^{**}(X) = \sum_{i=1}^{t}\beta_i^{**}\Phi^{(i)}(X), \tag{2.13}$$

where the coefficients α_k^*, α_k^{**} ($k \in K$) and β_i^*, β_i^{**} ($1 \le i \le t$) define convex linear combinations.

The following proposition is true by virtue of (2.1).

Theorem 2.5. *Suppose that* (2.10) *defines a set of transportation problems, where* $L^{n,m}$ *is a complete set of vector pairs, and expressions* (2.13) *specify such elements* $\big((\mathbf{A}^*,\mathbf{B}^*),\Phi^*(X)\big)$ *and* $\big((\mathbf{A}^{**},\mathbf{B}^{**}),\Phi^{**}(X)\big)$ *of* Q *that*

$$\min_{X\in M'(\mathbf{A},\mathbf{B})}\Big\{\Phi(X):\big((\mathbf{A},\mathbf{B}),\Phi(X)\big)\in Q\Big\}=\min_{X(\mathbf{A}^*,\mathbf{B}^*)\in M'(\mathbf{A}^*,\mathbf{B}^*)}\Phi^*(X(\mathbf{A}^*,\mathbf{B}^*))=\Phi^*,$$

$$\max_{X\in M'(\mathbf{A},\mathbf{B})}\Big\{\Phi(X):\big((\mathbf{A},\mathbf{B}),\Phi(X)\big)\in Q\Big\}=\max_{X(\mathbf{A}^{**},\mathbf{B}^{**})\in M'(\mathbf{A}^{**},\mathbf{B}^{**})}\Phi^{**}(X(\mathbf{A}^{**},\mathbf{B}^{**}))=\Phi^{**}.$$

(a) *If* $L^{n,m}=L^{n,m}(K;c)$ *and*

$$\min\Big\{\Phi^{(i)}(\overline{H}_k(\overline{\mathbf{E}}_k,\overline{\mathbf{F}}_k)):k\in K,\ 1\le i\le t\Big\}=\Phi^{(i^*)}(\overline{H}_{k^*}(\overline{\mathbf{E}}_{k^*},\overline{\mathbf{F}}_{k^*})),$$

$$\max\Big\{\Phi^{(i)}(\overline{H}_k(\overline{\mathbf{E}}_k,\overline{\mathbf{F}}_k)):k\in K,\ 1\le i\le t\Big\}=\Phi^{(i^{**})}(\overline{H}_{k^{**}}(\overline{\mathbf{E}}_{k^{**}},\overline{\mathbf{F}}_{k^{**}})),$$

then

$$(\mathbf{A}^*,\mathbf{B}^*)=c(\overline{\mathbf{E}}_{k^*},\overline{\mathbf{F}}_{k^*}),\quad \Phi^*(X)=\Phi^{(i^*)}(X),$$

$$\min_{X\in M'(\mathbf{A},\mathbf{B})}\Big\{\Phi(X):\big((\mathbf{A},\mathbf{B}),\Phi(X)\big)\in Q\Big\}=c\Phi^{(i^*)}(\overline{H}_{k^*}(\overline{\mathbf{E}}_{k^*},\overline{\mathbf{F}}_{k^*}))=\Phi^*;$$

$$(\mathbf{A}^{**},\mathbf{B}^{**})=c(\overline{\mathbf{E}}_{k^{**}},\overline{\mathbf{F}}_{k^{**}}),\quad \Phi^{**}(X)=\Phi^{(i^{**})}(X),$$

$$\max_{X\in M'(\mathbf{A},\mathbf{B})}\Big\{\Phi(X):\big((\mathbf{A},\mathbf{B}),\Phi(X)\big)\in Q\Big\}=c\Phi^{(i^{**})}(\overline{H}_{k^{**}}(\overline{\mathbf{E}}_{k^{**}},\overline{\mathbf{F}}_{k^{**}}))=\Phi^{**},$$

where $M'(\mathbf{A},\mathbf{B})=\widetilde{M}(\mathbf{A},\mathbf{B};c)$.

(b) *If* $L^{n,m}=L^{n,m}(c;K)$, *then the domain of the functional* $\Phi(X)\in\mathfrak{N}^{n,m}(\{X\})$ *consists of sets* $M'(\mathbf{A},\mathbf{B})=\widetilde{M}(\mathbf{A},\mathbf{B})$ *of uniform matrices, where* $(\mathbf{A},\mathbf{B})\in L^{n,m}$, *and the coefficients in expressions* (2.13) *for vector pairs* $(\mathbf{A}^*,\mathbf{B}^*)$ *and* $(\mathbf{A}^{**},\mathbf{B}^{**})$ *and functionals* $\Phi^*(X)$ *and* $\Phi^{**}(X)$ *satisfy the relations*

$$\alpha_k^*,\ \alpha_k^{**}\ge 0,\quad k\in K,\quad \sum_{k\in K}\alpha_k^*=1,\quad \sum_{k\in K}\alpha_k^{**}=1,$$

$$\beta_i^*,\ \beta_i^{**}\ge 0,\quad 1\le i\le t,\quad \sum_{i=1}^{t}\beta_i^*=1,\quad \sum_{i=1}^{t}\beta_i^{**}=1,$$

$$\sum_{k\in K}\alpha_k^*e_1^{(k)}\le 1,\quad \sum_{k\in K}\alpha_k^*f_1^{(k)}\le 1,$$

$$\sum_{k\in K} \alpha_k^{**} e_1^{(k)} \le 1, \quad \sum_{k\in K} \alpha_k^{**} f_1^{(k)} \le 1.$$

Remark 2.3.

(a) By proposition (a) of Theorem 2.5, the coefficients of convex linear combinations in (2.13) are

$$\alpha_k^* = \begin{cases} 1, & k = k^*, \\ 0, & k \in K, \quad k \neq k^*, \end{cases} \qquad \alpha_k^{**} = \begin{cases} 1, & k = k^{**}, \\ 0, & k \in K, \quad k \neq k^{**}. \end{cases}$$

These expressions define truncated uniform polyhedrons $\widetilde{M}(c\overline{\mathbf{E}}_{k^*}, c\overline{\mathbf{F}}_{k^*}; c)$ and $\widetilde{M}(c\overline{\mathbf{E}}_{k^{**}}, c\overline{\mathbf{F}}_{k^{**}}; c)$, each containing only one matrix.

(b) The right-hand side in (2.12) can be optimized under the constraints stated by proposition (b) of Theorem 2.5 by solving a simple mathematical programming problem [6, 7, 11].

Example 2.4: Application of Theorem 2.5.

Consider the linear forms

$$\Phi^{(1)}(X) = -3x_{11} - 2x_{12} + 2x_{21} + x_{22},$$

$$\Phi^{(2)}(X) = x_{11} + 3x_{12} + 3x_{21} - 2x_{22},$$

where $X = (x_{ij})$ $(1 \le i, j \le 2)$, as defined on the matrix set $\{X : X/2 \in M(2,2)\}$. The set $L^{2,2}(K^{2,2}; 2)$ of vector pairs defined by these matrices is complete. Assume that $\Phi_\beta(X) = \beta\Phi^{(1)}(X) + (1-\beta)\Phi^{(2)}(X)$, where $0 \le \beta \le 1$.

The extremal vector pairs and matrices that make up the fundamental polyhedrons $L(2,2)$ and $M(2,2)$ were listed in Example 1.1. Expression (2.12) has the form

$$\Phi_\beta(X) = 2\left[\beta\sum_{k=0}^{5}\alpha_k\Phi^{(1)}(\overline{H}_k(\overline{\mathbf{E}}_k, \overline{\mathbf{F}}_k)) + (1-\beta)\sum_{k=0}^{5}\alpha_k\Phi^{(2)}(\overline{H}_k(\overline{\mathbf{E}}_k, \overline{\mathbf{F}}_k))\right],$$

and the set of transportation problems in (2.10) is written as

$$Q = \left\{((\mathbf{A}, \mathbf{B}), \Phi_\beta(X)) : \mathbf{A}, \mathbf{B})/2 \in L(2,2),\ 0 \le \beta \le 1\right\}.$$

By Theorem 2.5, the optimal uniform transportation models in Q are determined by calculating the values of the functionals $\Phi^{(i)}(X)$ $(1 \le i \le 2)$

for the extremal matrices $\overline{H}_k(\overline{\mathbf{E}}_k, \overline{\mathbf{F}}_k)$ of the fundamental system $\overline{E}_{2,2}$. The indices i and k of the minimal and maximal values determine the optimal transportation problems. We denote the linear functional $\Phi^{(i)}(X)$ of an extremal matrix $\overline{H}_k(\overline{\mathbf{E}}_k, \overline{\mathbf{F}}_k)$ by $\Phi_k^{(i)}$, where $1 \le i \le 2$ and $0 \le k \le 5$. Recalling Example 1.1, we obtain

$$\Phi_0^{(1)} = 0, \quad \Phi_1^{(1)} = -3, \quad \Phi_2^{(1)} = -5, \quad \Phi_3^{(1)} = -1, \quad \Phi_4^{(1)} = -3, \quad \Phi_5^{(1)} = -2;$$

$$\Phi_0^{(2)} = 0, \quad \Phi_1^{(2)} = 1, \quad \Phi_2^{(2)} = 4, \quad \Phi_3^{(2)} = 4, \quad \Phi_4^{(2)} = 7, \quad \Phi_5^{(2)} = 5.$$

It can be readily shown that $k^* = 2$, $i^* = 1$, $k^{**} = 4$, and $i^{**} = 2$:

$$\min_{i,k} \Phi_k^{(i)} = -5 = \Phi_2^{(1)}, \quad \max_{i,k} \Phi_k^{(i)} = 7 = \Phi_4^{(2)}.$$

Thus, we single out the transportation models

$$\left((2\overline{\mathbf{E}}_2, 2\overline{\mathbf{F}}_2), \Phi^{(1)}(X)\right), \left((2\overline{\mathbf{E}}_4, 2\overline{\mathbf{F}}_4), \Phi^{(2)}(X)\right) \in Q.$$

In the first model,

$$\min_{X \in \widetilde{M}(\mathbf{A},\mathbf{B};2)} \left\{\Phi_\beta(X) : ((\mathbf{A}, \mathbf{B}), \Phi_\beta(X)) \in Q\right\} =$$
$$\min_{X \in \widetilde{M}(2\overline{\mathbf{E}}_2, 2\overline{\mathbf{F}}_2;2)} \Phi^{(1)}(X) = 2\Phi^{(1)}(\overline{H}_2(\overline{\mathbf{E}}_2, \overline{\mathbf{F}}_2)) = -10.$$

In the second model,

$$\max_{X \in \widetilde{M}(\mathbf{A},\mathbf{B};2)} \left\{\Phi_\beta(X) : ((\mathbf{A}, \mathbf{B}), \Phi_\beta(X)) \in Q\right\} =$$
$$\max_{X \in \widetilde{M}(2\overline{\mathbf{E}}_4, 2\overline{\mathbf{F}}_4;2)} \Phi^{(2)}(X) = 2\Phi^{(2)}(\overline{H}_4(\overline{\mathbf{E}}_4, \overline{\mathbf{F}}_4)) = 14.$$

This completes the first step of analysis in this example.

Next, consider the complete set $L^{2,2}(4; K^{2,2})$ of vector pairs. The corresponding functional (2.12) is

$$\Phi_\beta(X) = 4\Big(\beta \sum_{k=0}^{5} \alpha_k \Phi_k^{(1)} + (1-\beta) \sum_{k=0}^{5} \alpha_k \Phi_k^{(2)}\Big).$$

The values of $\Phi_k^{(i)}$ $(1 \le i \le 2,\ 0 \le k \le 5)$ have just been calculated. Further calculations yield

$$\Phi_\beta(X) =$$
$$4\left(\alpha_1 + 4\alpha_2 + 4\alpha_3 + 7\alpha_4 + 5\alpha_5 - 4\beta\alpha_1 - 9\beta\alpha_2 - 5\beta\alpha_3 - 10\beta\alpha_4 - 7\beta\alpha_5\right),$$

where

$$X \in \left\{X(\mathbf{A},\mathbf{B}) : (\mathbf{A},\mathbf{B})/4 \in L(2,2),\ a_1,\, b_1 \le 4,\ X(\mathbf{A},\mathbf{B}) \in \widetilde{M}(\mathbf{A},\mathbf{B})\right\}.$$

Note that since a_1, $b_1 \le 4$, $\mathbf{A}$, $\mathbf{B} \in \overline{R}^2_+$, and $a_1+a_2 = b_1+b_2$ in this example, we have $(\mathbf{A},\mathbf{B})/4 \in L(2,2)$.

Referring again to Example 1.1, we construct the following set of constraints for optimization of the functional $\Phi_\beta(X)$:

$$\begin{cases} \alpha_k \ge 0, \quad 0 \le k \le 5, \\ \alpha_0 + \alpha_1 + \alpha_2 + \alpha_3 + \alpha_4 + \alpha_5 = 1, \\ \alpha_1 + 2\alpha_2 + \alpha_3 + 2\alpha_4 + 2\alpha_5 \le 1, \\ \alpha_1 + \alpha_2 + 2\alpha_3 + 2\alpha_4 + 2\alpha_5 \le 1, \\ 0 \le \beta \le 1. \end{cases}$$

Now, it is obvious that the minimal and maximal values of $\Phi_\beta(X)$ are attained when $\beta = 1$ and $\beta = 0$, respectively:

$$\Phi_1(X) = 4(-3\alpha_1 - 5\alpha_2 - \alpha_3 - 3\alpha_4 - 2\alpha_5),$$

$$\Phi_0(X) = 4(\alpha_1 + 4\alpha_2 + 4\alpha_3 + 7\alpha_4 + 5\alpha_5).$$

Let us examine the functionals $\Phi_1(X) = \Phi^{(1)}(X)$ and $\Phi_0(X) = \Phi^{(2)}(X)$ separately.

We write the expression for $\Phi_1(X)$ with variables arranged in nonascending order:

$$\Phi_1(X) = 4\Big(-5\alpha_2 - 3(\alpha_1 + \alpha_4) - 2\alpha_5 - \alpha_3\Big).$$

Applying the constraints for α_k ($0 \le k \le 5$), we see that the functional $\Phi_1(X)$ is minimized when $\alpha_0 = \alpha_2 = 1/2$ and $\alpha_1 = \alpha_3 = \alpha_4 = \alpha_5 = 0$, and the minimal value is $\min \Phi_1(X) = -10$. The corresponding vector pair is

$$(\mathbf{A}^*,\mathbf{B}^*) = 4\alpha_2(\overline{\mathbf{E}}_2,\overline{\mathbf{F}}_2) = 2\left((2,0),(1,1)\right) = \left((4,0),(2,2)\right) = (2\overline{\mathbf{E}}_2, 2\overline{\mathbf{F}}_2).$$

Consequently, the optimal transportation model for minimization in Q is $((2\overline{\mathbf{E}}_2, 2\overline{\mathbf{F}}_2), \Phi^{(1)}(X))$:

$$\min_{X\in\widetilde{M}(\mathbf{A},\mathbf{B})} \left\{\Phi_\beta(X) : ((\mathbf{A},\mathbf{B}), \Phi_\beta(X)) \in Q\right\} = \min_{X\in\widetilde{M}(2\overline{\mathbf{E}}_2,2\overline{\mathbf{F}}_2)} \Phi^{(1)}(X) = -10.$$

The corresponding uniform polyhedron is degenerate: $\left|\widetilde{M}(2\overline{\mathbf{E}}_2, 2\overline{\mathbf{F}}_2)\right| = 1$.

The expression for $\Phi_0(X)$ with variables arranged in nonascending order is

$$\Phi_0(X) = 4\Big(7\alpha_4 + 5\alpha_5 + 4(\alpha_2 + \alpha_3) + \alpha_1\Big).$$

Applying the constraints for α_k $(0 \le k \le 5)$, we see that the functional $\Phi_0(X)$ is maximized when $\alpha_0 = \alpha_4 = 1/2$ and $\alpha_1 = \alpha_2 = \alpha_3 = \alpha_5 = 0$, and the maximal value is $\Phi_0(X) = 14$. Therefore,

$$(\mathbf{A}^{**}, \mathbf{B}^{**}) = 4\alpha_4(\overline{\mathbf{E}}_4, \overline{\mathbf{F}}_4) = 2\left((2,1),(2,1)\right) = \left((4,2),(4,2)\right) = (2\overline{\mathbf{E}}_4, 2\overline{\mathbf{F}}_4).$$

Consequently, the optimal transportation model for maximization in Q is $\left((2\overline{\mathbf{E}}_4, 2\overline{\mathbf{F}}_4), \Phi^{(2)}(X)\right)$:

$$\max_{X \in \widetilde{M}(\mathbf{A},\mathbf{B})} \left\{\Phi_\beta(X) : \left((\mathbf{A},\mathbf{B}), \Phi_\beta(X)\right) \in Q\right\} = \max_{X \in \widetilde{M}(2\overline{\mathbf{E}}_4, 2\overline{\mathbf{F}}_4)} \Phi^{(2)}(X) = 14.$$

The corresponding uniform polyhedron is given by the formula

$$\widetilde{M}(2\overline{\mathbf{E}}_4, 2\overline{\mathbf{F}}_4) = \left\{X(2\overline{\mathbf{E}}_4, 2\overline{\mathbf{F}}_4) = \begin{pmatrix} 2+\varepsilon & 2-\varepsilon \\ 2-\varepsilon & \varepsilon \end{pmatrix} : 0 \le \varepsilon \le 1\right\}.$$

The results presented in this chapter are based on [8–10].

References to Chapter 5

[1] Alekseev V. M., Tikhomirov V. M., and Fomin S. V., *Optimal'noe upravlenie* (Optimal Control), Nauka, 1979.

[2] Glazman I. M. and Lyubich Yu. I., *Konechnomernyi lineinyi analiz* (Finite-Dimensional Linear Analysis), Moscow: Nauka, 1969.

[3] Gol'shtein E. G. and Yudin D. B, *Zadachi lineinogo programmirovaniya transportnogo tipa* (Transportation Problems in Linear Programming), Moscow: Nauka, 1969.

[4] Dunford N. and Schwartz J. T., *Linear Operators, Part 1: General Theory,* New York: Interscience, 1958. Translated under the title *Lineinye operatory (obshchaya teoriya),* Moscow: Mir, 1962.

[5] Emelichev V. A., Kovalev M. M., and Kravtsov M. K., *Mnogogranniki, grafy, optimizatsiya* (Polyhedrons, Graphs, and Optimization), Moscow: Nauka, 1981.

[6] Karmanov V. G., *Matematicheskoe programmirovanie* (Mathematical Programming), Moscow: Nauka, 1969.

[7] Kuznetsov A.V., Sakovich V. A., and Kholod N. N., *Vysshaya matematika. Matematicheskoe programmirovanie* (Higher Mathematics: Mathematical programming), Minsk: Vysshaya Shkola, 1994.

[8] Mironov A. A., Geometry of the Points in the Space R^n that Can Be Realized in a Graph, *Usp. Mat. Nauk,* 1977, vol. 32, no. 6, pp. 231–232.

[9] Mironov A. A. and Tsurkov V. I., Approximation and Decomposition by Means of Extremal Graphs, *Zh. Vychisl. Mat. Mat. Fiz.*, 1993, vol. 33, no. 2, pp. 283–298.

[10] Mironov A. A., Uniform Generalized Graphs, *Dokl. Akad. Nauk*, 1996, vol. 351, no. 4, pp. 465–468.

[11] Moder J. and Elmaghraby S. S., *Handbook of Operations Research*, New York: Van Nostrand, 1978. Translated under the title *Issledovanie operatsii. Metodologicheskie osnovy i matematicheskie metody*, vol. 1, Moscow: Mir, 1981.

[12] Trius E. B., *Zadachi matematicheskogo programmirovaniya transportnogo tipa* (Transportation Problems in Mathematical Programming), Moscow: Sovetskoe Radio, 1967.

Index

MIX
Papier aus verantwortungsvollen Quellen
Paper from responsible sources
FSC® C105338

If you have any concerns about our products,
you can contact us on
ProductSafety@springernature.com

In case Publisher is established outside the EU,
the EU authorized representative is:
Springer Nature Customer Service Center GmbH
Europaplatz 3, 69115 Heidelberg, Germany

Printed by Libri Plureos GmbH
in Hamburg, Germany